T0251684

CLOUD COMPUTING

Methodology, Systems, and Applications

CLOUD COMPUTING

Methodology, Systems, and Applications

Edited by
Lizhe Wang • Rajiv Ranjan
Jinjun Chen • Boualem Benatallah

CRC Press
Taylor & Francis Group
Boca Raton London New York

CRC Press is an imprint of the
Taylor & Francis Group, an **informa** business

CRC Press
Taylor & Francis Group
6000 Broken Sound Parkway NW, Suite 300
Boca Raton, FL 33487-2742

© 2012 by Taylor & Francis Group, LLC
CRC Press is an imprint of Taylor & Francis Group, an Informa business

No claim to original U.S. Government works

Printed in the United States of America on acid-free paper
Version Date: 20110808

International Standard Book Number: 978-1-4398-5641-3 (Hardback)

Visit the Taylor & Francis Web site at
http://www.taylorandfrancis.com

and the CRC Press Web site at
http://www.crcpress.com

To my beloved mother, to whom I owe too much.

– Lizhe

To my parents and beloved wife, who are the pillars of my life.

– Rajiv

To my parents, to whom I owe too much.

– Jinjun

To my parents, without whom I would not be where I am today.

– Boualem

Contents

List of Figures

List of Tables

Foreword

Clouds have emerged as an important computing paradigm and are quickly dominating all aspects of the IT landscape. In fact, most organizations today are considering Cloud services as part of their IT roadmap, as users or providers or both.

Clouds are characterized by on-demand access to different levels of computing utilities, an abstraction of unlimited resources, and a usage-based payment model where users essentially "rent" virtual resources (or capabilities) and pay for what they use. Underlying these cloud services are, typically, consolidated and virtualized data centers that provide virtual machine (VM) containers hosting applications from large numbers of distributed users.

Clouds and the notion of computing as a service provides opportunities for exploring potentially interesting cost-benefit trade-off (for example, between capital and operational costs) models for operational flexibility and agility, approaches for reducing environmental impact, etc., and are leading to interesting economic and business models. Furthermore, integrating these public cloud platforms with more traditional data centers and Grids provides opportunities for on-demand scale-up, scale-out and scale-down, and allows service offering and applications capabilities to no longer be constrained by local infrastructure limitations.

Clearly, such a seductive paradigm can have a significant impact on a range of application in academia, industry and government. However, transitioning to the Cloud mind-set, integrating clouds into the currently computing infrastructure, and moving current applications and their operations to Clouds present a non-trivial challenge and often require new paradigms and practices at all levels. Furthermore, these challenges often go far beyond just the technical aspects. For example, cultural, legal, regulatory and social challenges are quickly overshadowing technical challenges. While certain usage modes are natural to Clouds, others are not and, as a result, the Cloud paradigm must coexist and complement other usage modes, such as, for example, high-performance computing. It is critical that these challenges be fundamentally understood and addressed.

I would like to congratulate the editors of this volume for putting together such a comprehensive and timely collection. This book brings together chapters authored by internationally recognized researchers, describing leading research efforts that are trying to get to the heart of many of the above listed challenges, with an overall goal to provide a comprehensive overview of the state-of-the-art of this emerging research area, from fundamental concepts to more specific technologies and application use cases. As a result, I do believe that this book will provide a tremendous resource to students, researchers and practitioners, and have a significant impact on this important and growing field.

Manish Parashar, Ph.D.
Professor, Department of Electrical & Computer Engineering
Rutgers, The State University of New Jersey

Preface

Cloud computing is the latest evolution of computing, where IT resources are offered as services. The hardware and software systems that manage these services are referred to as Infrastructure as a Service (IaaS) and Platform as a Service (PaaS), while actual applications managed and delivered by IaaS and PaaS are referred to as Software as a Service (SaaS). The combination of virtualization, IaaS, and PaaS hold the potential to revolutionize software application (SaaS) life-cycle management. If Cloud computing is properly applied within an overall IT strategy, it can help small and medium business enterprises (SMEs) and governments lower their IT costs, by taking advantage of economies of scale and automated IT operations, while at the same time optimizing investment in in-house computing infrastructure. The benefit of such an environment is efficiency and flexibility, through creation of a more dynamic computing enterprise, where the supported functionalities are no longer fixed or locked to the underlying infrastructure. This offers tremendous automation opportunities in a variety of computing domains including, but not limited to, e-Government, e-Research, high-performance computing, web hosting, social networking, multi-media, and e-Business.

The goal of this book is to give a comprehensive overview of the state-of-the-art of this emerging research area, which many believe to be the next platform for provisioning and delivering SaaS applications in various computing domains. The book also envisions future research topics and directions. This book is expected to serve as the most important reference and a milestone for research on Cloud computing since 2007, when the term "Cloud computing" was coined.

The book is organized into three parts. Part I focuses on the fundamentals of cloud computing and includes chapters that present the insight, concepts, methodologies, taxonomies, and architectures related to existing and future software systems and architectures. In Chapter 1, A. Kalapatapu and M. Sarkar provide an insight into evolution of Cloud computing and its origins, and direction. In Chapter 2, A. Edlund and I. Livenson present a comprehensive study on how Cloud computing can be leveraged for accelerating the innovation and business values for IT startups. In Chapter 3, R. Teckelmann, A. Sulistio, and C. Reich present a detailed study on interoperability issues of Cloud infrastructures (IaaS). Chapter 4, written by C. Baun and M. Kunze, gives an overview about the most popular Cloud services and categorizes them according to their organizational and technical realization. L. Atzori, F. Granelli, and A. Pescapé in Chapter 5 study the networking solutions for engineering Cloud computing architectures. In Chapter 6, A. Dastjerdi and R. Buyya provide a detailed taxonomy and survey of QoS management and service selection methodologies in Cloud computing environments. A detailed study on open-source software frameworks that are aimed at managing Cloud infrastructures (IaaS) is presented in Chapter 7 by P. Sempolinski and D. Thain. In Chapter 8 H. Khazaei, J. Mišić, and V. B. Mišić present an analytical model for performance evaluation of Cloud server farms, and demonstrate the manner in which important performance indicators such as request waiting time and server utilization may be assessed with sufficient accuracy. Chapter 9, written by A. Celesti, F. Tusa, M. Villari, and A. Puliafito investigates the new business advantages of a futuristic worldwide InterCloud ecosystem, considering evolutionary degree of the current infrastructures and the possible scenarios on which the InterCloud federation could be built.

Part II compiles the contributions on the research theme of Cloud computing functionality and service provisioning. In Chapter 10, S. Nepal, S. Chen, and J. Yao discuss the issue of building elastic and secure Cloud storage services—TrustScore, which allows end-users to scale their storage space to meet the ever-increasing need while improving utilization and manageability. A. Goscinski, M. Brock, and P. Church in Chapter 11 investigate what challenges exist when one attempts to use Clouds for high performance computing (HPC) and they further demonstrate the effectiveness of HPC in Clouds through broad benchmarking and through a new prototype system used to satisfy the needs of HPC clients. Chapter 12, written by E. Jiménez-Domingo, Á. Lagares-Lemos, and J. M. Gómez-Berbís, addresses the issue of architecting multitenancy application stacks in Clouds. V. Emeakaroha et al. in Chapter 13 present an innovative framework – LoM2HiS, for the mapping low level resource metric to high-level SLA parameters. In Chapter 14, E. Caron et al. examine Cloud computing IaaS and PaaS providers from three important resource management issues including load balancing, auto-scaling, and monitoring. Monitoring of IaaS and SaaS components is fundamental to ensure QoS guarantees in Cloud computing environments. To this end, G. Katsaros et al. in Chapter 15 analyze the requirements of an efficient monitoring mechanism for Cloud environments. To handle peak loads, M. Mattess et al. present an approach in Chapter 16 that leases Cloud infrastructure services in an opportunistic way. Energy efficient management of Cloud computing infrastructures that consists of many power-hungry components such as processors, memory modules, and network devices is an important issue. In Chapter 17, Y. Choon et al. present energy-efficiency models for provisioning and migrating applications in Cloud environments. Lack of security, privacy, and trust of data are the major hindrances in the path to moving sensitive applications to public Cloud Computing environments. S. Hamouda and J. Glauert in Chapter 18 present a detailed taxonomy on how these issues are currently handled in context of Clouds.

Part III focuses on specific case studies and applications that are deployed and provisioned over Cloud infrastructures. This part of the book also includes discussion on possible research directions. In Chapter 19, J. Shi discusses the fundamentals of architecting applications on Clouds with a focus on three objectives: maximal application survivability, unlimited application performance scalability and zero data losses. To develop a software framework that can provide a common access to all existing Cloud-based services (IaaS, PaaS, and SaaS) is challenging. F. Moscato et al. in Chapter 20 address this challenge by defining a common ontology for Cloud service negotiation and establishment. Cloud environments offer promising platforms for deploying and managing elastic RDBMS systems. L. Zhao, S. Sakr, and A. Liu in Chapter 21 explore the recent advancements in web scale data management on Clouds. In order to allow cost-effective scaling of videoconference systems, the use of Cloud computing appears to be a viable choice, mainly due to its elasticity and infinite resource capacity. To this end, J. Cervino et al. in Chapter 22 describe a new service that delivers a videoconferencing system through Clouds; further, they also analyze the cost and resource usage related to this case study. L. Carril et al. in Chapter 23 present a case study on designing a computing and storage service for planning radiotherapy treatment using Cloud infrastructures. In Chapter 24 K. Fletcher and X. Liu present a methodology to analyze Cloud security requirements and develop policies to deal with both internal and external security challenges. Integration of traditional HPC Grids with Cloud infrastructures is going to play a pivotal role for future scientific studies and experiments. H. Kim et al., in Chapter 25, experimentally investigate, from an applications perspective, interesting usage of nodes and scenarios for integrating HPC Grids and Clouds, and how they can be effectively enabled using an autonomic scheduling system. In Chapter 26, J. Kaylor, K. Läufer, and G. Thiruvathukal propose a file system, called RestFS, as a connector abstraction for flexible resource and service composition. Finally Y. Wei et al. present the Aneka Cloud computing application platform, in Chapter 27.

The compilation of this book has been possible because of the efforts and contributions of many individuals. Firstly, the editors would like to thank all contributors for their immense hard work in putting together excellent chapters that are informative, comprehensive, rigorous and, above all, timely. The editors would like to thank the team led by Nora Konopka, Amy Blalock and Brittany Gilbert, at Taylor & Francis Group/CRC Press for patiently helping us put this book together. Lastly, we would like to thank our families for their support throughout this journey and would like to dedicate this book to them.

Contributors

Luigi Atzori
Department of Electric and
    Electronic Engineering
University of Cagliari
Cagliari, Italy

Rocco Aversa
Dip. di Ingegneria dell'Informazione
Second University of Naples
Naples, Italy

Christian Baun
Steinbuch Centre for Computing
Karlsruhe Institute of Technology
Karlsruhe, Germany

Ivona Brandic
Vienna University of Technology
Vienna, Austria

Ivan Breskovic
Vienna University of Technology
Vienna, Austria

Michael Brock
School of Information Technology
Deakin University
Geelong, Australia

Rajkumar Buyya
Cloud Computing and Distributed
    Systems (CLOUDS) Laboratory
Department of Computer Science
    and Software Engineering
The University of Melbourne
Melbourne, Australia

Eddy Caron
LIP Laboratory
University of Lyon
Lyon, France

Luis M. Carril
CESGA
Galicia, Spain

Antonio Celesti
Faculty of Engineering
University of Messina
Messina, Italy

Javier Cerviño
ETS de Ingenieros de
    Telecomunicación
Universidad Politécnica de Madrid
Madrid, Spain

Shiping Chen
Information Engineering Lab
CSIRO ICT Centre
Marsfield, New South Wales

Philip Church
School of Information Technology
Deakin University
Australia

Amir Vahid Dastjerdi
The University of Melbourne
Melbourne, Australia

Frédéric Desprez
LIP Laboratory
University of Lyon
Lyon, France

Rubén Díaz
CESGA
Galicia, Spain

Beniamino Di Martino
Second University of Naples
Dip. di Ingegneria dell'Informazione
Naples, Italy

Schahram Dustdar
Vienna University of Technology
Vienna, Austria

Åke Edlund
PDC, KTH Royal Institute of
    Technology and SWEDACC – the
    Swedish Seed Accelerator
Stockholm, Sweden

Yaakoub El-Khamra
Texas Advanced Computing Center
The University of Texas at Austin
Austin Texas

Vincent C. Emeakaroha
Vienna University of Technology
Vienna, Austria

Fernando Escribano
ETS de Ingenieros de
    Telecomunicación
Universidad Politécnica de Madrid
Madrid, Spain

Carlos Fernández
CESGA
Galicia, Spain

Kenneth Kofi Fletcher
Missouri University of Science and
    Technology
Rolla, Missouri

Georgina Gallizo
High Performance Computing Center
Stuttgart, Germany

Saurabh Kumar Garg
The University of Melbourne
Melbourne, Australia

John Glauert
Department of Computer Science
University of East Anglia
Norwich, United Kingdom

Andrés Gómez
CESGA
Galicia, Spain

Juan Miguel Gómez-Berbís
Universidad Carlos III de Madrid
Madrid, Spain

Andrzej Goscinski
School of Information Technology
Deakin University
Geelong, Australia

Fabrizio Granelli
DISI (Department of Information
    Engineering and Computer
    Science)
University of Trento
Trento, Italy

Sara Hamouda
Department of Computer Science
University of East Anglia
Norwich, United Kingdom

Shantenu Jha
Center for Computation and
    Technology and Department of
    Computer Science
Louisiana State University
Baton Rouge, Louisiana

Enrique Jiménez-Domingo
Universidad Carlos III de Madrid
Madrid, Spain

Abhishek Kalapatapu
San Diego State University
San Diego, California

Dileban Karunamoorthy
Cloud Computing and Distributed
    Systems (CLOUDS) Laboratory
Department of Computer Science
    and Software Engineering
The University of Melbourne
Melbourne, Australia

Dilkushan T. M. Karunaratne
School of Information Technologies
University of Sydney
Sydney, Australia

Gregory Katsaros
High Performance Computing Center
Stuttgart, Germany

Joseph Kaylor
Department of Computer Science
Loyola University
Chicago, Illinois

Hamzeh Khazaei
University of Manitoba
Winnipeg, Manitoba, Canada

Hyunjoo Kim
Center for Autonomic Computing
Department of Electrical &
    Computer Engineering, Rutgers
The State University of New Jersey
New Brunswick, New Jersey

Roland Kübert
High Performance Computing Center
Stuttgart, Germany

Marcel Kunze
Steinbuch Centre for Computing
Karlsruhe Institute of Technology
Karlsruhe, Germany

Ángel Lagares-Lemos
Universidad Carlos III de Madrid
Madrid, Spain

Konstantin Läufer
Department of Computer Science
Loyola University
Chicago, Illinois

Young Choon Lee
Centre for Distributed and High
    Performance Computing
School of Information Technologies
University of Sydney
Sydney, Australia

Anna Liu
University of New South Wales
Sydney, Australia

Xiaoqing (Frank) Liu
Missouri University of Science and
    Technology
Rolla, Missouri

Ilja Livenson
PDC, KTH Royal Institute of
    Technology and SWEDACC – the
    Swedish Seed Accelerator
Stockholm, Sweden

Zahara Martín-Rodríguez
CESGA
Galicia, Spain

Michael Mattess
The University of Melbourne
Melbourne, Australia

Michael Maurer
Vienna University of Technology
Vienna, Austria

Jelena Mišić
Ryerson University
Toronto, Ontario, Canada

Vojislav B. Mišić
Ryerson University
Toronto, Ontario, Canada

Francesco Moscato
Dip. di Studi Europei e Mediterranei
Second University of Naples
Naples, Italy

Carlos Mouriño
CESGA
Galicia, Spain

Adrian Muresan
LIP Laboratory
University of Lyon
Lyon, France

Surya Nepal
Information Engineering Lab
CSIRO ICT Centre
Marsfield, New South Wales

Manish Parashar
Center for Autonomic Computing
Department of Electrical &
    Computer Engineering
Rutgers, The State University of
    New Jersey
New Brunswick, New Jersey

Antonio Pescapé
DIS (Department of Computer
    Science and Systems
University of Napoli Federico II
Naples, Italy

Dana Petcu
Computer Science Department
Western University of Timisoara
Timis, Romania

Antonio Puliafito
Faculty of Engineering
University of Messina
Messina, Italy

Massimiliano Rak
Dip. di Ingegneria dell'Informazione
Second University of Naples
Naples, Italy

Christoph Reich
Department of Computer Science
Hochschule Furtwangen University
Furtwangen, Germany

Luis Rodero-Merino
LIP Laboratory
University of Lyon
Lyon, France

Sherif Sakr
University of New South Wales
Sydney, Australia

Joaquín Salvachúa
ETS de Ingenieros de
    Telecomunicación
Universidad Politécnica de Madrid
Madrid, Spain

Mahasweta Sarkar
San Diego State University
San Diego, California

Peter Sempolinski
Department of Computer Science
    and Engineering
University of Notre Dame
Notre Dame, Indiana

Justin Y. Shi
Temple University
Philadelphia, Pennsylvania

Karthik Sukumar
Manjrasoft Pty. Ltd.
Melbourne, Victoria, Australia

Anthony Sulistio
Department of Applications, Models
    and Tools
High Performance Computing Center
Stuttgart, Germany

Ralf Teckelmann
Department of Computer Science
Hochschule Furtwangen University
Furtwangen, Germany

Douglas Thain
Department of Computer Science
and Engineering
University of Notre Dame
Notre Dame, Indiana

George K. Thiruvathukal
Department of Computer Science
Loyola University Chicago
Chicago, Illinois

Irena Trajkovska
ETS de Ingenieros de
Telecomunicación
Universidad Politécnica de Madrid
Madrid, Spain

Francesco Tusa
Faculty of Engineering
University of Messina
Messina, Italy

Christian Vecchiola
Cloud Computing and Distributed
Systems (CLOUDS) Laboratory
Department of Computer Science
and Software Engineering
The University of Melbourne
Melbourne, Australia

Salvatore Venticinque
Dip. di Ingegneria dell'Informazione
Second University of Naples
Naples, Italy

Massimo Villari
Faculty of Engineering
University of Messina
Messina, Italy

Chen Wang
CSIRO ICT Center
Marsfield, New South Wales,
Australia

Tinghe Wang
High Performance Computing Center
Stuttgart, Germany

Yi Wei
Manjrasoft Pty. Ltd.
Melbourne, Victoria, Australia

Jinhui Yao
Information Engineering Lab
CSIRO ICT Centre
Marsfield, New South Wales,
Australia

Albert Y. Zomaya
Centre for Distributed and High
Performance Computing
School of Information Technologies
University of Sydney
Sydney, Australia

Liang Zhao
University of New South Wales
Sydney, Australia

Part I

Fundamentals of Cloud Computing: Concept, Methodology, and Overview

1

Cloud Computing: An Overview

Abhishek Kalapatapu and Mahasweta Sarkar

San Diego State University, San Diego, California

CONTENTS

1.1 Introduction

Cloud computing, with the revolutionary promise of turning computing into a 5th utility, after water, electricity, gas, and telephony, has the potential to transform the face of Information Technology (IT), especially the aspects of service-rendition and service management. Though there are myriad ways of defining the phenomenon of Cloud Computing, we put forth the one coined by NIST (National Institute of Standards and Technology). According to them, Cloud Computing is defined as *"A model for enabling convenient, on-demand network access to a shared pool of configurable computing resources (e.g., networks, servers, storage, applications, and services) that can be rapidly provisioned and released with minimal management effort or service provider interaction"* [84]. Loosely speaking, Cloud computing represents a new way to deploy computing technology to give users the ability to access, work on, share, and store information using the Internet [762]. The cloud itself is a network of data centers, each composed of many thousands of computers working together that can perform the functions of software on a personal or business computer by providing users access to powerful applications, platforms, and services delivered over the Internet. It is in essence a set of network enabled

3

services that is capable of providing scalable, customized and inexpensive computing infrastructures on demand, which could be accessed in a simple and pervasive way by a wide range of geographically dispersed users. The Cloud also assures application based Quality-of-Service (QoS) guarantees to its users. Thus, Cloud Computing provides the users with large pools of resources in a transparent way along with a mechanism for managing the resources so that a user can access it ubiquitously and without incurring unnecessary performance overhead. The ideal way to describe Cloud Computing then would be to term it as "Everything as a Service" abbreviated as XaaS [100]. Below, we sum up the key features of Cloud Computing:

- Agility – helps in rapid and inexpensive re-provisioning of resources.

- Location Independence – resources can be accessed from anywhere and everywhere.

- Multi-Tenancy – resources are shared amongst a large pool of users.

- Reliability – dependable accessibility of resources and computation.

- Scalability – dynamic provisioning of data helps in avoiding various bottleneck scenarios.

- Maintenance – users (companies/organizations) have less work in terms of resource upgrades and management, which in the new paradigm will be handled by service providers of Cloud Computing.

However, Cloud Computing doesn't imply that it consists of only one cloud. The term "Cloud" symbolizes the Internet, which in itself is a network of networks. Also, not all forms of remote computing are Cloud Computing. On the contrary, Cloud Computing is nothing but services offered by providers who might have their own systems in place. [18]

1.2   Cloud Computing: Past, Present, and Future

A hundred years ago, companies stopped generating their own power with steam engines and dynamos and plugged into the newly built electric grid. The cheap power pumped out by such electric utilities did not just change how businesses operated — it set off a chain reaction of economic and social transformations that brought the modern world into existence. Today, a similar revolution is under way. Hooked up to the Internet's global computing grid, massive information-processing plants have begun pumping data and software code into our homes and businesses. This time, it's computing (instead of electricity) that's turning into a utility. Nicholas Carr in his book *The Big*

Switch: Rewiring the world from Edison to Google [164] has finely portrayed the transition of Cloud from its past till its present form. Cloud Computing has integrated various positive aspects of different computing paradigms, resulting in a hybrid model that has evolved gradually over the years beginning in 1960 when *John McCarthy* rightfully stated that *"computation may someday be organized as a public utility"* [608].

"The Cloud" is based on the idea of generating computing facility on demand. Just the way one turns on a faucet to get water or plug into an electric socket on the wall to get electricity, similarly Cloud Computing intends to create a paradigm where most of the features and functions of stand-alone computers today can be streamed for a user over the Internet [764]. Further probing into the philosophy of cloud computing will reveal that the concept dates back to the era of "Mainframes," where resources (like memory, computational capabilities) of centralized powerful computers owned by large organizations were used/shared by several users over a small geographical local area. Today Cloud Computing boasts of an architecture where the powerful computers are replaced by supercomputers and perhaps even a network of supercomputers and the users are dispersed over vast geographic areas, accessing these computing resources via the Internet (network of networks). In the past, issues like dearth of bandwidth, perception, loss of control, trust and feasibility proved to be major road blocks in realizing the Cloud concept of service rendition. Today most of these challenges have been overcome, or countermeasures are in place to resolve the challenges. Faster bandwidth, virtualization and more particular skills around cloud type technologies help in realizing the Cloud Computing paradigm.

The concept of Cloud Computing — as nascent as it might appear — itself, has undergone significant evolution. The first generation of Cloud Computing which evolved along with the "Internet Era" was mainly intended for "e-business services." The current generation of Cloud services has progressed several steps to now include "IT as a Service" which can be thought of as "consumerized internet services." Using standardized, highly virtualized infrastructure and applications, IT can drive higher degrees of automation and consolidation, thus reducing the cost of maintaining existing solutions and delivering new ones. In addition, externally supplied infrastructure, software, and platform services are delivering capacity augmentation and a means of using operating expense funding instead of a heavy duty capital. [388]

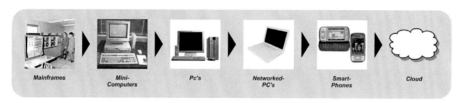

FIGURE 1.1
History of Computing.

Today, "the network is the (very big, very powerful) computer." The capabilities of the Cloud as a centralized resource can match to industrial scales. This implies that processing power involving thousands of machines embedded in a network has surpassed even the capabilities of the very high-performance supercomputers. By making this technology available through the network on an on-demand and as-needed basis, the Cloud holds the promise of giving individuals, businesses, organizations and governments around the world access to extraordinary computing power from any location and any device. It is a fact that data and information is growing at a break-neck pace. HP predicted that within three more years (by 2014) there will be more information produced and consumed than in the history of mankind. Thus, the next version of Cloud will enable access to information through services that are set in the context of the consumer experience. This is significantly different — it means that data will be separated from the applications — a paradigm where processes can be broken into smaller pieces and automated through a collection of services, woven together with access to massive amounts of data. It will eliminate the need for large scale, complex applications that are built around monolithic processes. Changes can be accomplished by refactoring service models, and integration achieved by subscribing to new data feeds. This will create new connections, new capabilities, and new innovations surpassing those that exist today. [388]. Thus it is envisioned that by 2020 most people will access

	Cloud 1 E-business Services			Cloud 2 IT as a Service		Cloud 3 Everything as a Service	
	1990	1995	2000	2005	2010	2015	2020
Primary forcing function		▪ Internet based supply chain integration and e-commerce		▪ Consumerized internet services ▪ Low cost IT		▪ Pervasive business and consumer services	
technology orientation		▪ Web based app design ▪ EAI & message bus Integration ▪ Internal protocols ▪ 3-tier architecture		▪ Web 2.0 & SCA app design ▪ Virtualization ▪ Cloudbased technology platforms		▪ Data oriented, context aware services ▪ Vertical and horizontal cloud ecosystems	
IT organization design		▪ Organized around technology domains ▪ Technology-centric		▪ Organized around service supply chain ▪ Service-centric		▪ Organized around value networks ▪ Service-centric	

FIGURE 1.2
Clouds Past, Present, and Future.

software applications online and share and access information through the use of remote server networks, rather than depending primarily on tools and information housed on their individual, personal computers. It is predicted that cloud computing will become more dominant than the desktop in the next decade [84]. In other words, most users will perform the majority of their computing and communicating activities through connections to servers that will be operated and owned by service-providing organizations. Most technologists and industrialists believe that cloud computing will continue to expand and come to dominate information transactions because it offers many advantages, allowing users to have easy, instant, and individualized access to tools and information they need wherever they are, locatable from any networked device. To validate this claim, the PEW INTERNET & AMERICAN LIFE PROJECT carried out a survey with a highly diverse population set. *71% of the survey takers believed that by 2020, most people won't do their work with software running on a general-purpose PC. Instead, they will work in Internet-based applications such as Google Docs, and in applications run from smart phones* [630]. However, quality of service guarantees, interoperability between existing working platforms and security concerns are some of the issues that still continue to plague the growing popularity of Cloud computing.

1.3 Cloud Computing Methodologies

Cloud Computing is based on two main techniques — (i) Service Oriented Architecture and (ii) Virtualization.

(i)*Service Oriented Architecture (SOA)*: Since the paradigm of Cloud computing perceives of all tasks accomplished as a "Service" rendered to users, it is said to follow the *Service Oriented Architecture*. This architecture comprises a flexible set of design principles used during the phases of system development and integration. The deployment of a SOA-based architecture will provide a loosely-integrated suite of services that can be used within multiple business domains. The enabling technologies in SOA allow services to be discovered, composed, and executed. For instance, when an end-user wishes to accomplish a certain task, a service can be employed to discover the required resources for the task. This will be followed by a composition service which will plan the road-map to provide the desired functionality and quality of service to the end-user. [579, 761]

(ii)*Virtualization*: The concept of virtualization is to relieve the user from the burden of resource purchases and installations. The Cloud brings the resources to the users. Virtualization may refer to *Hardware* (execution of software in an environment separated from the underlying hardware resources), *Memory* (giving an application program the impression that it has contiguous working memory, isolating it from the underlying physical memory implementation), *Storage* (the process of completely abstracting logical storage from physical storage), *Software* (hosting of multiple virtualized environments within a single Operating System (OS) instance), *Data* (the presentation of data as an abstract layer, independent of underlying database systems, structures and storage) and *Network* (creation of a virtualized network addressing space within or across network subnets) [486]. Virtualization has become an indispensable ingredient for almost every Cloud; the most obvious reasons being the ease of abstraction and encapsulation. Amongst the other important reasons for which the Clouds tend to adopt virtualization are:

(i) Server and application consolidation – as multiple applications can be run on the same server resources can be utilized more efficiently.

(ii) Configurability – as the resource requirements for various applications could differ significantly, (some require large storage, some require higher computation capability) virtualization is the only solution for customized configuration and aggregation of resources which are not achievable at the hardware level.

(iii) Increased application availability – virtualization allows quick recovery from unplanned outages as virtual environments can be backed up and migrated with no interruption in services.

(iv) Improved responsiveness – resource provisioning, monitoring and maintenance can be automated, and common resources can be cached and reused. [784]

In addition, these benefits of virtualization tend to facilitate the Cloud to meet stringent SLA (Service Level Agreement) requirements in a business setting which otherwise cannot be easily achieved in a cost-effective manner. Without virtualization, systems have to be over provisioned to handle peak load and hence waste valuable resources during idle periods.

1.4   The Cloud Architecture and Cloud Deployment Techniques

Geared with the knowledge of SOA and Virtualization, we now take a look at the overall Cloud architecture. From the end user's perspective, Figure 4.1 depicts a basic Cloud Computing architecture involving multiple components. Cloud architecture closely resembles the UNIX philosophy of involving multiple components which work together over universal interfaces [762]. Recall that the Cloud computing paradigm represents a Service oriented mechanism of managing and dispatching resources. Before we delve into studying the actual architecture of Cloud computing it will be beneficial to examine the possible characteristics that will be required to realize such a system. It is common knowledge that the architectural requirements of the Cloud will vary depending on the application for which the Cloud is being used. For instance, social networking applications like Facebook and Orkut will have a very differ-

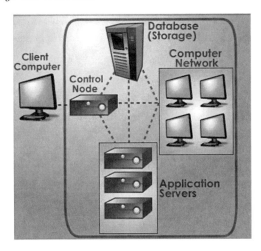

FIGURE 1.3
Basic Cloud Computing Architecture [86].

ent set of requirements, constraints and deliverables from the architecture in comparison to, say, a remote patient health monitoring application. However, some common architectural characteristics can still be identified. For instance, (i) the system should be scalable with the ability to include thousands to perhaps tens of thousands of members. (ii) It should be able to interoperate between various service requirements and effectively share resources amongst its users. (iii) The system should be easy to maintain and upgrade, maintaining user transparency during these processes. (iv) As outlined earlier, managing resources like servers and storage devices virtually, thereby creating a virtual organization, is absolutely crucial.

To mitigate the problem of designing customized Cloud architecture for each and every application and also to streamline the architecture design process of the Cloud, scientists resorted to the age old concept of a generalized "Layered approach" [163]. As with the classical 7-layer OSI model of data networks [759], the layered model in Cloud computing serves the same general purpose. Depending on the service requirement of an application, these layers are shuffled to create a customized architecture. The layered architecture adheres to the principle of Service Oriented Architecture (SOA) which forms the core of the Cloud computing paradigm. The components of a basic layered architecture are shown in the Figure 1.4, below [630], namely the Client, its required Services, the Applications that the Client runs, the Platform on which these applications run, the Storage requirement and finally the Infrastructure required to support the Client's computing needs.
We devote a few sentences on each component below.

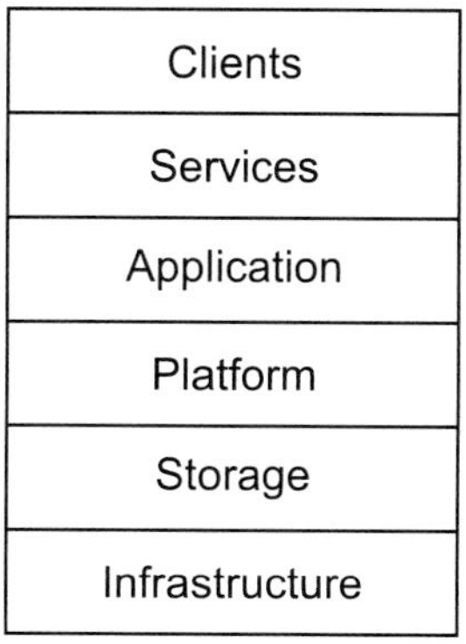

Clients
Services
Application
Platform
Storage
Infrastructure

FIGURE 1.4
Layered Architecture for a Customized Cloud Service [760].

Clients: the Clients of a Cloud comprise computer hardware and/or computer software that relies on the computational capability of the Cloud for application or service delivery. Examples include computers, mobile devices, operating systems, and browsers.

Services: this refers to the different service models made available by the Cloud like SaaS (Software-as-a-Service), IaaS (Infrastructure-as-a-Service) and PaaS (Platform-as-a-Service). This layer acts as a middleman between the user and the vast amount of resources accessible to the user. A resource includes "products, services and solutions that are delivered and consumed in real time over the Internet" [630]. Examples include location services and Search Engines among others.

Application: the Cloud enables resource management and user activity tracking from central locations rather than at each customer's site, enabling customers to access applications remotely via the Internet. Cloud application services deliver software as a service over the Internet, eliminating the need to install and run the application on the customer's own computer, thereby simplifying (almost eliminating) maintenance and support at the customer's end. Examples include Web Application and Peer-to-Peer computing.

Platform: it facilitates deployment of applications without the cost and complexity of buying and managing the underlying hardware and software layers. This layer delivers a computing platform and/or solution stack as a service, often consuming Cloud infrastructure and sustaining Cloud applications. Examples include Web Application Frameworks like Ruby on Rails and Web Hosting.

Storage: the storage layer consists of computer hardware and/or computer software products that are specifically designed for the storage of Cloud services. Computer hardware comprises huge data centers that are used for resource sharing. Examples include Amazon SImpleDB, and Nirvanix SDN

(Storage Delivery Network).

Infrastructure: this layer delivers computer infrastructure, typically a platform virtualization environment as a service. It includes management of virtual resources too. Rather than purchasing servers, software, data center space or network equipment, clients instead buy those resources as a fully outsourced service. Examples include Network Attached Storage and Database services.

The main advantage of such a layered architecture is the ease with which they can be modified to suit a particular service. The way in which these components interact leads to various architectural styles. There are two basic architectural styles on which most of the services are based. They are:

- **Outside-In:** This architectural style is inherently a top-down design emphasizing the functionality of the components. Implementing this style leads to a better architectural layering with various functionalities. It infuses more feasibility enabling better integration and interoperation of components.

- **Inside-Out:** This architectural style, on the other hand, is inherently a bottom-up design which takes an infrastructural point of view of the components. This style is more application oriented than service oriented. [631]

It is to be noted that incorporating new functionalities in a pre-existing architectural scheme is done in an incremental fashion. The ease of transforming an existing architecture into another, depends on the complexity of the architecture, the functionalities of the components and their integration. The vast landscape of the services and their growing complexity has lead to implementation of innovative architectural styles and several hybrid architectures. [221, 631]

Cloud Deployment Techniques

Cloud deployment is the manner in which a Cloud is designed to provide a particular service. Obviously these deployment methods will vary according to the way in which a Cloud provides service to the users. Thus, their deployment techniques are user specific [494]. For instance, a deployment technique might depend on the level of security commissioned for a particular user.

Figure 1.5 depicts the various Cloud deployment techniques which predominantly comprise (i) the Public deployment, (ii) the Private deployment or (iii) the Hybrid deployment. We discuss each of these deployment strategies briefly below.

(i) **Public Cloud:** it is the traditional mainstream Cloud deployment technique whereby resources are dynamically provisioned by third party providers who share them with the users and bill the users on a fine grained utility computing basis. It offers easy resource management, scalability and flexibility

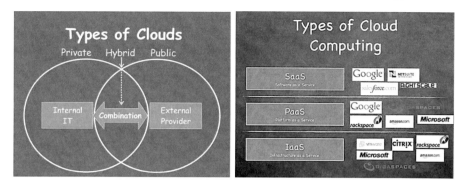

FIGURE 1.5
Cloud Deployment Techniques [325].

with an economical pay-as-you-go model which is extremely viable especially for small businesses. On the negative side, the user lacks visibility and control over the computing infrastructure. Since computing infrastructures are shared between various organizations, these Clouds face various security and compliance issues. Amazon's Web Services and Google's AppEngine are few examples of Public Clouds, also known as external Clouds. [630]

(ii) **Private Cloud:** in this Cloud deployment technique, the computing infrastructure is solely dedicated to a particular organization or business. These Clouds are more secure because they belong exclusively to a particular organization [494]. These Clouds are more expensive because one needs in-house expertise for their maintenance. Private Clouds are further classified based on their location as:

(a)*On-Premise Clouds* – these refer to Clouds that are for a particular organization hosted by the organization itself. Examples of such Clouds would include Clouds related to military services which have a considerable amount of confidential data.

(b)*Externally hosted Clouds* – these refer to Clouds that are also dedicated for a particular organization but are hosted by a third party specializing in Cloud infrastructure. These are cheaper than On-premise Clouds. Examples of such Clouds would be small businesses using services from VMware, Amazon etc. Such Clouds are also known as Internal Clouds. [631]

(iii) **Hybrid Cloud:** this deployment technique integrates the positive attributes of both the Public Cloud and Private Cloud paradigm. For instance, in a Hybrid Cloud deployment, critical services with stringent security requirements may be hosted on Private Clouds while less critical services can be

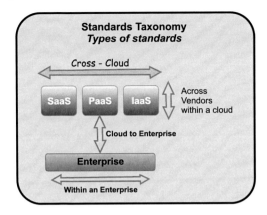

FIGURE 1.6
Types of Cloud Services [569].

hosted on the Public Clouds. The criticality, flexibility and scalability requirement of a service governs its classification into either the Public or Private Cloud domain. Each Cloud in the Hybrid domain retains its unique entity. However, they function synchronously to gracefully accommodate any sudden rise in computing requirements. Hybrid Cloud deployment is definitely the current trend amongst the major leading Cloud providers currently [630].

(iv) **Community Cloud:** this deployment technique is similar to Public Clouds with the only difference being the distribution of the sharing rights on the computing resources. In a Community Cloud, the computing resources are shared amongst organizations of the same community. So this Cloud covers a particular group of organizations, which have the same functionalities. For example, all Government organizations within the state of California may share the computing infrastructure on the Cloud to manage data related to citizens residing in California [608].

1.5 Cloud Services

The Cloud can provide us with a myriad service models and services. They include SaaS (Software as a Service), PaaS (Platform as a Service), HaaS (Hardware as a Service), DaaS ([Development, Database, Desktop] as a Service), IaaS(Infrastructure as a Service), BaaS (Business as a Service), FaaS (Framework as a Service), OaaS (Organization as a Service) amongst others [100]. However, Cloud Computing products can be broadly classified into three main Services (SaaS, PaaS and IaaS) which are showcased in Figure 1.6 along with their relationship to a user (Enterprise). The following section is an attempt to familiarize the reader with the several different Cloud Services that are currently rendered:

Infrastructure-as-a-Service (IaaS): This service provisions for hardware related services like storage, and virtual servers on a pay-as-you-go basis. The main advantage of IaaS is the usage of latest technology at all times with regard to computer infrastructure which allows users to achieve faster service. Organizations can use IaaS to quickly build new versions of applications or environments without incurring unnecessary purchase and configuration delay. On-demand scaling via resource virtualization and use-based billing makes IaaS competent enough for any kind of businesses. The major companies already providing IaaS are Amazon [57, 58], Rackspace, GoGrid, AT&T and IBM. [367]

Platform-as-a-Service (PaaS): PaaS offerings may include facilities for application design, application development, testing, deployment and hosting as well as application services such as team collaboration, web service integration and marshalling, database integration, security, scalability, storage, persistence, state management, application versioning, application instrumentation and developer community facilitation. These services may be provisioned as an integrated solution over the web, providing an existent managed higher-level software infrastructure for building particular classes of applications and services. The platform includes the use of underlying computing resources, typically billed similar to IaaS products, although the infrastructure is abstracted away below the platform. Major companies providing PaaS are Google's AppEngine [92], Microsoft Azure, and Force.com etc. [608, 630, 631]

Software-as-a-Service (SaaS): Provides specific already created applications as fully or partially remote services. Sometimes it is in the form of web-based applications and other times it consists of standard non-remote applications with Internet-based storage or other network interactions. It allows a user to use the provider's application using a thin client interface. Users can access a software application hosted by the Cloud vendor on payper-use basis [527]. It is a multi-tenant platform. The pioneer in this field has been Salesforce.com offering online Customer Relationship Management (CRM) space. Other examples are online email providers like Google's Gmail and Microsoft's hotmail, Google docs and Microsoft's online version of office called BPOS (Business Productivity Online Standard Suite). [16, 608, 630, 631]

Other than the above services David Linthicum has described a more granular classification of services which includes:

Storage-as-a-Service (SaaS): Storage as a Service is a business model that helps a smaller company or individual in renting storage spaces from a large company. Storage as a Service is generally seen as a good alternative for a small or mid-sized business that lacks the capital budget and/or technical personnel to implement and maintain their own storage infrastructure. SaaS is also being promoted as a way for all businesses to mitigate risks in disaster recovery, provide long-term retention for records and enhance both business continuity and availability. Examples include Nirvanix, Cleversafe's dsNET etc.

Database-as-a-Service (DbaaS): It constitutes delivery of database software and related physical database storage as a service. A managed service, offered on a pay-per-usage basis that provides on-demand access to a database for the storage of application data is what constitutes DbaaS. Examples include Amazon, Force.com etc.

Information-as-a-Service (IfaaS): Information as a service accepts the idea that data resides within many systems and repositories. Its main function is to standardize the access of data by applying a standard set of transformations to the various sources of data thus enabling service requestors to access the data regardless of vendor or system. Examples include IBM, Microsoft etc.

Process-as-a-Service (PraaS): Refers to a remote resource that's able to bind many resources together, either hosted within the same Cloud computing resource or remote, to create business processes. These processes are typically easier to change than applications, and thus provide agility to those who leverage these process engines that are delivered on-demand. Process-as-a-service providers include Appian Anywhere, Akemma, and Intensil.

Integration-as-a-Service (InaaS): Integration-as-a-Service includes most of the features and functions found within traditional Enterprise Application Integration technology, but delivered as a service. Integration-as-a-Service takes the functionality of system integration and puts it into the Cloud, providing for data transport between the enterprise and SaaS applications or third parties. Examples include Amazon SQS, OpSource Connect, Boomi, and Mule OnDemand.

Security-as-a-Service (SeaaS): Delivers core security services remotely over the Internet like anti-virus, log management etc. While typically the security services provided are rudimentary, more sophisticated services are becoming available such as identity management. Security-as-a-Service providers include Cisco, McAfee, Panda Software, Symantec, Trend Micro and VeriSign.

Management/Governance-as-a-Service (MaaS): Provides the ability to manage one or more Cloud services, typically simple things such as topology, resource utilization, virtualization, and uptime management. Governance systems are becoming available as well, such the ability to enforce defined policies on data and services. Management/governance-as-a-service providers include RightScale, rPath, Xen, and Elastra.

Testing-as-a-Service (TaaS): These systems have the ability to test other Cloud applications, Web sites, and internal enterprise systems, and do not require a hardware or software footprint within the enterprise. They also have the ability to test services that are remotely hosted. SOASTA is one of the many Testing-as-a-Service providers [608, 631].

1.6   Cloud Applications

Cloud Applications are not only developed with a business perspective but also take into account activities oriented towards socializing and sharing information. This information may be as basic as checking news headlines or more sensitive in nature, such as searches for health or medical information. Thus Cloud Computing is often a better alternative than local servers handling such applications.

Virtualization is the main basis of Cloud Computing, thus it is fully enabled with virtual appliances. A virtual appliance is an application which has all its components bundled and streamlined with the operating system [762]. The main advantage of a Cloud Computing application is that the provider can run various instances of an application with minimum labor and expense. A Cloud service provider needs to anticipate a few issues before launching its application in the Cloud computing environment. Keeping in mind the issues an application must be designed to scale easily, tolerate failures and include management tools [388]. We discuss these issues in the following section.

- *Scale:* Application in a Cloud environment needs to have maximum scalabilities and to ensure this, one should start building the application in the simplest manners avoiding complex design patterns and enhancements. The next step would be to split the functions of an application and integrate them loosely. The most important step in ensuring on demand scalability of an application is sharding, which can be described as splitting up the system into many smaller clusters instead of scaling the single system up so as to serve all users.

- *Failures:* Any application due to one or the other reason is bound to face failure at some point of time. To tolerate failures, an application must operate in an asynchronous fashion and one should spread the load across multiple clusters so that the impact of failure gets distributed. The best way to tolerate failures would be testing the application for all kinds of failure scenarios, and also users should be aware of the real cost incurred if an application faces any kind of failure.

- *Management Tools:* Having a proper management tool helps in automating the application configuration and updates, thereby reducing management overhead. The management system helps in not just minimizing economic expenditures but also leading to optimized usage of resources. The most

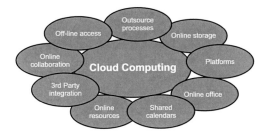

FIGURE 1.7
Cloud Applications [569].

difficult and expensive problem that can be handled with a proper management tool is variability [795]. This helps in delivering an application that can boast of consistent performance.

Applications in Cloud Computing have been designed keeping in mind the various types of services offered by it. Cloud Computing has impacted people's perception of applications over the Internet. The applications of Cloud computing are streamlined with each and every field of sciences. In Figure 1.7 we have highlighted a few areas in which Cloud Computing has shown great potential for developing various applications which have fostered the growth of businesses and enterprises.

These applications can be categorized into broad areas like datacenters and storage, security, military applications, the health industry, platforms and software usages, applications with high performance computing resources, and last but not the least, the growing need for virtual environments. [221, 494]

1.7 Issues with Cloud Computing

So far we have focused on the promises that the Cloud holds for users and applications alike. However, there are issues that need to be resolved before this technology can be exploited to its maximum potential [238]. In the section below we enumerate and discuss the burning issues which might deter the phenomenal growth of Cloud Computing technology.

(i)Security Issues: Security is as much of a concern in Cloud computing as it would be in any other computing paradigms. Cloud Computing can be vaguely defined as outsourcing of services, which in turn causes users to lose significant control over their data. There is always a risk of seizure associated with the public Clouds. For instance, an organization sharing data in an environment

where other organizations are doing the same is always under the threat of compromising the security and privacy of its data if any other organization in the shared scheme happens to violate the security protocols [390]. Moreover, in a virtualized environment one needs to consider the security of not just the physical host but also the virtual machine. This is because if the security of a physical host is compromised, then automatically all virtual machines face security threat and vice versa. Since the majority of services in Cloud computing are provided using web browsers, there are many security issues related with it as well [392]. Flooding is also a major issue where an attacker sends huge amounts of illegitimate service requests which cause the system to run slow thereby hampering the performance of the overall system. Cloud networks stand the potential threat of both Indirect Denial of Service attacks and Distributed Denial of Service attacks.

(ii)Legal and Compliance Issues: Clouds are sometimes bounded by geographical boundaries. Provision of various services is not location dependent but because of this flexibility Clouds face Legal & Compliance issues. These issues are related mainly to the vendors though they still affect the end users. These issues are broadly classified as functional (services in the Clouds that have legal implications for both service providers and end users), jurisdictional (where governments administer laws to follow) and contractual (terms and conditions). Issues include (a) **Physical Location of the data** referring to where the data is physically located and if a dispute occurs, which jurisdiction will help in resolving it (b) **Responsibilities of the data** where if a vendor is hit by a disaster will the businesses using its services be covered under insurance (c) **Intellectual Property Rights** which deals with the way trade secrets are maintained [732].

(iii)Performance and QoS Related Issues: For any computing paradigm performance is of utmost importance. Quality of Service (QoS) varies as the user requirements vary. One of the critical QoS related issues is the optimized way in which commercial success can be achieved using Cloud computing. If a provider is not able to deliver the promised QoS it may tarnish its reputation [238]. Since Software-as-a-Service (SaaS) deals with provision of softwares on virtualized resources, one faces the issue of Memory and Licensing constraints which directly hamper the performance of a system.

(iv)Data Management Issues: The main purpose of Cloud Computing is to put the entire data on the Cloud with minimum infrastructure requirements for the end users. The main issues related to data management are scalability of data, storage of data, data migration from one Cloud to another and also different architectures for resource access [127]. Since data in Cloud computing even includes high confidential information it is of utmost importance to manage these data effectively. There has been an instance where an online storage service called The Linkup got shut down after losing access to as much as 45% of its customers. While transferring data, i.e., data migration, in a Cloud has to be done very carefully as it could lead to bottlenecks at each

and every layer of the network model, as huge chunks of data are associated with the Cloud [762].

(v)Interoperability Issues: The Cloud computing interoperability idea was conceived by Reuven Cohen. Reuven Cohen is founder and Chief Technologist for Toronto based Enomaly Inc. Vint Cerf, who is a co-designer of the Internet's TCP/IP standards and widely considered a father of the Internet, spoke about the need for data portability standards for Cloud computing. Companies such as Microsoft, Amazon, IBM, and Google all own their independent Clouds but they lack interoperability amongst them. Each service provider has its own architecture, which caters to a specific application. To make such uniquely distinct Clouds interoperate is a non-trivial problem. The lack of standardized protocols in the domain of Cloud computing further makes interoperability a challenge. The key issues hampering implementation of interoperable Clouds are the large scale access and computational capabilities of the Clouds, resource contention and the dynamic nature of the Cloud. However, interoperability amongst the various Clouds would only add to the value of this technology, making it more widely accessible, fault tolerant, and thereby robust.

1.8   Cloud Computing and Grid Computing: A Comparative Study

Cloud computing should not be confused with Grid Computing, Utility Computing and Autonomic Computing. Grid computing consists of clusters of loosely coupled, networked computers acting in concert to perform very large tasks. Utility computing packages computing resources as a metered service. Autonomic computing stresses self management of resources. Grids require many computers, typically in the thousands, and commonly use servers, desktops, and laptops. Clouds also support non-grid environments [284]. The differences between Utility computing and Cloud computing are crucial. Utility computing relates to the business model in which application, infrastructure, resources, hardware and/or software are delivered. In contrast, Cloud computing relates to the way we design, build, deploy and run applications that operate in a virtualized, shared-resource environment accompanied by the coveted ability to dynamically grow, shrink and self-heal. Thus the major aspects which separate Cloud Computing from other computing paradigms are user centric interfaces, on-demand services, QoS guarantees, autonomous system organization, scalability and flexibility of services [369].

In the mid 1990s, the term "Grid" was coined to describe technologies that would allow consumers to obtain computing power on demand. Ian Foster and others posited that by standardizing the protocols used to request com-

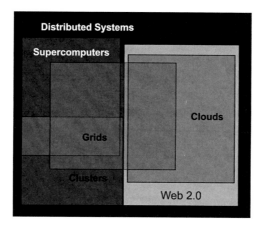

FIGURE 1.8
Cloud and Grid Computing [283].

puting power, we could spur the creation of a Computing Grid, analogous in form and utility to the electric power grid. Figure 1.8 showcases the overlap and distinctions amongst several cutting edge computing technologies [369]. Web 2.0 covers almost the whole spectrum of service-oriented applications, whereas Cloud Computing lies mostly on the large-scale computing domain. Supercomputing and Cluster computing have been more focused on traditional non-service applications. Grid Computing overlaps with all these fields and is generally considered to be catering to smaller scaled computing requirements than the Supercomputers and the Clouds [284].

Half a decade ago, Ian Foster gave a three point checklist to help define what is, and what is not a Grid. We present his checklist below:
1.) Coordinates resources that are not subject to centralized control,
2.) Uses standard, open, general-purpose protocols and interfaces, and
3.) Delivers non-trivial qualities of service.

Although the third point holds true for Cloud Computing, neither point one nor two is applicable for today's Clouds. The vision for Clouds and Grids are similar but the implementation details and technologies used differ considerably. In the following section, we discuss the differences in these two technologies based on their architecture, business model, and resource management techniques.

Architecture
Grids provide protocols and services at five different layers as identified in the Grid protocol architecture shown in Figure 1.9.

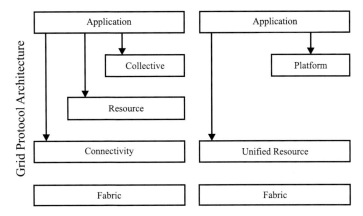

FIGURE 1.9
Cloud Architecture vs. Grid Architecture [283].

At the *fabric layer*, Grids provide access to different resource types such as computation, storage and network resource, code repository, etc. The *connectivity layer* defines core communication and authentication protocols for easy and secure network transactions. The *resource layer* defines protocols for the publication, discovery, negotiation, monitoring, accounting and payment of sharing operations on individual resources. The *collective layer* captures interactions across collections of resources and directory services. The *application layer* comprises whatever user applications are built on top of the above protocols and APIs environments. [284, 370]

In contrast, the Cloud architecture presents a four-layer architecture composed of (i)Fabric (ii)Unified resource (iii) Platform and (iv)Application layers. (It is to be noted that there are other paradigms of cloud architecture as well. We however, choose to focus on the 4-layer architecture for our discussion).

The Fabric layer contains the raw hardware level resources, such as computational resources, storage resources, and network resources. The Unified Resource Layer contains resources that have been abstracted/encapsulated (usually by virtualization) so that they can be exposed to upper layer and end users as integrated resources, for instance, a virtual computer/cluster, a logical file system, a database system, etc. The Platform layer adds a collection of specialized tools, middleware and services on top of the unified resources to provide a development and/or deployment platform. Finally, the Application layer contains the applications that run in the Clouds. Clouds in general provide services at three different levels, namely IaaS, PaaS, and SaaS, although some providers can choose to expose services at more than one level. [284]

Resource Management

With regard to resource management we compare Clouds and Grids using the following models:

(i)Computational Model: Most Grids use a batch-scheduled computational model, in which a local resource manager (LRM), such as PBS, Condor and SGE manages the computational resources for a Grid site. Users submit batch jobs via GRAM that occupies a certain amount of resources for a stipulated period of time [787]. In contrast to the Grid's paradigm of allocating dedicated resources to specific "jobs" for a stipulated amount of time with a scheduling algorithm governing the process, the Cloud computing model takes a resource sharing approach by all users at all times [371]. The goal here (in Cloud computing) is to achieve low queuing latency for resource access. Moreover, this design should allow latency sensitive applications to operate natively on Clouds, although ensuring the required level of QoS is likely to be one of the major challenges for Cloud Computing, especially as the Cloud scales to incorporate more users. [284]

(ii)Data model: The importance of data has caught the attention of the Grid community for the past decade; Data Grids have been specifically designed to tackle data intensive applications in Grid environments, with the concept of virtual data playing a crucial role. Virtual data captures the relationship between data, programs, and computations and prescribes various abstractions that a data grid can provide. For instance, (a) location transparency where data can be requested without regard to data location, a distributed metadata catalog is engaged to keep track of the locations of each piece of data (along with its replicas) across grid sites, and privacy and access control are enforced; (b) materialization transparency where data can be either recomputed on the fly or transferred upon request, depending on the availability of the data and the cost to re-compute [378]. There is also representation transparency where data can be consumed and produced no matter what their actual physical formats and storages are, data are mapped into some abstract structural representation and manipulated in that way. In contrast, the Cloud computing paradigm is centralized around Data.Cloud Computing and Client Computing will coexist and evolve hand in hand, while data management (mapping, partitioning, querying, movement, caching, replication, etc.) will become more and more important for both Cloud Computing and Client Computing with the increase of data-intensive applications. [163]

(iii)Virtualization: Grids do not rely on virtualization as much as Clouds do, but that might be more due to policy issues and having each individual organization maintain full control of their resources (i.e., not by virtualization). However, there are efforts in Grids to use virtualization as well, such as Nimbus, which provides the same abstraction and dynamic deployment capabilities. A virtual workspace is an execution environment that can be deployed dynamically and securely in the Grid. In addition, Nimbus can also provision

a virtual cluster for Grid applications (e.g., a batch scheduler, or a workflow system), which is also dynamically configurable — a growing trend in Grid Computing. [284]

(iv)Monitoring: Another challenge that virtualization brings to the Clouds is the potential difficulty in retaining fine grained control over the monitoring of resources. Although many Grids (such as TeraGrid) also enforce restrictions on what kind of sensors or long-running services a user can launch, Cloud monitoring is not as straightforward as in Grids, because Grids in general have a different trust model in which users via their identity delegation can access and browse resources at different Grid sites, and Grid resources are not highly abstracted and virtualized as in Clouds. Monitoring can be argued to be less important in Clouds, as users interact with a more abstract layer that is potentially more sophisticated. This abstract layer could respond to failures and quality of service (QoS) requirements automatically in a general-purpose way irrespective of application logic. In the near future, user-end monitoring might be a significant challenge for Clouds, but it will become less important as Clouds become more sophisticated and more self-maintained and self-healing.

(v)Programming Model: Grids primarily target large-scale scientific computations, so they must scale to leverage large number/amount of resources, and we would also naturally want to make programs run fast and efficiently in Grid environments. Programs must also run correctly, so reliability and fault tolerance must be considered. Clouds (such as Amazon Web Services, Microsoft's Azure Services Platform) have generally adopted Web Services APIs where users access, configure and program Cloud services using predefined APIs exposed as Web services, and HTTP and SOAP are the common protocols chosen for such services. Although Clouds adopted some common communication protocols such as HTTP and SOAP, the integration and interoperability of all the services and applications remain the biggest challenge as users need to tap into a federation of Clouds instead of a single Cloud provider. [519]

(v)Application model: Grids generally support many different kinds of applications, ranging from high performance computing (HPC) to high throughput computing (HTC). HPC applications are efficient at executing tightly coupled parallel jobs within a particular machine with low-latency interconnects and are generally not executed across a wide area network Grid. On the other hand, Cloud computing could in principle cater to a similar set of applications. The one exception that will likely be hard to achieve in Cloud computing (but has had much success in Grids) are HPC applications that require fast and low latency network interconnects for efficient scaling to many processors. As Cloud computing is still in its infancy, the applications that will run on Clouds are not well defined, but we can certainly characterize them to be loosely coupled, transaction oriented (small tasks in the order of millisec-

onds to seconds), and likely to be interactive (as opposed to batch scheduled as they are currently in Grids).

(vi)Security Model: Clouds mostly comprise dedicated data centers belonging to the same organization, and within each data center, hardware and software configurations and supporting platforms are in general more homogeneous as compared with those in the Grid environments. Interoperability can become a serious issue for cross-data center and cross-administration domain interactions. Imagine running your accounting service in Amazon EC2 while your other business operations are on Google infrastructure. Grids, however, are built on the assumption that resources are heterogeneous and dynamic, and each Grid site may have its own administration domain and operation autonomy. Thus security has been engineered in the fundamental Grid infrastructure. The key issues considered are: single sign-on, so that users can log on only once and have access to multiple Grid sites; this also facilitates accounting and auditing, delegation (so that a program can be authorized to access resources on a user's behalf and it can further delegate to other programs), privacy, integrity and segregation. Moreover, resources belonging to one user cannot be accessed by unauthorized users and cannot be tampered with during transfer. In contrast, currently, the security model for Clouds seems to be relatively simpler and less secure than the security model adopted by the Grids. Cloud infrastructure typically relies on Web forms (over SSL) to create and manage account information for end-users, and allows users to reset their passwords and receive new passwords via emails in an unsafe and unencrypted communication. Note that new users could use Clouds relatively easily and almost instantly, with a credit card and/or email address. To contrast this, Grids are stricter about security.

Business models:

In a Cloud-based business model, a consumer pays the provider on the basis of resource consumption, akin to the utility companies charging for basic utilities such as electricity, gas, and water. The model relies on economics of scale in order to drive prices down for users and profits up for providers. Today, Amazon essentially provides a centralized Cloud consisting of Compute Cloud EC2 and Data Cloud S3. The former is charged based on per instance-hour consumed for each instance type and the latter is charged by per GB-Month of storage used. In addition, data transfer is charged by TB /month data transfer, depending on the source and target of such transfer. The business model for Grids (at least that found in academia or government labs) is project-oriented, in which the users or community represented have a certain number of service units (i.e., CPU hours) they can spend.

Thus Clouds and Grids share a lot of commonality in their vision, architecture and technology, but they also differ in various aspects such as security, programming model, business model, computational model, data model, applications, and abstractions. We believe a close comparison such as this can help the two communities understand, share, and evolve infrastructure and technology within and across, and accelerate Cloud computing to leap from early prototypes to production systems. [360]

1.9 Conclusion

Cloud Computing plays a significant role in varied areas like e-business, search engines, data mining, virtual machines, batch oriented scientific computing, online TV amongst many others. Cloud computing has the potential to become an integral part of our lives. Examples include, (i) the Cloud operating system which provides users with all the basic features of an operating system like data storage and applications; (ii) mapping services which help users in finding routes to various places; (iii) Telemedicine applications for collecting data of a patient and calling emergency services in dire need.

As an increasing number of businesses move toward Cloud based services, issues like interoperability, security, portability, migration and standardized protocols are proving to be critical concerns. For instance, the need for higher transparency in scheduling tasks with guaranteed QoS is proving to be a challenging issue in the area of data management over the Clouds. In addition, the service driven model of Cloud Computing often leads to concerns and queries regarding the Service Level Agreement (SLA) of enterprises.

The research and business community alike are coming up with new and innovative solutions to tackle the numerous issues that are making its way as Cloud Computing is shaping itself as the Future of Internet. The potentials are endless and the sky is the limit as we watch "a computing environment to elastically provide virtualized resources as a service over the Internet in a pay-as-you-go manner" [631] bloom and blossom to its maximum potential.

APPENDIX: Comparisons of different Cloud computing technologies
In this section, we present a comparative data-set that showcases the various solution and service strategies of some of the major Cloud service providers today, like Amazon Web Services, GoGrid, FlexiScale, Google, Nimbus and a few others. Detailed description of this data set is available at [608].

TABLE 1.1

Outages in Different Cloud Services

Vendor	Service and outage	Outage Duration
Microsoft	Malfunction in Windows Azure	22 hours
Google	Gmail and Google Apps	2.5 hours
	Google search outage due to programming error	40 Mins
	Gmail site unavailable due to outage in contacts system	1.5 hours
	Google App engine partial outage	5 hours
Amazon	Authentication overload single bit error leading to protocol blowup	2 hours 6–8 hours
Flexiscale	Core network failure	18 hours

TABLE 1.2

Comparison of Different Cloud Computing Technologies and Solution Provider

Feature	Amazon Web Services	GoGrid	Flexiscale
Computing Architecture	Provides IaaS.Gives client API's to manage their Infrastructure	Provides IaaS.Designed to deliver a guaranteed QoS level and reconfigures itself depending on demand.	Provides IaaS.Functions similar to GoGrid's architecture but allows multi -tier architectures.
Load Balancing	Round-Robin Load Balancing, HAproxy.	F5 Load Balancing. Algorithms used are round-robin, sticky session, SSl least connect,source address.	Automatic Equalization of server load within clusters.
Fault Tolerance	System alerts automatically, does failover and resyncs to last known state.	Instantly scalable and reliable file level backup service.	Full Self-Service.
Interoperability	Supports horizontal interoperability	Working towards interoperability.	Working towards interoperability.
Storage	S3 and SimpleDB	First connects each server to private network and then uses transfer protocols.	Persistent storage based on a fully virtualized high end SAN/NAS back end.
Security	Type-I, Firewall, SSl, Access Control list.	No guarantee of security.	Customers have their own VLAN.
Programming Framework	Amazon Machine Image, Amazon MapReduce.	Uses REST-like Query interface and supports Java, Python and Ruby.	Supports C, C++, Java, PHP, Pearl and Ruby.

TABLE 1.3

Features of Different PaaS and SaaS Providers

Feature	Google App Engine	Azure	Force.com
Computing Architecture	Google's geo-distributed architecture	Platform is hosted on Microsoft datacenters. Provides an OS and a set of developer's cloud services.	Facilitates multi-tenant architecture allowing a single application to serve multiple customers.
Load Balancing	Automatic scaling and Load Balancing	Built in hardware Load Balancing. Containers are used as load balancer.	Load Balancing among tenants.
Fault Tolerance	Automatically pushed to a number of fault tolerant servers.	On Failure SQL data services will automatically begin using another replica of container.	Self-Management and Self-Learning.
Interoperability	Supports interoperability between platforms of different vendors and programming languages.	Supports interoperability between platforms.	Application level integration between different clouds.
Storage	Proprietary database (Big Table distributed storage)	SQL Server Data Services (SSDS).	Database deals in terms of relationship fields.
Security	Google's Secure Data Connector (SDC).SDC uses TLS based server authentication and uses RSA/128-bit or higher AES CBC/SHA.	Security Token Service (STS) creates a security assertion markup language token according to rule.	SysTest SAS, 70 Type II.
Programming Framework	MapReduce framework supporting Python and Java	Microsoft NET.	Apex for database service and supports NET, Apache Axis (Java, C++).

TABLE 1.4

Comparisons of Different Open Source-Based Cloud Computing Services

Feature	Eucalyptus	OpenNebula	Nimbus
Computing Architecture	Can configure multiple clusters in a Private Cloud	It is based on Haizea scheduling. Focuses on efficient, dynamic and scalable management of VM's within private clouds.	It has a client side cloud computing interface to Globus enabled TeraPort cluster. Nimbus Context Broker combines several deployed VM's into "turnkey" virtual cluster.
Load Balancing	Simple Load Balancing cloud controller	Nginx server is used as Load Balancer with round-robin or weighted selection mechanism.	Triggers self configuring virtual clusters.
Fault Tolerance	Separate clusters within a cloud reduce the risk of correlated failure.	Daemon is restarted and a persistent database backend is used to store host and VM info.	Checks worker nodes periodically and performs recovery operation.
Interoperability	Multiple private clouds use the same backend infrastructure.	Interoperable between intra cloud services.	Standards "rough consensus and working code."
Storage	Walrus is used.	SQLite3 is used as a backend component. Persistent storage for ONE data structures.	GridFTP and SCP.
Security	WS-security for authentication, cloud controller generates the public/private key.	Firewall, virtual private network tunnel.	PKI credential required, works with GRID proxies VOMS, shibboleth custom PDP's.
Programming Framework	Supports Hibernate, Axis2 and Java.	Supports Java and Ruby.	Supports Python and Java.

2

Cloud Computing and Startups

Åke Edlund and Ilja Livenson

PDC, KTH Royal Institute of Technology, Stockholm, Sweden
SICS Startup Accelerator, Stockholm, Sweden

CONTENTS

2.1 Introduction

Early stage companies, typically referred to as startups, usually have very small resources to play with and, at the same time, a strong demand for flexibility and scalability. To be able to develop business ideas into services quickly, and at the same time be able to adapt to the customer feedback is vital for these companies. Cloud computing brings the ability to scale both

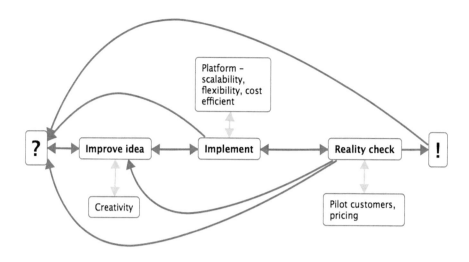

FIGURE 2.1
The Iterative Process of Innovation.

technologically and business-wise using a pay-as-you-go paradigm, allowing the users to concentrate more resources and time on their ideas. The match between startups and cloud computing was identified early, making many of the startups early adopters of this paradigm shift.

In this chapter we look more closely at how cloud computing accelerates the highly iterative innovation cycle for startups (see Figure 2.1). A number of examples is given; we also discuss how cloud computing changes the overall landscape for startups and investors in startups.

2.2 Time to Market

During the development, from the very first idea to a product or a service, time to market is often considered as one of the key components for success. Even if the value of being the very first to a new market is hard to evaluate [49, 473], the ability to adapt rapidly to competition and customer need is not. Figure 2.2 shows a historical perspective on the increasing intensity of the competition. Quickly launching early prototypes for customer feedback is very useful for choosing the right path to a new service. This is one of the strong features of the cloud computing concept, to be able to do rapid and adaptive development to explore new markets. A very similar process occurs also in larger corporations, especially in research and development departments but

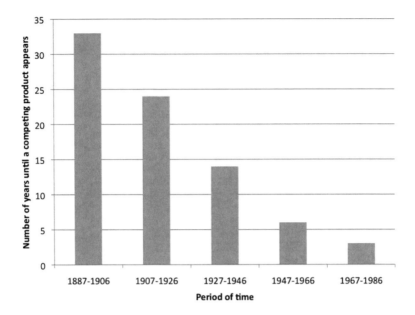

FIGURE 2.2
Over the Years the Time It Takes for a Competing Product to Be Released to Market Shrank Dramatically.

also in the interaction with customers and partners. The main difference is that in larger companies development more often relies on hybrid clouds, consisting of a combination of internal infrastructure and services together with public cloud offerings.

Failing is not a bad thing, this is a part of the learning process. Not being able to reach the market in a quick way can mean running a failed company longer than necessary. Using flexible and scalable technologies, the startups can quickly change direction and try new angles. Let us now look more closely at the essential cloud characteristics [541] and their implications on startups.

2.3 Cloud Computing Implications

2.3.1 On-Demand Self-Service

Description: The user can — without any human interaction with the cloud service provider — provision computing capabilities, such as server time and network storage, in an on-demand self-service manner.

Implications for startups: the company can easily and swiftly get needed IT infrastructure in place. Especially for smaller newer companies, and even more so for not-yet-started companies, negotiating sales contracts is not their strongest side. In addition, the need of server time and network storage is highly unpredictable for early stage companies, making the on-demand self-service characteristics of cloud computing even more attractive. One early example: Yieldex [781] during startup phase used Amazon Wed Services to demo the capabilities of their publishing service for investors with a total cost of 40 USD for the first month. This was made possible by allocating cloud resources for the actual meetings, and releasing them directly after the meetings, with no human interaction, and on-demand.

2.3.2   Broad Network Access

Description [541]: Capabilities are available over the network and accessed through standard mechanisms that promote use by heterogeneous thin or thick client platforms (e.g., mobile phones, laptops, and PDAs).

Implications on startups: Using cloud services and distribution platform for mobile clients a whole new field of services arises. With this delivery chain, the smallest company can grow overnight into a much larger one by offering services in a scalable way. The most famous examples include the Apple App Store distribution platform and Android applications that often rely on the Google App Engine-based backend.

2.3.3   Resource Pooling

Description [541]: The provider's computing resources are pooled to serve multiple consumers using a multi-tenant model, with different physical and virtual resources dynamically assigned and reassigned according to consumer demand. There is a sense of location independence in that the customer generally has no control or knowledge over the exact location of the provided resources but may be able to specify location at a higher level of abstraction (e.g., country, state, or datacenter). Examples of resources include storage, processing, memory, network bandwidth, and virtual machines.

Implications on startups: Resource pooling is one of the reasons why public IaaS can be more cost-effective than owning own infrastructure. However, unless a startup is building a service on top of a private cloud, the price tag for a certain cloud service is much more important.

2.3.4   Rapid Elasticity

Description [541]: Capabilities can be rapidly and elastically provisioned, in some cases automatically, to quickly scale out, and rapidly released to quickly scale in. To the consumer, the capabilities available for provisioning often appear to be unlimited and can be purchased in any quantity at any time.

Implications on startups: Through rapid elasticity, the company can quickly adapt its service to address the customer demands. This results in a cost effective scalable business model very useful for both small and medium companies. From services built directly on IaaS, Animoto [90] is the more well-known example, porting its photo presentation application to Facebook, generating a large peak in usage. Animoto was prepared for this, using RightScale and Amazon to handle the peak in an economical way. Dropbox [243] and other storage services sell space on demand in an elastic way avoiding large overhead in capacity.

2.3.5 Measured Service

Description [541]: Cloud systems automatically control and optimize resource use by leveraging a metering capability at some level of abstraction appropriate to the type of service (e.g., storage, processing, bandwidth, and active user accounts). Resource usage can be monitored, controlled, and reported providing transparency for both the provider and consumer of the utilized service.

Implications on startups: Being able to calculate the cost of a certain business transaction is very useful for making decision, for example, for establishing a price-list for the end customer. Resource usage metrics of the cloud services make this process much easier, as they can be directly converted into monetary values and service level agreements (SLAs).

2.4 Changes to the Startup Ecosystem

Cloud computing opens up a new way of launching startups. By creating scalable business models with consumption based pricing, the startups can evolve with a higher degree of cost control than earlier. Customer needs and behavior are hard to predict, which means a high risk of developing wrong or too costly services. With a higher level of agility, companies get a better control of cost versus revenue.

In *The Future of Web Startups* [Graham 2007] Paul Graham lists a number of changes to the startup ecosystem, changes that since then have come true — especially for Paul Grahams own Y Combinator [779]. More startups are launched, with faster turn-around from testing to the next step (continuation or end). Many of these startups are web based, most of them rely on cloud computing, many with cloud computing as part of their offering.

2.4.1   Lowering the Barrier to Entrance

As a result of the availability of low cost cloud computing services, mature open source software stacks, high-quality connectivity and novel mobile service platforms, more entrepreneurs start web-based companies in a shorter time. For similar reasons, many new-formed companies can operate further than before without the need for external investments. This results in a change in the overall investment chain, with a shift of control to the favor of the entrepreneurs.

By Figure 2.3 we try to give an idea of the implications on the startup-investor ecosystem. Cloud computing together with a number of earlier changes in IT (open source, network, mobility, commodity low cost hardware) lowers the cost for startups to start. Due to the quick launch through cloud computing, the entrepreneurs get quick feedback on their ideas — as well as the investors get quick feedback on their investments. This lowers the risk and need of initial capital, enabling more investments and startups to launch. At the same time startups can now develop longer before involving external capital, if involving it at all.

All the above mentioned features of cloud computing are very appealing also from the investors point of view. Investors, for example, business angels, venture capitalists, the entrepreneurs themselves, do not have to make heavy IT infrastructure investments at the early stage of the companies, which has been the case for at least a decade. They can get relatively quick feedback and only later, once the company matures, an option of purchasing their own infrastructure could be considered, for security reasons or total cost minimization. It also means that if at some point a certain startup fails to meet expectations, shutting it down is as easy as stopping the virtual machines, without the hassle of the IT infrastructure leftovers.

2.4.2   Seed Accelerator Programs

In Figure 2.3 we also point at the investor's point of view: they now need to evaluate more startups if they want to get involved early. This is not possible in most venture capital firms due to low staffing, and new models are needed. One of these is to get involved in a so called seed accelerator.

The seed accelerator program [42, 190, 779] is a new model of funding and assisting startup companies. The model is especially suitable for quick-starting web and media based technology and service businesses with a relatively low entry barrier. Cloud computing is the key for many of these startups.

Y Combinator [779] was the first seed accelerator, specializing in web services. Y Combinator invites technology focused teams to develop their service within the seed accelerator for a short period of time, typically 12 weeks, twice a year. During this stay the teams evolve, from idea to early stage company, exposed to the network of investors and services (marketing, sales, legal). Successful examples from Y Combinator includes Dropbox [243], Zencoder

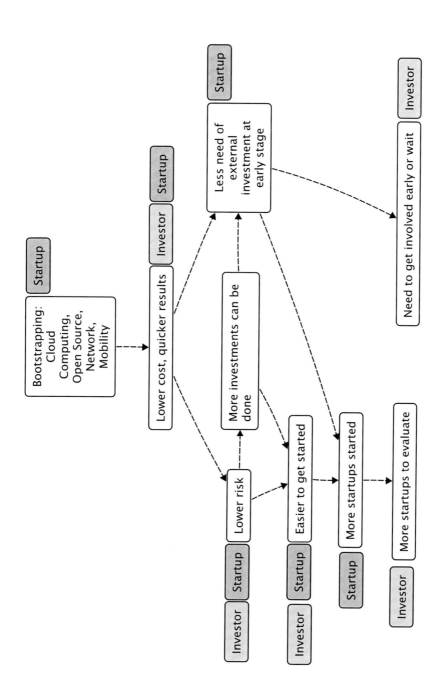

FIGURE 2.3
Evolution of a Startup Ecosystem to Today's Cloud-Based Startups and How It Affects Startups and Investors.

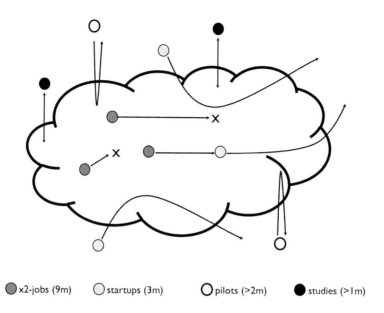

○ x2-jobs (9m)   ○ startups (3m)   ○ pilots (>2m)   ● studies (>1m)

FIGURE 2.4
The SICS Startup Accelerator, Interacting with External Cloud Based Companies and Developing Cloud Focused Startups in ×2-jobs.

[791], and Heroku [347] — all based on cloud computing. In mid-2010 about 25% of the teams had found investors during the 12 weeks stay in the seed accelerator. In exchange for this service Y Combinator takes 2 to 10% of the company. Through seed accelerators, venture capital companies can sometimes get an early view of the upcoming startups. In the Y Combinator case, Sequoia Capital [640] is a partner backing the overall investment process.

2.4.2.1   SICS Startup Accelerator Projects and Services

Y Combinator has inspired a number of new seed accelerators, for example, Aalto Venture Garage [42] and SICS Startup Accelerator [683]. In the SICS Startup Accelerator case (see Figure 2.4) focus is on active interaction also with external cloud based companies, including larger corporations:

- ×2 job is a 9 month long MSc thesis work (or on the same level), with the goal of starting companies from the outcome of the work. These ×2 jobs are in close collaboration with local innovation support organizations in the region.

- Incubator support projects, where SICS Startup Accelerator is supporting incubators on the technology side, e.g., architecture and design.

- Corporation support projects, where SICS Startup Accelerator is supporting larger size companies on innovation in the cloud area.

- Corporation analysis work for larger organizations and companies.

The number of seed accelerators has been increasing rapidly in the last years to accommodate a growing demand for such services.

2.4.3 Cloud Innovation Platforms

The Cloud Innovation Platform is one of the services that a Seed Accelerator could offer. It comprises a set of prepared and tested ready-to-go recipes to even further accelerate the development of startups.

The Cloud Innovation Platform consists of a common set of cloud services available for testing and development — a set of cloud services that will be constantly improved and extended, based on the feedback from its users. The Cloud Innovation Platform addresses two main issues: the basic scalable IT functionality needed for implementation of the startup's idea, typically mimicking and incorporating IaaS offerings, both private and public; and a specialized functionality for novel media and content delivery, payments and accounting services, and modern programming models.

A Cloud Innovation Platform can be used both for development of the new services and education in the best practices of cloud computing usage, for example, for the enterprise clients considering migration of the internal IT systems to the cloud.

A very important requirement for the Cloud Innovation Platform is its interoperability with other cloud offerings. Ability to move away is a startup-friendly approach that most of the public Platform-as-a-Service offerings, for example, Google App Engine or Microsoft Azure, lack.

2.5 Evolution of the Cloud-Based Company

Like any living organism, startups change over time. They can grow into something bigger, get "eaten" by another company or simply die. Figure 2.5 shows the most typical stages of the startup development.

The usage patterns of cloud computing change with the size and profitability of the company. For example, early stage companies with very little financial capabilities often have no other reasonable option but to go with a public cloud offering for all of their needs. Large and wealthier companies with higher demands on security could choose a more restrictive solution by creating a

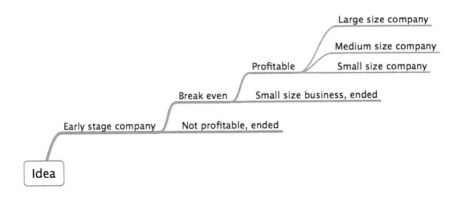

FIGURE 2.5
Evolution of a Startup.

hybrid or private cloud for some of the more mission critical tasks. Figure 2.5 depicts a conservative approach of the companies of different sizes to satisfying certain goals.

2.5.1　Costs and Risks

Balancing out costs and risks is a complicated but necessary step in making a certain business decision, for example, in uptaking a certain set of cloud services. One of the key selling points of the public cloud providers is that using their offer is more optimal if you take into account all of the cost components. Whether it is actually true often comes down to a specific customer situation. In some cases, a potential cloud customer possesses something that can make using out-of-the box cloud computing a bad decision. Some of the more important decision modifiers include availability of server administration know-how, usage of the data regulated by a certain policy, and existence of the in-house infrastructure.

Below we list and explain major cost inflicting factors of the IT infrastructure from the perspective of the consumer of the cloud resources. We also provide a brief introduction of the risks associated with using cloud computing.

2.5.1.1　Cost Components

Consumed resources

The charging model of the cloud providers often includes a pay-as-you-go model. That means that they break down the service offering into many components and charge you according to the monitored consumption. Typical components include virtual machine uptime, average storage used in a month,

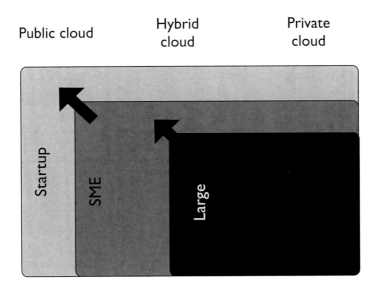

FIGURE 2.6
Adoption of Cloud Deployment Models by Companies of Various Sizes and
the Trend toward Wider Cloud Adaption.

inbound and outbound network. Quite often there are also more services offered that can be used for creating applications running in the cloud: message
queues, databases, identity providers, content distribution networks and so
on. While from the sales point of view these services are independent from
each other — for example, you can use AWS EC2 with your own data storage
rather than using AWS S3 — but in practice this often leads to a considerable
overhead, for example, increase in network traffic.

The pricing policies are also quite diverse. Apart from the basic pay-as-you-
go one, there are policies that offer you service at a reduced quality for a
cheaper price. This is the case for AWS S3 Reduced Redundancy Storage that
has a lower SLA on the durability due to a smaller number of replicas of the
data being stored. AWS also offers reduced price in case of a certain upfront
commitment — AWS EC2 Reserved Instances — and a market-based bidding
system for the computational resources — AWS EC2 Spot Instances.

In case of a small resource consumption these policies are not really important.
However, for a larger consumption, and also in case of a sustainable business,
it is definitely worth taking a look at potential cost optimizations by switching
to another pricing policy.

Administrative overhead

Depending on the level of cloud computing offering (IaaS, PaaS, SaaS) there
are different needs for administrating services. While buying an Infrastructure-

as-a-Service offering does relieve you of a need to manage hardware, network and application administration is still required. This can be quite expensive unless you have that know-how inside the company. For PaaS and SaaS the overhead is smaller and the required skills are typically closer to the consumer of this service.

License costs

The Cloud computing paradigm was quite disruptive with respect to the standard software licensing models, as it essentially requires the ability to be able to rent licenses. Moreover, if a company has already bought a set of licenses, it can easily be the case that they cannot be used for running in a cloud, as it is outside the domain of a company, or that they are linked to a number of working machines, which makes the flexibility of the cloud not really usable. One of the solutions to this problems was creation of the virtual appliance marketplaces. They provide means for the software vendors to publish their products as specific software package or even full virtual machine image, so that once it is running, the client is paying extra for the uptime of the virtual machine that goes directly to the software vendor. Examples include GoGrid Exchange [307] and Amazon DevPay [235].

2.5.1.2   Risks

Security is often being named as one of the main obstacles on the way to cloud computing adoption [251]. Most of the time it means that the risks connected with such a decision are either too high or too complicated to evaluate. While the concept of outsourcing IT-processes is not new at all, and so the concept is familiar, there are several new threats introduced by the cloud computing that both startups and more mature companies should consider. Cloud Security Alliance [211] has produced several documents addressing these new threats, mostly from the security and technological points of view [212]. For most of the startups, however, functionality is more important than the security guarantees. Picking a cloud provider with the most comfortable set of services, and trusting it, is the most popular way to go. While acceptable behavior in the beginning, when the overhead of the security analysis is too high, risks should definitely be more thoroughly analyzed once the startup has started to grow.

Apart from the technological risks, there are also risks related to the way companies operate. Nowadays almost all of the companies generate and store a lot of data. This leads to a risk of vendor lock-in, even if the interfaces are compatible and transition could be smooth. Data storage inflicts a constantly raising cost as most of the time companies tend to save as much information as possible. This makes storage cost optimization a very important problem. However, even if you find a cheaper option for data storage, the "momentum" of the existing data can be a problem — transferring petabytes of data is neither fast nor cheap.

TABLE 2.1
Importance of IT Requirements for Companies of Various Sizes.

	Startup	SME	Large company
Scalability	High	Average	Low (stability is more important)
Security	Low	Average	High
Risk analysis	Low	Average	High
Cost optimization	Low	Average	High
Vendor lock-in	Low	Average	High

2.5.1.3 Evolution of Requirements

With the development of a company, its requirements for supporting IT infrastructure change to reflect priorities of a company. Table 2.1 lists major changes in a company's requirements based on its size. For example, while for a small company the ability to scale their business quickly is very important, for medium size, and especially larger ones, stability and predictability are of a higher value.

2.6 Summary

Cloud computing characteristics map to the needs of startups very well. So well that a new group of cloud specific startups is now rapidly evolving. On one hand, one could argue that it is just the same active group that has quickly picked up a new technology, but on the other hand the number of startups evolving and how fast they can get started and how far they can go on low funding is a game changer. Startups have always been very good at bootstrapping, getting as far as possible on no or very little funding. But now, with cloud computing and high quality open source software, better and less expensive network, a more open mobile market, and more evolved customer base, the bootstrapping can take you very far, possibly all the way to a self supported profitable business.

For investors the market is also changing, creating a need for very early stage, close-to-the-founder, technology knowledgeable services — the seed accelerators. The seed accelerators act as an early investor — helping the startups with technology decisions — and at the same time helping future investors in the identification of interesting objects.

3

A Taxonomy of Interoperability for IaaS

Ralf Teckelmann

Department of Computer Science
Hochschule Furtwangen University, Germany

Anthony Sulistio

Department of Applications, Models and Tools
High Performance Computing Center Stuttgart (HLRS), Germany

Christoph Reich

Department of Computer Science
Hochschule Furtwangen University, Germany

CONTENTS

The idea behind cloud computing is to deliver Infrastructure-, Platform- and Software-as-a-Service (IaaS, PaaS and SaaS) over the Internet on an easy pay-per-use business model. However, current offerings from cloud providers are based on proprietary technologies. As a consequence, consumers run into a risk of a vendor lock-in with little flexibility in moving their services to other providers. This can hinder the advancement of cloud computing to small- and medium-sized enterprises. In this chapter, we present our work in outlining the motivations and current trends of achieving interoperability, especially in the area of IaaS. More specifically, this work delivers a comprehensive taxonomy as a guideline for cloud providers to enable interoperability within their cloud infrastructures. Thus, this taxonomy discusses important topics of IaaS, such as access mechanism, virtual appliance, security, and service-level agreement.

3.1 Introduction

According to National Institute of Standards and Technology (NIST), there are five essential characteristics of cloud computing, i.e., *on-demand self-service, broad network access, resource pooling, rapid elasticity* and *measured Service* [509]. *On-demand self-service* means that customers are able to obtain computing capabilities without human interaction from a service provider. *Broad network access* defines the need for network-based access and standardized mechanisms in order to facilitate access through heterogeneous platforms.

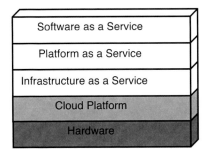

FIGURE 3.1
Layers of Cloud Computing.

Resource pooling describes providers' resources as pools of different physical and virtual resources dynamically assigned to consumers. The assignment and reassignment are based on a multi-tenant model in order to achieve high utilization. The possibility of rapid scaling up or down of provisioned capabilities is referred to as *rapid elasticity*. Moreover, the impression of infinite resource capabilities that are obtainable in a large quantity at any time is proposed. Finally, *measured service* means resources are automatically monitored, controlled, and if needed, optimized using metering mechanisms appropriate to the resource type. Furthermore, the according utilization is transparently traceable for customers and cloud providers.

The service models of cloud computing consist of *Infrastructure as a Service (IaaS), Platform as a Service (PaaS)* and *Software as a Service (SaaS)* [509], as shown in Figure 3.1. They differ in the complexity of the provisioned capability, and how customers can access or use them. IaaS is marked through the provision of basic computing capabilities for customers. In PaaS, users have control only of their applications, and to a certain extent the configuration of hosting environments. In the SaaS model, customers just use a certain application running on a cloud infrastructure.

3.1.1   Motivation

Currently, offerings from first generation clouds, such as Amazon Web Services (AWS) [63], Rackspace [594], and Flexiant's FlexiScale [273], are based on proprietary middleware [588], thus, resulting in isolated environments. This isolation obstructs the further advancement of cloud computing. New models, e.g., federations, are not feasible without interoperability [611]. The same applies to the enforcement of Service-Level Agreements (SLA) in a hybrid cloud and the management of logically-grouped virtual machines (VMs).

The aforementioned issues lead to a series of disadvantages for customers, such as vendor lock-in and high migration costs. To address these, standardization efforts have to take place in order to support further developments in the

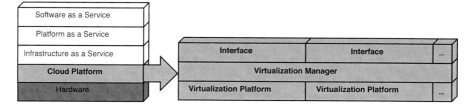

FIGURE 3.2
The Cloud Platform Layer.

second generation of clouds. Standardized exchange mechanisms and interfaces are crucial in order to facilitate interoperability [490]. It is also important that the standardization process should apply to the evolution of cloud.

As cloud computing becomes popular, myriads of offerings are announced. However, the first wave of these offerings are IaaS-based products like Amazon Elastic Compute Cloud (EC2) [63], Rackspace and FlexiScale. PaaS and SaaS solutions are later offered on top of IaaS. Figure 3.2 illustrates this relationship, which includes the *Cloud Platform* layer and its underlying hardware. Deriving from this stack, interoperability for IaaS has to be achieved first. In order to do so, standardization efforts have to focus on one layer below, i.e., the cloud platform layer, since it encapsulates all technologies used to provide virtual environment and provides functionalities to the upper layers. As shown in Figure 3.2, the cloud platform layer consists of several *Virtualization Platforms* that are responsible for the resource virtualization, a *Virtualization Manager* for managing VMs, and *Interfaces* that provide a controlled and uniform access to the underlying environment.

This chapter deals with technologies and mechanisms of the cloud platform layer specific to IaaS, in order to achieve interoperability between clouds. Only when the interfaces, documentation and standards are open and transparent, cloud services can be easily migrated to various platforms. To give an overview of related topics and important areas, this work presents a taxonomy. The taxonomy discusses all important issues to outline the needs and trends in current developments aiming for interoperability for IaaS. Important developments, such as the rise of Virtual Appliances (VAs) over VMs and the adoption of SLAs to clouds, are also considered in this taxonomy.

The rest of this chapter is organized as follows. In Section 3.2, the term *interoperability* is defined and discussed. Furthermore, the benefits of interoperability are pointed out. The core of this work is a taxonomy about IaaS interoperability and is presented in Section 3.3. Section 3.4 provides some related work. Finally, Section 3.5 concludes the chapter and gives future work.

3.2   Interoperability of Cloud Platforms

According to Oxford Dictionary of Computing, interoperability refers to "The ability of systems to exchange and make use of information in a straightforward and useful way; this is enhanced by the use of standards in communication and data format" [215]. This definition applies well to cloud computing, since it addresses the current state of cloud solutions. There are multiple systems or clouds that are able to use the information that they have, but are not able to 1) exchange and share them, and 2) understand the information from others. Only the second point is somewhat related to interoperability. There are two kinds of interoperability, i.e., *syntactic* and *semantic* [728]. Syntactic means that clouds are trying to communicate through standardized mechanisms, where interfaces, data formats and communication protocols are used to achieve interoperability. Semantic interoperability means that the information is not only exchanged but also interpreted in order to use it.

3.2.1   Benefits of Interoperable Clouds

The benefits of interoperability cover business/economy and technical issues. These issues are relevant to stakeholders, i.e., customers, developers, and cloud providers. From a business perspective, advantages to customers are: (i) no vendor lock-in, where a customer is no longer restricted to a single cloud provider; (ii) flexibility, where a customer is able to interact with others using various cloud offerings and distribute his/her applications across several providers; and (iii) cost reduction, where if a customer moves to a different provider, he/she can move major parts of the application without building it from scratch again. From a technical perspective, developers can take many advantages of using interoperable clouds:

- Monitoring of a distributed infrastructure scaled over several clouds can be achieved through a standardized communication mechanism.

- Consistent and uniform system management of multiple clouds. This addresses the current lack of having multiple information points, e.g., a local monitor for a private cloud, and Amazon CloudWatch [63] for virtual machines in the Amazon EC2.

- Secure verification and service discovery using standardized interfaces.

- SLA enforcement for the possibility of automatically reacting to a SLA violation.

- Fault tolerance by scaling over several clouds.

- Disaster recovery, where distributing information and components prevents complete data loss and reduces downtime.

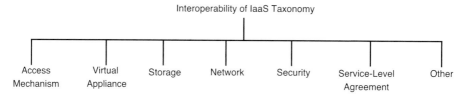

FIGURE 3.3
Interoperability of IaaS Taxonomy.

The above advantages provide cloud providers with a case for interoperability. One incentive for cloud providers is an increase in market share, as interoperability further drives wide-spread adoption by users and developers in using cloud computing. This translates to a growth in revenue. Another factor is to be able to provide users with a SLA guarantee, in case of hardware failure and/or security attacks. Using a federated cloud model as an example, users' VMs can be migrated to a different provider with minimal overhead, thus, reducing SLA violations.

3.3   Taxonomy of Interoperability for IaaS

Figure 3.3 gives an overview of the taxonomy categories, which can be seen as essential building blocks towards an interoperability for IaaS on the cloud platform level. The first category shown in Figure 3.3 is *Access Mechanism*, where several access types and important topics of service discovery are discussed. *Virtual Appliance* is the second category and deals with the interoperability of computational resources. Next is *Storage*, where it discusses storage organization and management as important areas for interoperability.

The fourth category shown in Figure 3.3 is *Network* that focuses on addressing issues regarding to high-level communications. Afterwards, *Security* is considered in this taxonomy. *Service Level Agreement (SLA)* is discussed next, especially in the area of architecture, template format, and monitoring of SLA objectives. Finally, the *Other* category is intended for topics like consensus and regulations, and audit standards that do not belong to other parts.

3.3.1   Access Mechanism

Access mechanisms have a major influence on interoperability since they define how a service can be accessed. Furthermore, they define how a service can be discovered. Two areas in access mechanisms are identified, i.e., *Types* and *Service Discovery*, as shown in Figure 3.4.

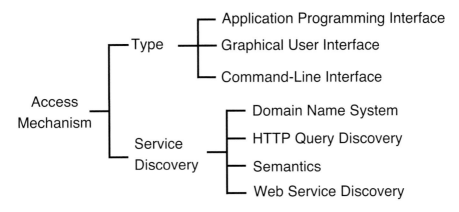

FIGURE 3.4
Access Mechanism Taxonomy.

3.3.1.1 Type

In an access mechanism, *Type* means different ways that a customer has access to services. In the first generation of clouds, common standard types are *Application Programming Interface (API)*, *Graphical User Interface (GUI)*, and *Command-Line Interface (CLI)*, as depicted in Figure 3.4.

Application Programming Interface

APIs offered by cloud providers can be divided into three architectural styles, i.e., *Web Service*, *HTTP Query*, and *Extensible Messaging and Presence Protocol (XMPP)*. Simple Object Access Protocol (SOAP) is a commonly-used specification for implementing web services, since it is a stateless message exchange protocol. The common underlying protocol is Hypertext Transfer Protocol (HTTP), but Remote Procedure Call (RPC) is also applicable. However, SOAP does not provide its own syntax within Web Services Description Language (WSDL), an XML-based language that provides a model for describing web services. Thus, it may lead to existing systems unable to communicate with each other due to using proprietary or implementation-specific solutions. HTTP Query provides an alternative mechanism to SOAP, where it uses either Remote Procedure Call (RPC) or Representational State Transfer (REST). Common RPC analogs are XML-RPC and JSON-RPC. XML-RPC is the precursor of SOAP and it uses XML to encode its calls, whereas JSON-RPC uses Java Simple Object Notation (JSON) as the data format. In contrast, REST is an architectural style. Although REST is exhaustively elaborate and explained in [271], no concrete definition is formalized. Thus, it leads to various interpretations on what RESTful means. This is reflected through APIs from several cloud providers, such as Amazon and Rackspace that are promising RESTfulness, but unable to communicate with each other.

XMPP is an XML-based protocol. The design goals of XMPP are messaging, presence and request-response services in a near real-time manner. Like other XML-based communications over HTTP, XMPP can be adopted to a programmatic client-to-cloud communication. However, XMPP suffers from the low adoption by cloud applications.

Command Line Interface

Many cloud providers offer tools and scripts to support CLI, e.g., Amazon's EC2 API tools or other third-party tools for AWS. In most cases, these tools are just implementations on top of existing APIs like REST or XML-RPC. Secure Shell (SSH) also becomes a standard tool for accessing Unix/Linux-based computational resources.

Graphical User Interface

The main role of GUI is to provide users with an easy access to a cloud offering. In general, access using GUI can be differentiated into *Remote Desktop Protocol (RDP)* and *Web Portals*. RDP is a proprietary protocol from Microsoft, and it is used to connect VMs that use the Windows operating system (OS) remotely. In contrast, web portals are used independent of OS. Thus, many cloud providers like Amazon and Flexiant offer access to their IaaS offerings via web portals.

3.3.1.2 Service Discovery

Figure 3.4 also shows various forms of service discovery like *Domain Name System (DNS)*, *HTTP Query*, *Semantics* and *Web Service Discovery*. These forms are discussed next.

Domain Name System

DNS Service Discovery (DNS-SD) describes a mechanism to discover services of a certain domain [182]. It is a convention for naming of resource-record types, thus, allowing the discovery of service instances under a certain domain with basic DNS queries. DNS-SD provides a discovery mechanism, which can be integrated into existing DNS structures without major changes. Thus, DNS-SD is an access mechanism independent of a discovery mechanism.

HTTP Query

It refers to the *HTTP Query* described in Section 3.3.1.1. If a web service follows the REST constraint of statelessness [271], a discovery request on a resource should return all related information including the available operations. Moreover, each service is self-describing and discoverable through simple

HTTP requests. However, it only refers to the discovery of a service's capabilities, not the service itself.

Semantics

It aims to discover services by finding a meaning in certain contexts. Resource Description Framework (RDF) and Web Ontology Language (OWL) are commonly-used semantic models. RDF is a general-purpose language for representing information on the Web, and uses XML for the information presentation. In contrast, OWL enables the description of connections between information. Thus, resources can be described using RDF and their relations through OWL in a machine-readable way. As a result, resources can be discovered not only by keywords, but also by meaning. This makes service discovery very flexible, because applications are no longer bound to or developed against a static set of particular services.

Web Service Discovery

There are four aspects to a web service discovery, i.e., *Registry, Federated, Language,* and *Multicast.* Universal Description, Discovery & Integration (UDDI) is an XML-based registry for classifying and locating web service applications. It uses WSDL for service description and SOAP for communication.

A federated discovery is a hybrid version of distributed and centralized ones [660]. Several autonomous discovery entities are federated, and each of them provides its information within the federation.

Web Service Inspection Language (WSIL) is a language that defines formats and rules on how information about services are made available [113]. A WSIL document abstracts different descriptions' formats through a uniform XML-based format. It provides a mapping of different descriptions to the according service through pointers. Because the WSIL document only holds references and no concrete descriptions, it is extensible and easy to process.

Web Services Dynamic Discovery (WS-Discovery) is a multicast discovery protocol for locating web services in local area networks [556].

3.3.1.3 Discussion

CLI and GUI are important tools for giving users access to cloud offerings, but they lack interoperability. The only exception is standard tools for remote access to computational resources, which are widely-accepted like SSH or RDP. CLI and GUI are mainly implemented on top of existing APIs.

These APIs are the main access points. They use proprietary XML, JSON or plain HTTP as data formats resulting in a high diversity of implementation details. To achieve interoperability for IaaS, best practices or other restrictive rules are necessary, especially when using XML or JSON to describe data, due to high extensibility and adaptivity. Thus, a consensus about the meanings of syntactic constructs has to be achieved. Furthermore, the appropriateness of existing approaches can be evaluated, and the inappropriate ones can be rejected in order to reduce the diversity.

In order to facilitate interoperability, APIs have to expose service discovery functionalities. A well-accepted approach is the combination of REST and HTTP Query service discovery of a certain domain. The advantages of this approach are ease of integration and simple discovery through well-known and well-defined technologies.

A federated service discovery mechanism can also help interoperability. A federation of mediators like WSIL offers the ability to encapsulate a broad band of services and offer request specific response. The approach of mediators also applies well with a semantic service discovery mechanism. An environment with services described in RDF is summarized using WSIL. Moreover, it can be discovered through the use of OWL, since it leverages the power of a semantic discovery, thus, leading to a more dynamic and flexible way. These technologies are also applicable in a RESTful architecture.

3.3.2   Virtual Appliance

The idea of *Virtual Appliance (VA)* is to deliver a service as a complete software stack installed on one or more VMs, as proposed by a Distributed Management Task Force (DMTF) [240]. A VA description is more than the XML-based summary of VM descriptions. Figure 3.5 shows the taxonomy of VA, where it consists of *Life Cycle*, *Virtualization Platform* and *Virtualization Manager*.

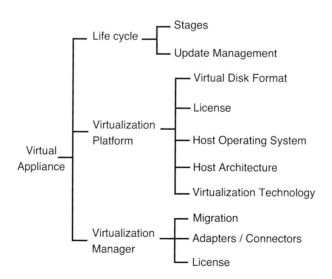

FIGURE 3.5
Virtual Appliance Taxonomy.

FIGURE 3.6
The Life Cycle of a Virtual Appliance [240].

3.3.2.1 Life Cycle

As shown in Figure 3.5, there are two important topics of the VA life cycle, i.e., *Stages* and *Update Management*.

Stages

According to DMTF, the life cycle of a VA consists of five stages, as shown in Figure 3.6. In this figure, it starts with the *Development* stage, where images are prepared and meta-data are assembled. Next is the *Package and Distribute* stage. Because VAs consist of multiple entities, they have to be assembled. However, packaging means nothing more than a collection of all components. The *Distribute* part is the step of making the package available by uploading it to the cloud. Then, the *Deploy* stage launches the VA. The *Management* stage manages the VA, such as performing pause, resume, and stop operations. Finally, in the *Retirement* stage, the VA is decommissioned and the relevant resources are released.

Update Management

Currently, running VMs need to be updated individually by the users for software upgrades or security fixes. The update management can also pose a significant problem to cloud providers, as they need to update all pre-configured images. For PaaS and SaaS providers, an update management needs to be done carefully so as not to avoid SLA violations that disrupt VMs' uptime and network performance. Therefore, an update management can become a complex problem. Currently, one viable solution is to move from a conventional software patching to a redeployment [681]. With this solution, the update procedure is being shifted into the development phase of the VA life cycle. Thus, VA developers or maintainers can update their images and test them before they are being packaged and deployed. As a result, the newly-deployed VA takes over the older one after its retirement or slowly replaces it.

3.3.2.2 Virtualization Platform

Figure 3.5 also shows the core characteristics of a virtualization platform, i.e., *Virtual Disk Format, License, Host Operating System (OS), Host Architecture,* and *Virtualization Technology*. Descriptions of each type are described next.

Virtual Disk Format

The current landscape of virtual disk formats (VDFs) is heterogeneous, like VMware's Virtual Machine Disk (VMDK), Microsoft's Virtual Hard Disk (VHD), and QEMU's qcow. However, most of them are proprietary. To advocate interoperability, the support of various VDFs is recommended.

License

Software license is one of the most complicated topics because there are many free and proprietary licenses existing nowadays. Virtualization platforms with a free license are preferred in order to support interoperability. Moreover, they are less restrictive, hence, enabling broader adoption and agreement.

Host Operating System (OS)

In terms of interoperability, it is important that the host OS supports VMs with a variety of guest OSs, if a virtualization manager is not capable of managing different platforms. Moreover, the host OS should allow paravirtualization of a guest OS to enhance the performance, in comparison to using full virtualization.

Host Architecture

Several host architectures were developed within the evolution of computing like SPARC (Scalable Processor ARChitecture) and x86. The x86 architecture with 64-bit registers (x86-64) is compatible to 32-bit ones, but not vice versa. Since a VM's guest architecture depends on the host architecture and most x86 processors are used in new PCs and servers, x86-64 should be the standard host architecture to achieve interoperability.

Virtualization Technology

It is an essential part of virtualization, because it enables the running of multiple VMs inside one host or server. Currently, there are four technologies used in today's virtualization, i.e., full, para, OS-level, and hardware-assisted. Full virtualization allows a VM to be running without any OS modifications, for example, a VM with Microsoft Windows OS running inside a Linux-based host OS. In contrast, paravirtualization modifies a guest OS to communicate with a host system to improve performance and efficiency. Hence, VMs and the host need to be running on the same OS.

OS-level virtualization has the appearance of a stand-alone system from the view of running applications. However, the host OS is actually shared by many guest OSs. Hardware-assisted virtualization is a hardware technology that supports virtualization by enhancing the communication between a guest OS and the underlying hardware. For example, Intel's VT and AMD's AMD-V processors have built-in support to enable a hardware-assisted virtualization.

To support interoperability, it is recommended that the cloud providers use a variety of virtualization technologies to address the different needs and requirements of their customers.

3.3.2.3   Virtualization Manager

A virtualization manager is responsible for managing VAs on physical hosts. For interoperability, an important task is to move the VAs to different hosts and/or cloud providers. Therefore, *Migration* is an essential task for the manager, as shown in Figure 3.5. Moreover, this figure shows other relevant interoperability issues like *Adaptors/Connectors* and *License*.

Migration

There are two kinds of migration, i.e., *cold* and *live*. For a cold migration, the VM is being shutdown and then moved to another host or server. This is a time- and resource-consuming process, where it has many disadvantages like increasing downtime and potentially violates SLAs. A second approach is a live migration that aims to avoid the aforementioned disadvantages. With this process, data need to be stored externally before the migration begins.

Some virtualization managers, such as VMware vSphere and Microsoft Hyper-V, are under closed source development and have restrictive licenses. Thus, they limit the ability to reproduce the same migration procedures and perform uniform benchmark tests [191]. Moreover, with various vendors using different live migration strategies, it would be a challenging task to achieve interoperability in this issue.

Adaptors/Connectors

The terms adaptor, connector, broker or driver are refer to the same thing, i.e., the integration of an API in a modular way that allows the translation of operations and commands from one API to the function set of another. This modular approach is adopted by multiple virtualization managers of second generation clouds like OpenNebula [667] and Eucalyptus [544].

The modular approach results from the need to achieve a certain level of interoperability through the integration of a variety of interfaces. Currently, only the proprietary Amazon EC2 API is gaining wide acceptance since interfaces from Open Grid Forum (OGF) [551] and DMTF are still in an early adoption state. However, in the future, the adoption of several APIs through adaptors should be a de facto requirement for a virtualization manager.

License

The same issues with a license for a virtualization platform, as discussed in Section 3.3.2.2, can be applied to a virtualization manager.

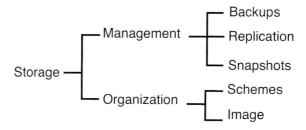

FIGURE 3.7
Storage Taxonomy.

3.3.2.4   Discussion

In this chapter, virtual appliances are introduced and presented as a new computational resource for cloud computing, instead of virtual machines. As mentioned earlier, the VA approach provides new possibilities. Thus, its adoption and further enhancement is recommended.

Due to heterogeneity in virtual disk formats, licenses, host and guest OSs, and virtualization technologies, only host architecture and virtualization manager can be used to achieve interoperability. The virtualization manager is capable of abstracting various virtualization platforms through an intelligent controller, which orchestrates the platform utilization and allows a heterogeneous environment. However, interoperability among virtualization managers is obscured by closed or proprietary sources, and license restrictions. Interoperability can only be achieved through the adoption of virtualization platforms and virtualization managers without such limitations. The adaptors/connectors concept can also help interoperability though a modular integration of various APIs.

3.3.3   Storage

Figure 3.7 shows two important topics in terms of storage interoperability, i.e., *Management* and *Organization*. The former addresses storage functionalities and discusses the need of their availability. The latter points out several kinds of storage organization, which are important to prevent incompatibility.

3.3.3.1   Management

Figure 3.7 also shows the three topics of a storage management, i.e., *Backup*, *Replication*, and *Snapshots*. Because these topics are sometimes mixed up or treated as alternatives, their differences are highlighted to avoid this confusion. Moreover, they have to be present in order to achieve interoperability of IaaS. The backup functionality is important to guarantee long-term preservation on a non-volatile storage media, either for data to be backed up or VM images to be protected against data loss. On the other hand, replication means that data are not only being copied to one location, but they are distributed to several sites. However, replicated data are not moved into a non-volatile storage media. Instead, they are copied to volatile storage for availability and faster access time. Finally, snapshots are full copies of data objects. Compared to backups, snapshots are not moved to non-volatile storage. Snapshots are also different from replications, because in the process only one copy is made to a single location and not to multiple ones.

3.3.3.2 Organization

Storage organization depends on the kind of data to be stored. Figure 3.7 also shows the areas of a storage organization, i.e., *Schemes*, and *Image*.

Schemes

There are three major schemes of storage organization: *Block Storage*, *File System*, and *Object Storage*. Block storage is a kind of storage where data are saved in blocks. A block has a certain size and consists of a structured sequence of bytes. Thus, block storage can be seen as a logical array of unrelated blocks, which are addressed by an index in the array called Logical Block Address (LBA) [260]. On the other hand, a file system is responsible for imposing a structure on the address space of disks, such that applications can refer to files' abstract names of data objects. Finally, object storage is a kind of abstract storage consisting of one or more object stores or devices [260]. An object store is a collection of objects, which consists of data objects and their metadata. With this approach, objects can be stored or modified via methods exposed through a standardized file-like interface. Everything below the method calls is encapsulated.

Image

Image formats have already been discussed in Section 3.3.2.2. Hence, this section focuses on their relationship to the underlying organization scheme. Hard disk images are one example already mentioned above. A hard disk image is a by-sector copy of a hard disk. The important point is that sectors are smaller than blocks, different from files, which are mostly larger. Therefore, the image could be placed on the block storage more efficiently. Furthermore, all information is within the image and no metadata have to be considered. In case of a VA image containing files [241], the usage of block storage is not efficient due to additional file sources.

Decoupling pairs of images and related data to the most appropriate scheme can be a solution, but only possible if the schemes are hidden behind intelligent object storage. In this scenario, block and file system storages are chosen based on the type of data. Therefore, interoperability can be achieved through the intelligent placement of data beneath an object storage layer.

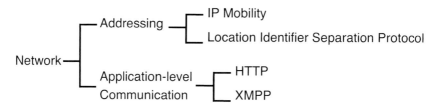

FIGURE 3.8
Network Taxonomy.

3.3.3.3   Discussion

Compatibility of underlying storage resources is required to allow the shift of data from one location to another while still retaining the possibility of an efficient usage. This issue is highlighted through the introduction of three storage organization schemes, and the statement about the relationship between various images and the underlying storage infrastructure. To avoid compatibility problems, the introduction of an intelligent object storage layer can be a solution. VM images and related data are decoupled and stored on most appropriate storage. Logical connections can be kept using metadata information. The mentioned functionalities can be supported from such an environment and fulfill interoperability requirements.

3.3.4   Network

Figure 3.8 shows two important topics in terms of network interoperability, i.e., *Addressing* and *Application-level communication*.

3.3.4.1   Addressing

A major problem for cloud computing is the need to have a reliable remote access to applications and their underlying VMs, even when they are being moved between subnets within a cloud or to others. Therefore, network addressing plays an important role for interoperability. As shown in Figure 3.8, *IP Mobility* and *Locator Identifier Separation Protocol (LISP)* aim to address the aforementioned problem.

IP Mobility

The possibility to keep its IP address during live migration is essential for a VM, otherwise it would directly lead to service downtime and SLA violations. Therefore, extended mechanisms need to be considered to allow seamless migration and facilitate interoperability. The mechanisms of an IP mobility are available for both IPv4 and IPv6.

IPv4 mobility is an extension of the IPv4 protocol [247]. The first component is the so called a *home agent*. This can be a network router that a VM is working on. When the VM moves, a so-called *care-of address* of the machine is deposited to the home agent. This is the address through which the VM is accessible in the new network. If the address is provided by the router, then this router is called *foreign agent* and has the function to forward data to the VM. When datagrams are sent to the migrated VM, the home agent tunnels them to the foreign agent. In the other direction, the foreign agent serves as a standard gateway.

IPv6 mobility implements a mechanism similar to IPv4. Care-of addresses and home agents are also used [398], but there are many differences resulting in incompatibility and the inability to inter-operate. The most important difference is that, because of mechanisms like proxy neighbor discovery and stateless or stateful auto-configuration, IPv6 mobility does not need foreign agents [398]. Furthermore, the communication routes differ, because IPv6 mobility allows bidirectional tunneling over the home agent, as well as direct communication to the care-of address for the other communicating entity.

Locator Identifier Separation Protocol

LISP aims to have a single IP address to be used for multiple purposes [126]. An IP address locates and identifies a host. The idea is to separate these functionalities into *Endpoint Identifiers (EIDs)* and *Routing Locators (RLOCs)* [263], where EIDs are hosts and RLOCs are router addresses. End-to-end communication is realized on EID-to-EID communication. EIDs are mapped to their according RLOCs through conversion algorithms resulting in a RLOC-based communication on the way between start and destination network. Therefore, LISP is placed on edge routers. This model works with both IPv4 and IPv6, and allows mobility over network barriers through the decoupling of ID and location. However, because of the placement of LISP on edge routers, the implementation and adoption is extremely difficult even if the model seems to be powerful. Deriving from this, further research is needed.

3.3.4.2 Application-Level Communication

Figure 3.8 also shows two communication protocols important in terms of interoperability, i.e., *HTTP* and *XMPP*. HTTP was already mentioned in the context of RESTful APIs in Section 3.3.1.1. However, in cloud computing, HTTP is the common application-level transport protocol, since HTTP mechanisms apply well to REST and SOAP.

XMPP has also been discussed previously, where it provides XML over TCP. Unfortunately, due to its low adoption in clouds and the deficiencies already pointed out in Section 3.3.1.1, XMPP is not an alternative to HTTP in terms of interoperability.

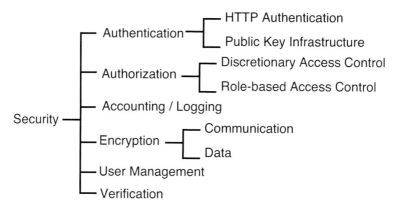

FIGURE 3.9
Security Taxonomy.

3.3.4.3   Discussion

Interoperability in networking is achieved through the dominant position of the IP protocol family. It is achievable even if IPv4 and IPv6 are incompatible, because there is cooperation among translation mechanisms. The ability of a continuous low-level communication is essential, but it is currently not possible among heterogeneous environments. A comprehensive solution through LISP does not seem feasible at the moment. Therefore, a homogenous addressing founding on IPv6 has to become standard in order to achieve an interoperability of cloud platforms. This is due to the exhaustion of the IPv4 address space, and the IPv6 mobility is a core component of the initial protocol design and has no later addition.

XMPP is currently not suitable for a high-level intercloud communication, since it is not widely-used by cloud applications. Therefore, HTTP can be seen as the only way to achieve interoperability in application-level communication. Nevertheless, the XMPP approach should be traced and reviewed against the applicability of REST to it.

3.3.5   Security

Figure 3.9 shows important security topics for a cloud interoperability. *Authentication*, *Authorization* and *Accounting/Logging* introduce the AAA model into this taxonomy. Next, *Encryption* mechanisms for communication and data are presented. Afterwards, important considerations in the area of *User Management* are pointed out. Finally, a description about a service discovery specific topic *Verification* is discussed.

3.3.5.1 Authentication

Authentication means a confirmation of a stated identity, and it is an essential security mechanism in clouds and other IT areas. In the following, two kinds of mechanism are presented, as shown in Figure 3.9.

HTTP Authentication

HTTP offers two authentication mechanisms, i.e., basic and digest access [285]. The basic authentication is founded on a Base64 encoding of user ID and password within the authorization header field. It is not secured since Base64 is an encoding not an encryption mechanism. Digest access authentication addresses this problem through MD5-based encryption for the password and several other values, but not the username. Unfortunately, due to the well-known weakness of MD5, this mechanism can not be treated as secure. Moreover, these mechanisms are not interoperable with each other.

Public Key Infrastructure

Public Key Infrastructure (PKI) is a foundation for security mechanisms that allows the use of public keys and certificates, instead of logins and passwords [755]. Therefore, PKI provides an opportunity to establish a trustable relationship between an individual or an organization to its credentials (key or certificate), which can be used by other mechanisms to verify an identity and authenticate it. In terms of interoperability, four components of PKI are important, i.e., *Trust Model*, *Cross Certification*, *Algorithms*, and *X.509*.

There are several trust models available for PKIs. One model is the hierarchical model. There, one or more entities, called certification authorities (CAs), build a hierarchy above the credentials of an end-user. In this model, a CA guarantees the confidentiality of the CAs or end-user credentials below and can be verified against a next higher CA if necessary. Another model is called *Web of Trust*. Because no higher authorities are present, confidentiality has to be verified on each end-user characteristic. In case of Pretty Good Privacy (PGP), other end-users acknowledge an identity and the consumer decides if he/she trusts these acknowledgements. A third model is the *Peer-to-Peer* model were each entity establishes trust with each other entity through individual trust negotiation.

Cross certification is an extension of the hierarchical model. It gives entities of one PKI the chance to verify the confidentiality of entities belonging to other PKIs. The idea is quite simple; if one CA trusts another, an entity trusting one of them can also trust the other one.

In cryptography, the requirement on replacing algorithms is recommended, because of the need to be able to substitute an algorithm with a stronger one if it got broken. Thus, it is necessary that communicating systems can negotiate an algorithm they are capable of. On the other hand, X.509 is an International

Telecommunication Union (ITU) standard for important PKI components like public key certificates, certificate revocation lists, and attribute certificates.

3.3.5.2 Authorization

Authorization means an allocation of resources or rights according to the user's credentials after the user has proven to be the one he/she stated. Mainly two models, *discretionary* and *role-based access control* are discussed, as shown in Figure 3.9.

Authorization in clouds is achieved through Discretionary Access Control (DAC). Amazon and Rackspace are examples of using DAC in their public clouds. The concept of separation of users and objects hides behind this term [232]. The access in this case is controlled through lists, named Access Control Lists (ACLs). These lists consist of mappings of usernames and functions that they are able to execute on the according object. The mappings are determined by the owner of the object. This model has a great potential for interoperability, because only the ACL format has to be uniform to achieve interoperability. The function terms do not have to be unified, since they are only applied to a specific environment.

Role-based Access Control (RBAC) is based on the assumption that access control decisions are determined by the role of a user [266]. This can be by his/her duties, responsibilities or qualification. In RBAC, functions are dedicated to users according to their roles, resulting in a concept of different roles with different function sets. Unprivileged users, privileged users, database administrators or network administrators are examples of roles. Another difference from DAC is that these roles are not determined by users or owners, instead they are specified by the system. Currently, only one cloud API, VMware's vCloud with vExpress, implements the RBAC approach. A multi-tiered RBAC separates user and administrative functions. However, RBAC is not compatible with DAC, since the roles of RBAC systems can differ.

3.3.5.3 Accounting/Logging

In conjunction to security, accounting/logging means the record of the amounts of events and operations, and the saving of information about them. In terms of security, the importance of accounting data lies in their availability sustaining trust and compliance. Logs are central information sources for common system and fault analysis. Interoperability between clouds can be seriously affected if no such information is available and there are no present mechanisms to access them.

The concrete information could be made available using Information Technology Infrastructure Library (ITIL)'s *Incident Logging* [379]. This ITIL process defines several parameters which have to be present in every event record. Furthermore, it proposes multi-level categorization for the identification of event importance. Event-based logging would also apply well to a monitoring solution to be discussed later in Section 3.3.6.3. The influence of ITIL and its well-defined best practice guidelines could lead to acceptance and adoption. As a result, trust and interoperability can be achieved.

3.3.5.4 Encryption

Figure 3.9 also shows two important topics of encryption, i.e., *Communication* and *Data*. Security between endpoints through communication encryption is as important as endpoint security. Especially through the close relationship of REST and SOAP to HTTP, Secure Socket Layer (SSL) or its new version Transport Layer Security (TLS), become the standard encryption protocol for communication over unsecure networks. Virtual Private Networks (VPNs) and SSH are other common mechanisms providing secure communication per default. For Windows-based systems, RDP also provides a secure mechanism. Data encryption is a topic mostly all cloud providers distance themselves from. Currently, the only agreement made is that existing encrypted data can be stored [350]. The problem of encryption methods is the lack of knowledge of their practicality in cloud computing, due to its scale. Mechanisms like `encfs`, Linux Unified Key Setup (LUKS) or `dm-crypt` are feasible for local systems with lower scale data size (in terms of Gigabyte), and are subjects of open licenses. However, their feasibility on cloud scenarios is not yet proven.

3.3.5.5 User Management

In the first generation of clouds, data regarding a user profile are mostly limited to login name, password, e-mail address and credit card number. The implementation of a secure authentication environment like a PKI raises the data amount as a direct consequence. This potentially enlarges the profile through certificates, stored public keys and signatures, and more personal or organizational data necessary for such security mechanisms.

A lot of operations in the cloud are based on user interactions that require credentials. Thus, how are profiles treated in case of federation? One solution is to leverage a Single Sign-On (SSO) technology with Shibboleth, where it allows a single user credential to access and use various services across multiple layers [657]. In Shibboleth, special components like *Identity Providers (IdPs)* and *Service Providers (SPs)* are introduced to build an environment which is also able to federate with other SSO systems. A user or system account has to authenticate if he/she wants to use a service, and has to communicate with the SP. Then, the SP verifies the provided credential against information it gets from the IdP. If the user is authenticated, credentials are placed on his/her side; then the SP can be used to authenticate user interactions without disturbing the user. The process can also be realized between SPs and IdPs of federated partners. Moreover, the amount of data retrieved between IdP, SP and the user can be configured, thus improving the security of the user data.

3.3.5.6   Verification

Verification is a security requirement for the identification of services in conjunction with the service discovery [126]. Verification means the ability to decide if a service or application can trust others [350]. If it can not, then interoperability will be hindered, because services are not able to prove their intentions as stated. Thus, the communication will be rejected due to security issues. In order to avoid this, verification through certificates or signatures is needed. Every service of a domain owning a certificate or being signed with a verifiable signature can be assumed as trustable. For example, RDF or WSIL documents could be signed easily. Overall, verification introduces security into the service discovery process and into the interaction of communicating services. This strengthens interoperability as the requester can assure itself about the authenticity of its partner.

3.3.5.7   Discussion

HTTP authentication does not provide a secure mechanism, so it is not applicable commercially. PKIs, especially hierarchical ones that use well-known and secure mechanisms are recommended to achieve interoperability in the short run, and with the support of cross certification or SSO in the long run. X.509, as a standard certification format used in mostly all PKIs, also provides interoperability. Moreover, certificates can be used to verify more than just users. Services and service entry points can be provided with certificates as well.

Only DAC becomes widely accepted since RBAC plays a smaller role. Interoperability between DAC and RBAC is not achievable due to different underlying concepts. Even if RBAC provides several advantages compared to DAC, its adoption is not as common as DAC. This leads to interoperability through the broad adoption of ACL-based DAC for authorization.

Accounting/logging enables trust and compliance through the availability of event and operation traceability. Interoperability between logging facilities can be achieved through APIs providing information and hiding the concrete infrastructure. However, their presence is important to achieve the willingness to inter-operate. The functionalities exposed through APIs should follow ITIL guidelines, since these best practices are widely accepted.

The two topics, communication and data encryption, are also pointed out in this section. However, data encryption lacks adoption within cloud computing. User management in federated scenarios is also an important topic, because personal data have to be stored, but not transmitted. If an operation needs to be billed, information will be needed by the partners. This process can be arranged through the utilization of federated SSO infrastructures. The operating principle of Shibboleth and the resulting possibilities were highlighted previously. The proposed approach also applies well to the authentication and authorization mechanisms, because of the usage of common strong security and web standards. A critical point is the difference from the underlying

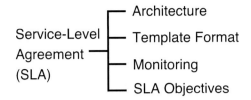

FIGURE 3.10
Service-Level Agreement Taxonomy.

attribute-based authorization. Further research efforts are necessary at this point.

Programmatic service discovery is another important feature for automated service delivery and consumption. Verification secures the communication between services through the introduction of verifiability of the confidentiality of their opponents. This strengthens the idea behind interoperable systems. Therefore, verification mechanisms have to be adopted in form of signing the service discovery endpoints with verifiable certificates.

3.3.6 Service Level Agreement

Service Level Agreements (SLAs) are ubiquitous in modern IT systems, and thus, they have to be discussed in terms of interoperability [126] [350] [606] [567]. Figure 3.10 shows four important topics, i.e., *Architecture, Template Format, Monitoring* and *SLA Objectives.*

3.3.6.1 Architecture

Web Service Agreement Specification (WS-A) is the standard for SLA management in web service environments, and its web service-based interface called *agreement layer* [87]. While WS-A is XML-based and exposes a rich function set, RESTful is preferred since it is minimal, with extensible function sets of Virtual Execution Environment Management Interface (VMI).

3.3.6.2 Template Format

Several template formats for SLAs were developed in the past years, for example IBM's Web Service Level Agreement (WSLA) or SLA Definition Language (SLAng). One important property is that these template formats are machine-readable documents, where they are the electronic representations for the purpose of automated management. In the first generation cloud systems, SLAs have predefined metrics and values for SLA Objectives (SLOs), and they are mostly not machine readable [513]. This results in one-round agreement, where a customer can accept or reject an agreement. The possibility to negotiate the metrics and values of service level objectives on runtime allows multiple round agreements and is more dynamic.

One template format developed in an extensible way with such scenarios in mind is WS-A, which is an OGF standard. Similar to WSLA and SLAng, WS-A offers a language to describe a SLA template and all of its components, and a basic XML-based schema for automated management. Moreover, WS-A specifies a protocol advertising the capabilities of service providers and creating agreements based on creational offers, and for monitoring agreement compliance at runtime [87].

In order to achieve interoperability, it is necessary to agree upon a single template format. The possibility to process documents electronically is crucial for automated management and the extensibility to dynamic negotiation is required also. WS-A seems to be a good choice; however, cloud providers have not yet started to offer machine-readable SLAs. Deriving from the choice of XML as the description language for WSLA, SLAng and WS-A, an XML-based template format seems to be in favor.

3.3.6.3   Monitoring

Monitoring of SLAs is important for both cloud providers and customers. A provider wants to ensure that the provision of resources is according to the SLA and no liability arises. On the other hand, a customer wants to ensure the adherence of cloud providers as mentioned in the contract. A way to monitor SLA is to use event-based dynamic monitoring infrastructures capable of associating low-level metrics with Key Performance Indicators (KPIs) of SLOs. Thresholds can be determined dynamically and re-negotiated between parties having already accepted them in case of events forcing a change [199]. In terms of interoperability, a dangerous scenario arises when high static thresholds of SLOs can not be satisfied by any cloud providers. The consequence is permanent SLA violations and the inability to move to other clouds, because the SLA requirements could not be met. Moreover, an insufficient event monitoring of static threshold monitoring systems, which are not capable of re-negotiation, can stress interoperability. The result are possibly more critical SLA violations from shift delays and unavailability. Overall, an event-based dynamic monitoring system fits better to the dynamic nature of clouds, dynamic SLOs, and machine-readable and negotiable SLAs.

3.3.6.4   SLA Objectives

SLA Objectives (SLOs) are the core component of a SLA, because they directly influence the interoperability of cloud systems. In a scenario where a customer wants to change his/her cloud provider, the absence of SLOs can make it difficult to compare the old SLA with a new one, for example. In order to achieve interoperability, dynamic negotiation and re-negotiation are in favor [199] [513] [87]. This approach applies well to a modular and independently-extensible SLA management entity and a dynamic extensible machine-readable SLA template format previously discussed. For SLOs, dynamic negotiation means the participating parties communicate and negotiate SLOs, implied values, and related information, to which only relevant SLOs are of concern.

3.3.6.5 Discussion

APIs are crucial to interoperability for exposing SLA management functionalities. Furthermore, the separation of a SLA managing entity strengthens its interoperability through exchangeability and extensibility. In addition, a modular architecture that supports dynamic SLA negotiations avoids interoperability issues, since SLA terms can be negotiated during runtime. Assuming that machine-readable documents become common, a sole-accepted SLA template format is essential to achieve interoperable systems, since current existing formats are not compatible. WS-A seems to be a good choice, due to its extensibility through XML and orientation for more complex negotiation processes, and an OGF standard.

The danger in SLA monitoring for interoperability arises from characteristics of monitoring systems using a static low-level threshold to realize high-level SLOs. Not only permanent or more critical SLA violations can occur, depending on the monitoring systems behavior, the cumbering of interoperability through unsatisfied SLOs can be a consequence as well. Furthermore, dynamic monitoring systems apply better to the dynamic nature of clouds and other SLA-related mechanisms like dynamic SLO negotiation.

KPIs are well-known and should be present in cloud computing. However, mechanisms for dynamic negotiation are still required. Interoperability of SLOs can be achieved through the negotiation of KPI and all other relevant information and their aggregation to SLOs. Measures can be controlled by using a dynamic threshold-based monitoring system, which is encapsulated as a SLA service. This service again provides the necessary functionality in form of an API to the cloud provider and customers. Such a system would address all issues and achieve interoperability.

3.3.7 Other

Besides the mostly functional topics discussed for this taxonomy, other topics can have major input on interoperability as well. The topic on *Consensus* highlights the importance for conformity of initiatives, companies and communities. Some compliance issues with potential cumbering influence on interoperability are discussed in *Regulation and Auditing Standards*.

3.3.7.1 Consensus

As already pointed out and shown in the previous sections, standard compliant implementations and de-facto standard tools play a decisive role in cloud computing. Interoperability can be achieved if standards get adopted and certain tools accepted. Standard Development Organizations (SDOs) and companies try to achieve this by founding on alliances and cooperation with other initiatives or companies. Open and academic communities are also part of these alliances. However, a broad consensus about standards and best practices is necessary between company- and community-driven projects to avoid the old battle of open source communities against closed source companies. The first generation of clouds lacked of interoperability, due to the absence of standards and best practices. The second generation should not fail, since differentiating business interests are resulting in the creation of a smaller number of still not interoperable cloud specifications for protocols and APIs.

3.3.7.2   Regulation and Auditing Standards

Regulations can also have major influence on interoperability, because governmental laws or regulations could restrict or prohibit interoperability. For example, storing data in the cloud may elicit various federal and state privacy and data security law requirements, such as the US Health Insurance Portability and Accountability Act, and EU Data Protection Directive [669]. Thus, privacy and data security laws present a significant challenge for cloud providers to comply with.

Auditing standards can also play important roles for interoperability [350]. For example, ITIL provides an auditable best practices catalog for IT Service Management (ITSM) [379]. However, in cloud computing, IT resources are no longer solely in users' own data center. Therefore, it is at the discretion of a cloud provider to follow standards.

3.4   Related Work

Other publications focus on specific problems, such as requirements [567], security [469], semantic-based [195], or mainly express some considerations [490] and use cases [239]. However, several publications analyze the cloud computing paradigm through the establishment of taxonomies [606] [588] [126]. This has the advantage to be able to structure the components and define their borders, comprehensiveness and influences on other topics.

Rimal et al. [606] focus on a high-level architectural observation of cloud functionalities and capabilities. They use these criteria to compare a few cloud providers and their offerings. However, they do not focus on IaaS and interoperability. This makes their taxonomy more general. Moreover, they focus only on existing cloud offerings, but not on current developments and standards.

Prodan and Osterman [588] also utilize a taxonomy to summarize the elements in a cloud environment, but from a general perspective. This is reflected in the analysis of several cloud providers and web hosting providers who are against their proposed taxonomy. Similarly, the difference to our work lies in the specialization on IaaS, interoperability and the analysis of different technologies.

On the other hand, Bernstein et al. [126] focus on an intercloud communication by discussing multiple problems and providing solutions to them. The result of their work is a set of protocols, which can be utilized to achieve an intercloud communication. In their work, the use of open standards is essential in order to achieve interoperability. However, this approach focuses only on the communication part, but does not consider current trends or other important topics, such as virtual appliance, security, storage and SLA.

3.5 Conclusion and Future Work

This work presented a taxonomy of important categories and topics in the area of interoperability for Infrastructure as a Service (IaaS). Firstly, a detailed explanation of cloud computing is presented. Then, the term interoperability is defined and elaborated, and its benefits to stakeholders, i.e., customers, developers and cloud providers, are explained.

The taxonomy itself spans many essential cloud computing topics, such as access mechanism, virtual appliances (VAs), security, Service-Level Agreements (SLAs), and general recommendations. Moreover, a comprehensive scope of topics is considered and hot topics are mentioned. The depth and width of the categories are varied, but this taxonomy outlines a detailed picture of the significant characteristics. However, there are several times mentioned in this chapter, where lack of current solutions are identified. Thus, further research is warranted.

As for future work, VA life cycle management, network addressing issues and SLAs need to be studied in more detail. Moreover, the use of the Information Technology Infrastructure Library (ITIL) to provide interoperability is an interesting area to explore. Finally, further work is needed to compare existing cloud standards and offerings with the topics and criteria presented in this taxonomy.

4

A Taxonomy Study on Cloud Computing Systems and Technologies

Christian Baun

Steinbuch Centre for Computing, Karlsruhe Institute of Technology, Germany

Marcel Kunze

Steinbuch Centre for Computing, Karlsruhe Institute of Technology, Germany

CONTENTS

Building on compute and storage virtualization, and leveraging the modern web, cloud computing provides scalable, network- centric, abstracted IT infrastructures, platforms, and applications as on-demand services that are billed by consumption [121]. The key advantages of cloud computing are flexibility, scalability, and usability. These features originate in the combination of virtualization technologies with flexible and scalable web services, and lead to cost saving [97].

There are various deployment and delivery models for cloud services. This chapter gives an overview about the most popular cloud services and categorizes them according to their organizational and technical realization.

Afterwards, the different existing services, applications and tools for the customers' work with the cloud infrastructure and storage services are discussed.

4.1 Deployment Models

The existing cloud computing services differ with respect to organization and quality of the offering. Three different deployment models of clouds exist: Public clouds, private clouds, and hybrid clouds.

4.1.1 Public Cloud

In a public cloud, the service provider and the service consumer belong to different organizations. Public clouds always implement commercial business models and only the actual resource usage is being accounted for. Although public clouds offer a very good value due to their supply side economy of scale, there is the concern that sensitive data might leave the customer's environment. Another concern is the possibility of a vendor lock-in [456]: It might be difficult to quickly access and transfer the data in case of an urgent need. A possible solution here is to use independent cloud services that are compatible to the APIs and features.

There are popular success stories of companies that used public cloud services to complete complex tasks in a short time at a low budget, such as *The New York Times*[12] and Animoto[3].

Back in 2007 *The New York Times* processed 4 TB of data containing their scanned articles from 1851 until the 1980s. The task was to convert the scans of 11 million news stories into PDF format for online distribution. One hundred virtual Linux servers, run by the Amazon public cloud, only needed 24 hours to convert all images at an expense of 500 dollars.

Animoto offers an online service for the creation of videos out of a repository of pictures and music. In April 2008 the service was referred to on Facebook and immediately the number of active users increased from 25,000 to about 250,000 within three days. At peak time, within a single hour, up to 20,000

---

[1]http://open.blogs.nytimes.com/2007/11/01/
self-service-prorated-super-computing-fun/

[2]http://open.blogs.nytimes.com/2008/05/21/
the-new-york-times-archives-amazon-web-services-timesmachine/

[3]http://animoto.com

people signed up and tried to render a video with the Animoto service. Thanks to the elasticity of the cloud services the company was able to ramp up the number of (virtual) servers to 450 in a short time following the demand.[4] Another example of a heavy user of public cloud services is the social network game developer Zynga[5]. The company develops browser-based games that are hosted on 12,000 virtual servers.[6]

4.1.2   Private Cloud

The services of a private cloud are always operated by the same organization the consumer belongs to. The motivation for building and running a private cloud may be security [610] and privacy concerns. However, it is difficult to reach the economy of scale, the availability and security level of a certified professional public cloud service provider. Lots of private cloud solutions to build up infrastructure, platform and storage services are open source but solutions that are licensed under proprietary terms also exist.

Two examples that enable the customers to integrate public cloud resources in the form of a virtual private cloud into their local IT infrastructure are the Amazon Virtual Private Cloud (VPC)[7] which is a part of the Amazon Web Services (AWS)[8] and the Google Secure Data Connector[9] for the Google App Engine[10].

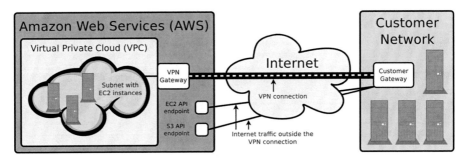

FIGURE 4.1
Amazon Virtual Private Cloud (VPC) ©2010 Amazon.

---

[4]http://blog.rightscale.com/2008/04/23/animoto-facebook-scale-up/
[5]http://zynga.com
[6]http://www.slideshare.net/davidcchou/scale-as-a-competitive-advantage/
[7]http://aws.amazon.com/vpc/
[8]http://aws.amazon.com
[9]http://code.google.com/securedataconnector/
[10]http://appengine.google.com

4.1.3   Hybrid Cloud

In a hybrid cloud, services of public and private clouds are combined. The public cloud services in a hybrid cloud can be leveraged to satisfy peak loads or to spread redundant data inside the cloud to achieve high availability. A special instantiation of the hybrid cloud is the so-called virtual private cloud where resources of a service provider are transparently mapped into a customer network by use of a network tunnel.

4.2   Delivery Models

Besides the different deployment models, four different service delivery models exist to characterize service instances. Figure 4.2 gives an overview on the different cloud service groups. It includes popular public cloud service offerings as well as private cloud solutions that are Open Source.

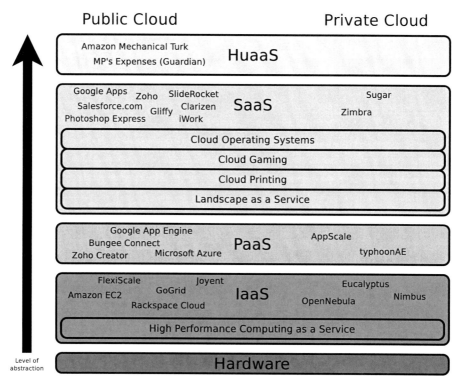

FIGURE 4.2
Different Public and Private Cloud Services.

4.2.1   Infrastructure as a Service

Infrastructure as a Service (IaaS) allows hosted virtual instances of servers without the need to physically access the hardware. The clients have full administrative privileges inside their server instances and are also allowed to define their own level of networking security. The number and type of supported operating systems inside an IaaS depends on the service provider. IaaS in principle allows to virtualize a complete datacenter and transfer it into the cloud. The most popular public cloud IaaS offering is the Amazon Elastic Compute Cloud (EC2)[11]. Popular competitors are GoGrid[12], Joyent[13], FlexiScale[14] and Rackspace Cloud[15]. Open Source IaaS solutions that are popular in industry and science projects are Eucalyptus[16], Nimbus[17] and OpenNebula[18]. All these projects are open source. Two proprietary private cloud IaaS solutions are the Elastic Computing Platform from Enomaly[19] and the quite new project Nimbula Director that was released into beta in December 2010 from the start-up company Nimbula[20] that was founded by a team of ex-Amazon developers that developed EC2. Nimbula is an IaaS and works with the Xen hypervisor [115] and KVM [436]. The solution uses its own Nimbula API and has no compatibility with EC2 or other cloud services from the AWS. EC2 enables the customers to run virtual Linux, Windows Server and Open Solaris instances inside Amazon's data centers based on Xen hypervisor technology. While the user data inside an instance is lost after its termination, customers may save important data inside persistent Elastic Block Store (EBS)[21] volumes.

The Amazon Simple Storage Service (S3)[22] is a storage service for web objects. It allows to store virtual machine images and to provide easily handled web based storage for applications. The functionality of EC2 and it's API may be considered a de facto standard and a lot of the existing Open Source solutions try to be more or less compatible to the EC2 API (see Table 4.1). Eucalyptus and Nimbus both include dedicated S3-compatible storage services that are called Walrus for Eucalyptus and Cumulus for Nimbus.

Each instance of an IaaS obtains an internal and an external DNS name that are created dynamically during startup. Both names are lost when an instance is terminated. In order to realize a permanently available service, it is possible to create elastic IPs and assign them to the instances. An elastic IP can be

---

[11]http://aws.amazon.com/ec2/
[12]http://www.gogrid.com
[13]http://www.joyent.com
[14]http://www.flexiant.com/products/flexiscale/
[15]http://www.rackspacecloud.com
[16]http://www.eucalyptus.com
[17]http://www.nimbusproject.org
[18]http://www.opennebula.org
[19]http://www.enomaly.com
[20]http://nimbula.com
[21]http://aws.amazon.com/ebs/
[22]http://aws.amazon.com/s3/

TABLE 4.1
Private Cloud Solutions that are Open Source and Their Level of Compatibility to the Amazon Web Services

| Name | AWS APIs Implemented | | |
	EC2	S3	EBS
Abiquo (abiCloud)	—	—	—
CloudStack	subset	—	—
Enomaly ECP	—	—	—
Eucalyptus	full support	full support	full support
Nimbus	subset	subset	—
OpenECP	—	—	—
OpenNebula	subset	—	—
Tashi	under dev.	—	—

remapped to another instance at any time. The elastic IP concept is provided by all public and private cloud IaaS offerings and solutions.

Instances can be launched on the basis of different machine images and instance types. The images represent templates of different systems like Windows, Linux etc. The instance types define the number of cores, architecture, main memory and storage provided, I/O performance and thus the price per hour. With the different instance types and images, the customers of an IaaS have the freedom to create an arbitrary IT infrastructure on-demand and scale it in an elastic way.

The underlying infrastructures and components of public cloud services remain usually unclear for the users. As an example of a private cloud infrastructure service, Figure 4.3 shows the structure and components of a Eucalyptus IaaS. The main components are the Cloud Controller (CLC), Cluster Controller (CC) and Node Controller (NC) and the storage services Walrus and Storage Controller (SC) [546] [547]. The NC runs on every node in the cloud as well as a Xen hypervisor or KVM and provides information about free resources to the CC. The CC needs to run inside every site of the cloud, schedules the distribution of virtual machines to the NC and collects (free) resource information. The CLC runs exactly one time inside the private cloud and collects resource information from the CC. The CLC operates like a meta-scheduler in the cloud too and controls the database with the user and resource information. Figure 4.3 shows the structure of Eucalyptus including CLC, CC, NC, Walrus and SC.

4.2.1.1　High Performance Computing as a Service

A more specialized instance of IaaS ls High Performance Computing as a Service (HPCaaS). An important feature of high performance computing is the need for networking with low latency and high bandwidth. To archive

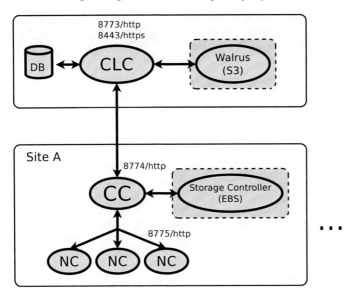

FIGURE 4.3
Structure of a Eucalyptus Private Cloud IaaS.

this goal, the machine instances need to be located next to each other physically. The feature of grouping nodes is supported by OpenNebula. Several cloud service providers offer solutions for HPCaaS. Examples are the Amazon Cluster Compute Instances[23], Gridcore Gompute[24], Penguin Computing on Demand[25], Sabalcore HPC on Demand[26] and UNIVA UD UniCloud[27].

4.2.2 Platform as a Service

A Platform as a Service (PaaS) is a scalable runtime environment for developers to host web applications. PaaS represents a development and execution environment to support a single or few programming languages. The target audience is software developers and end users that very often are able to consume the services in a corresponding marketplace. Due to the fact that the infrastructure is abstracted away, a PaaS intrinsically allows scaling from a single server to many. The end user has no need to care about operating system maintenance and installation of application software packages. There is almost no administrative overhead in the process of service delivery.

---

[23]http://aws.amazon.com/hpc-applications/
[24]http://www.gompute.com
[25]http://www.penguincomputing.com
[26]http://www.sabalcore.com
[27]http://www.univa.com/products/unicloud.php

Popular public cloud PaaS offerings are Google App Engine[28], Bungee Connect[29], Zoho Creator[30] and Windows Azure[31]. The Google App Engine allows to run web applications that are written in Python or Java inside Google's data centers. Bungee Connect has its own C-style programming language called Bungee Sky, and Zoho Creator is focused on the creation of online database applications with HTML forms. While PaaS applications are usually run in a sandboxed environment, it is an outstanding feature of Windows Azure that it grants access to the underlying infrastructure, too.

It is possible to implement and host a private cloud PaaS with e.g., AppScale[32] [33] [185] [147] and typhoonAE[34]. Both solutions are compatible to the Google App Engine API. To provide the same functionality and API these private clouds both use the publicly available Google development server as a fundamental platform. It is a limitation of all PaaS that the users are bound to given compiler versions and features of the environment.

4.2.3 Software as a Service

In a Software as a Service (SaaS) environment, software or applications are operated by a provider and can be consumed as a utility. In general, no software has to be installed at the local site and the services are accessed by use of a web browser. However, there needs to exist a trust relationship to the provider regarding privacy and availability of the service because not only the application itself, but also the data are stored at the provider site.

Popular SaaS offerings are Google Apps[35], Salesforce.com[36], Gliffy[37], Clarizen[38], Zoho[39], SlideRocket[40], Adobe Photoshop Express[41] and Apple iWork.com[42]. For running a private SaaS, Sugar[43] and Zimbra[44] are two of the existing solutions.

The SaaS category includes not only simple web applications, but also more complex services such as Cloud Printing, Cloud Gaming or Cloud Operating Systems.

---

[28]http://code.google.com/appengine/
[29]http://www.bungeeconnect.com
[30]http://creator.zoho.com
[31]http://www.microsoft.com/azure
[32]http://appscale.cs.ucsb.edu
[33]http://code.google.com/p/appscale/
[34]http://code.google.com/p/typhoonae/
[35]http://www.google.com/apps/
[36]http://www.salesforce.com
[37]http://www.gliffy.com
[38]http://www.clarizen.com
[39]http://www.zoho.com
[40]http://www.sliderocket.com
[41]http://www.photoshop.com/express
[42]http://www.apple.com/iwork/
[43]http://www.sugarcrm.com
[44]http://www.zimbra.com

4.2.3.1   Cloud Gaming

Cloud based gaming services enable users to play 3D video games without the need to operate specific hardware or operating systems locally. Players furthermore get rid of installing and registering the games. The advantages for the game developers and publishers are the increased number of potential customers, and that it's impossible to make or use a pirate copy of games with a Cloud Gaming service. The games are run with the infrastructure at the provider's site. The graphics and sound output is streamed to the user's device while input commands are sent back to the provider and interpreted there. To use these services, a broadband internet connection is necessary, with low latency. Popular Cloud Gaming service providers are OnLive[45], Gaikai[46] and Otoy[47].

4.2.3.2   Cloud Print

Cloud Print[48] is a part of Google's new ChromeOS[49] strategy and a new concept for accessing printers from any applications and from any device with any operating systems in the world without the need to install any driver. Such a service is important due to the growing number of portable networked devices like smart phones and tablet computers. These lack drivers for printers and sometimes even a printing subsystem. It is the idea of Google Cloud Print to send the print jobs to a service that processes the jobs and redirects them to a printer.

The printer can be a network printer compatible to this service or a legacy printer that requires a proxy software running on the host it is connected to. Cloud Print is an open standard to get rid of the need to collect and maintain print drivers. HP announced it will support Cloud Print and offers a series of inexpensive and compatible printers.

4.2.3.3   Cloud OS

Cloud operating systems or web desktops are browser based operating systems that provide a virtual desktop running in a web browser. All applications, user data, and configuration are stored by a provider, and the computing takes place remotely. Examples for well-known cloud operating systems are eyeOS[50], Icloud[51], Online OS[52] and Cofio[53].
It is a major advantage of these services that a desktop in the cloud can be

---

[45]http://www.onlive.com
[46]http://www.gaikai.com
[47]http://www.otoy.com
[48]http://code.google.com/apis/cloudprint/
[49]http://www.chromium.org/chromium-os
[50]http://www.eyeos.org
[51]http://www.icloud.com
[52]http://oos.cc
[53]http://www.cofio-hosting.com

accessed from anywhere in the world by use of a web browser. There is no need to fear security issues like malware and viruses. The availability may be better compared to personal computers administrated by the users themselves. Drawbacks are the speed of the user interface which is limited by the network connection and the web technologies in use. Cloud Operating Systems also don't provide the same level of rich features compared to traditional desktop systems.

4.2.4   Landscape as a Service

A special use case of SaaS is Landscape as a Service (LaaS). The aim of such a service is to provide complex, non-multitenancy software like SAP R3. LaaS makes an entire SAP system landscape available as a set of services. Offers like these enable companies to outsource their complete datacenter including hardware, software and maintenance. Providers of LaaS are companies like Fluid Operations[54] or Zimory[55] together with T-Systems.

4.2.5   Humans as a Service

Humans as a Service (HuaaS) follows the principle of crowdsourcing by using the intelligence of the crowd in the web. Whenever a computer is not very well suited to fulfill the task at hand and time is a problem, crowdsourcing may be a good solution. Large amounts of human work force can be leveraged for a small amount of money.

The most popular marketplace for HuaaS is the Amazon Mechanical Turk[56] (see Figure 4.4). Typical tasks that are offered here are translation jobs, ratings, and matching products with product categories.

4.3   Cloud Resource Management

To interact with cloud services that provide SaaS and HuaaS the users need nothing more than an ordinary browser. The management of cloud platform services like the Google App Engine or compatible private cloud solutions is done with the tools the provider of these services makes available to the users. The existing cloud infrastructure and storage service management tools can be classified by four categories:

---

[54]http://www.fluidops.com
[55]http://www.zimory.com
[56]http://www.mturk.com

FIGURE 4.4
Amazon Mechanical Turk.

- Command-line tools like the AWS tools delivered by Amazon, the Euca2ools[57] of Eucalyptus, GSutil[58] of Google Storage and s3cmd[59]

- Locally installed management applications with graphical user interface like EC2Dream[60], Gladinet Cloud Desktop[61] and Cyberduck[62]

- Firefox browser extensions like ElasticFox[63], Hybridfox[64] and S3Fox[65]

- Online services such as the AWS Management Console[66] offered by Amazon, the Google Storage Manager[67] and Ylastic[68]

---

[57]http://wiki.debian.org/euca2ools/
[58]http://code.google.com/p/gsutil/
[59]http://s3tools.org/s3cmd
[60]http://www.elastdream.com
[61]http://www.gladinet.com
[62]http://cyberduck.ch
[63]http://sourceforge.net/projects/elasticfox/
[64]http://code.google.com/p/hybridfox/
[65]http://www.s3fox.net
[66]http://aws.amazon.com/console/
[67]https://sandbox.google.com/storage/
[68]http://www.ylastic.com

A trade-off analysis of the available cloud management solutions is shown in the Tables 4.2, 4.3, 4.4 and 4.5.

4.3.1 Command-Line Tools

The command-line AWS tools offered by Amazon only support their own public cloud offerings. The AWS tools are a collection of a large number of command-line applications (e.g., `ec2-create-volume`, `ec2-delete-keypair`, `ec2-describe-images`, `ec2-start-instances`,...) that are written in Java and enable the customers to access the whole functionality of services like EC2, EBS and ELB that are part of the AWS. The Euca2ools of the Eucalyptus project and GSutil of Google Storage support both public and private cloud infrastructure services.

The Euca2ools are inspired by the AWS tools. These command-line applications have a similar naming schema (e.g., `euca-create-volume`, `euca-delete-keypair`, `euca-describe-images`, `euca-run-instances`,...) compared to the AWS tools and largely accept the same options and environment variables. The implementation is made with Python and the Python library boto[69] which is an interface to cloud services that are interface compatible with the AWS. The GSutil is also implemented with Python and boto.

TABLE 4.2
Command-Line Tools to Interact with Cloud Services

	Euca2ools	AWS tools	s3cmd	GSutil
License	BSD	Apache v2.0	GPLv2	Apache v2.0
Charge	none	none	none	none
Support for EC2	yes	yes	no	no
Support for S3	no	no	yes	yes
Support for EBS	yes	yes	no	no
Support for ELB	no	yes	no	no
Eucalyptus	yes	yes	yes	no
Walrus	no	no	yes	yes
Nimbus	yes	yes	no	no
Cumulus	no	no	yes	no
OpenNebula	yes	yes	no	no
CloudStack	no	no	no	no
Google Storage	no	no	no	yes

The command-line tool s3cmd works with S3, as well as the S3-compatible private cloud storage services Walrus and Cumulus. An advantage of s3cmd is that it is an easy to use command-line interface for S3. With just a few

----

[69]`http://code.google.com/p/boto/`

options, all important functions of S3 and interface compatible cloud storage services can be used by the customers.

A drawback of all command-line tools is that they require a local installation and therefore may lack ease of use because there is no graphical user interface. The Euca2ools and GSutil both require a Python runtime environment. The AWS tools need the Java runtime edition to be installed.

4.3.2 Locally Installed Management Applications with GUI

EC2Dream is a management application for EC2, EBS, RDS and Eucalyptus. The software is open source and runs with Windows, Mac OS X and Linux. Cyberduck is an open source client for various file transfer protocols and cloud storage services. The application runs with Windows and Mac OS X and it supports directory synchronization and multiple languages. The proprietary application Gladinet Cloud Desktop also supports various different cloud storage services and integrates them into the Windows Explorer interface.

TABLE 4.3
Locally Installed Management Applications with Graphical User Interface to Interact with Cloud Services

	EC2Dream	**Cloud Desktop**	**Cyberduck**
License	Apache v2.0	proprietary	GPLv3
Charge	none	$59.99/year	none
Support for EC2	yes	no	no
Support for S3	no	yes	yes
Support for EBS	yes	no	no
Support for ELB	yes	no	no
Eucalyptus	yes	no	no
Walrus	no	no	yes
Nimbus	no	no	no
Cumulus	no	no	no
OpenNebula	no	no	no
CloudStack	no	no	no
Google Storage	no	yes	yes

4.3.3 Browser Extensions

The existing browser extensions only work with the Firefox browser and not other popular browsers like Internet Explorer, Google Chrome, Safari or Opera. The customers need to install and maintain the extension on their local computer, a fact that somehow does not reflect the cloud paradigm very well. Recent versions of ElasticFox, Hybridfox can interact with EC2, the storage services EBS and Eucalyptus infrastructure services. The extension S3Fox (see Figure 4.5) is only capable of working with S3.

TABLE 4.4
Browser Extensions to Interact with Cloud Services

	ElasticFox	Hybridfox	S3Fox
License	Apache v2.0	Apache v2.0	proprietary
Charge	none	none	none
Support for EC2	yes	yes	no
Support for S3	no	no	yes
Support for EBS	yes	yes	no
Support for ELB	no	no	no
Eucalyptus	yes	yes	no
Walrus	no	no	no
Nimbus	no	no	no
Cumulus	no	no	no
OpenNebula	no	no	no
CloudStack	no	no	no
Google Storage	no	no	no

4.3.4   Online Services

When using online services the customers have the freedom to use any operating system and a wide range of browsers. However, Amazon's AWS Management Console (see 4.6) is in line with just Amazon's cloud service offerings. It is impossible for the users to configure it in a way to manage other cloud infrastructure or storage services. The same holds for the Google Storage Manager (see 4.7) that cannot be configured to work with any other cloud service besides Google Storage.

Ylastic is a commercial service from a company that offers just a management tool and no cloud infrastructures or storage services. Ylastic offers support for most AWS cloud services and Eucalyptus infrastructures but it is not possible to manage other compatible infrastructures e.g., Nimbus or OpenNebula. As the access keys are stored with the provider of the management service, the customer also has to trust the provider of the management tool regarding privacy and security.

The KOALA and the AWS Console are compatible with any up to date browser. The Google Storage Manager and Ylastic cannot be used with the Internet Explorer.

The software service KOALA[70] (Karlsruhe Open Application (for) cLoud Administration) is designed to manage all cloud services that are compatible to

---

[70]http://koalacloud.appspot.com/

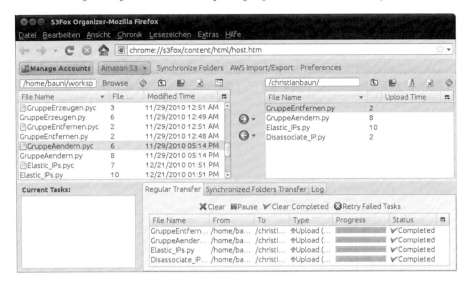

FIGURE 4.5
S3Fox Browser Extension.

the AWS API (see figure 4.8). KOALA[71] has been developed in the framework of the Open Cirrus™ cloud computing testbed that is sponsored by HP, Intel and Yahoo! [160] The service not only supports the public cloud services EC2, S3, EBS and Google Storage, but also private cloud services based on Eucalyptus, Nimbus or OpenNebula, and Walrus and Storage Controller from the Eucalyptus project can be used with this management solution. In addition, it is possible to create and manage elastic load balancers with Amazon's Elastic Load Balancing (ELB) service.

KOALA itself is a software service developed for the Google App Engine. Besides public hosting, it can be run as well inside private cloud platforms based on AppScale or typhoonAE. Operating KOALA inside a private cloud allows it even to host the management tool inside the same managed cloud that it controls. A customized version of KOALA for Android and iPhone/iPod touch is also included, which makes it easy to control cloud infrastructures and storage services with mobile devices.

A collection of alternative management solutions are optimized for mobile phones and similar devices and are run inside the e.g., Apple or Android platforms in a native way Ylastic's iPhone and Android App, the iAWSManager[72], Elasticpod[73] and directThought[74].

---

[71]http://code.google.com/p/koalacloud/
[72]http://www.iawsmanager.com
[73]http://www.elasticpod.com
[74]http://www.directthought.com

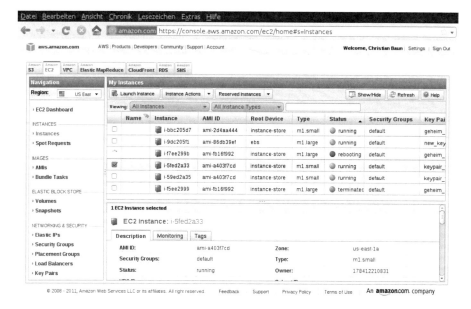

FIGURE 4.6
AWS Management Console.

4.4   Conclusion

The growing number of public and private cloud services, as well as tools to work with these services keep the cloud market in a dynamic condition. Big players like Amazon and Google as well as the open source projects and start-up companies like Cloudera[75] with their Hadoop distribution and Nimbula continuously improve their solutions, products and tools. This dynamic situation leads to new interesting services like the new Amazon Cluster GPU instances that are equipped with two Quad-Core Intel Xeon-X5570 Nehalem CPUs and include two Nvidia Tesla-M2050 GPUs each. Such new services give more opportunities to the customers to use IT resources in an elastic and favorable way without the need to buy and run their own hardware.

---

[75]http://www.cloudera.com

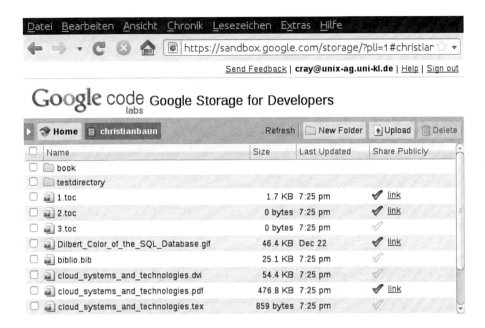

FIGURE 4.7
Google Storage Manager.

TABLE 4.5
Online Services to Interact with Cloud Services

	KOALA	**AWS Console**	**Ylastic**	**Storage Manager**
License	Apache v2.0	proprietary	proprietary	proprietary
Charge	none	none	$25/month	none
Support for EC2	yes	yes	yes	no
Support for S3	yes	yes	yes	no
Support for EBS	yes	yes	yes	no
Support for ELB	yes	yes	yes	no
Eucalyptus	yes	no	yes	no
Walrus	yes	no	yes	no
Nimbus	yes	no	no	no
Cumulus	no	no	no	no
OpenNebula	yes	no	no	no
CloudStack	no	no	no	no
Google Storage	yes	no	no	yes

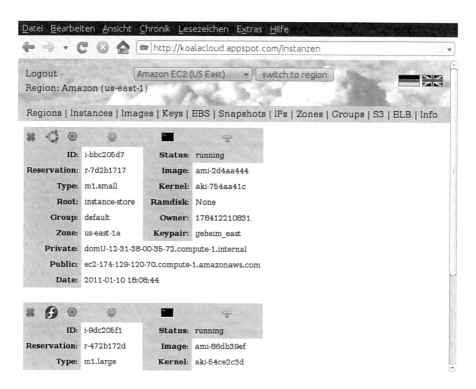

FIGURE 4.8
KOALA Cloud Management Console Showing the Users Instances.

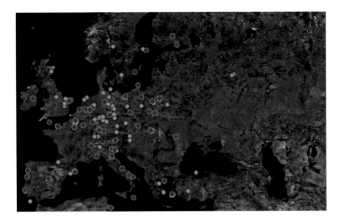

FIGURE 5.4
PlanetLab Europe Nodes on the Map (Source: http://www.planet-lab.eu/).

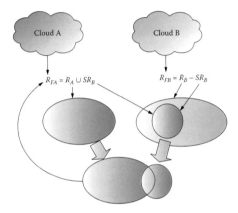

FIGURE 9.3
Example of Resource Sharing Scenario, Where a Cloud Borrows the Resources Shared by Another Federated Cloud with Exclusive Access.

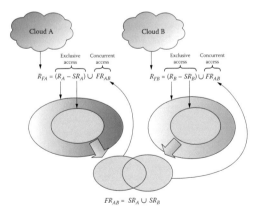

FIGURE 9.4
Example of Resource Sharing Scenario, Where Two Clouds Share a Subset of Their Resources with Concurrent Access.

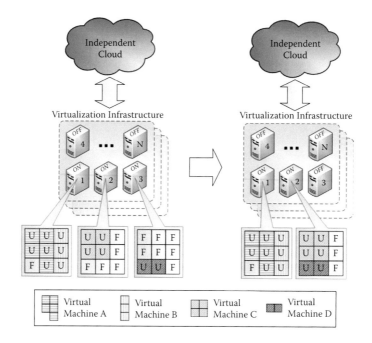

FIGURE 9.7
Example of Power Saving Scenario of Stage-1.

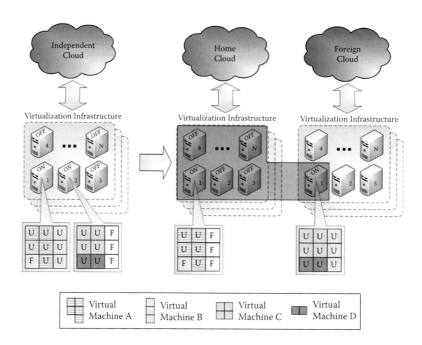

FIGURE 9.8
Example of Power Saving Scenario of Stage-3 (Intracloud/Intercloud).

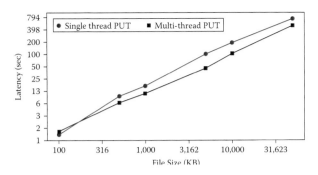

(a) latency comparison for put

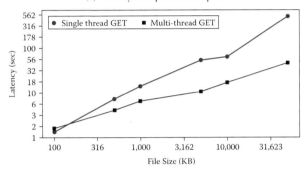

(b) latency comparison for Get

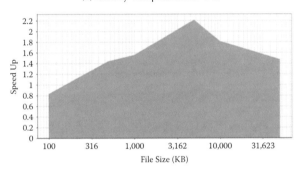

(c) speed up of using multi-threads

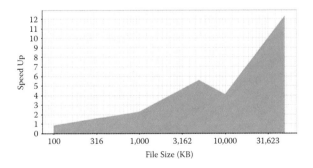

(d) speed up using multi-thread

FIGURE 10.8
Performance Improvement: Single Thread (Amazon) vs. Multi-Thread SSS.

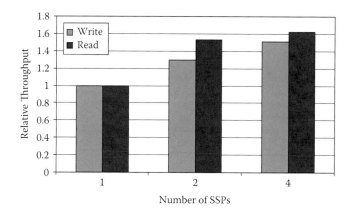

(a) Using Single SSP Parameters

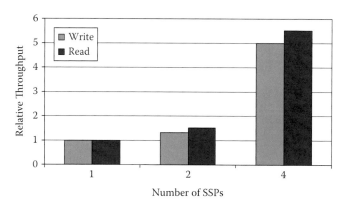

(b) Using Multiple SSPs Parameters

FIGURE 10.10
SDSI Benchmarking for Multiple SSPs.

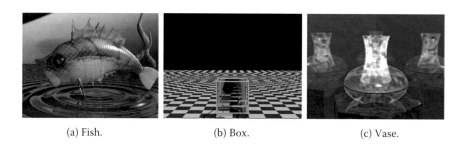

(a) Fish.   (b) Box.   (c) Vase.

FIGURE 13.6
Example of Images for Each of the Three Animations.

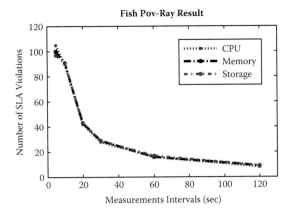

(a) Fish.

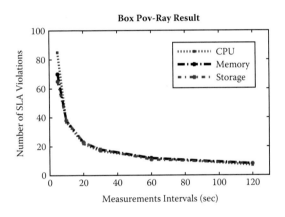

(b) Box.

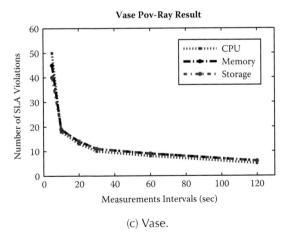

(c) Vase.

FIGURE 13.9
POV-Ray Experimental Results.

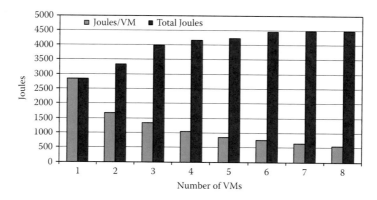

FIGURE 17.8
Energy Reduction in VM Consolidation.

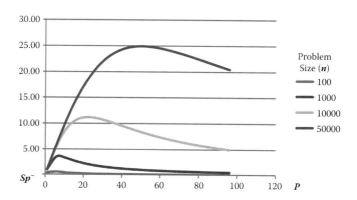

FIGURE 19.13
Parallel Performance Map of Matrix Multiplication.

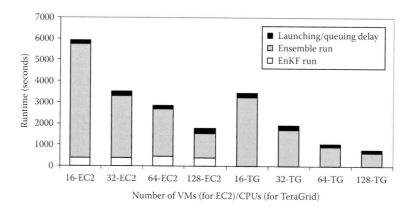

FIGURE 25.4
Baseline TTC for EC2 and TeraGrid for a 1-Stage, 128 Ensemble-Member EnKF Run. The First 4 Bars Represent the TTC as the Number of EC2 VMs Increase; the Next 4 Bars Represent the TTC as the Number of CPUs (nodes) Used Increases.

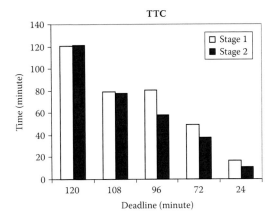

(a) Time To Completion

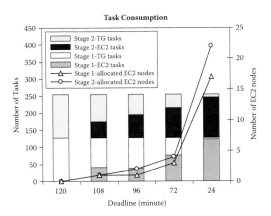

(b) Task consumption

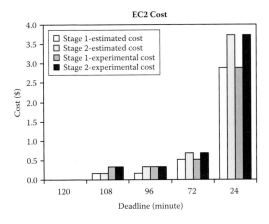

(c) EC2 cost

FIGURE 25.8

Results from Baseline Experiments (without adaptivity) but with a Speci- fied Deadline. We Run the EnKF with 2 Stages, 128 Ensemble Members and Limit EC2 Instance Types to m1.small and c1.medium. Tasks are Completed within a Given Deadline. The Shorter the Deadline, the More EC2 Nodes are Allocated.

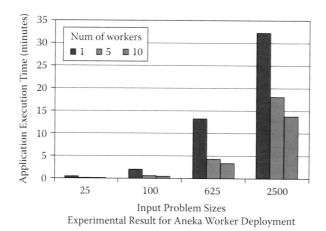

Experimental Result for Aneka Worker Deployment

FIGURE 27.20
Experimental Result Showing the Execution Time for Running the Mandel-brot Application on Aneka Worker Deployment.

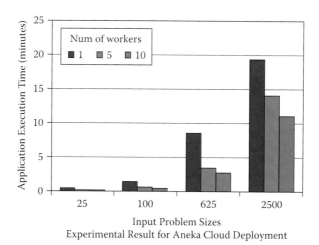

Experimental Result for Aneka Cloud Deployment

FIGURE 27.21
Experimental Result Showing the Execution Time for Running the Mandel-brot Application on Aneka Cloud Deployment.

5

A Network-Oriented Survey and Open Issues in Cloud Computing

Luigi Atzori

DIEE (Dept. of Electric and Electronic Engineering) – University of Cagliari, Italy

Fabrizio Granelli

DISI (Dept. of Information Engineering and Computer Science) – University of Trento, Italy

Antonio Pescapé

DIS (Dept. of Computer Science and Systems) – University of Napoli Federico II, Italy

CONTENTS

Cloud computing represents an emerging paradigm in the framework of ICT, describing a computing model where business applications are allocated to a combination of connections, and software and services are accessed through a web browser over a network, known as "The Cloud." This permits access to power computing through a variety of entry points and eliminates the need for organizations to install and run heavy duty applications on their computers; the data and software themselves are retrieved "on demand" like a utility service. In general, no unified and quantitative approach to analyze and design cloud computing architectures is available in the literature, as well as specific supporting technologies to control the QoS and ensure predictable performance.

The chapter is aimed at studying networking solutions for "engineering" Cloud Computing architectures, with the final goal to provide a framework for enabling analysis, design and performance improvement of such architectures.

5.1   Introduction

Cloud computing represents an emerging paradigm in the framework of ICT, describing a computing model where business applications are allocated to a combination of connections, and software and services are accessed through a web browser over a network, known as "The Cloud" [282]. This enables access to power computing through a variety of entry points and eliminates the need for organizations to install and run heavy duty applications on their computers; the data as well as software itself is retrieved "on demand" like an utility service [237, 563, 752]. As a matter of fact, cloud computing constitutes a specialized distributed computing paradigm, with the following features:

1. It is massively scalable and provides economy of scale;

2. it can be encapsulated as an abstract entity offering different level of service to customers outside the Cloud (i.e., the Internet, or any communication network);

3. services can be dynamically configured (e.g., via virtualization) and delivered on-demand;

4. it supports energy saving.

As in several well-known cases (such as peer-to-peer), diffusion of this paradigm was driven by software developers and service providers rather than research, leading to the existence of already available solutions — including Google App Engine[1] and Google Docs, Yahoo! Pipes[2], web Operating Systems

---

[1]ttp://appengine.google.com
[2]http://pipes.yahoo.com/pipes/

(G.ho.st, etc.), without the support from research studies related to architecture and protocol design, performance analysis, dimensioning, and similar issues.

Indeed, the main open challenges related to the cloud computing paradigm include:

1. data portability, lack of trust and privacy issues;

2. QoS (Quality of Service) control or guarantee, as Clouds will grow in scale and number of users and resources will require proper management and allocation;

3. increase of data-intensive applications will put a heavy burden on the communication infrastructure;

4. difficulty in fine-control over resources monitoring, as the layered architecture makes it difficult for an end user (but also for a developer or an administrator) to deploy his own monitoring infrastructure;

5. virtualization itself represents both an advantage (it provides the necessary abstraction to unify fabric components into pool of resources and resource overlays) as well as a disadvantage (reliable and efficient virtualization is required to meet SLA requirements and avoid potential performance penalties at application level).

To the best of the knowledge of the authors, in general, no unified and quantitative approach to analyze and design cloud computing architectures as well as specific supporting technologies to enable to control the QoS and ensure predictable performance are available in the literature.

In this framework, this chapter is aimed at identifying and analyzing solutions for engineering Cloud Computing architectures, with the final goal to provide a description of open issues, available solutions and ongoing activities on the subject.

The chapter is organized as follows. Section 5.2 presents a brief overview to cloud computing and related issues; Section 5.3 illustrates the main network-oriented challenges related to this new computing model, while Section 5.4 concludes the chapter with final remarks.

5.2   A Brief View of Cloud Computing

As cloud computing encompasses several disciplines related to ICT, it is not possible to provide a comprehensive state-of-the-art. Therefore, in the following, we only provide a review of the main subjects within this wide topic.

5.2.1   Cloud Computing Architectures

The paper by Buyya, Yeo and Venugopal [151] presents a modern vision of computing based on Cloud computing. Different computing paradigms promising to deliver the vision of computing utilities are presented, leading to the definition of Cloud computing and the architecture for creating market-oriented Clouds by leveraging technologies (such as VMs). Chappell[3] provides an introduction on cloud computing, starting from a brief classification of the current implementations of cloud computing, identifying three main categories. Wang and Von Laszewski [744] reveal how the Cloud computing emerges as a new computing paradigm which aims to provide reliable, customized and QoS guaranteed dynamic computing environments for end-users. Their paper analyzes the cloud computing paradigm from various aspects, such as definitions, distinct features, and enabling technologies. It considers the Hardware as a Service (HaaS), the Software as a Service (SaaS) and the Data as a Service (DaaS) visions, and the main functional aspects of cloud computing. While these works provide different definitions and visions of the Cloud, all rely on a common architecture which is shown in Figure 5.1 and that is made of the following components: the Fabric Layer which encompasses the hardware resources; the Unified Layer, which is aimed at making available to the upper layer a uniform set of services to make use of the hardware resources; the Platform, which is the operating system for the management of the system; and the Application layer, which includes the applications provided by the Cloud.

The main distinctions between Cloud Computing and other paradigms (such as Grid computing, Global computing, Internet Computing), are the following:

1. User-centric interfaces.

2. On-demand service provisioning.

3. QoS guaranteed offer.

4. Autonomous System.

5. Scalability and flexibility.

5.2.2   Cross-Layer Signaling

Tighter cooperation among protocols and functionalities at different layers of the protocol stack (especially between application-level clouds and networking infrastructure) is envisaged as an enabling framework for (i) supporting fine control over network and computation resources, and (ii) providing APIs for network-aware applications.

As far as cross-layering is concerned, several cross-layer approaches have been

---

[3]Chappell David (August 2008). A Short Introduction to Cloud Platforms. David Chappell & Associates. Retrieved on 2008-08-20.

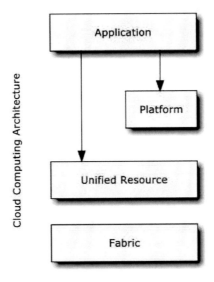

FIGURE 5.1
A Reference Cloud-Computing Layered Model.

proposed in the literature so far, focusing on specific problems, providing ad-hoc solutions and raising relevant issues concerning implementation of different solutions within the TCP/IP protocol reference model. Indeed, coexistence and interoperability represent a central issue, especially in the cloud computing scenario, leading to the need for a common cross-layer signaling architecture. Such an architecture should provide the implementation of the cross-layer functionalities as well as a standard way for an easy introduction of cross-layer mechanisms inside the protocol stack. Kliazovich et al.[4] present some possible cross-layer signaling architectures that can be proposed based on ideas available in the literature, even if it must be underlined that none is currently available in the form of an enabling framework to study and fully exploit the advantages of a reasonably wide range of cross-layer solutions.

5.2.3 Services and Traffic Analysis and Classification

Service characterization [476, 477] and the related traffic classification and analysis represent important steps in the entire framework of service management and delivery [141]. Since the service characterization phase depends on the particular service scenario, low level view (network traffic level) allows to adopt well established techniques proposed in literature, and at the

---

[4]D.Kliazovich, M.Devetsikiotis, F.Granelli, "Formal Methods in Cross layer Modeling and Optimization of Wireless Networks: State of the Art and Future Directions", in Heterogeneous Next Generation Networking: Innovations and Platform, IGI, 2008.

same time a preferred way to analyze the behavior of the applications. Traffic classification represents a powerful tool for understanding the characteristics of the traffic relying on the network. Today the classification approaches used are port-based and payload-based. Such techniques were initially considered very reliable, such to be used to build reference data in the evaluation of novel classification approaches[5] [573]. Because of the increasing problems (e.g., privacy or unreliability), in the last few years researchers have proposed several classification techniques that do not need access to packets content, while they are commonly based on the statistical analysis of traffic patterns and on machine-learning. The explosion of high-quality scientific literature in this field [105, 125, 411, 522] testifies to the great interest in researching novel and accurate techniques for traffic classification, which find application in several networking areas. It has been demonstrated that statistical and machine-learning techniques can achieve high degrees of accuracy, and that they appear to be the most promising approaches to face problems like protocol obfuscation, encapsulation, and encryption. Despite the great effort in traffic classification and analysis, the literature lacks the studies considering traffic generated by cloud computing applications.

5.2.4   QoS Technologies and Management Issues

QoS management has to be provided in a consistent and coordinated fashion across all layers of enterprise systems, ranging from enterprise policies, applications, middleware platforms, down to the network layers. In addition, a comprehensive set of QoS characteristics in categories like performance, reliability, timeliness, and security must be managed in a holistic fashion. This is the approach that has been followed by most significant works in this field in the last few years. Efforts are being made in various areas, for example, SOA, application-oriented networking, and autonomic computing[6], to achieve a scalable, secure, and self-managed service delivery framework to shorten the time-to-market of new Internet applications, as well as lower the management costs of service providers [181]. A prototype for QoS management to support Service Level Management for global enterprise services is presented in [741]. These works try to define a solution with a holistic view of the QoS management problem, starting from an architectural point of view and identifying standards, analyzing interoperability issues and defining workflows. Other studies focus more on the networking issues, that is, how service differentiation is fulfilled while transmitting the content from an end to another end of the network on the basis to high-level QoS targets. [89] specifically focuses on the issues and complexities on merging WDM and IP technologies and concentrates on making the core network efficient for transporting differentiated service traffic, adaptive to changes in traffic patterns, and resilient against

---

[5]L7-filter, Application Layer Packet Classifier for Linux. http://l7-filter.sourceforge.net.
[6]ITU-T Rec M.3060/Y.2401, "Principles for the Management of the Next Generation Networks," Mar. 2006.

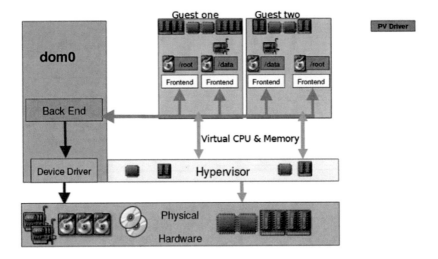

FIGURE 5.2
Xen Paravirtualization Architecture (source: RedHat).

possible failures. [102] deals with the bandwidth management in NGN with particular attention to the Differentiated-Service-Aware Traffic Engineering in Multiprotocol Label Switching networks.

5.2.5 Virtualization

Virtualization mechanisms [586] find the right employment in the cloud computing context. The Linux VServer[7] technology is a soft partitioning concept based on Security Contexts, which permits the creation of many independent Virtual Private Servers (VPS) that run simultaneously on a single physical server at full speed, efficiently sharing hardware resources. FreeBSD Jail[8] is the virtualization mechanism used in Emulab in order to support multiple experiments running concurrently on the same physical node. OpenVZ[9] is an operating system-level virtualization technology built using GNU/Linux. It gives the ability to run multiple isolated system instances, called Virtual Private Servers or Virtual Environments. Xen [116] is a paravirtualization system developed by the University of Cambridge. Xen provides a VM monitor for x86 processors that supports execution of multiple guest operating systems at the same time. VIOLIN [623] is a shared distributed infrastructure formed by federating computation resources from multiple domains. Usher [506] is a VM

---

[7]http://linux-vserver.org
[8]http://www.freebsd.org/doc/en/books/handbook/jails.html
[9]http://openvz.org

management system designed to impose few constraints upon the computing environment under its management.

5.3   Research Challenges for Engineering Cloud Computing Architectures

The focus of the chapter is the identification and analysis of solutions for engineering Cloud Computing architectures, with the final goal to provide a description of open issues, available solutions and ongoing activities on the subject. The term "Cloud" refers to the characteristic of accessing data and services from different access technologies through a transparent network (i.e., the Internet). The term "engineering" defines the novelty of the approach, and clearly identifies the focus on the novel perspective of enabling quantitative analyses and effective design of solutions covering issues related to the support of cloud computing services, including protocols and architectures, QoS/QoE, interoperability, SLAs. As a consequence, the chapter refers to a vertical vision of the paradigm across the layers of the protocol stack and on the performance-oriented integration of the functionalities at different layers.

Indeed, besides the existence of actual cloud computing applications, still they don't address issues related to the underlying transport infrastructure and how to interact and collaborate with it — focusing on universal access rather than performance or predictability.

The next sections provide a brief description of the main issues associated with each challenge.

5.3.1   Assuring the Target QoS/QoE

With the aim of engineering the network so that the target QoS and QoE (Quality of Experience) requirements in a distributed architecture are met, the DS-TE (DiffServ aware Traffic Engineering) architecture represents one of most appropriate solutions for the transport of the relevant Cloud Computing application traffic flows. It manages the QoS in a scalable, flexible and dynamic way and allows for performing Traffic Engineering (TE) in a differentiated service environment by applying routing constrains with class granularity [454]. In a DS-TE domain, TE policies are performed on a per-class basis through DiffServ classification. This goal is achieved by introducing three new concepts:

- Class Type (CT): is a set of traffic trunks with the same QoS requirements. In a DS-TE domain, no more than eight CTs can be set up, on the basis of traffic trunks CoS (Class of Services) values.

- TE-Class: is a combination of a CT and a preemption value, defined as

$< CT, p >$. It allows traffic trunks belonging to the same CT to be forwarded over different LSPs (Label Switched Path) at different priorities. Preemption allows high-priority LSPs to be routed through paths in use by low-priority LSPs, which are then terminated or re-routed. TE-Classes are defined by assigning one or more preemption values to each CT. The maximum number of TE-Class is eight and the belonging of a packet to a TE-Class arises from the EXP bits, which is a field in the MPLS header.

- Bandwidth Constraint (BC) model: specifies the amount of links bandwidth that can be used to route LSP belonging to each CT. To this, there are appropriate Bandwidth Constraints defined for each CT and each link.

Regardless of which BC model is adopted, the resulting network performance and resource utilization depend on both the aggregation of the CoS in the implemented CTs and the setting of the BCs. The bandwidth constraints over all the network links have a direct impact on the performance of the constrained-based routing [412], which then heavily influences the call block probability and resulting QoS. It also affects the frequencies of the preemption occurrences, which have to be kept as low as possible [225]. BCs' setting, together with the selected aggregation of the traffic into the active CTs, are also major tasks to control the end-to-end performance in term of delay, losses and jitter. In fact, the belonging of a traffic flow to a certain CT determines the priority of the relevant packets with respect to the others. Additionally, the amount of packets with higher priority depends on the BCs set for the high priority classes. For this reason the setting of the BC and aggregation of the CoS in CTs are problems that have to be jointly addressed. Indeed, traffic flows with dissimilar features characterize the Cloud Computing scenario, such as: bulk data transfers between data-storage centers, short control messages between distributed applications, constant real-time flows for multimedia communications.

Herein, we present a generic DS-TE bandwidth management procedure, which aims to configure the BC Model in terms of the effective network requirements when providing Cloud Computing services. Accordingly, the system adopts the solution of implementing a single algorithm to achieve both Class Type classification and BC definition. This approach performs the two tasks in an interdependent way, optimizing the output of the one in terms of the solution obtained for the other and vice versa.

As shown in Figure 5.3, the bandwidth management setup works on the basis of the following input information: a reference set of key performance indicators (KPIs), which allow for characterizing the service requirements and evaluating QoS network performance; a quality profile, which defines the services classification into CoSs; and the profile of the forecasted ingress traffic. This information, together with the BC model adopted in the network, are the input to a "what-if analysis" to examine network performance and resource utilization at varying CT classifications and BC settings. Note that while the

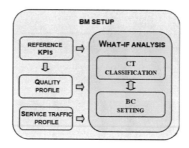

FIGURE 5.3
Setup of Bandwidth Management.

setting of the BC model may vary from link to link, CT classification has to be unique for the whole cloud network.

5.3.1.1   Key Performance Indicators (KPIs)

The Key Performance Indicators provide a quantitative and objective solution to compare the obtained quality performance with the desired ones and are used for both traffic classification and network analysis. Quality can be evaluated from the point of view of the end-user, as perceived quality which refers to the experience in the use of a service, or from the point of view of the operators, as offered quality which refers to the policies of service provisioning, the sustained costs and the capacity to maintain service availability. It can even be intrinsic if it refers to technical network performance parameters, such as delay, jitter, packet loss, and throughput. It can be specified either quantitatively or qualitatively.

Several QoS definitions have been proposed in the literature[10] [638]. However, the following are those that are most often selected: IP Packet Transfer Delay (IPTD); IP Packet Delay Variation (IPDV); IP Packet Loss Ratio (IPLR); IP Packet Error Ratio (IPER); and Spurious IP Packet Rate (SIPR).

5.3.1.2   Quality Profile

The definition of the quality profile consists in identifying the QoS requirements and performing DiffServ classification for the set of services S required in the cloud computing scenario. Service characterization and classification rely on a set of KPIs. Let P be the number of selected reference KPIs; the i-th service within S is associated to a vector that specifies its quality requirements:

$$< S >_i = [\Delta KPI^S_{1,i}, \Delta KPI^S_{2,i}, \Delta KPI^S_{3,i}, ..., \Delta KPI^S_{P,i}]$$

Each element of this vector defines the threshold for the allowed values for each

---

[10]ITU-T Rec. G.1540, "IP Packet transfer and Availability Performance Parameters," Dec. 2002.

KPI. On the basis of these values, each service is classified into a specific CoS according to the DiffServ model. The 14 standard CoSs [412] can be adopted or new CoSs can be defined by the network operator. One or more services can be classified in the same CoS. As a result of this procedure another vector of KPI is obtained:

$< CoS >_j = [\Delta KPI^c_{1,j}, \Delta KPI^c_{2,j}, \Delta KPI^c_{3,j}, ..., \Delta KPI^c_{P,j}]$

This vector contains the KPI threshold values which the j-th CoS is able to satisfy. They correspond to the intersection of quality requirements of all services in the j-th CoS. The cardinality of the set CoS can be lower than the number of services to be provided.

5.3.1.3   Service Traffic Profile

In accordance with the IETF rules, the BC model specifications have to be defined link by link; as a consequence, the proposed DS-TE bandwidth management procedure needs to perform a traffic prediction for each link in the network. This prediction can be obtained through a network analysis by considering the following inputs which are available to the operator:

- cloud network topology;

- bandwidth required by each cloud computing service, B^S_i;

- number of users $U^{a,b}_i$ of service i accessing the network at node a and ending at node b.

To estimate the traffic load per link, it is necessary to consider the paths between each pair of ingress-egress routers. For the generic edge routers (a, b), where a and b are the ingress and egress nodes, respectively, the load for service i is equal to:

$$C^{a,b}_i = B^S_i U^{a,b}_i \tag{5.1}$$

This traffic spreads over the set of available paths from a to b. The distribution of the traffic through the available paths is evaluated through an empirical simple algorithm, which distributes the traffic according to the path length. In particular, an inverse linear relationship between the traffic load and the path length can be adopted. Only the disjoint Z paths no longer than two times the shortest one are considered in this process and the traffic is distributed according to the length of each path p_z. The traffic load from a to b along the z-path can be assumed to be equal to:

$$c^{a,b}_{i,z} = C^{a,b}_{i,z} \frac{1}{P_z \sum_{x=1}^{Z} 1/p_x} \tag{5.2}$$

From this distribution, the total load per link and per service is computed.

Note that the proposed algorithm is quite simple. The choice has been driven by the fact that at this stage it is just required to find enough resources from the source to the destination to satisfy the bandwidth demands for the different services. For this purpose, the use of complex routing procedures would be useless [412]. These are instead adopted when addressing LSP setup requests.

At this stage the maximum length among all the paths traversing each link needs to be computed. This is used to compute local KPI thresholds from the end-to-end KPI thresholds in $< CoS >_j$, as discussed in the following.

5.3.1.4   What-if Analysis

This procedure is intended to provide a solution to the problem of mapping the cloud computing services into the appropriate Class Type and to find out the optimal setting of the bandwidth constraints. Herein, appropriateness and optimality are defined according to the KPI constraints (for each service) and in terms of resource utilization and/or saving.

The first step is the detection of a possible CT classification, which is performed by evaluating the $< CoS >$ KPI vectors which were defined during the quality profile definition phase. The possible mappings from CoSs to CTs can be obtained in two possible ways:

- activating a CT for each CoS (since the maximum number of activable CTs is 8, this solution is possible only if the number of CoSs is lower than 8).

- grouping more CoSs in the same CT. In this case, the bandwidth allocation benefits from reducing the number of CTs at the expense of a lower efficiency in terms of QoS requirements satisfaction. The allowed combinations of CoSs are those which satisfy the following conditions:

  - at least three CTs are defined: CT2 for expedited traffic, CT1 with intermediate guarantees and CT0 for best effort services;

  - the priority order defined by DiffServ classification is respected (only consecutive CoSs are grouped).

If W is the cardinality of the set $\{CoS\}$, the resulting total number of $\{CT\}$ classifications is:

$$H = \sum_{v=3}^{V} \frac{(W-1)!}{v!(W-v)!} \tag{5.3}$$

where $V = W$ if $W <= 8$ and $V = 8$ otherwise.

Each k-th CT of the considered classification needs to satisfy the quality parameters of the encompassed CoSs so that another vector of KPIs $<CT>$ is defined as the intersection of all corresponding $< CoS >_j$. These represent the KPI thresholds for all the services included in each CT.

Once all the possible aggregations have been defined, these are evaluated by computing a gain function that takes into account both KPI gain (GKPI) and bandwidth gain (GBDW). GKPI evaluates to which extent the considered solution allows for keeping each KPIs lower than the desired thresholds $< CoS > j$. GBDW provides a measure of the amount of bandwidth that is still free for each CT when applying the proposed solution. While the meaning and the objectives of both these functions are clearly defined, the exact formula to be used depends on the specific scenario under investigation. This analysis is performed for all other $\{CT\}_h$ until $h = H$. Then, the optimal CT classification and the optimal BCs for each link are identified by selecting the solution with the highest gain function among all the evaluated combinations.

5.3.2 Service Characterization and Traffic Modeling

A suitable framework for performance-oriented analysis of cloud computing architectures and solutions is needed, which in turn implies the introduction of an effective characterization of the services and their footprint on the transport infrastructure, and in general the definition of service requirements and their dependence on the number of users. In general, an open issue is related to the quantification of a service, since especially in the case of cloud computing a service represents a multi-dimensional load on the computing and networking infrastructure, both in terms of computational load and of network load (including bandwidth, requirements, etc.). The idea here proposed is to instrument cloud computing platforms with traffic classification features useful to both cloud traffic and applications classification. This permits one to clearly understand (i) the traffic received/generated by the cloud; (ii) the applications responsible for a specific portion of traffic (e.g., heavy hitters); (iii) the relationships between traffic flows and cloud applications. Therefore, there is a strong need for new methodologies and techniques for service characterization and for understanding the impact of cloud computing services on the network. We propose a two step approach:

- The first step of this analysis is the study of the interactions among the entities of the considered distributed applications, that is, service characterization. This task provides insights related to the process of the composition and use of a service. This step will study the cloud computing services and their characterization from the following viewpoints: (i) interactions among the several entities of the considered service; (ii) interactions between users and services; (iii) interactions between users and networks.

- The second step is the characterization of the traffic generated by cloud computing applications: after the "high-level" analysis of the previous step, in this second step the target of the traffic analysis and footprinting stage is the understanding of the dynamics associated to cloud computing services and final objectives are: (i) to gain knowledge of the traffic generated by these services; (ii) to study techniques and methodologies for the identification and classification of the traffic generated by cloud computing services;

(iii) to improve the support for such new services; (iv) to guarantee their reliability and proper use; (v) to better plan their evolution; (vi) to develop new network architectures for better supporting them.

5.3.3   Signaling

Introducing specific signaling structures and protocols is required to enable tighter cooperation between cloud computing applications and the underlying transport infrastructure. In this framework, the cloud architecture could benefit from the definition of suitable built-in monitoring primitives aimed at providing information about the number of users, SLAs, network context and resources. In the following, possible cross-layer signaling architectures are briefly described:

- *Interlayer Signaling Pipe* allows propagation of signaling messages layer-to-layer along with packet data flow inside the protocol stack in bottom-up or top-down manner. Signaling information propagates along with the data flow inside the protocol stack and can be associated with a particular packet incoming or outgoing from the protocol stack. Packet headers or packet structures are two techniques considered for encapsulation of signaling information and its propagation along the protocol stack.

- *Direct Interlayer Communication* introduces signaling "shortcuts" performed out of band allowing non-neighboring layers of the protocol stack to exchange messages, without processing at every adjacent layer, thus allowing fast signaling information delivery to the destination layer. Despite the advantages of direct communication between protocol layers this approach is mostly limited by request-response action, while more complicated event-based signaling should be adapted. To this aim, a mechanism which uses callback functions can be employed. This mechanism allows a given protocol layer to register a specific procedure (callback function) with another protocol layer, whose execution is triggered by a specific event at that layer.

- *Central Cross-Layer Plane* implemented in parallel to the protocol stack is the most widely proposed cross-layer signaling architecture. Typically, it is implemented using a shared bus or database that can be accessed by all layers. Parameter exchanged between layers is standardized and performed using well-defined layer interfacing modules, each of which exports a set of IP functions.

- *Network-Wide Cross-Layer Signaling* represents a novel approach, allowing network-wide propagation of cross-layer signaling information adding another degree of freedom in how cross-layer signaling can be performed. Implementation of network-wide cross-layering should include a combination of signaling techniques (like packet headers, standalone messages, or feedback functions) depending on signaling goals and the scope (at the node or in the network) the cross-layer signaling is performed.

The choice of a specific signaling architecture should be driven by quantitative analysis in terms of communication requirements, resulting overhead, interoperability, etc. A specific framework should be identified to address such issue and provide a design tool for cloud computing system engineers.

5.3.4 Overlay and Performance Optimization

As dynamic adaptation and optimization will represent the central paradigm of a cloud computing platform, the issue represents a core problem. Optimization of cloud computing systems, with specific emphasis on supporting effective scalability, dependability, and reliability of the solutions under a dynamic perspective, involves several interdisciplinary areas.

As all distributed applications, cloud computing involves the definition of a virtual topology or "overlay" in terms of logical connections among the servers, services, clients and other entities in the application domain. The virtual topology is then mapped onto the network topology, which is the actual communication infrastructure. A specific issue which raises in this process is whether such mapping operation between the two topologies should be blind or resource-aware. Similar issues, even if related to computational time optimization, were faced in grid computing platforms — leading to the definition of specific functionalities such as the Network Weather Service[11]. Clearly, efficient mapping implies suitable signaling architectures (see previous section).

In general, performance optimization in a cloud computing scenario represents a multi-dimensional problem, depending on several variables (network context, operating procedures) and aimed at different but joint optimization goals (response time, execution time, interactivity, etc.). As a consequence, relevant works in the framework of scientific literature on Multi-Objective Optimization (MOO) should be considered and customized and adapted to the cloud computing scenario.

Finally, efficient runtime resource utilization at computational and network levels puts a strong emphasis on the aspect of measurements in order to provide the necessary knowledge about the operating context. In this scenario, measurement-based approaches should be considered in order to support the theoretical benefits deriving from the chosen MOO strategies. Again, relevant works are available in the literature, but not specifically tailored for cloud computing platforms.

5.3.5 Interoperability and Testing

Current architectures for cloud computing offer limited possibilities to the user such as flat prices for resources leasing, proprietary interfaces for services

---

[11] http://nws.cs.ucsb.edu

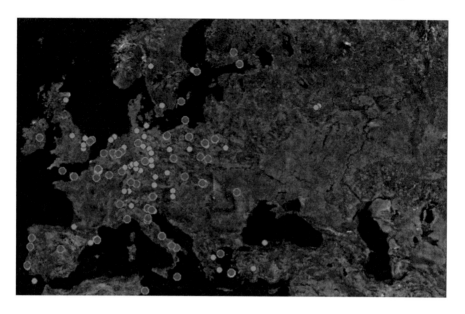

FIGURE 5.4
PlanetLab Europe Nodes on the Map (Source: http://www.planet-lab.eu/).

access and the lack of portability for the applications, making it very difficult for a user to migrate applications from one service platform to another.
In the future, it will be very useful to have a standard for interoperability
among the different cloud architectures. This would allow the growth of new
applications and the development of a market based approach to cloud computing. This can be studied thanks to distributed experimental facilities and
testbeds based on the concept of federation. It is being widely used for network
testbeds. In the framework of PlanetLab[12], a peer-to-peer federation between
PlanetLab Central and PlanetLab Europe (see Figure 5.4) has been successfully established thanks to the ONELAB European Project[13]. Just to provide
the reader a quantitative dimension of these testbeds, PlanetLab currently
consists of 1132 nodes at 517 sites (as of October 2010).
DETER[14] testbed is built by implementing a sort of federation between several independent EMULAB-based testbeds[15]. Likewise, a similar federation
effort is being developed with the objective of running experiments across
different ORBIT[16] wireless testbeds. A more challenging step in this process
consists in further extending the concept of federation across heterogeneous

---

[12]http://www.planet-lab.org/
[13]http://www.onelab.eu/
[14]http://www.isi.edu/deter/
[15]http://www.emulab.net/
[16]http://www.orbit-lab.org/

testbeds. Federation between PlanetLab and EMULAB is currently investigated. Federation is also addressed by the ONELAB2 project (part of the FIRE initiative[17]), a follow-up of ONELAB started in September 2008.

Future architectures for cloud computing should offer the possibilities to adopt federation approaches to integrate distributed testbeds in a unique experimental facility defining a framework for the interoperability among architectures.

5.4   Conclusions and Final Remarks

Cloud computing represents an emerging paradigm in the world of distributed computing. Nevertheless, it provides relevant challenges in the areas of networking and system engineering in general. The chapter proposed a brief overview of available platforms for cloud computing, focusing on current and prospective networking issues to be considered by network engineers in order to control and optimize performance of such computing services. Some of those are already considered in the literature, even if in different domains, while others remain completely uncovered — and therefore represent problems where competences related to networking enter the scenario of cloud computing.

---

[17]http://cordis.europa.eu/fp7/ict/fire/

6

A Taxonomy of QoS Management and Service Selection Methodologies for Cloud Computing

Amir Vahid Dastjerdi

The University of Melbourne, Australia

Rajkumar Buyya

The University of Melbourne, Australia

CONTENTS

A problem that has become one of the recent critical issues in service computing is automating web services selection. Web service selection has been applied in different computing paradigms namely as SOA, and Grid. A number of approaches with a variety of architectures and algorithms have been proposed to resolve the problem. The aim of this chapter is to create a comprehensive taxonomy for web service selection solution, and apply this taxonomy to categorize selection approaches. In addition, the survey and taxonomy results are presented in a way to identify gaps in this area of research. Furthermore, it indicates how web service selection approaches in SOA and Grid can share their contributions to tackle selection problems in Cloud.

6.1   Introduction

A web service [14] is a piece of software interface that can be invoked over the Internet and can be roughly viewed as a next-generation successor of the Common Object Request Broker Architecture (CORBA) [15] or the Remote Procedure Call (RPC) [763] technique. The main benefits of web services are interoperability, ease of use, reusability and ubiquitous computing. Currently, an increasing number of companies and organizations implement their applications over the Internet. Thus, the ability to select and integrate inter-organizational and heterogeneous services on the Web efficiently and effectively at runtime is an important step toward the development of Web service applications. Recently, researches study how to specify (in a formal and expressive enough language), compose (automatically), discover, select and ensure the correctness of web services [689]. The typical architecture of web services includes three roles, namely service requestor, service broker and service provider. Once the requestor sends a service request to the broker, a matched service provider should be sent back by the broker from a published service metadata repository. When the broker finds a set of services which all satisfy the functional requirements of a service request, then the crucial issue is how quality of service (QoS) requirements should be used to select the most proper service.

Web service selection has been applied in different computing paradigms namely as SOA, and Grid. That is because a resource can be represented by web services and web service selection approaches can help assigning each request in the queue to the most proper resource in a fair manner. In addition, recently Cloud Computing has emerged as a promising paradigm, in which the platform and even the infrastructure are represented as service, therefore Cloud selection plays a significant role and will attract lots of attention. Considering the popularity of Cloud computing, it is highly possible to conduct redundant researches in this area. There are many works that in the context of SOA and Grid web service selections, can share their contributions to tackle selection problems in the Cloud. Consequently, the aim of this taxonomy is:

- To identify state-of-the-art challenges in web service selection and extract a general model for service selection.

- To classify works on how they are modeling QoS attributes and how they approach selection including investigation of their advantages and disadvantages.

- To investigate what are new challenges in web service selection particularly in the context of Cloud computing, and what can be inherited works in the context of Grid and SOA.

6.2   General Model of Web Service Selection

Based on a survey of web service selection literatures, a general model for service selection has been extracted. The model consists of two parts. The first part describes steps that have to be considered for QoS management, and the second part is deals with the process of web service selection based on provided QoS preferences and service descriptions.

6.2.1   QoS Management

As depicted in Figure 6.1, several steps have to be taken into account for QoS management. In general, a QoS management approach for QoS-aware Web service selection can include following phases:

1. **Identifying Roles**: Provider and client (requestor) are the two basic roles in all service selection problems. The objective of selection is determined based on the defined roles in the problem. Selection solutions usually try to maximize provider, requestor, or both profits. In some problems, other roles such as monitoring party or other third parties can be identified.

2. **QoS Modeling**: QoS can be described in service requestor's preferences to express their expectations more. It also can be included in a web service advertisement when there are different WS providers that present diverse versions of services to answer varying requirements of their customers. The requestor's requirements and the service offerings have both functional and complex non-functional aspects, which need to be expressed and then for evaluation purposes of service and request, matched against each other [785]. Moreover,

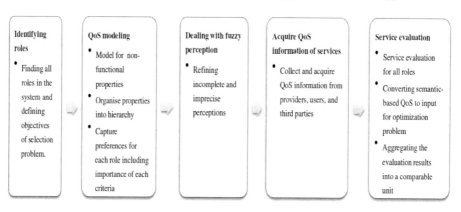

FIGURE 6.1
QoS Management Process.

the model should give the user the ability to express which QoS criteria are more important for them and a way to define the relations between QoS criteria. For example, one user may prefer a service with better response time compared to a service with a lower price and the other may like the cheaper service better. In addition to the model, it increases the transparency if we place QoS properties into hierarchical structure. For example, in the hierarchy throughput and quality, properties are performance aspects, while security and privacy are both safety aspects. Usually work uses ontology to create the hierarchy [713, 715].

3. **Taking Care of Customers' Fuzzy Perceptions**: Dealing with fuzzy perceptions of customers is another important task in managing QoS preferences of requestors. That is because brokers usually face user's perceptions which are incomplete and imprecise [745] (due to complexity and mistrusted information).

4. **Collecting QoS Information**: In this phase, a proper approach to collect and acquire QoS information is needed. Even though some works [707] believe service consumers are responsible for development of QoS information, others [450, 737, 747] assume that service providers are supposed to offer QoS information along with their service descriptions.

5. **Aggregating the Evaluation Results into a Comparable Unit**: This phase can include the transferring of semantically described QoS to a proper input for an optimization problem [295]. It is essential to aggregate QoS criteria and sub-criteria scores to gain a final score for the service. In this step, a suitable aggregation method needs to be selected [785].

6.2.2   Process of Web Service Selection

After web services QoS values have been obtained, they are used as an input for a selection approach to find the most preferable web services. Most of researches suggest the following selection phases:

- Formulating and modeling the problem is done in this first phase which includes finding constraints and objectives for the selection problem. For example, in a work done by D. Tsesmetzis et al. [715] the bandwidth has been considered as a constraint and the objective is to minimize the cost.

- Next, the selection problem is tackled by a proper optimization or decision making technique which suits the modeled problem.

6.3   Taxonomy

6.3.1   Taxonomy of Web Service Selection Context

Web service selection has been investigated in different computing paradigms and most importantly for Grid, and SOA. This section shows how characteristics of the problem in each realm differ from the others significantly. Major objectives of works in each context are summarized and illustrated in Figure 6.2.

6.3.1.1   Grid Computing

Grid Computing aims to enable resource sharing and coordinated problem solving in dynamic, multi-institutional virtual organizations [279, 280]. The goal of such a paradigm is to enable federated resource sharing in dynamic and distributed environments. Therefore, QoS management and selection work in this context is mainly focused on load balancing [149] and fair distribution of resources among service requests. Some efforts [91, 737] present the problem of selection as a challenge of finding proper services to form the composition which can maximize objective function.

Load Balancing Xiaoling et al. [747] mentioned that because web services are highly dynamic and distributed, it is likely that loads on service providers are not distributed symmetrically. Consequently, the work approaches the problem by applying the queuing theory. Their idea is using queuing theory to assign requests to service providers that have the least load. Hence, the system obtains all providers which can serve the request and choose the one which has the minimal expected waiting time. The Expected Waiting Time (EWT) is calculated based on the formula below:

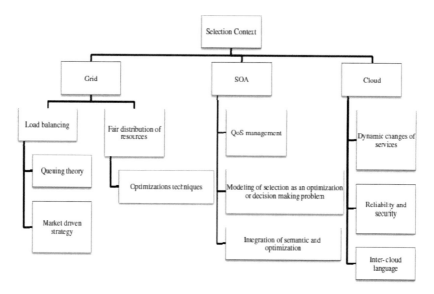

FIGURE 6.2
Selection Contexts.

$$EWT = E(w) = \frac{1}{\mu}\left(\frac{\rho}{1-\rho}\right) = \frac{\lambda}{\mu(\mu-\lambda)} \qquad (6.1)$$

where λ is the arrival rate, which specifies the number of requests the provider served in the time unit. This can be obtained by:

$$\lambda = \frac{NUM}{elapstime} \qquad (6.2)$$

In the equation elapstime is calculated from the point when the web service provider is started, and Num is number of requests have been satisfied during elapstime. Performance evaluation for the approach shows the selection strategy reduces total waiting time in general and works better than random selection when the provider's capacity is limited and low. Most service selection strategies only take care of requestor's interests, which brings heavy loads to best service providers, however Wancheng et al. [737] uses the price-demand rule in the commodity-market to improve load balancing. The paper discusses that using best-effort strategy to serve requesters leads to an unbalanced system because of greedy competition for the best services. So in this approach, service requesters and providers are considered as buyers and sellers in a commodity-market to adjust their supply and demand [149]. In addition, the selection strategy offers "enough" QoS as a substitute for "best" QoS. The evaluation model for web service selection introduces the relationship between price of attributes and value of attributes instead of just simply using an additive weighting method for aggregation of QoS properties. It models the service selection as a 0–1 multi- dimensional knapsack problem. Furthermore, the enhancement to the service selection model was made by the use of the web service's price to adjust their supply and demand. The demands of web services are determined by their invocation rate. Therefore, it increases the price when the invocation rate is high and vice versa to obtain a balanced system. The work has not considered requestor's profit in the web service selection model which can be considered as its shortcoming.

Fair Distribution of Resources Tapashree Guha et al. [321] proposed an algorithm, which enables optimal selection on the Grid while considering fair distribution of resources among requestors in a condition that several clients should be persuaded concurrently. First, the paper analyzed traditional matchmaking algorithms [484]. Those algorithms take the first request and then map it to the service with the highest score. And then they select the second best for the second request in the list. The major drawback of this sort of algorithm is that requests placed later in the queue are mapped to services with poorer match scores. One way to tackle this problem is applying a constraint-satisfaction-based matchmaking algorithm (CS-MM). The algorithm effectiveness appears when several requests are assigned to a single service which can only serve one request at a time. Therefore, the algorithm selects the service for a request, which has the lower, second highest match score. Consequently, the algorithm can solve traditional matchmaking problems in a fair manner. However, to achieve better results when we are processing a large number of concurrent requests for particular services the Multiple Objective based Particle Swarm Optimization Algorithm using Crowding Distance (MOPSO-CD) was adopted. The algorithm is swarm based artificial intelligence algorithm [198, 354, 568] and was inspired from the nature of bird flocking. The experimental results in the paper show although the MOPSO-CD taking considerably more time to be executed; however, it can produce a far more accurate solution than CS-MM.

With the aim of fair distribution of Grid resources (services) among requests, Ludwig et al. [485] presented a service selection approach for Grid computing based on a Genetic Algorithm (GA) and compares its performance with traditional matchmaking methods. First, the paper proposes a scenario in which many service requests coming into the system have to be served at the same time. There are five quality of service criteria considered as: execution duration, price, reputation, reliability, and availability. Next, it compares matchmaking and genetic algorithm approaches. The matchmaking algorithm uses a weighting approach for selection and selects the service with highest score for each request. The main weakness of this method is that requests later in the list are very likely to map to services having worse match scores. To tackle this problem the paper applies a NSGA-II genetic algorithm to find the solution for mapping the request to services in a fair manner. The genetic algorithm has been tested for several populations and the average matching score has been compared with matchmaking algorithms. Results show that if proper population size is chosen for NSGA-II approach, it can deliver better performance compared to traditional matchmaking algorithms. However, the work has solved the fairness problem by introducing a new problem of finding a proper population size for the GA.

6.3.1.2   Service Oriented Architecture

In this paradigm the main concern is QoS management (including QoS description languages for services and user preferences), definition of QoS ontology and optimization of multi-criteria web service selection. For this purpose the semantic web service has been applied by many works [295, 709] in SOA to increase expressiveness, flexibility, and accuracy by applying ontology for representing QoS properties. Semantically described QoS information then has to be converted to comparable units to be proper inputs for optimization-based solutions [295] for selection. In the following we show how the challenges mentioned have been addressed.

Quality of Service Description Toma et al. [708] discuss various approaches and their advantages and disadvantages toward modeling QoS. Initially, the work starts with discussing three main approaches to processing QoS in SOA namely as: combined broker, separate QoS-broker, and direct negotiation. The combined broker approach [642] is an extended version of UDDI in which a broker extended to be able to process QoS information. In the second method [502, 740] the devoted broker is responsible for processing QoS. And the third one requires provider and requestor to agree on service level agreement [482, 712]. Subsequently, the paper talks about current supports of QoS in WSMO with the help of non-functional properties. WSMO proposes a number of non-functional properties for each element in a service description. The description of QoS is based on the Dublin Core Metadata Initiative [753]. Finally, the article offers three techniques to extend WSMO/WSML support for QoS. The first method specifies each QoS property through the use of relations in WSML. Therefore, there is no need to add separate vocabulary for QoS. The second technique is to define a new concept for QoS. And through the appropriate property, we can include relationship between non-functional properties. The third approach is to model them like capabilities in WSML. The main disadvantage of this approach is the need for extending WSML syntax. Among the three, the third approach seems to be the most suitable one, as it can provide the full support of QoS. However, this approach has to be developed from scratch to deliver clear syntax and proper ontology for reasoning.

Integration of Semantic and Optimization In addition, with the emergence of semantic web services, some of the works in this context concentrate on filling the gap between semantic-based solutions and optimization-based solutions [295]. Furthermore, some efforts [91,478,737] describe the selection problem as a part of service composition problem.

J. Garcia et al. [295,296] present a framework to transfer the user preferences which are in the form of semantic web services into an optimization problem which aims to select the best web service. The paper argues that, semantic web services define ontology which helps web services to be discovered and composed automatically. As they are mostly using description logic for those purposes, they are infertile in dealing with QoS based selection. The reason is QoS based selection approaches lead to optimization problems which cannot be solved by applying description logic. Authors build their selection on their previous efforts [294] which apply semantically described utility function to define user requirements. They use the work to offer a selection approach that transforms user preferences into an optimization problem. That optimization problem can be tackled using various techniques, like constraint programming or dynamic programming. The selection approach consists of the following phases. In the first phase, QoS values are retrieved from the semantic description of web services. Proceeding to the next stage, the pervious phase results are linked to the user preferences which express utility functions for related QoS criteria. In the last stage, an XSL transformation has been applied to the mentioned utility function to acquire the specification of desired optimization problem. The optimization problem chosen for this work is the Constraint Satisfaction Optimization Problem (CSOP) [621]. Moreover, any other optimization techniques can be supported by designing a proper XSL style sheet. And then web services are ranked and selected using CSOP.

6.3.1.3 Cloud Computing

Rajkumar Buyya and colleagues' definition of the Cloud highlighted main aspects of the Cloud, namely as dynamic scalability, and SLA negotiability. According to their definition [152] "A Cloud is a type of parallel and distributed system consisting of a collection of interconnected and virtualized computers that are dynamically provisioned and presented as one or more unified computing resource(s) based on service-level agreements established through negotiation between the service provider and consumers." Therefore, in a Cloud environment, users' preferences are changing dynamically [281], therefore selection strategy has to match them dynamically to proper services. At the same time services in the Cloud also change dynamically, for example based on the market circumstances the price of services may jump up as demand for the service is growing.

Even though there are a few works [337,793] which presented service selection in Cloud context, however, to the best of our knowledge they are mostly in the early stage and not considered Cloud specific characteristics and QoS dimensions. In the context of a Cloud, selection objective is not fair distribution of resources between requestors. That is because in a Cloud resources can be considered as infinite [514]. Instead, the Cloud environment has emphasized QoS dimensions such as reliability, cost (including data transfer cost), and security [281]. Cloud customers are typically concerned about the reliability of Cloud providers' operations. The reason is by migrating to a Cloud, they move their information and services out of their direct control [399].

In the Cloud environment, as SLA plays an important role, reliability can be measured based on SLA. Consequently, SLA has to contain a set of goals (for example certain network and system performance), which determines acceptable performance of services. By monitoring of SLA we can find out to which extent the provider has deviated from promised SLA for each criteria and then calculate reliability. Therefore, monitoring services in the Cloud are crucial for determining QoS criteria such as reliability.

Moreover, in a heterogeneous environment such as a Cloud, it is difficult to enforce syntax and semantics of QoS description of service. Therefore, the first step towards QoS modeling in a Cloud would be building an inter-Cloud language. Furthermore, service selection in a Cloud has to be done based on requestor context and with the consideration of all roles (such as user, service creator, and provider) and their objectives.

The rest of the chapter aims to survey various strategies that have been taken for QoS modeling and service selection in detail, and shows which one can be applied to tackle Cloud computing issues.

6.3.2   Web Service QoS Modeling Taxonomy

Different approaches have been offered in literature to respond to requirements explained in each phase in Section 1.2.1. In fact, selected works can be classified based on how they tackle those issues. For example, currently there are three ways to define QoS attributes for web services namely, extended Universal Discovery, Description and Integration (UDDI), and semantic web services. However, UDDI and Web Services Description Language(WSDL) [442] cannot support modeling of QoS properties of web services; therefore, works [122,598] such as UDDIe [642] and web service level agreement [483] were proposed to enrich web services with QoS properties. In following subsections more classifications for QoS management is given and their summary illustrated in Figure 6.3.

6.3.2.1   User Preferences and QoS Criteria Relation and Tendency Modeling

When service requestors express their expectations from services, they identify functional and non-functional (QoS) characteristics of required services. In addition, they have to identify which of the QoS criteria are more important than the others. A simple way to do this is to ask requestors to give weight to

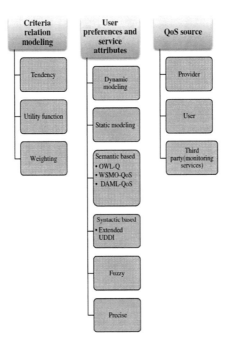

FIGURE 6.3
Web Service QoS Modeling Taxonomy.

each criterion. Using weights to achieve a decision matrix is one of the primary ways of modeling importance of criteria in user preferences. This approach has been applied by many works [264, 498, 746] as it is simple and computationally efficient. However, the major drawback concerning this approach is the complexity of finding proper weight coefficients in real world application. Furthermore, requestors' QoS preferences in terms of tendencies [746] have to be considered. For example, it has to be defined whether a parameter value is more desirable for a particular user when it is smaller or greater. There can be some other tendencies assigned to Qos properties like "exact" and "close," as presented by Vuong Xuan Tan [713].

Moreover, works in this area have focused on modeling relations between QoS criteria. For example, Tsesmetzis et al. [715] discussed the importance of QoS consideration for service providers and requestors which can help providers attract more customers and make them able to increase their infrastructure utilization. Moreover, the work creates a QoS vocabulary ontology in OWL [642] with the maximum height of two, to reduce the complexity. The QoS vocabulary covers a wide range of non-functional properties from performance to security and reliability. It also applies a standard generic model [564] for defining association between QoS attributes and the approach for measuring them.

A comprehensive work on the relation modeling is done by Hong Qing [784]. In the work a mechanism for ranking of web services using logic scoring preferences (LSP) [244] and ordered weighted averaging (OWA) [162,265] is offered. The work has found out that current ranking algorithms are ignoring relations between individual criteria and the simple arithmetic metric they used is incapable of representing logic relations such as simultaneity and replaceability. The work claims those drawbacks can be addressed by adopting LSP and OWA. The work makes use of LSP, which was originally developed for solving hardware selection, and considers the relation between criteria of selection such as reparability, simultaneity, and mandatory-ness. However, since the work is based on LSP, it cannot deal with selection problems with many QoS criteria.

Utility function recently has been used [296,445,450,621] and it is said [296] to be the most appropriate way to model user preferences expressively. Therefore, work in a Cloud also can adopt it for Service selection. Utility function is a normalized function and shows which values of QoS criteria are more preferable. For the selection of the best alternative, all of the utility functions — each of them belongs to a QoS attribute — have to be aggregated to compute the global utility value. J. Garcia et al. [296] offer a new approach of ranking based on the description of user preferences in the form of a utility function. The work presents novel hybrid architecture for service ranking integrated in the Web Service Modeling Ontology (WSMO) [224,615]. Therefore, user requirements and preferences are modeled using the Web Service Modeling Language (WSML), adding support to utility functions. When we deal with several non-functional properties, each utility function has to be associated with a relative weight. Consequently, to solve a multi-criteria ranking problem, the user preference value calculated as weighted composition of the associated utility function values. User preferences definitions are inserted as part of a goal in the form of WSML. Figure 6.4 shows a sample of a utility function extracted from their work. As depicted by the Figure 6.4 highest utility value is reflected by that function if the price is lower than 60 and it is falling down when the price increases.

In addition to those methods, Tran et al. [713] adopted (Analytic Hierarchy Process) AHP [626] for QoS criteria relation modeling and service selection. AHP consists of three main phases: problem breakdown, comparative evaluation, and priority composition. In the first phase, each problem is broken done to three elements — overall goal, its criteria and sub-criteria. Next, pairwise comparisons for all criteria and sub-criteria will be done to obtain their relative importance for decision making and subsequently all solutions will be ranked locally applying those sub-criteria. In final stage, all relative local ranks of solutions will be combined to obtain the overall rank for the solution. Since the method is using pairwise ranking its performance decreases when the selection problem consists of large number of criteria. Moreover, it increases flexibility

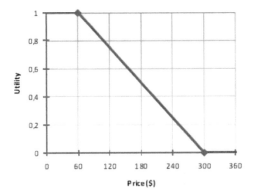

FIGURE 6.4
Price Utility Function [296].

of the ranking algorithm by allowing two options of mandatory and optional QoS constraints. It offers more options (exact and close) for tendency characteristics of QoS criteria in addition to negative and positive options offered by other works [296, 709]. The exact option shows that the property value should be equal to the defined value by request. And the close option says that closer value to the requested value is more desirable than the others.

Furthermore, Wang et al. [745] presents a novel resolution process for determining the linguistic weight of QoS criteria. First, it creates the framework for evaluation QoS criteria, which summarizes QoS requirements of web services from decision makers' opinions and market surveys. Next, it determines the importance of criteria by aggregating all groups of participants' opinions. It can be said that the approach is a complement to the works [222, 686] which helps to select web services in market places based on QoS.

6.3.2.2   QoS Source: Provider, User, or Third Party

The majority of works made an assumption that QoS information has been supplied by providers along with services description. However, some believe [745] that not all provides are willing to provide the related QoS information for comparison, or that they are likely to advertise their services exaggeratedly. And that's why we have to consider consumers' feedback on their experience of using web services to determine QoS values. In addition, there are values that cannot be determined by users or providers such as network related QoS information, reliability, and trust, which is usually evaluated by a third party (monitoring services). To collect QoS information Zhenyu et al. [478] built an approach based on distributed agents. The main contribution of the work is building a procedure for processing the quality of web service from multiple locations. The article presents a distributed approach to acquire the QoS data from users in different locations by the help of agents scattered in the network.

6.3.2.3   Context-Aware

There are cases when QoS information of web services vary based on the requestors' context. The context-aware QoS information let service providers or third parties offer QoS information for web service based on the service requestor context. Hong Qing et al. [784] adopts an in-context project [12, 13] for providing dynamic context information. This information includes location, budget of users, and availability of services. Therefore, the web service selection is based on reasoning based on the context data. In addition, a selection approach offered by Lamparter et al. [450] is context sensitive as it adopts utility function policies to model context dependent user preferences. For example, there might be a case where web service which is selected as the best service in the list is not available in a particular location; therefore selecting it for the requestor in that context is not acceptable.

In Cloud commuting service selection, context information also plays a key role. The first is context of users and providers can affect performance, client location information can be used by selection as the basis for determining which data center location is closest and thus can provide service with less latency and perhaps higher throughput. There are some restrictions applied by law for deploying on Clouds in particulate Geographic locations. For example, according to DPA, Clouds located outside the European Union are unlawful. As Papakos et al. [565] discussed, another important QoS information (for service selection in Cloud) to be considered is the user's device. As they have mentioned, requirements of a client with a mobile device can change because of changes in the context of the device. These status changes encompass hardware resources, environmental variables, or user preferences. Binding to a service offering different quality of service levels from the ones required may lead to excess consumption of mobile resources such as battery life, as well as unnecessarily high provision costs. Therefore, the paper proposes VOLARE middleware that dynamically selects Cloud service requests.

6.3.2.4   Dynamic versus Static Modeling of User Preferences and Web Service QoS Attributes

The value of QoS property for a web service can remain constant and therefore identified once, or it can be updated regularly. On the other hand, user preferences also can be specified once or can change during interaction with the system. For example, a service response time can rely heavily on network traffic. Therefore, it can have a short response time at a moment in a day and then can increase dynamically to a certain level when it is not any longer available.

Lampar [450] argued the shortcomings of current works like the WS-agreement [88] in modeling of users' dynamic preferences on service configuration. And then it shows how those could be tackled by a mechanism which efficiently declares non-functional attributes such as price. In order to succeed in that, the service configuration is modeled using utility function policies [426, 451] which are presented in the form of OWL ontology. Another achievement of this work is finding out that performance crucially depends on the way web service offers and requests are modeled. Consequently, if service selection is runtime, one way to increase the performance would be limiting the expressiveness of biding language.

In context of the Cloud, Andrzej et al. [793] have taken the advantage of state-full web services to deal with dynamic changes of Cloud services. They proposed a higher layer of abstraction which provides selection based on QoS criteria values that describe dynamically the state and characteristics of Cloud Services (cluster as a service). In addition, they have implemented their proposed solution, and through several tests, it was proved that the proposed technology is feasible.

6.3.2.5   Semantic-Based versus Syntactic-Based

There are two ways to describe entities of a web service, namely as semantic-based or syntactic-based (extended UDDI). The syntactic-based approach uses numerical and key word values for web service QoS properties. Although an extended version of UDDI can encompass the QoS information of web services, it is not machine understandable, hence not suitable for automatic selection. The automation selection enables service provider and requestor to be decoupled, which means they don't have to be aware of each other before execution phase. In addition, different service providers and users can apply a variety of models for describing QoS attributes, and then it is essential to acquire a solution to understand different QoS representations. That solution which covers mentioned drawbacks is semantic web service which can increase expressiveness, flexibility, and accuracy by applying ontology for representing QoS properties. A number of QoS ontologies have been proposed for web service which mainly developed based on three semantic languages, namely as OWL-Q [445, 629], WSMO-QoS [503, 709, 746], and DAML-QoS [798].

OWL-Q is built on OWL-S [498] language and has the strength of not only modeling and measuring static QoS attributes but also dynamic properties. Nonetheless, it doesn't offer a way of defining QoS criteria importance and whether they are mandatory or optional. WSMO-QoS is yet another promising approach for QoS modeling of web services. It offers an upper level ontology which can describe each attribute of a web service in detail. Moreover, a new category of properties with the name of non-functional has been added to Goal and semantic service description. The new category provides attributes for describing QoS type, metric, dynamicity level, tendency, and importance. Tran et al. [713] offers comprehensive OWL-based QoS ontology (include diverse data type) and an AHP-based ranking algorithm to dynamically rank services at different levels of provider and consumer requirements.

Finally, DAML-QoS developed as a complement to DAML-S to be capable of specifying QoS ontology. More specifically, the ontology consists of three layers — QoS profile, QoS property, and QoS metric. The QoS-profile layer is developed mainly to provide advertisement, and request ontology for providers and requestors. The QoS property deals with name, and domain, and range of QoS attributes can be defined in DAML-QoS. And finally, the QoS metric layer is offered for measuring QoS metrics and computing their values.

In Cloud computing environment semantic of services and data can be a first step towards building an inter-Cloud language. As we have discussed in our previous work [220] in a heterogeneous environment such as Cloud, it is difficult to enforce syntax and semantics of service description (including QoS description) and users' requirements. Therefore, applying symmetric attribute-based matching between requirements and request is impossible. To tackle that we proposed an advertisement approach for Cloud providers by modeling Cloud services into WSMO. Then we offered ontology-based matchmaking to help users deploy their application on the most proper IaaS providers based on their QoS preferences when both sides are not using the same notation to describe their services and requirements.

6.3.2.6   Fuzzy versus Precise Preferences

As Wang et al. [745] believe, the complex nature of QoS made the service consumers' perception fuzzy. Therefore, it proposes new selection algorithms based on MaX-Min-Max composition of Intuitionistic Fuzzy Set (IFS) under the vague information. The work expects fuzzy perception of service consumers and providers. Therefore, it suggests a fuzzy multi-criteria decision making solution with following aspects:

- Capable of handling imprecise preferences of service consumers.

- A clear weighting strategy for QoS criteria.

- A QoS aware service ranking ability.

The article uses the QoS criteria presented by the W3C working group in 2003. In addition, it mentions the fact that not all providers are willing to provide the related the QoS information for comparison. And that is why we have to consider works that use consumers' feedback on their experience of using web services. As mentioned earlier, the web service selection approach in this work is based on IFS. IFS was introduced in 1989 by Atanassov [101] and it can be considered as a generalization of the concept of fuzzy set which is effective in dealing with vagueness, in addition, this type of selection problem can be modeled by Fuzzy multi-objective and multi-level Optimization [801].

6.3.2.7   Identifying Roles in the Problem

It is important to determine the selection algorithm goal. Is it working to maximize provider profit, or taking care of requestors' interests or both? There are some works [713] which are flexible enough to act for both parties. In details, they give weight to objectives of requestors and providers in the utility function. For example, by giving more weight to providers in the utility function they are working in favor of providers. Nevertheless, before this step all the roles in the problem and their objectives have to be clearly determined and then a decision has to be made on the importance of each role goal. A summary of surveyed approaches in different context and their applicability to Cloud Computing is shown in Table 6.1.

TABLE 6.1
Selection Works in Different Contexts and Their Applicability to Cloud Computing

Criteria	Grid	SOA	Cloud
Semantic based		[295, 708, 709]	Yes
Load balancing		[737, 747]	No
Fair distribution of resources	[321, 485]	[747]	No
maximize user or provider profit	[149, 715]	[450, 737]	No
Integration of semantic and optimization		[295, 715]	Yes
Considering reliability		[383]	Yes
Provider as QoS source	[295, 708, 709]		No
Consumers as QoS source		[450, 707]	No
Third Party as QoS source(monitoring service)		[220]	Yes
Selection in Service Composition contex		[91, 737]	Yes
Utility based function for preferences		[295, 296, 450]	Yes
Context aware		[450, 478, 565, 784]	Yes
Dynamic modeling	[450, 793]		Yes

6.3.3   Taxonomy of Web Service Selection Approach

As it is illustrated by Figure 6.5, selection work mainly has utilized two types of approach for web service selection, namely as optimization and decision making. Decision-making can be defined as process of identifying and choosing alternatives based on values and goals of decision makers. Therefore, the assumption is that, there are many candidates to be chosen and the aim is to select the one that best fits our goal, desires, and constraints. This process can be depicted as Figure 6.6. When there is one single criterion, then selection can be made by identifying the alternatives with the best value of that criterion. However, when there are several criteria as it is depicted in Figure 6.6, we have to know which criteria have higher priorities to users. And this is where the decision making techniques offer approaches such as AHP to help users assign the comparative importance to those criteria. In the case of multiple criteria and a small number of explicitly given alternatives, and when there is no existing scale of measurement for the criteria, the problem can be solved using decision making approaches such as the Analytical Hierarchy process (AHP) and the multi-attribute utility theory (MAUT).

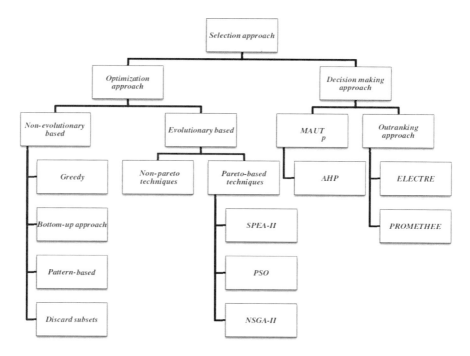

FIGURE 6.5
Web Service Selection Approaches.

However, if given alternatives are many or implicit, multiple criteria optimization techniques can be applied. These techniques can be categorized into evolutionary-based and non-evolutionary based techniques. Evolutionary-based techniques (for multi-objective problem) are based on the Pareto solution, which is an economic concept and applied for the condition when a better value for an attribute can only happen once the value of at least one other attribute gets worse. AI solutions, such as the Non-Dominated Sorting Genetic Algorithm-II (NSGA-II), Strength Pareto Evolutionary Approach 2 (SPEA-2), and particle swarm optimization (PSO) are among the most scientifically used techniques in this area. In the following each of these selection approaches (from decision making to optimization) will be discussed in detail.

FIGURE 6.6
Process of Decision Making.

6.3.3.1 Decision Making Formulation and Analytical Hierarchy Process(AHP)

The problem of multi-criteria decision making (MCDM) is a favorite method for expressing decision making problems which can be considered in the following form: C1,...,Cm as criteria, A1,Ě,An as alternatives, W1,Ě,Wm which are weights assigned to criteria, a Matrix which its aij element shows the score of alternative Aj against criteria Ci, and Xi,Ě, Xn are aggregative score of alternative Ai. MCDM approaches can be classified into two main categories namely, Multi-Attribute Utility Theory (MAUT) and outranking approach. MAUT is benefited from applying utility function which has to be maximized. Moreover, it allows the complete payoff between criteria that can show relative importance of each attribute in Alternatives evaluation. In literature, the Analytical Hierarchy process (AHP) is one of the most applied methods in the MAUT category, therefore in this section we will investigate it more. The AHP method was suggested by Satty in 1998 and is based on a pairwise comparison of criteria to determine their weights in utility function. From other work in this category, a distance-minimizing approach [299] can be named.

The major contribution of AHP is to convert subjective assessments of relative importance to numerical values or weights. The methodology of AHP is based on pairwise comparisons of the criteria by asking the question of "How important is criterion Ci compare to criterion Cj?" The answer to this question determines weights for criteria. Figure 6.7 depicts the process of choosing the best service provider when there are four criteria (cost reliability, security, and trust) to be considered. The answers to questions can be one of the following: After the pairwise comparison as it can be seen from Figure 6.7, relative importance has been given to each criterion. In next step, and similar questions have to be asked to evaluate the performance scores for alternatives on the subjective criteria. And then based on the result of this phase we can rank alternatives and select the one with the highest rank.

The next category in decision making approaches is named Outranking which was introduced by Roy in 1968, and the idea behind it can be simply defined as follows. In this approach all alternatives are compared in a way that alternative Ai can outrank Aj if on the majority of criteria Ai performs as well as Aj,

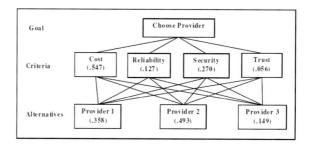

FIGURE 6.7
Choosing a Provider Using AHP.

TABLE 6.2
Major Scale of Pairwise Comparisons

Scores	Response to the question
1	Equal importance or preference.
3	Moderate importance or preference of one over another.
5	Strong or essential importance or preference.
7	Very strong or demonstrated importance or preference.
9	Extreme importance or preference.

at the same time as it can achieve an sufficiently acceptable score in other criteria. ELECTRE [272] and PROMETHEE [138] are among the most famous approaches in this category.

6.3.3.2   Optimization Methods

Optimization methods seek the most suitable service, which is usually the one which maximizes or minimizes one or several criteria, such as cost, deployment time, etc. The problem can be complicated when we consider more than one criterion. Furthermore, there are some constraints in selection problems which can be imposed by service requestors or consumers. By considering constraints in selection, the definition of optimization can be rewritten as "finding of the most suitable services for the clients or providers, which maximizes or minimizes one or several criteria and still adhere to the constraints." For example, assume that the best service is the one with minimum cost, highest availability, and least deployment time when there is a limitation for providers to serve the requestor [715, 737] due to bandwidth limitation. Then, the problem of selection can be formulated as the "Selective Multiple Choice Knapsack Problem" (SMCKP) which is a kind of NP-hard problem. The work in areas of selection has faced the problem in different ways, some trying to find the optimal solution [582], others aimed at finding the semi-optimal solution by offering suitable heuristics [91,485,512,745]. Dominant approaches in this area can be classified in two main categories — non-evolutionary and evolutionary optimization methods — which will be investigated in the following.

Non-Evolutionary Optimization Method Four classes of selection approach in this category are: pattern based, selection using discarded subset results, bottom-up selection, and the greedy. This classification is done by Jaeger et al. [383]. The objective is to solve the problem of multi-criteria selection. It compares several algorithms for selecting the best web service candidates. They have considered four QoS categories introduced by Zeng et al. and Menasce [511,792]. The categories are namely: execution time, cost, reputation and availability. After that, an approach for aggregating QoS [385] of individual web services was applied. At the comparison stage the work applied the Simple Additive Weighting (SAW) approach which was extracted from the context of Multiple Criteria Decision making (MCDM) [358]. They have compared algorithms' performances, and the results are reported as follows:

- A greedy selection is not able to consider constraints. Instead it can find the candidate that scored the highest among all the other candidates.

- Bottom-up approach [384] relies on a fact that the selection problem for composition (selection in composition) of web services shows similarities to Resource Constrained Project Scheduling Problem (RCPSP). In RCSP, project is divided into individual tasks, and then each task has to be assigned to an available worker to complete the whole project in a way that meets the constraints, such as time. Bottom-up selection results in the second worse QoS; however, the computation effort is negligible.

- Pattern-Based selection [316] which considers each composition pattern separately and then tries to find the best assignment. This approach offers the best achieved QoS compared to all heuristic approaches. The computational effort of this selection is reasonable, and depends on the composition structure.

- The selection by discarding subsets is a kind of backtracking-based algorithm which uses a search tree consisting of nodes, each representing a possible pair of a candidate and task. It results in the best QoS possible and also meets the constraints.

Evolutionary Multi-Objective Optimization Method Evolutionary Multi-objective Optimization methods are based on the principle of natural selection, which is also called the survival of the fittest and originally characterized by Charles Darwin [219]. Since evolutionary approaches have shown desirable potential for solving optimization problems, this class of search strategy has been utilized for multi-objective optimization from mid-1980s. Evolutionary multi-objective optimization is in fact a combination of the evolutionary computation and traditional multiple criteria decision making. Evolutionary approaches follow two major concepts. The first concept is the competition for reproduction which is called selection. And second one mimics the ability of producing new generation by mutation, which is also called variation.

A considerable number of evolutionary multi-objective optimization (EMOO) techniques have been developed in recent years [197, 719]. In an attempt to discuss the most important approaches proposed, we decided to classify these techniques using the following scheme of Non-Pareto Techniques and Pareto-based Techniques such as NSGA, NPGA, three of the most applied methods in the web selection literature as NSGA, SPEA, and PSO are briefly explained.

NSGA Non-dominated Sorting Genetic Algorithm (NSGA) [675] was proposed by Srinivas and Deb. The algorithm modifies the ranking procedure originally proposed by Goldberg [309] based on several layers of classifications of the individuals. NSGA was highly computational intensive and had several other drawbacks which led to the rise of [227] NSGA-II. In the first step NSGA-II constructs a space of solutions, then performs sorting based on non-domination level, applies a crowded-comparison operator to create a new pool of offspring. As it applies a fast non-dominated sorting approach which has O(MN2) computational complexity, M is the number of objectives and N the population size. The algorithm is capable of outperforming many other genetic optimization algorithms [227]. That is mainly because it applies the crowded-comparison operator.

SPEA The Strength Pareto Evolutionary Algorithm (SPEA) is presented by Zitzler and Thiele [802]. The method is a result of integrating different EMOO techniques. It has the unique character of archiving non-dominated solutions already found in order not to lose certain portions of the current non-dominated front due to random effects. For ranking (calculating the strength) of non-dominated solutions an approach similar to MOGA has been used. In addition, fitness of individuals is calculated based on the strengths of all non-dominated solutions in archive which can dominate it. Moreover, for maintaining diversity a method called the "average linkage method" [523] has been used. The 0/1 knapsack problem [582, 802] is one of the main fields SPEA has been applied to.

PSO The particle swarm optimization (PSO) [424] was built by Kennedy and Beernaert and inspired by the flocking and swarm behavior of birds and insects. Every solution in PSO is represented by a particle. In the first phase a number of particles are generated with random positions and velocities. In the next step, each particle flies through the search space with the velocity constantly updated based on two important factors, first by its best position, and second by the position of the best particle in a problem space which corresponds to the cooperative effect of the particles in optimization searching. Therefore it guides particles toward the global best position found so far. On the other hand, in order to control the ability of the particles to search and be restricted within the search space boundary, a maximum velocity vector Vmax was introduced. The PSO algorithms are mainly used in scheduling problems and New PSO alternatives are constantly being developed to improve scheduling performance [321].

6.4 Future Directions and Conclusion

This chapter introduced a general model for QoS-aware service selection which consists of QoS management, and service selection. Furthermore, selection works in different context of Cloud, Grid, and SOA have been reviewed in order to find out what could be inherited from works in other paradigms and what has to be done uniquely (considering special characteristics of Cloud Computing) for Cloud Computing. Below, a summary of future directions is given:

- Cloud services have specific characteristics and QoS dimensions which have to be defined, and then approaches for measuring those QoS criteria (reliability, security, and trust) have to be discovered. For example, methods to evaluate reliability and trust of providers from user feedback can be further studied.

- In modeling user preferences, there is opportunity to explore how to capture relative importance of criteria from requestors or other decision makers in selection problems.

- In addition, when selection aim is to find the best Individual services to form a composition, the general utility function which is given for a whole composition can be decomposed to provide a utility function for each individual service in the composition. Techniques for the decomposition of utility functions in this area can be investigated.

- In Cloud computing, dynamic modeling of service status and user demand and preferences is an essential task which can be further enhanced.

- In addition, work can focus on identifying roles and parties in the decision making process to ensure success of selection in maximizing all engaging parties' profits.

- Building an inter-Cloud language using Semantic web service for modeling services and data is another promising field to be considered.

- Moreover, researchers can investigate and identify Cloud users' context attributes (users' device characteristics and location) and study their effects on the performance of Cloud service deployment.

7

An Introduction to Open-Source IaaS Cloud Middleware

Peter Sempolinski

Graduate student in the Department of Computer Science and Engineering at the University of Notre Dame, Indiana

Douglas Thain

Associate Professor in the Department of Computer Science and Engineering at the University of Notre Dame, Indiana

CONTENTS

Some parts of this chapter are a summary and update on some of the topics covered in a previous work by the same authors [639].

7.1   Introduction

Much of the focus of cloud computing in recent years has been on the various offerings of commercial cloud providers. For example, Amazon's EC2 cloud [641] and Microsoft's Azure [109] cloud have received a great deal of attention. However, for a wide variety of reasons, open-source software has its own place in cloud computing as well. Open-source frameworks have a unique function in that they allow an organization to construct a "private" cloud environment. Such a system usually falls into what is commonly described as "infrastructure-as-a-service" [743], in which a user requests some number of virtual machines (a.k.a. VM), which are hosted somewhere on the cloud's own resources. The key difference is that open-source cloud middleware allows any individual or organization to build their own cloud to host their own VM, without relying on an external, commercial cloud provider.

There are numerous reasons why some individual or organization may to want to have their own cloud. To name a few:

- Private clouds allow organizations to process potentially sensitive information "in-house."

- If utilization of the private resources is high enough, a small private cloud might be cheaper then purchasing from a commercial cloud.

- Virtual machines of private clouds can be tuned to a particular organization's resources or software tools.

- Research organizations may want to experiment directly with the cloud technology itself.

- For whatever proprietary reason, an organization wishes to customize its cloud.

As these are only a few of the reasons why open-source clouds can be useful, it is clear that the notion of a private cloud is a significant part of an overall understanding of cloud computing. This chapter is intended for those who wish to understand the main, general ideas of open-source cloud computing. It could also be used as a starting point for someone who wishes to construct their own private cloud, or for anyone who wishes to understand the issues involved.

In this summary, we describe several open-source software frameworks that are designed to host infrastructure-as-a-service. These include Eucalyptus [548], OpenNebula [552], and Nimbus [539]. Then, we briefly describe other open-source projects (Condor [700], xCAT [769]) that, while they can host VM in a similar way, were not originally designed for that purpose. More importantly, we will describe the overall structure of an open-source cloud, enumerating the parts that all open-source clouds must have. We wish to emphasize that a cloud computing system is composed of many pieces and one of the consistent challenges of building and managing a private cloud is understanding how the parts fit together. Throughout, we will be describing various common pitfalls and difficulties faced in building a private cloud. We will conclude after naming some future challenges and opportunities that we anticipate in the field.

Some of the open-source softwares cited in this chapter also have "enterprise" variations, in which support, improvements and guidance can be purchased. We make note that, while it is true that purchasing such an enterprise version is a viable option for some organizations, our focus is on the pure open-source version of any software we refer to.

7.2 Previous Work

In the field of cloud computing, there have been many works that compare various parts of the cloud computing landscape and the various toolkits available [666] [338] [607] [462]. Some of these comparisons, like ours, are focused on open-source software [171] [250]. These works are able to provide some insights into the overall cloud landscape, and some of the features of various open-source and commercial cloud offerings. We note, however, that many of the projects referenced in these papers are active, growing projects, so lists of features can often become outdated very quickly.

Rather than duplicate the above work, this chapter will mainly be intended as an introduction for persons who wish to understand the current state of open-source clouds, with a specific focus on the challenges and issues related to building and maintaining a generic private cloud. There will be less of a focus on raw feature set, or any other information that might become quickly outdated.

7.3 Components of an Open-Source Cloud

A typical open-source cloud computing framework, hosting VM as an infrastructure-as-a-service cloud, is comprised of many components which must work together. Understanding the construction of such a system requires an awareness of all of these parts.

7.3.1 Summary of Parts

In general, most private cloud providers must configure several main components:

- The Underlying Physical Hardware: To build a cloud, some physical hardware is needed. It is important that this hardware have the processor and memory capacity to host numerous virtual machines. Also, network access and disk space must be accounted for.

- The Underlying Operating System: All of the components to build a private cloud must run on some operating system. The selection of the operating system will be critical for determining the compatibility issues that a private cloud will face between its various components.

- The Hypervisor: Also known as a Virtual Machine Monitor, or VMM, the choice of underlying hypervisor plays a huge role in the stability and performance of a cloud system.

- Disk Storage: In addition to storing the binaries for the running software, disk storage must be allocated for the disk image files that will form the basis for the virtual hard disks of the VM. This includes storing both the template disk images for future VM as well as the current disk images for running VM.

- User Front-End: In some way, users must be able to request and access VM. Moreover, the front-end must be tuned to the usability and security requirements of the cloud provider and user.

- The Cloud Controlling Software: This component ties all of the parts together and coordinates them. Open-source options for this part include Eucalyptus, OpenNebula and Nimbus.

Of course, there are many more parts to a private cloud, especially those parts internal to the cloud controlling software. This includes things such as user or VM databases or scheduler software. These, however, tend to be bundled tightly with the cloud controller itself and can be considered part of that unit. We note the above major components since these are usually what the cloud provider has to spend the most time setting up, configuring, customizing and tuning.

7.3.2 Recurring Considerations

In constructing a private, open-source cloud, there are several ideas that must be kept in mind:

- As stated before, an open-source cloud cannot be regarded as a single component. Rather, a private cloud is constructed out of a **Complete Cloud Computing Software Stack**.

- In order to develop and maintain an open-source cloud, **Compatibility** must be constantly kept in mind. All of the hardware and software components must be able to talk to each other. In practice, we found that insuring such compatibility required numerous configuration tweaks to get a working system.

- The cloud middleware of Eucalyptus, OpenNebula and Nimbus are **Active Projects**. Each group has a very active user mailing list. In our experience with all three we found that, for every one of them, some problem which we had encountered was later fixed in a subsequent version.

- Since this is open-source software, administrators can **Customize**. The ability to mix and match components, tweak configurations, or even write new code, allows private cloud providers to tune their systems to their own needs.

- If one is part of an organization, one might encounter a certain level of **System Limitations** determining some software configurations. A private cloud provider must conform their system to their organization's expectations regarding network configuration, available hardware, and underlying software. Maintaining working ties with other system administrators in the organization is critical to a successful cloud system.

- Organizations wishing to construct a private cloud must be aware of **User Requirements**. Fortunately, since private clouds are customizable, there is a great deal of opportunity to conform the cloud to what users want. Some considerations are security, ease-of-use and the ability to interface with previously used systems.

- The construction and maintenance of a private cloud is an exercise in **Systems Administration**. All of the best practices for administering robust, stable systems apply in this setting as well.

7.4 Open-Source Cloud Implementations

There are several open-source projects available for constructing an infrastructure-as-a-service private cloud. These include Eucalyptus, OpenNebula, and Nimbus. Since these are active projects under consistent development, the features and performance of these projects change over time. However, there are aspects of each of these projects which change less often, such as their overall aim and guiding philosophy. Below, we summarize the differences between these three projects. This comparison is a summary of previous work by us [639]. In table 1, we have a short summary of the main comparisons. For our comparison, we primarily examined Eucalyptus version 1.6.2 Nimbus 2.5, and Open Nebula 1.4.0, while emphasizing the more static features.

7.4.1   Eucalyptus

The quickest way to summarize Eucalyptus is that it is an open-source imitation of Amazon's EC2. Their published design specification describes the system's structure [548]. Eucalyptus also maintains its website and mailing list [258], and has published work on some of their recent versions [692] [545]. The front-end provided by Eucalyptus, called euca2ools, is designed to share compatibility with Amazon's EC2 front-end tools and the EC2 API. Similarly, Eucalyptus implements a distributed storage system called "Walrus" which is designed to be compatible with Amazon's S3 storage.

Internally, Eucalyptus is designed to be extensible enough to support large numbers of machines and users. Eucalyptus has a particularly decentralized design, permitting multiple "clusters" of machines in a single cloud. Putting together numerous clusters, each with its own cluster controller, can be a way of achieving a large private cloud deployment. Moreover, Eucalyptus provides a user-management web interface, allowing the cloud administrator to keep track of a large database of users and giving new users the ability to "signup" for the cloud. Moreover, Eucalyptus very cleanly hides the internal working of the cloud from users.

In its literature, Eucalyptus describes itself as an "industry" cloud. This, given the features of the software, makes sense. A private company with a large number of machines and users can easily take advantage of the scaleability of the system, and the ability to coordinate many users. Moreover, compatibility with EC2 makes the system compatible with numerous pre-existing tools. However, the focus on EC2 compatibility means that more limits are placed on the customizability of the system as compared with the other toolkits.

7.4.2   OpenNebula

OpenNebula's most distinctive feature is the very high degree of customizability that is given to administrators and users. This is especially the case with regard to the user front-end. OpenNebula's available interfaces include a command line on the cloud's head node that users login to, an XML-RPC API, front-ends for OGF OCCI and EC2 API and more. Moreover, using one of these API to build one's own custom front-end is a fairly straightforward task. [552]

If the cloud administrator wishes to use the default, command-line interface on the head node, users specify the VM configuration in a formatted file. This file allows for almost any configuration of memory, disk, network and more. Of course, such customizability has the drawback of leaving users with the ability to request invalid (or destructive) configurations of VM. But, the option of providing a more protected front-end to users is also available.

Internally, a great deal of customization is built around the underlying file system. OpenNebula can employ one shared file system (such as NFS or any distributed file system) for all files OpenNebula. This can, of course, pose security and bottlenecking issues, which, depending on the file system used, might limit the scale of the system. Or, OpenNebula can use SCP to copy needed files (mostly VM disk images) from one machine to another. (This is discussed in more detail below, under disk storage.)

The wide range of customizations allow OpenNebula to do a great number of things, but can also mean that there are more avenues to make mistakes, requiring either more user knowledge or a configuration to protect users. Overall, many of the default configurations lend themselves toward smaller, more private deployments. Of course, OpenNebula is so customizable that everything depends on the individual setup.

7.4.3 Nimbus

The Nimbus project describes itself in terms of "cloud computing for science" with a fairly clear emphasis on that end [539]. Nimbus is also affiliated with Globus, and many of the available front-ends employ X509 credentials for authentication. There is a pretty clear emphasis on collaboration. In addition to being the subject of numerous research projects such as [494], Nimbus allows for interaction with other clouds API, such as interfaces for the WSRF and EC2 protocols. Also, Nimbus has recently implemented (version 2.5) an Amazon S3 compatable storage called Cumulus. Of course, Nimbus also has its own client front-end interface..

There is a real emphasis on both openness and ease-of-use. With regard to ease-of-use, Nimbus includes a number of startup scripts and example configuration files in an effort to quickly get users and administrators into a working system. While Nimbus has a certain level of customizability, that customizability is (deliberately, according to the documentation) built around certain defaults that are incorporated so as to allow for quick initial setups. Moreover, this customizability is built around a fairly large community of related projects, furthering their goal of collaboration. This leads to some unique properties of the cloud components. For example, the aforementioned affiliation with Globus tools leads to a heavy use of proxy certificates, in the style of other Globus projects. We note, however, that Nimbus does not require these certificates, but also allows other means of authentication depending on the interface that is used.

We also note, that, due to the emphasis on "openness," Nimbus does not have an "enterprise" variant (unlike OpenNebula and Eucalyptus). As such, much of the development of Nimbus code is done through the academic community, resulting in numerous affiliated projects and papers [421] [352] [420].

7.4.4 Others

There are other open-source softwares that, while not originally designed for infrastructure-as-a-service, can be used to host virtual machines in a similar

TABLE 7.1

A Brief Summary of 3 Major Open-Source Clouds

Cloud:	Overall Focus:	Customizability:
Eucalyptus:	Open-source implementation of Amazon EC2	Some, but maintains EC2 "look"
OpenNebula:	Private cloud with high level of custom tuning	Almost everything
Nimbus:	"Science Cloud" with open collaboration	Numerous interchangeable parts
Cloud:	Unique Aspects:	Tailored To:
Eucalyptus:	Web management of user access	Enterprise level private cloud
OpenNebula:	Underlying SCP or NFS storage	Smaller scale or heavily custom clouds
Nimbus:	Design allows many interchangeable components, no enterprise variant	Research Community

way, provided that they are set up properly. We mention these for completeness' sake.

One such system is Condor [700] [701]. Condor is, of course, not originally designed to be an IaaS cloud. Rather, it is designed for "cycle scavenging," that is, giving users the ability to run jobs on currently idle machines. However, it is possible to set up a job such that the job is to run a VM through the "virtual" universe. This, however, requires setting up at least some of the machines in the Condor pool with VMM, disk space, network bridging and all of the components that running VM require. Then those machines must advertise their VM capabilities to the Condor pool. While this is possible, this is a newly added feature to Condor that is not, yet, heavily documented. Lastly, Condor performs much of its scheduling and user arbitration by killing running jobs to try them again later or on a different machine. As such, in Condor, running VM are prone to simply die and disappear, which is in direct contrast to the scheduling of an IaaS cloud.

Another system that can be used to manage VM is xCAT [769]. xCAT stands for Extreme Cloud Administration Toolkit, and is primarily designed for remote system administration of a large number of physical machines. That is, it can be used to quickly set up the operating system of numerous physical machines, all built around the same template. This has wide application as an assistant for systems administration for large groups of machines. However, with special configurations, xCAT can administer virtual machines in the same way. However, this is subject to the same limitations as xCAT. First, xCAT can only administer a fairly small subset of OS. Second, since it is an administration toolkit, there is no provision to allowing end-users to request arbitrary VM, unlike an IaaS cloud.

7.5 A Cloud Builder's Checklist

There are several practical issues that a builder of a private cloud should be aware of before beginning:

- **How is the Network set up?** In our experience, one of the more frustrating aspects of building a private cloud is making sure that the network works. We find that three things are helpful. First, whoever is building and maintaining a private cloud should clearly understand the various components of a TCP/IP network, including DHCP, DNS, virtual NIC and similar components. Second, the cloud controller should be very clear on what limitations his or her organization places on the network setup. This includes, in particular, an awareness of the firewall setup and a clear understanding of the available IP address ranges of the organization. Third, the administrator of the private cloud and the administrator of the organization's network should be working close together.

- **Hardware Inventory and Expectations:** The person building the cloud should have a clear notion of what hardware is currently available and what hardware will be made available in the future. In particular, the disk space, memory, network bandwidth and number of cores should all be noted, as well as whether or not the hardware has virtualization extensions. One of the key things to avoid is a mismatch in capacity. For example, if one has enough cores for 60 VM, then one must also have enough disk space to store the virtual hard disks for 60 VM. Moreover, any expectations that the organization has for the hardware or software must be noted as well. For example, a certain OS might be the company standard.

- **User Privileges:** Many of the functions needed to perform a successful install of an open-source cloud require a careful understanding of the UNIX permissions used to perform the various tasks. Root privileges are typically needed for at least some parts of the install on all machines. However, not all tasks performed by various private clouds use root. Being aware of what privilege level should be used to run what commands can prevent frustrating errors.

- **User Expectations:** The selection of Eucalyptus, OpenNebula and Nimbus probably depends on user expectations more than any other factor. In our experience, the most important factor was user familiarity with various toolkits, API and interfaces. For example, Eucalyptus imitates EC2, which might be ideal if the users have already done work with EC2. On the other hand, if the users are very unfamiliar with any cloud toolkits, then it might be worth the time to construct a customized front-end to simplify the interface. In any case, being aware of what users can and can't do is critical to a successful private cloud setup.

- **Be Ready To Tinker:** In our experience, before getting a successful setup with any of the common open-source clouds, we were required to tinker with configurations throughout the software stack. The best advice we can give is to locate **all** of the relevant log files for virtualization, user permissions, and the cloud software itself and check them frequently. Of course, it also helps to keep good notes.

7.6 The Cloud Computing Software Stack

As we discussed before, the best way to understand open-source cloud computing is as a collection of components, all of which are, to some degree, customizable. In Figure 1, we have a generic picture of how the pieces fit together. There is one (or, sometimes, more than one) head node, used to control the actions of several cloud nodes, on which VM are run. In order for all of this to happen, several hardware and software components must work in concert. It is the role of the cloud middleware, (Eucalyptus, OpenNebula and Nimbus) to provide that coordination.

7.6.1 Hardware

The first aspect of constructing a private cloud is the underlying hardware. Obviously, hardware for running virtual machines must be robust for all of the things normally needed to run virtual machines. (High memory, disk space, reliability, etc.) Most pertinently, many of these cloud computing frameworks require the hardware to have virtualization extensions (Intel-VT [698] or AMD-V [66]) to run. This is related to an old issue with virtualization in general. A paper by Goldberg outlines the properties required for a processor to run virtual machines [587]. For Intel architectures, virtualization extensions are needed, in most open-source virtual machine settings, to have the necessary virtualization properties. Otherwise, the virtualization is usually limited to paravirtualization, which is slower. This issue is mostly evident in one of two settings. If an organization is repurposing older hardware for its private cloud, these extensions might not be available. Or, the extensions might be there, but often must be manually enabled in the BIOS of the chipset to be used. Apart from this issue of the virtualization extensions, the hardware needs to be good, reliable hardware for virtual machines.

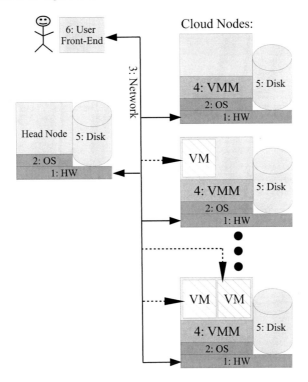

FIGURE 7.1
These are some of the parts that have to be coordinated to build a private cloud. 1) Underlying Hardware 2) Underlying OS 3) Physical and Virtual Network 4) VMM or Hypervisor 5) Disk Storage 6) User Front-End. In an open-source cloud, middleware software is installed on each machine to coordinate the various pieces.

Hardware RAID setups are also a hardware feature that might be useful, depending on the desired reliability of the cloud. We also recommend that any machine for any cloud purpose be filled with as much RAM as possible, based on our experiences with virtualization technology.

7.6.2 Operating System

Once hardware is set up, the operating system is the next logical component. Since all of the other components run on whatever OS the cloud builder chooses, compatibility of components is often dictated by the OS chosen. For the cloud controllers Eucalyptus, OpenNebula and Nimbus, each of them is compatible with a number of Linux operating systems. (The open-source frameworks rely on open-source OS, as expected.) However, they must also face the compatibility challenge. For each of these frameworks, online advice is available for how to deal with the various quirks of many various Linux distributions. Also, taking advantage of the active mailing lists of each project is often helpful in this context. Nearly every combination of OS and other software is going to produce tricky, sometimes time-consuming, configuration issues in open-source clouds, since these are developing projects. However, since these are active, developing projects, our experience is that help is usually available in some form.

The underlying hardware and OS form the basis of the open-source cloud before it is constructed. The hardware dictates the cloud's capacity, and the OS dictates many aspects of the behavior of the system. However, the good news is that many of the open-source cloud systems are constructed so that they can run on a variety of systems. Moreover, since this is open-source software, it is usually an option to compile the cloud software on your own distribution, if needed.

7.6.3   Network

There are two main aspects of the network to be aware of. First, there is the physical network of all of the machines involved in the private cloud. Second, there is the virtual network of the virtual machines. In our experience, network configuration is the most challenging aspect of setting up a private cloud, as the physical and virtual networks must both be accommodated.

Typically, a private cloud uses a program such as iptables or bridge-utils to create a virtual network device for each VM. The cloud controller calls these programs when building a VM to create the device and supply a MAC address. The tricky part is in the MAC to IP address mapping. There are several factors to keep in mind. First, what is the firewall filtering? It is not uncommon for organizations to have rules regarding filtering dynamically allocated IP addresses, for security purposes. The cloud controller needs to be certain that there are at least some addresses that can be used for VM. Second, those IP address must be correctly registered with the DNS and routers being used by the organization. Otherwise, the VM cannot be accessed. Third, like most physical machines the IP address for the VM must be assigned using a DHCP server. The dhcpd can be located in many places, depending on the configuration, but however it is setup, there must not be a conflict with the wider network. Some examples include: (1) a dhcpd handled separately from the cloud, (2) The cloud automatically configuring and setting up a single dhcpd of it own, or (3) each running cloud node having a dhcpd for the VM on it. With respect to the various places that a dhcpd might be, the various networking modes offered by the various clouds dictate how this must done. The dhcpd configuration and cloud configuration must agree, both in where the dhcpd is located and in the MAC to IP pairings the dhcpd will provide.

In practice, we found that the most elegant and flexible way for a private cloud to be set up is for the cloud controller to have exclusive control over some IP address space. This typically requires configuring routers such that there is a single subnet with an address range exclusively for the private cloud. Then, the cloud controller can assign part of the address range to physical machines and part to virtual machines. Then, depending on the expected configuration of the cloud, dhcpd and (possibly) dns, can be set up for these addresses.

If giving the cloud controller exclusive control over an address range is not an option, due to the configuration of the rest of the network, then it is probable that the network administrator and the cloud administrator will have to spend time together, putting together a configuration to meet both their needs.

7.6.4 Hypervisor

The hypervisor, also known as the Virtual Machine Monitor or VMM, is the software, on each cloud node, that starts and runs VM. Typically, VMM are divided into Type I and Type II VMM [311]. For open-source hypervisors, the two most widely used in open-source clouds are Xen and KVM. [233] [117]. Recently, there has been some interest in some of the open-source clouds for using VMware (a commercial product) as well [729]. In the future, Oracle VirtualBox might also be used [555].

In essence, the hypervisor dictates the properties of the virtual machines that are allowed to run. Most basically, not all OS can be run on all hypervisors. More to the point, Xen and KVM (open-source hypervisors) are mainly designed for running Linux VM. Xen, in particular, requires that both the physical machine and the virtual machine use a special Xen-enabled Linux kernel. The various hypervisors have different options for how disk images are packaged. In order to use a particular hypervisor, the disk images for the running VM must conform to the expectations of the hypervisor in question. As such, if users are supplying their own VM images, they must be knowledgeable about the underlying hypervisor as well.

(KVM can also run Windows VM, but we have not seen much experimentation with this in the context of open-source clouds, as of yet, probably due to the licensing issues, and technical issues of activation, of running multiple, proprietary Windows VM.)

We briefly make note of the package called libvirt. Libvirt is used by various open-source cloud systems as an intermediary between the cloud software and the hypervisor. We make brief note of this fact because the configuration of libvirt is often key to resolving difficulties in getting any open source cloud to run and, second, the libvirt log file is a very useful debugging resource for any open-source cloud provider. Lastly, we again note that libvirt, xen, and kvm are all active projects, with new versions being released periodically.

7.6.5 Disk Image Storage

Running a virtual machine requires some files to store the virtual hard disks of that machine. This is true for VM, whether or not they are in a cloud. For clouds, there are three sets of virtual disks. First, there are virtual disk templates for future VM. Second, there are the virtual disks for currently running VM. Third, there are (optionally) stored virtual disks of previously running VM that have been saved for future use. Of course, the most important thing for a private cloud is to insure that there is sufficient disk space and that the hard disks are reliable. In our experience, we found that some form of hardware RAID for disk storage was useful for insuring against system failure. Eucalyptus and Nimbus both take a similar approach to the storage of files. They each employ a distributed storage mechanism that implements the API of Amazon's S3 storage. For Eucalyptus it is called Walrus and for Nimbus it is called Cumulus. The main feature of this is that individual users can upload their own VM images, (or even share them with each other), for future use. Images to be run are (usually) specified from the user's set of stored images. One important advantage to this scheme, apart from compatibility with the common S3 standard, is that it decentralizes the storage of disk images, hence the idea of "distributed" storage.

OpenNebula takes a different approach, which is more customizable but also tends to be more centralized. The two main setups, (though, as said before, it is very customizable) are SCP and NFS. In SCP, the template disk images are stored on the front-end and are copied over to wherever they are needed, to start VM. In NFS, all the OpenNebula related files are stored on a Network File System server (or other shared file system), and all files, running VM and template VM disk images, are accessed over the NFS. These setups have the advantage of being fairly easy to understand, (and if desired, customized), but are very centralized. For SCP, the front-end can be a bottleneck, and requires enough disk space for all template disk images. For NFS, the file server is a potential bottleneck, and, depending on the openness of the network, a security vulnerability.

7.6.6   User Front-End

In a certain sense, the front-end for the user is one of the most interesting parts of the idea of open-source clouds. This is mainly due to the customizability that is available. Since there is, as of right now, no universally accepted "standard" interface for private clouds, many of the open-source clouds provide multiple API, such as EC2, WSRF, OGF OCCI and more. Moreover, more front-ends are being implemented as time goes on, often as third-party projects. Since there are so many options, the decision as to what kind of front-end should be used depends heavily on the requirements of users. That is, it would be unwise to force users to adopt one standard, when they are familiar with another. One example is that, since Nimbus is affiliated with Globus, many (though not all) of the front-ends used with Nimbus use security certificates in the style of Globus, with which users may or may not be familiar.

We also note that security plays a big role in what front-end is used. This includes preventing users from maliciously causing problems, but also for preventing users from accidentally causing problems as well. This is a key contrast between Eucalyptus and OpenNebula. While Eucalyptus has numerous tools for user management, and cleanly separates users from the physical resources, OpenNebula can allow users to login to the head node to create VM. That particular OpenNebula interface allows basically anything about the VM to be specified by the user. For OpenNebula, (at least with the login interface) the advantage is customizability for the user; for Eucalyptus, the advantage is that it is more difficult for users to ask for an invalid (or dangerous) configuration. (Of course, OpenNebula also has safer front-ends available as well.)

We note that tools for managing user allotments of VM resources are not yet heavily developed. That is, preventing users from using more than their "fair share" of resources is not yet a standard feature. For some cloud settings, this is fine, as an attentive administrator can simply remove users who abuse the system. However, in a setting where more fine-grained control of user privileges is required, a custom front-end is needed. We found that OpenNebula, in particular, works well for this, by having custom PHP web front-end route commands to OpenNebula's XML-RPC interface. However, it is also possible to do something similar to this with Eucalyptus and Nimbus as well, using one of several available API.

Regardless of the situation, the configuration of the front-end is the key means by which private cloud providers can adjust the experience for the users. This includes user-friendliness, ease-of-use, security, resource allocation, and more. A private cloud administrator who wishes to do anything unique will, in all likelihood, have to spend some time tweaking the front-end for that purpose. Lastly, while the front-end must account for the means by which VM are created and removed, the front-end must also handle the security of allowing users ssh access to their own VM. In our experience, this second aspect was not always satisfactory. There are four main approaches used: (1) Users maintain ssh keys which the cloud software inserts into the VM prior to the first boot of a VM. (This is common for Eucalyptus and Nimbus.) The downside is that users must maintain their own ssh private keys and understand them. (2) The VM disk image, provided by the user, has a root password the user knows. The downside is that this makes it harder for users to share disk images with each other, as the password must be known. (3) A custom front-end is built to manipulate the password and sudoers file of the VM to grant access to specified users. This has the downside of requiring a custom front-end to do it. (4) The cloud gives vnc access to VM. This has the downside of being much slower, not a standard feature of most open-source clouds, and being very difficult to secure. In any case, some facility must be available for users to get into their own VM.

7.7   Future Opportunities

In order to complete this introduction, we make note of several developments that we expect to occur in open-source cloud computing in the near future:

- One current drawback of these open-source clouds is that updating the cloud software and restoring VMs from backups is an unwieldy task, usually requiring a shutdown of the private cloud. We expect that more progress will be made to facilitate clean recovery of crashed or updated cloud systems.

- We expect that more attention will be paid to scheduling and resource allocation for private cloud resources. Since private clouds do not, typically, use money as an arbiter for resource access, there is an open research area here regarding how to fairly allocate VM resources, depending on the needs of the private cloud. We expect that many different approaches will be proposed. (We note a very recent proposed feature to Nimbus, allowing for "spot instances" similar to those used by Amazon [671] which automatically lease cloud resources to waiting users when demand drops below a certain level.)

- As the various software pieces are developed, we expect that there will be tighter abstractions on the interfaces between various components, especially for the libvirt layer. The benefit of such cleaner abstractions is that they allows for more heterogeneity lower in the software stack and, therefore, a wider array of possible cloud configurations. For example, a cleaner abstraction at the libvirt layer would make it easier to have a single private cloud hosted on machines with different VMM or other underlying software.

- In general, we expect more research into combining things. There is already research combining clouds with each other [612], dispatching excess private cloud demand to a commercial (or other) cloud [495], and more. We expect more interest in combining private clouds with other batch computing systems to produce more useful systems.

- Security is always a topic of interest. One of the reasons that we stated that private clouds are of interest is that some organizations cannot risk placing their data in a public cloud. Private clouds, however, also have security challenges. Some areas of interest include: securing data in distributed cloud file storage, balancing secure user authentication with ease-of-use and investigating security vulnerabilities in network bridges. We expect that continued research will both expose and fix vulnerabilities in private clouds.

- As we mentioned before, support for API standards and underlying VMM software continues to be implemented in the various open-source cloud projects. We expect this to continue.

- And, lastly, we further expect that continued improvements in xen, kvm, libvirt, Eucalyptus, OpenNebula, Nimbus and the other key cloud components will continue to improve the performance of open-source clouds and broaden their use and influence well into the future.

7.8   Conclusion

For many individuals or organizations, open-source clouds are a useful option for supporting a private cloud infrastructure. Numerous software projects are available to enact such an endeavor, including Eucalyptus, OpenNebula, Nimbus and others. However, construction of such a private cloud requires a careful selection of the software involved at all levels of the cloud computing software stack. This includes the underlying hardware, operating systems, the physical and virtual networks, the hypervisor, the storage for virtual disk images and the software of the user front-end. Moreover, each of these components must be compatible with the others. In this chapter, we have summarized these major components and several of the common challenges involved in open-source clouds.

Fortunately, the major players in the open-source cloud computing community, such as Eucalyptus, OpenNebula and Nimbus are vibrant, growing projects, with active support communities. This allows newcomers to the construction and maintenance of private clouds avenues of support for problems encountered. Due to the maturity of these projects, and the usefulness of the software they provide, we believe that the community of open-source cloud providers and users will remain an active and innovative fixture in the future cloud computing community.

7.9   Acknowledgments

We are grateful to Paul Brenner and Stephen Bogol for providing hardware resources and technical support during our research. Furthermore, we are grateful for the work being done by the Eucalyptus, OpenNebula and Nimbus projects, as well as their assistance in solving many of the problems we encountered.

8

Cloud Computing: Performance Analysis

Hamzeh Khazaei

University of Manitoba, Winnipeg, Manitoba, Canada

Jelena Mišić

Ryerson University, Toronto, Ontario, Canada

Vojislav B. Mišić

Ryerson University, Toronto, Ontario, Canada

CONTENTS

Cloud computing is a computing paradigm in which different computing resources, including infrastructure, hardware platforms, and software applications, are made accessible to remote users as services. Successful provision of infrastructure-as-a-service (IaaS) and, consequently, widespread adoption of cloud computing, necessitates accurate performance evaluation that allows service providers to dimension their resources in order to fulfil the service level agreements with their customers. In this chapter, we describe an analytical model for performance evaluation of cloud server farms, and demonstrate the manner in which important performance indicators such as request waiting time and server utilization may be assessed with sufficient accuracy.

8.1 Introduction

Significant innovations in virtualization and distributed computing, as well as improved access to high-speed Internet, have accelerated interest in cloud computing [720]. Cloud computing is a general term for system architectures that involve delivering hosted services over the Internet. These services are broadly divided into three categories: Infrastructure-as-a-Service (IaaS), which includes equipment such as hardware, storage, servers, and networking components that are made accessible over the Internet); Platform-as-a-Service (PaaS), which includes computing platforms—hardware with operating systems, virtualized servers, and the like; and Software-as-a-Service (SaaS), which includes sofware applications and other hosted services [291]. A cloud service differs from traditional hosting in three principal aspects. First, it is provided on demand, typically by the minute or the hour; second, it is elastic, since the user can have as much or as little of a service as they want at any given time; and third, the service is fully managed by the provider; the user needs little more than a computer and Internet access. Cloud customers pay only for the services they use by means of a customized service level agreement (SLA), which is a contract negotiated and agreed upon between a customer and a service provider: the service provider is required to execute service requests from a customer within negotiated quality of service(QoS) requirements, for a given price.

Due to the dynamic nature of cloud environments, diversity of users' requests and time dependency of load, providing expected quality of service while avoiding over-provisioning is not a simple task [772]. To ensure that the QoS perceived by end clients is acceptable, the providers must exploit techniques and mechanisms that guarantee a minimum level of QoS. Although QoS has multiple aspects such as response time, throughput, availability, reliability, and security, the primary aspect of QoS considered in this work is related to response time [742].

Cloud computing has been the focus of much research in both academia and industry, however, implementation-related issues have received much more attention than performance-related ones; here we describe an analytical model for evaluating the performance of cloud server farms and verify its accuracy with numerical calculations and simulations. We assume that any request goes through a *facility node* and then leaves the center. A facility node may contain different computing resources such as web servers, database servers, and others, as shown in Fig. 8.1. We consider the time a request spends in one of those facility nodes as the response time; response time does not follow any specific distribution. Our model is flexible in terms of cloud center size and service time of customer requests. We model the cloud environment as an $M/G/m$ queuing system which indicates that inter-arrival time of requests is exponentially distributed, the service time is generally distributed and the number

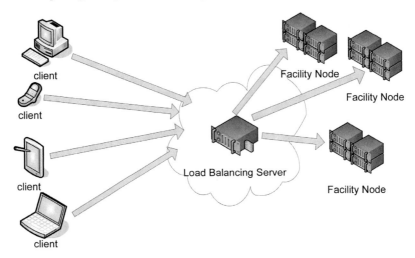

FIGURE 8.1
Cloud Clients and Service Provider.

of facility nodes is m. Also, due to the the nature of the cloud environment (i.e., it is a service provider with potentially many customers), we pose no restrictions on the number of facility nodes. These two characteristics, general service time and large number of nodes, have not been adequately addressed in previous research.

8.2 Related Work

Cloud computing has attracted considerable research attention, but only a small portion of the work done so far has addressed performance issues, and the rigorous analytical approach has been adopted by only a handful among these. In [782], the authors studied the response time in terms of various metrics, such as the overhead of acquiring and realizing the virtual computing resources, and other virtualization and network communication overhead. To address these issues, they have designed and implemented C-Meter, a portable, extensible, and easy-to-use framework for generating and submitting test workloads to computing clouds. Most of the research related to cloud computing has dealt with implementation issues, while performance-related issues have received much less attention.

In [772], the authors consider a cloud center which is modeled as the classic open network; they obtained the distribution of response time based on assumption that inter-arrival time and service time are both exponential. Using the distribution of response time, they found the relationship among the maximal number of tasks, the minimal service resources and the highest level of services.

Theoretical analyses have mostly relied on extensive research in performance evaluation of $M/G/m$ queuing systems [135, 353, 433, 488, 520, 543, 560, 691, 706, 778]. As solutions for distribution of response time and queue length in $M/G/m$ systems and $M/G/m/m + r$ can't be obtained in closed form, suitable approximations were sought. However, most of these provide reasonably accurate estimates of mean response time only when the number of servers is comparatively small as well as small *coefficient of variation* of service time, CV, (less than unity), but fail for large number of servers and higher CV. Approximation errors are particularly pronounced when the offered load ρ is small, and/or when both the number of servers m and the CV of the service time, are large [135, 433, 706].

A closed form expression for the blocking probability in $M/G/m/m + r$ based on the exact solution for finite capacity exponential queues was proposed in [662]. There are essentially two problems of interest in this paper; the first is how to estimate the blocking probability and the second problem concerns the allocation of buffers so that the loss/delay of blocking probability will be below a specific threshold. The building block of this approach is the exact solution of the $M/M/m/m + r$ queuing systems so this approach is more likely to be suitable for service time distributions for which the CV does not exceed one.

An approximations for the mean queue length in the $M/G/m/m + r$ queue without deriving the distribution of number of tasks in the system is proposed in [543]. Moreover, their methods were given in transform and required lots of computation in order to be evaluated.

In [775], the cloud center was modeled as an $M/M/m/m + r$ queuing system, which has been used to compute the distribution of response time. Inter-arrival and service times were both assumed to be exponentially distributed, and the system had a finite buffer of size $m + r$. The response time was broken down into waiting, service, and execution periods, assuming that all three periods are independent, which is unrealistic, based on their own argument.

For an $M/G/m/m + r$ queue there is no explicit formula for probability distribution of the number of tasks in the system except in a few special cases: if $G = M$, $r = 0$ or $m = 1$, then the exact and close form of distribution of tasks in the system are attainable; M denotes the exponential cumulative distribution function. However, it is quite difficult to obtain explicit formula for the probability distribution of the number of tasks in the system in a general case.

The author in [435] proposed a transform-free approach for a steady-state queue length distribution in an $M/G/m$ system with finite waiting space; his approach was given in an explicit form and hence its numerical computation is easier than that for previous approximations [353, 705]. Although the approach is exact for $M/M/m/m+r$, $r = 0$ and reasonably accurate for a general case, apparently the method is suitable for a small number of servers only.

In [434], the author considered the problem of optimal buffer designing for $M/G/m$ in order to determine the smallest buffer capacity such that the rate of lost tasks remains under a predefined level. Here, Kimura used the same approach in [435] in order to approximate the blocking probability, and then applied the approximate formula to the buffer design problem. Based on convex order value, Kimura concluded that the higher the order of convexity for service time, the bigger the optimal buffer size.

As a result, the former approaches are not directly applicable to performance analysis of cloud computing server farms, where the number of servers is huge and service request arrival distribution is not generally known.

8.3 The Analytical Model

We model a cloud server farm as an $M/G/m$ queuing system which indicates that the inter-arrival time of requests is exponentially distributed, the service times of customers' requests are independent and identically distributed random variables with a general distribution whose service rate is μ; both μ and CV, the coefficient of variation defined as standard deviation divided by the mean, are finite.

An $M/G/m$ queuing system may be considered as a Markov process which can be analyzed by applying the embedded Markov chain technique. The embedded Markov Chain techique requires selection of Markov points in which the state of the system is observed. Therefore we monitor the number of the tasks in the system (both in service and queued) at the moments immediately before the task request arrival. If we consider the system at Markov points and number these instances 0, 1, 2, ..., then we get a Markov chain [438]. Here, the system under consideration contains m servers, which render service in order of task request arrivals.

Task requests' arrival process is Poisson. Task request interarrival time A is exponentially distributed with rate to $\frac{1}{\lambda}$. We will denote its Cumulative Distribution Function (CDF) as $A(x) = Prob[A < x]$ and its probability density function (pdf) as $a(x) = \lambda e^{-\lambda x}$. The Laplace Stieltjes Transform (LST) of interarrival time is $A^*(s) = \int_0^\infty e^{-sx} a(x) dx = \frac{\lambda}{\lambda+s}$.

Task service times are identically and independently distributed according to a general distribution B, with a mean service time equal to $\bar{b} = \frac{1}{\mu}$. The CDF of the service time is $B(x) = Prob[B < x]$, and its pdf is $b(x)$. The LST of service time is $B^*(s) = \int_0^\infty e^{-sx} b(x) dx$.

Residual task service time is time from the random point in task execution until the task completion. We will denote it as B_+. This time is necessary for our model, since it represents time distribution between task arrival z and departure of the task which was in service when task arrival z occured. It can be shown as well that probability distribution of elapsed service time (between start of the task execution and next arrival of task request B_- has the same probability distribution [690].

The LST of residual and elapsed task service times can be calculated in [690] as

$$B_+^*(s) = B_-^*(s) = \frac{1 - B^*(s)}{s\bar{b}} \tag{8.1}$$

The offered load may be defined as

$$\rho \triangleq \frac{\lambda}{m\mu} \tag{8.2}$$

For practical reasons, we assume that the system never enters saturation, which means that any request submitted to the center will get access to the required facility node after a finite queuing time. Furthermore, we also assume each task is serviced by a single server (i.e., there are no batch arrivals), and we do not distinguish between installation (setup), actual task execution, and finalization components of the service time; these assumptions will be relaxed in our future work.

8.3.1 The Markov Chain

We are looking at the system at the moments of task request arrivals — these points are selected as Markov points. A given Markov chain has a steady-state solution if it is ergodic. Based on conditions for ergodicity [438] and the above-mentioned assumptions, it is easy to prove that our Markov Chain is ergodic. Then, using the steady-state solution, we can extract the distribution of the number of tasks in the system as well as the response time.

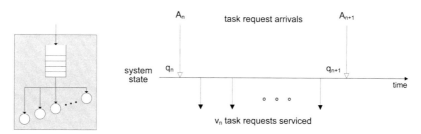

FIGURE 8.2
Embedded Markov Points.

Let A_n and A_{n+1} indicate the moment of nth and $(n + 1)^{th}$ arrivals to the system, respectively, while q_n and q_{n+1} indicate the number of tasks found in the system immediately before these arrivals; this is schematically shown in

Fig. 8.2. If v_{n+1} indicates the number of tasks which are serviced and depart from the system between A_n and A_{n+1}, the following holds:

$$q_{n+1} = q_n - v_{n+1} + 1 \qquad (8.3)$$

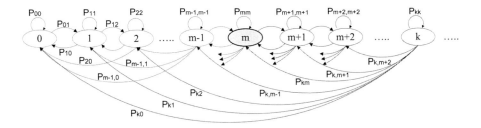

FIGURE 8.3

State–Transition–Probability Diagram for the $M/G/m$ Embedded Markov Chain.

We need to calculate the transition probabilities associated with this Markov chain, defined as

$$p_{ij} \triangleq Prob\left[q_{n+1} = j | q_n = i\right] \qquad (8.4)$$

i.e., the probability that $i + 1 - j$ customers are served during the interval between two successive task request arrivals. Obviously for $j > i + 1$

$$p_{ij} = 0 \qquad (8.5)$$

since there are at most $i+1$ tasks present between the arrival of A_n and A_{n+1}. The Markov state-transition-probability diagram as in Fig. 8.3, where states are numbered according to the number of tasks currently in the system (i.e those in service and those awaiting service). For clarity, some transitions are not fully drawn, especially those originating from states above m. We have also highlighted the state m because the transition probabilities are different for states on the left- and right-hand side of this state (i.e., below and above m).

8.3.2 Departure Probabilities

Due to ergodicity of the Markov chain, an equilibrium probability distribution will exist for the number of tasks present at the arrival instants; so we define

$$\pi_k = \lim_{n \to +\infty} Prob\left[q_n = k\right] \qquad (8.6)$$

From [690], the direct method of solution for this equilibrium distribution requires that we solve the following system of linear equations:

$$\pi = \pi \mathbf{P} \qquad (8.7)$$

where $\pi = [\pi_0, \pi_1, \pi_2, \ldots]$, and $\mathbf{P}$ is the matrix whose elements are one-step transition probabilities p_{ij}.

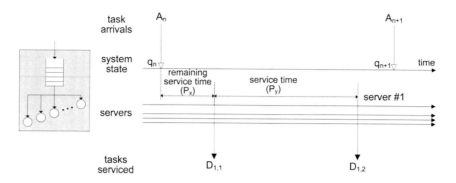

FIGURE 8.4
System Behavior between Two Arrivals.

To find the elements of the transition probability matrix, we need to count the number of tasks departing from the system in between two successive arrivals. Consider the behavior of the system, as shown in Fig. 8.4. Each server has zero or more departures during the time between two successive task request arrivals (the inter-arrival time). Let us focus on an arbitrary server, which (without loss of generality) could be the server number 1. For a task to finish and depart from the system during the inter-arrival time, its remaining duration (residual service time as defined in (8.1)) must be shorter than the task inter-arrival time. This probability will be denoted as P_x, and it can be calculated as

$$P_x = Prob\,[A > B_+] = \int_{x=0}^{\infty} P\{A > B_+|B_+ = x\,\}P\{B_+ = x\}$$
$$= \int_0^{\infty} e^{-\lambda x}dB_+(x) = B_+^*(\lambda) \tag{8.8}$$

Physically this result presents the probability of no task arrivals during residual task service time.

In the case when an arriving task can be accommodated immediately by an idle server (and therefore queue length is zero) we have to evaluate the probability that such a task will depart before next task arrival. We will denote this probability as P_y and calculate it as:

$$P_y = Prob\,[A > B] = \int_{x=0}^{\infty} P\{A > B|B = x\,\}P\{B_+ = x\}$$
$$= \int_0^{\infty} e^{-\lambda x}dB(x) = B^*(\lambda) \tag{8.9}$$

However, if queue is non-empty upon task arrival, the following situation may happen. If between two successive new task arrivals a completed task departs from a server, that server will take a new task from the non-empty queue. That task may be completed as well, before the next task arrival, and if the queue is still non-empty a new task may be executed, and so on until either the queue gets empty or a new task arrives. Therefore the probability of $k > 0$ job departures from a single server, given that there are enough jobs in the queue can be derived from expressions (8.8) and (8.9) as:

$$P_{z,k} = B_+^*(\lambda)(B^*(\lambda))^{k-1} \qquad (8.10)$$

note that $P_{z,1} = P_x$.

Using these values we are able to compute the transition probabilities matrix.

8.3.3 Transition Matrix

Based on our Markov chain, we may identify four different regions of operation for which different conditions hold; these regions are schematically shown in Fig. 8.5, where the numbers on horizontal and vertical axes correspond to the number of tasks in the system immediately before a task request arrival (i) and immediately upon the next task request arrival (j), respectively.

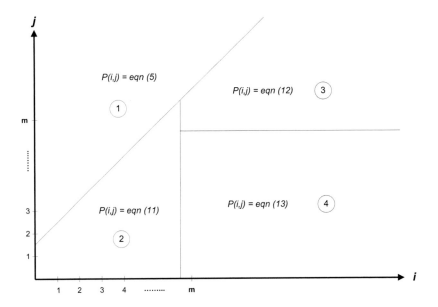

FIGURE 8.5
Range of Validity for p_{ij} Equations.

Regarding the region labeled 1, we already know from Eq. 8.5 that $p_{ij} = 0$ for $i + 1 < j$.

In region 2, no tasks are waiting in the queue, hence $i < m$ and $j \leq m$. In between the two successive request arrivals, $i + 1 - j$ tasks will complete their service. For all transitions located on the left side of state m in Fig. 8.3, the probability of having $i + 1 - j$ departures is

$$p_{ij} = \binom{i}{i-j} P_x^{i-j}(1 - P_x)^j P_y + \binom{i}{i+1-j} P_x^{i+1-j}(1 - P_x)^{j-1}(1 - P_y)$$

$$\text{for } i < m, j \leq m \tag{8.11}$$

Region 3 corresponds to the case where all servers are busy throughout the inter-arrival time, i.e., $i, j \geq m$. In this case, all transitions remain to the right of state m in Fig. 8.3, and state transition probabilities can be calculated as

$$p_{ij} = \sum_{s=\phi}^{\sigma} \binom{m}{s} P_x^s (1 - P_x)^{m-s} P_{z,2}^{i+1-j-s}(1 - P_{z,2})^s \tag{8.12}$$

$$\text{for } i, j \geq m$$

In the last expression, the summation bounds are $\sigma = min\,[i + 1 - j, m]$ and $\phi = min\,[i + 1 - j, 1]$.

Finally, region 4, in which $i \geq m$ and $j \leq m$, describes the situation where the first arrival (A_n) finds all servers busy and a total of $i - m$ tasks waiting in the queue, which it joins; while at the time of the next arrival (A_{n+1}) there are exactly j tasks in the system, all of which are in service. The transition probabilities for this region are

$$p_{ij} = \sum_{s=1}^{\sigma} \binom{m}{s} P_x^s (1 - P_x)^{m-s} \binom{\eta}{\alpha} P_{z,2}^{\psi}(1 - P_{z,2})^\zeta \beta \tag{8.13}$$

$$\text{for } i \geq m, j < m$$

where we used the following notation:

$$
\begin{aligned}
\sigma &= \ min\,[m, i + 1 - j] \\
\eta &= \ min\,[s, i + 1 - m] \\
\alpha &= \ min\,[s, i + 1 - j - s] \\
\psi &= \ max\,[0, i + 1 - j - s] \\
\zeta &= \ max\,[0, j - m + s] \\
\beta &= \ \begin{cases} 1 & \text{if } \psi \leq i + 1 - m \\ 0 & \text{otherwise} \end{cases}
\end{aligned}
\tag{8.14}
$$

8.4   Numerical Validation

The steady-state balance equations outlined above can't be solved in closed form, hence we must resort to a numerical solution. To obtain the steady-state probabilities $\pi = [\pi_0, \pi_1, \pi_2, ...]$, as well as the mean number of tasks in the system (in service and in the queue) and the mean response time, we have used the probability generating functions (PGFs) for the number of tasks in the system:

$$P(z) = \sum_{k=0}^{\infty} \pi_z z^k \qquad (8.15)$$

and solved the resulting system of equations using Maple 13 from Maplesoft, Inc. [492]. Since the PGF is an infinite series, it must be truncated for numerical solution; we have set the number of equations to twice the number of servers, which allows us to achieve satisfactory accuracy (as will be explained below), plus the necessary balance equation

$$\sum_{i=0}^{2m} \pi_i = 1. \qquad (8.16)$$

The mean number of tasks in the system is, then, obtained as

$$E[QS] = P'(1) \qquad (8.17)$$

while the mean response time is obtained using Little's law as

$$E[RT] = E[QS]/\lambda \qquad (8.18)$$

We have assumed that the task request arrivals follow the gamma distribution with different values for shape and scale parameters; however, our model may accommodate other distributions without any changes. Then, we have performed two experiments with the variable task request arrival rate and coefficient of variation CV (which can be adjusted in the gamma distribution independently of the arrival rate).

To validate the analytical solutions we have also built a discrete even simulator of the cloud server farm using the object-oriented Petri net-based simulation engine Artifex by RSoftDesign, Inc. [620].

The diagrams in Fig. 8.6 show analytical and simulation results (shown as lines and symbols, respectively) for the mean number of tasks in the system as functions of the offered load ρ, under different number of servers. Two different values of the coefficient of variation, $CV = 0.7$ and 0.9, were used; the corresponding results are shown in Figs. 8.6(a) and 8.6(b). As can be seen, the results obtained by solving the analytical model agree very well with those obtained by simulation.

The diagrams in Fig. 8.8 show the mean response time, again for the same range of input variables and for the same values of the coefficient of variation. As above, solid lines correspond to analytical solutions, while different symbols correspond to different number of servers. As could be expected, the response time is fairly steady up to the offered load of around $\rho = 0.8$, when it begins to increase rapidly. However, the agreement between the analytical solutions and simulation results is still very good, which confirms the validity of our modeling approach.

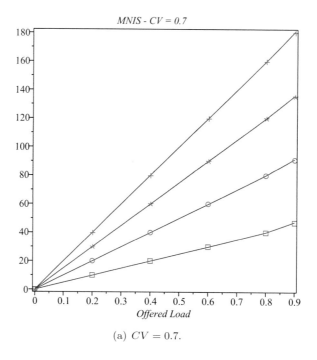

(a) $CV = 0.7$.

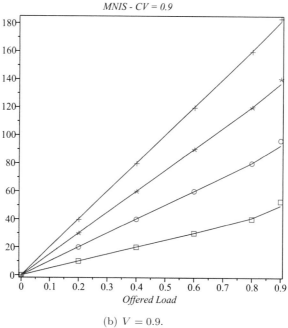

(b) $V = 0.9$.

FIGURE 8.6
Mean Number of Tasks in the System: $m = 50$ (denoted by squares), 100 (circles), 150 (asterisks), and 200 (crosses).

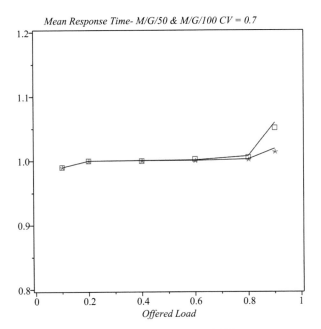

(a) Results for $CV = 0.7$, $m = 50$ and 100 servers.

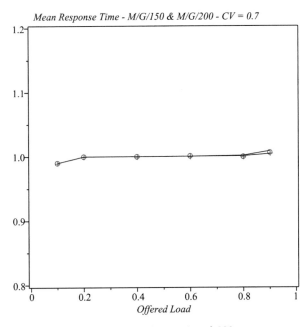

(b) Results for $CV = 0.7$, $m = 150$ and 200 servers.

FIGURE 8.7
Mean Response Time $CV = 0.7$, $m = 50$ (denoted by squares), 100 (asterisks), 150 (circles), and 200 (crosses).

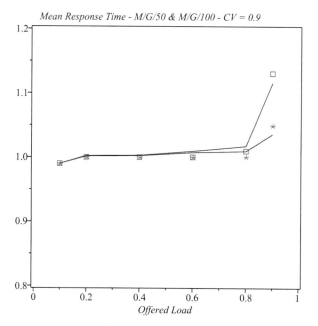

(a) Results for $CV = 0.9$, $m = 50$ and 100 servers.

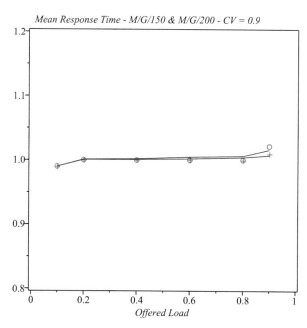

(b) Results for $CV = 0.9$, $m = 150$ and 200 servers.

FIGURE 8.8

Mean Response Time for $CV = 0.9$, $m = 50$ (denoted by squares), 100 (asterisks), 150 (circles), and 200 (crosses).

8.5 Conclusions

Performance evaluation of server farms is an important aspect of cloud computing which is of crucial interest for both cloud providers and cloud customers. In this chapter we have proposed an analytical model for performance evaluation of a cloud computing center. Due to the nature of the cloud environment, we assumed general service time for requests as well as a large number of servers; in the other words, our model is flexible in terms of scalability and diversity of service time. We have further conducted numerical experiments and simulation to validate our model. Numerical and simulation results showed that the proposed method provided a quite accurate computation of the mean number of tasks in the system and mean response time.

The model could be extended for burst arrivals of requests or a kind of task including several subtasks; examining other types of distributions as service time which are more realistic in cloud computing area, e.g., Log-Normal distribution and looking in to the facility node and breaking down the response time into several components such as setup, execution, return and clean up time could be another dimension of extension. The authors will address all these issues in future work.

8.6 Glossary

Markov Process: In probability theory and statistics, a Markov process, named after the Russian mathematician Andrey Markov, is a time-varying random phenomenon for which a specific property (the Markov property) holds. In a common description, a stochastic process with the Markov property, or memorylessness, is one for which conditional on the present state of the system, its future and past are independent.

Markov Chain: A Markov chain is a random process with the Markov property, i.e., the property, simply said, that the next state depends only on the current state and not on the past.

Embedded Markov Chain Technique: One method of finding the stationary probability distribution, π, of an ergodic continuous-time Markov process, Q, is by first finding its embedded Markov chain (EMC). Strictly speaking, the EMC is a regular discrete-time Markov chain, sometimes referred to as a jump process. Each element of the one-step transition probability matrix of the EMC, S, is denoted by p_{ij}, and represents the conditional probability of transitioning from state i into state j.

9

Intercloud: The Future of Cloud Computing. Concepts and Advantages

Antonio Celesti

Faculty of Engineering, University of Messina, Italy

Francesco Tusa

Faculty of Engineering, University of Messina, Italy

Massimo Villari

Faculty of Engineering, University of Messina, Italy

Antonio Puliafito

Faculty of Engineering, University of Messina, Italy

CONTENTS

9.1   Introduction

Nowadays, cloud computing is considered an emerging computation paradigm able to pursue new levels of efficiency in service delivering, and could represent, at the same time, a tempting business opportunity for IT operators of increasing their revenues. Until now, the trend of the cloud computing ecosystem has been characterized by the steady rising of hundreds of independent, heterogeneous cloud providers, managed by private subjects, yielding various types of cloud-based services to their clients (e.g., IT societies, organizations, universities, desktop and mobile end-users, etc.). Currently, most of such clouds can be considered as "islands in the ocean of the cloud computing" and do not presents any form of federation. At the same time a few clouds are beginning to use the cloud-based services of other clouds, but there is still a long way to go toward the establishment of a worldwide Intercloud ecosystem including thousands of cooperating clouds. In such a perspective, the latest trend toward cloud computing is dominated by the idea to federate heterogeneous clouds. This means not to think about independent private clouds any more, but to consider a new Intercloud scenario where different clouds, belonging to different administrative domains, interact with each other, sharing and gaining access to physical resources, and becoming themselves at the same time both "users" and "resource providers."

This chapter aims to investigate the new business advantages of a futuristic worldwide Intercloud ecosystem, considering evolution degree of the current infrastructures and the possible scenarios on which the Intercloud federation model could be applied. In particular, a discussion about the requirements for the federation will be provided, together with an overview of the involved technological concerns. More specifically, several new business advantages of such a futuristic scenario will be analyzed, focusing on the requirements needed for the establishment of a High Cooperation Federation among cloud platforms. Moreover, an overview of the major future research challenges will be provided.

The chapter is organized as follows. Section 9.2 introduces the general meaning of the term federation, then focuses specifically on its implications in the cloud, and presents a terminology useful for classifying the status of evolution of clouds as far as regards the federation. Some of the existing middlewares for implementing private, public and hybrid clouds are described, focusing on their abilities for sharing and accessing resources using specific interfaces and protocols. Subsequently, the discussion will point out the new perspective of Intercloud: after a presentation of the Intercloud concepts, which includes both the actors and their interactions, a series of possible scenarios will be reported, each associated to different models for sharing and accessing resources among different clouds. The description will continue with the presentation of a possible solution for achieving the Intercloud federation: a theoretical and practical approach, based on the execution of three subsequent phases, will be proposed for building up the federation. As will be deeply explained later, such an approach will consist of the execution, by the clouds which want to belong to a Intercloud, of three different phases: Discovery, Match-Making and Authentication. Section 9.6 will present some details about the technological solutions that could be employed for implementing the aforementioned three-phase approach. Such a section will provide some details about the Extensible Messaging and Presence Protocol (XMPP) [380] as support for the discovery, eXtensible Access Control Markup Language (XACML) [1] for addressing the Match-Making phase, and Security Assertion Markup Language (SAML) [2] for implementing an authentication mechanism based on the Identity Provider/Service Provider model. Finally, Section 9.7 will present some conclusions on the proposed idea, also opening a debate on the new possible challenges and research topics regarding the cloud federation.

9.2   Federation: From the Political World to the IT

The concept of federation has always had both political and historical implications: the term refers, in fact, to a type of system organization characterized by a joining of partially "self-governing" entities united by a "central government." In a federation, each self-governing status of the component entities is typically independent and may not be altered by a unilateral decision of the "central government."

More specifically, looking at the political philosophy, the federation refers to the form of government or constitutional structure known as federalism and can be considered the opposite of the "unitary state." The components of a federation are in some sense "sovereign" with a certain degree of autonomy from the "central government": this is why a federation can be intended as more than a mere loose alliance of independent entities [3].

Until now, the cloud ecosystem has been characterized by the steady rising of hundreds of independent and heterogeneous cloud providers, managed by private subjects which yield various services to their clients. Using this computing infrastructure it is possible to pursue new levels of efficiency in delivering services (SaaS, PaaS, and IaaS) to clients such as IT companies, organizations, universities, generic single end-user which can range from desktop to mobile users, and so on. For shortness, in the rest of the chapter, we will also refer to these services with the term *aaS.

Even though such an ecosystem includes hundreds of independent, heterogeneous clouds, many business operators have predicted that the process toward interoperable federated Intracloud/Intercloud scenarios will begin in the near future. We imagine a scenario where different clouds, belonging to different administrative domains, could interact with each other, becoming themselves both "users" and resource providers at the same time. Obviously the interaction and cooperation among the entities of this scenario might be complex and needs to be deeply investigated: this is why the term "federation" is also in the IT world and cloud computing.

As it has been claimed in [130], the evolution of the cloud computing market is hypothesized to evolve according to the following three subsequent stages:

- Stage-1 "Independent Clouds" (now) — cloud services are based on proprietary architectures, islands of cloud services delivered by mega-providers (this is what Amazon, Google, Salesforce and Microsoft look like today);

- Stage-2 "Clouds using *aaS of other clouds" — over time, some cloud providers will leverage cloud services from other providers. The clouds will be proprietary islands yet, but the ecosystem will start;

- Stage-3 "Intracloud/Intercloud" — smaller, medium, and large providers will federate themselves to gain economies of scale, an efficient use of their assets, and an enlargement of their capabilities.

The current gap between the infrastructural development of stage-1 and stage-2-3 clouds providers, is reflected by the cloud infrastructure developments in US and Europe. While US cloud landscape is dominated by the mega scale providers like Amazon, Azure and Google, in Europe the cloud providers are generally Telcos developing their private clouds and Enterprises which offer their infrastructures to specific groups of users with specific needs.

Even though the idea of creating federated infrastructures seems to be very profitable, bridging such a gap could not be straightforward: on one hand, in fact, highly scalable infrastructures are required to comply with the varying load, software and hardware failures using cloud federation scenarios. On the other hand, autonomic managed infrastructures are required to adapt, manage and utilize cloud ecosystems in a efficient way. Furthermore for building up an interoperable heterogeneous federated Intercloud environment, clouds have to cooperate together accomplishing trust contexts for providing new business opportunities such as cost-effective assets optimization, power saving, on-demand resources provisioning, delivery of new types of *aaS, etc.

9.2.1   The Existing Cloud Models and Solutions

In the following, we analyze the current state-of-the-art in Cloud Computing, paying attention to the existing middleware implementations, evaluating their main features, and focusing on their ability to create interconnections and federations.

Nowadays, together with the monolithic independent mega providers which employ proprietary and closed implementation to build their clouds, some open source solutions are arising. Such environments mainly address the requirements of organizations to build their own IaaS clouds using their internal computing resources [664]. This is the concept on which the private clouds are based, i.e., infrastructures provided by an organization offering a dedicated operating environment with a high trust level. In such a scenario, the computing infrastructure is owned by a single customer that controls the applications being executed. The main aim of these clouds is not to provide and sell computing capacity over the Internet through publicly accessible interfaces, but to give local users a flexible and agile private infrastructure to run services within their administrative domains.

Together with the concept of private clouds, the possibility of supporting a hybrid cloud model by adding to the local infrastructure more computing capacity coming from external public clouds, is currently emerging. Furthermore, a private/hybrid cloud could also expose remote access to its resources over the Internet using remote interfaces, such as the web service interfaces that, for instance, Amazon Elastic Compute Cloud (Amazon EC2) [4] uses. This trend could be associated as a first attempt of creating a "loosely coupled" federated cloud infrastructure, according to the aforementioned stage-2 definition.

In order to better understand the state of the evolution (related to the federation capabilities) of the existing middlewares implementing the private/hybrid cloud model, it could be useful to explain how such middlewares are logically organized and which features they implement. The task can be addressed keeping in mind a stack similar to the one reported in Fig. 9.1. As the figure shows, three different levels exist: the lowest one named *Virtual Machine Manager*, the second one named *Virtual Infrastructure Manager* and finally the highest one named *High-Level Cloud Manager*).

- The Virtual Machine Manager (VMM) can be generally associated to the Hypervisor running on top of the Operating System;

- The Virtual Infrastructure Manager (VIM) (at the lowest layer in the picture) which acts as a dynamic orchestrator of Virtual Environments (VEs);

- The Cloud Manager (CM) (at the highest level in the picture) which accomplishes the security, contextualization and federation issues.

The Virtual Machine Manager provides an abstraction layer to the VIM. It interacts with the Hypervisor running on top of the Operating System (OS) of each server composing the cloud's datacenter, and enables the capability of deploying Virtual Machines (VMs) on the physical hardware where the VMM is running. Most of the cloud middlewares mainly exploit this layer, adding some new features.

On the one hand, the main purpose of the middlewares implementing the VIM, essentially refers to the ability of setup VMs (preparing disk images, setting up networking, and so on) regardless of the underlying VMM (Xen, KVM, VMware, Virtual Box, Virtual PC, etc.). Most of these solutions concentrate their features on the previous tasks and do not provide the ability of communicating with external clouds.

On the other hand, projects which can be considered as CM (they mainly deal with the highest layer of the stack of Fig. 9.1) are able to transform existing infrastructure into an IaaS cloud with cloud-like interfaces. However, although these tools are fully functional with respect to providing cloud-like interfaces and higher-level functionality for security, contextualization and external cloud interaction, their VI management capabilities are limited and lack features for the support to specialized VIM solutions.

Such an analysis demonstrates that a middleware able to address all the main issues regarding cloud computing, such as low-level VIM management and high-level features implementation (i.e., a middleware which spans the two levels of the previous stack) does not exist right now. This leads to the need of finding a trade-off when choosing an open-source solution, depending on the specific requirements that should be satisfied. The cloud ecosystem offers middleware belonging to the first category such as OpenQRM [5], OpenNebula [667] and CLEVER [717]; the other category comprises middlewares specialized in providing high-level features (external interfaces, security and contextualization) such as Globus Nimbus [590], Eucalyptus [544] and RESERVOIR [6].

Regarding the current trend about the implementation of "federation capabilities" within such middlewares, it is strictly related to the logical level on which they belong: considering the VIM middlewares, only the more recent versions of OpenNebula are trying to support a hybrid cloud to combine local infrastructure with public cloud-based infrastructure, enabling highly scalable hosting environments. OpenNebula provides Cloud interfaces to expose its features for virtual machine, storage, and network management. Regarding the middlewares dealing with Cloud Management, Nimbus provides interfaces to support VM management functions. These interfaces are based on the Web Services Resource Framework (WSRF) [7] set of protocols; there is also an alternative implementation based on Amazon EC2 Web Service Description Language (WSDL). Nimbus is designed also to support additional client-side interfaces. The workspace service of the middleware uses Grid Service Infrastructure (GSI) to authenticate and authorize VM creation requests. Among others, it allows a client to be authorized based on Virtual Organization role information contained in the VOMS credentials and attributes obtained via GridShib [8].

CLEVER is a cloud-enabled virtual environment which specifically aims at the design of a VI management layer for the administration of private cloud infrastructures. Differently from the other middleware existing in the literature, CLEVER also provides simple and easily accessible interfaces for enabling the interaction of different "interconnected" computing infrastructures and thus the ability of deploying VMs on these heterogeneous clouds. The concept of interface is also exploited for integrating security, contextualization, VM disk image management and federation features made available from higher level software components.

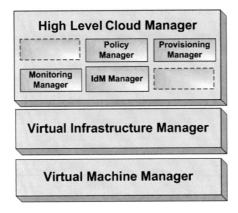

FIGURE 9.1
The Stack: The Logical Organization of Private/Hybrid Cloud Reference Architectures.

9.2.2 The New Perspective of Intercloud

Nowadays lots of clouds are independent and isolated platforms able to provide various types of services belonging to three main categories: IaaS, PaaS, and SaaS. The users of such services can be multiple, including IT companies, organizations, universities, and generic single end-users which can range from desktop to mobile users. In these clouds there is not any concept of federation and each one uses only its own datacenter. At the same time, some cloud platforms such as Amazon are beginning the transition to Stage-2, supplying several services to other clouds, according to *a priori* business and economic agreements. Instead, as far as it is concerned the transition to Stage-3 is related to the achievement of the "Intracloud" and "Intercloud" scenarios, things are not so easy.

Doing a parallelism with the Internet, which is considered a network of networks, the Intercloud can not be simply considered cloud of clouds. As the networks composing the Internet are governed by the policies which define access control rules, also the clouds composing the Intercloud need to be regulated by policies defining their relationships. But the definition of such relationships is not simple because the Intercloud can theoretically consider infinite possible scenarios depending on the business model which the involved clouds want to accomplish. Behind these considerations takes place the concept of federation, which is an indispensable requirement for the establishment of relationships between clouds. The Intercloud vision can be achieved with strong federation technologies providing gateways between different clouds and their internal datacenters. A "Cloud Federation" can be defined as a relationship between two or more independent homogeneous/heterogeneous cloud platforms which establish a trusted federation agreement in order to benefit a particular form of business advantage. This latter definition is quite generic and does not specify which type of relationship may be established between two clouds. In fact, there are many types of possible cloud federation models. As already introduced, a few cloud platforms have stage-2 federation utilities for the use of Amazon EC2 web services. Regarding federation mechanisms of stage-3 in an Intercloud perspective, nowadays, one of the few cloud communities which have faced the federation topics is the one of the European Project RESER-VOIR [6]. Nevertheless, despite RESERVOIR support federation mechanisms among different datacenters (i.e., sites in the RESERVOIR terminology) running the same RESERVOIR platform, no utilities for Intercloud federation with other cloud solutions have been planned yet. In order to clarify the ideas, we basically distinguish between two major federation categories:

- **Low Cooperation Federation**. Clouds establish agreements in order to use the *aaS offered by other clouds (the basis of the stage-2).

- **High Cooperation Federation**. Clouds establish agreements in order to use the storage and computing capabilities of the virtualization infrastructures of other clouds along with the capabilities of their own virtualization infrastructure. In this way clouds gain an elastic enlargement of their virtualization power for the arrangement of their own *aaS, theoretically without resource limitations.

The low cooperation federation involves a relationship among clouds for the use of a *aaS. This type of federation takes place at high-level involving the interaction between the Cloud Manager layers of the considered clouds. Instead, high cooperation federation involves the establishment of closer relationships among two or more loosely coupled cloud stacks. More specifically, such a type of federation takes place both at high-level involving the interaction between the Cloud Manager layers of the considered clouds, and at low-level also involving their own Virtual Infrastructure Manager layers according to specific federation agreements. Nowadays federation for the use of cloud services is beginning to take place in several cloud providers. For example, let us think of Eucalyptus and Open Nebula, which trust the Amazon cloud supporting the Amazon EC2 web services to launch instances of virtual environments with a variety of operating systems.

This attests that the evolution of cloud computing is passing at the stage-2. Instead, the high cooperation federation may be the key element for the near future transition to the stage-3 of the cloud computing evolutionary line. In fact, this could be the basis for new variants of the current cloud computing paradigm represented by futuristic Intracloud and Intercloud scenarios. Figure 9.2 summarizes the evolutionary line of cloud computing.

According to what already mentioned, and doing another parallelism with the classic concepts of Intranet and Internet, it is possible to introduce the concepts of "Intracloud" and "Intercloud." The Intracloud can be defined as an interconnection of federated clouds distributed over a circumscribed area and built over either a Local Area Network (LAN) or a Metropolitan Area

	Stage	Description	Federation Type	Current Evolution State
	1	Independent clouds	None	Completed
	2	Clouds using the *aaS of other clouds	Low Cooperation Federation	Partially completed
	3	Intracloud/Intercloud	High Cooperation Federation	To be planned

(Left vertical label: Cloud Computing Evolutionary Line)

FIGURE 9.2
Possible Evolutionary Line of Cloud Computing.

Network (MAN). Instead, the Intercloud can be defined as an interconnection of federated clouds distributed around the world which communicate through a Wide Area Network (WAN), i.e., Internet.

The concept of Intercloud implies several more evolved and complex scenarios than the current ones where clouds are independent and isolated from each other, or where they merely use the services provided by other clouds. In fact, the Intercloud scenario involves more than the simple provisioning of services from a cloud to another, and therefore it needs to rely on closer relationships between clouds for the establishment of federated environments. But in order to establish such "closer relationships," achieving the high cooperation federation, several issues concerning both the compatibility among heterogeneous cloud platforms and the management of federated cloud platforms need to be addressed. More specifically, it is required to identify how to establish and manage a high cooperation federation, maintaining at the same time each cloud independent from each other.

An indispensable requirement is that the federation agreements do not have to affect all the assets of the clouds having relationships with other clouds. In fact, the federation has to respect the local management and policies, but at the same time it has to enable the involved clouds to control and use part or the whole range of the resources lent by other clouds. Typically these resources are storage and computational capabilities which clouds use for the allocation of *aaS. More specifically, using such resources the new emerging clouds generally arrange the virtualization infrastructures used to allocate virtual machines and other virtual assets for the composition of *aaS(s) which are orchestrated by cloud providers and delivered to clients. The access to the federated resources is regulated by *a priori* federation agreements which define how each federated cloud can "borrow" the resources of other clouds and how each federated cloud can "lend" the resources to other federated clouds. For example, a possible pay-per-use business scenario could include clouds renting their storage and computing resources to other federated clouds which pay them for the utilization of such resources, according to time-based utilization charges. These forms of resource borrowing and lending are tied to the fact that clouds meet or do not meet the requirements of other clouds in terms of both management and compatibility policies.

9.3 Intercloud Resource Sharing Models

The Intercloud allows to enlarge the amount of resources of each involved cloud providers, allowing clouds to hold a theoretically inexhaustible tank of resources and count on an elastic virtualization infrastructure without barriers for the arrangement and delivery of several *aaS. However, it is still unclear how to establish a low-level sharing of resources between cloud platforms in order to set up a federation driven by clear rules, policies, and agreements. For the sake of simplicity, from now on with the term "resource" we will indicate a generic computing or storage capability (e.g., the number of free CPUs and the amount of free disk space in a server running a hypervisor) provided by the data center of each cloud. Typically these resources are used by the virtualization infrastructure (i.e., a set of servers running hypervisors) of each cloud for the allocation of virtual environments including: virtual machines running guest operating systems, virtual network devices or other types of virtualized assets which are orchestrated and composed for the arrangement of *aaS(s). The entity which is responsible to both manage the virtualization infrastructure and to arrange, maintain, and deallocate *aaS(s) is generally a cloud software solution which acts as orchestrator of the whole cloud platform, also satisfying the user requests and guaranteeing the respect of the policies governing the relationships within the cloud.

Basically we can identify two main resource sharing scenarios among federated clouds belonging to an Intracloud/Intercloud. In the first one, a cloud does not share anything but uses the sets of resources shared by other federated clouds. In the second one, two or more clouds share their own sets of resources, and each one has concurrent access to the set of shared resources of the other clouds. In order to clarify the ideas let us consider two simplified scenarios, each one including two clouds: cloud A and B. We define $R_A = \{ra_1, ra_2, \cdots, ra_n\}$ the set of resources held by cloud A so that $\forall ra_i, ra_j; i, j, n \in N; i \leq n; j \leq n; ra_i \neq ra_j$. We define $R_B = \{rb_1, rb_2, \cdots, rb_n\}$ the set of resources held by cloud B such that $\forall rb_i, rb_j; i, j \in N; i \leq n; j \leq n; rb_i \neq rb_j$. The conditions $ra_i \neq ra_j$ and $rb_i \neq rb_j$ mean that each resource is unique. This can be simply clarified thinking about virtual machines: when these latter have to be instantiated, it is needed to reserve part of CPU, memory, and disk space capabilities of the physical hypervisor server where such a virtual machine is hosted. These capabilities have to be exclusively assigned to a virtual machine in order to not affect the performance of other virtual machines. For better explaining the two resource sharing scenarios, in Figures 9.3 and 9.4 we are going to consider Venn diagrams.

Scenario 1 (Figure 9.3). Cloud A does not share resources but at the same time it uses both its own resources and the resources shared by cloud B. So we define with $SR_B \subseteq R_B$ the subset of resources shared by cloud B. The

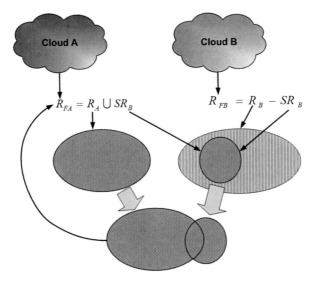

FIGURE 9.3
Example of Resource Sharing Scenario, Where a Cloud Borrows the Resources Shared by Another Federated Cloud with Exclusive Access.

inclusion relation between SR_B and R_B specifies that cloud B can share with other federated clouds either part of its own resources or all its resources (in case we would have $SR_B = S_R$). Therefore, the total available resources for cloud A after the federation with cloud B can be indicated by the union of the set R_A of resources (held by cloud A and placed in its own virtualization infrastructure) with SR_B, representing the subset of the shared resources of cloud B placed in the cloud B's virtualization infrastructure. So the resources of the federated cloud A can be indicated by the set $R_{FA} = R_A \cup SR_B$, where $R_A \cap SR_B = \emptyset$, instead the resources of the federated cloud B can be indicated by the set $R_{FB} = R_B - SR_B$ also called with the term completion of R_B relative to SR_B.

Scenario 2 (Figure 9.4). Clouds A and B share part of their own resources with each other, so we define $SR_A \subseteq R_A$ the subset of shared resources by cloud A and $SR_B \subseteq R_B$ the subset of shared resources by cloud B. Also in this scenario both the inclusion relations $SR_A \subseteq R_A$ and $SR_B \subseteq R_B$ indicate that the two involved clouds can share either part of their own resources or all their resources. Considering the federation between clouds A and B, we indicate $FR_{AB} = SR_A \cup SR_B$, where $SR_A \cap SR_B = \emptyset$, the set of the federated resources available for both clouds A and B. Therefore, due to the federation, Cloud A has only exclusive access to the set of its resources given by $R_A - SR_A$ (completion of R_A relatively to SR_A), and cloud B has only exclusive access to its set of resources given by $R_B - SR_B$ (completion of R_B relatively to SR_B), and at the same time they also have concurrent access

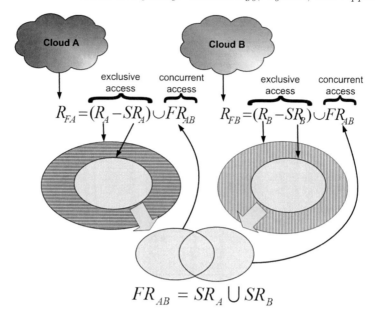

FIGURE 9.4
Example of Resource Sharing Scenario, Where Two Clouds Share a Subset of
Their Resources with Concurrent Access.

to FR_{AB}. So, after the federation, the resources available for cloud A are
indicated by the set $R_{FA} = (R_A - SR_A) \cup FR_{AB}$, whereas the resources
available for cloud B are indicated by the the the set $R_{FB} = (R_B - SR_B) \cup FR_{AB}$,
where $FR_{AB} \in R_{FA}, FR_{AB} \in R_{FB}$. It is important to notice that clouds A
and B have to consider the shared resources indicated by the set FR_{AB} as
an elastic tank whose capabilities vary according to the rate of utilization of
each involved cloud. For example, if at a given time t cloud A uses the 70% of
the shared resources for the temporary enlargement of its own capabilities, at
time t_1 cloud B will have the availability of the 30% of the shared resources
for the enlargement of its own capabilities. However, this does not prevent, for
example, that at time t_2 cloud B might use the 90% of the shared resources.

9.4   Advantages and New Business Opportunities

The Intercloud undoubtedly involves various new business scenarios for both
federated clouds and their users. The business scenario of stage-1 follows a
pay-per-use model: the cloud's users (e.g., IT companies, organizations, universities, generic single end-users ranging from desktop to mobile users) pay

the cloud for the use of an *aaS. The most obvious example of a commercial cloud adopting such a model is represented by Amazon EC2. In fact, Amazon adopts various types of pay-per-use forms including pay per hour, pay per data transferred, pay per GB-month of provisioned storage, pay per 10.000 GET requests, etc. Such forms of payment are the same which begin to be also used in stage-2 scenarios. Here, the only difference with respect to stage-1 is represented by the nature of the cloud's clients which could be also other cloud providers. Besides adopting "cloud's client"/"cloud provider" pay-per-use relationships for the utilization of *aaS(s), stage-3 also includes other forms of pay-per-use relationships between the federated cloud providers forming the Intracloud/Intercloud environment for the "renting" and "borrowing" of storage and computing resources. In order to better explain such a futuristic scenario, we distinguish among cloud's client, home cloud, and foreign cloud:

- **Cloud's client**. An IT company, organization, university, generic single end-user ranging from desktop to mobile users or a cloud provider using the *aaS supplied by a target "home cloud" according to a pay-per-use model.

- **Home cloud**. A cloud provider which receives *aaS instantiation requests from its clients. Each home cloud for the arrangement, composition, and delivery of such services can use the computing and storage resources of its own virtualization infrastructure along with the resources borrowed from foreign clouds according to a pay-per-use model.

- **Foreign cloud**. A cloud provider which lends its storage and computing resources to home clouds according to a pay-per-use model. More specifically, a foreign cloud reserves part of its own virtualization infrastructure for a home cloud, so that the home cloud can logically count on an elastic virtualization infrastructure whose capabilities are greater than the capabilities of its own physical virtualization infrastructure. Therefore, even though the virtual environments and services of a home cloud are logically placed in its virtualization infrastructure, in reality they can be physically placed in parts of the virtualization infrastructure lent by foreign clouds.

Therefore, the Intercloud scenario triggers two pay-per-use relationships: the first one takes place between the cloud's client and the home cloud, whereas the second one takes places between the home cloud and the foreign cloud. In addition, the whole Intercloud scenario becomes more complicated if we consider thousands of interconnected cloud providers which could be at the same time both home cloud and/or foreign cloud. The Intercloud embodies more than the traditional pay-per-use model. Nowadays most of data centers do not use 100% of their resources. Hence, the Intercloud on one hand allows to elastically increase the virtualization capabilities of clouds and, on the other hand to enable clouds to rent their computational and storage capabilities when their virtualization infrastructures are partially or totally unused. Let us think of two clouds placed in different time zones, e.g., the cloud placed in a time zone where it is morning might use the shared resources of another federated cloud placed in another time zone where it is night. There is no limit to the possible business scenarios which can take place in an Intracloud/Intercloud environment. Such scenarios include virtualization capability enlargement, resource optimization, provisioning of distributed *aaS, power saving and so on. In order to clarify these ideas, in the following we present an overview of some of the aforementioned scenarios.

9.4.1   Capability Enlargement and Resource Optimization

Commonly, commercial clouds host virtual environments in their own virtualization infrastructure in order to arrange *aaS for their clients. Unfortunately, clouds do not have infinite resources. There are three main situations which can take place when a cloud temporarily runs out of storage and computing capabilities:

- The cloud is not able to satisfy any further *aaS instantiation requests.

- The cloud is not able to satisfy the Service Level Agreements (SLA) *a priori* established with the client concerning the utilization terms of the *aaS.

- The cloud is not able to satisfy a SLA modification request sent by its client who needs to increase the capabilities and features of its *aaS.

All three situations imply that, for a period of time, the cloud is not able to satisfy the requests of its clients as long as some of its resources will not be released. This can have a serious negative economic impact, especially for small and medium clouds. Such a problem can be solved using the Intercloud, as each cloud operator is able to transparently enlarge and optimize its own resource capabilities, increasing the number of instantiable virtual environments, also balancing the workload so that clouds can never deny services or a request from their clients. In order to better explain this idea, we consider the scenario depicted in Figure 9.5 where the home cloud is already at stage-2 because it is able to provide services to other clouds (top part of the Figure). Moreover, the home cloud is also at stage-3, in fact, when it realizes that its virtualization infrastructure has run out of capabilities. In order to continue providing services to its clients, it decides to federate itself with foreign clouds A and B. The home cloud, besides hosting virtualization environments inside its own virtualization infrastructure, is also able to host virtual machines inside the foreign clouds A and B virtualization infrastructures, enlarging the amount of its available virtualization resources (See Figure 9.5, bottom part). Therefore, although the virtualization resources rented to the home cloud are physically placed within the virtualization infrastructures of foreign clouds A and B, they are logically considered to be resources of the home cloud virtualization infrastructure.

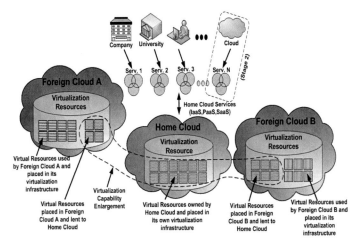

FIGURE 9.5
Example of Virtualization Capability Enlargement and Resource Optimization Scenario.

9.4.2 Provisioning of Distributed *aaS

Nowadays, according to the size of a virtualization infrastructure, we can distinguish among small, medium, and large clouds. Small and medium clouds are held by small and medium size companies; large Clouds are held by large size companies (e.g., Amazon Europe or RESERVOIR). These latter are able to provide new sorts of cloud services we named "Distributed IaaS, PaaS, and SaaS" (generalizing "D*aaS"). With D*aaS we indicate *"a distributed cloud service composed of a set of VMs spread over a wide geographical area, orchestrated in order to achieve a target purpose, and provided on-demand to a cloud client to meet his business needs."* In this perspective the Intercloud can give also to both small and medium clouds the opportunity to offer D*aaS to their clients. By means of the Intercloud, small and medium clouds, acting as home clouds, can enlarge their virtualization capabilities using the virtualization infrastructures of other foreign clouds (small, medium, or large) for a given business purpose, becoming as competitive as large clouds. In this scenario the reason driving a home cloud to be part of the Intercloud is that it needs to have several resources placed in given geographical locations, in order to arrange and provide D*aaS composed of several VMs, hosted in different foreign clouds placed in particular geographical areas. Figure 9.6 depicts a scenario where the home cloud provides D*aaS to its clients. More specifically the D*aaS is arranged by the home cloud aggregating and orchestrating the virtual machines lent by different foreign clouds placed in different European cities, for example Rome, Madrid, and London. An example of DIaaS can be a Content Delivery Network (CDN) including mirrors of a main web server in different cities, a DPaaS aggregating the data sent by different sensor net-

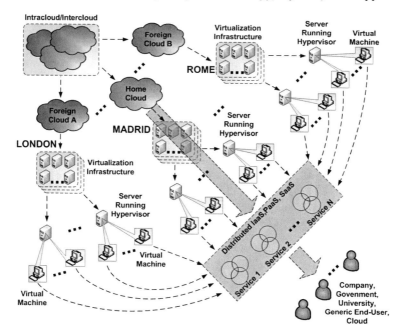

FIGURE 9.6
Example of Distributed *aaS Provisioning Scenario.

works placed in particular geographical areas offering API for development of applications for the territory monitoring, DSaaS accessing to data placed in different countries that for the local laws can not pass their boundaries, and so on.

9.4.3   Power Saving

Even though cloud computing has been thought to be for the optimization of the resources of datacenters, considering an independent cloud (i.e., the stage-1 of the evolutionary line of the cloud ecosystem) there could be situations where a part of the resources of a cloud remains unused although there is high energy consumption. In such a scenario, virtual machine migration along with green computing strategies allow clouds to reduce the energy consumption of their own datacenters. Moreover, considering Intracloud/Intercloud scenarios the use of extra resources, although small in quantity, might both maximize the exploitation of the cloud's virtualization infrastructure and reduce the energy consumption. This is possible by means of several appropriate virtual machine migrations, from one server to another placed inside the virtualization infrastructure of the home cloud itself, and/or to another server placed inside the virtualization infrastructure of one or more foreign clouds.

Figure 9.7 depicts a typical scenario where an independent cloud arranges its virtual machines in order to reduce the energy consumption of its virtualization infrastructure. The independent cloud has a virtualization infrastructure

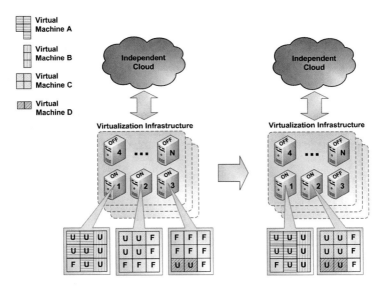

FIGURE 9.7
Example of Power Saving Scenario of Stage-1.

including several servers, each one running an hypervisor. More specifically, servers 1, 2, and 3 are turned on, whereas the other ones are turned off. For simplicity we assume each server has the same computing and storage capabilities and that each one has nine units of virtualization resources. With this latter generic term we can indicate for example either a target utilization rate of a CPU, or a target amount of disk space of the server. In addition we assume that one or more virtualization resource units are needed for the instantiation of a virtual machine. After several virtual machine allocations, resizing, movements, and deallocations due to the compositions and decompositions of *aaS(s), the virtualization infrastructure is arranged as depicted in Figure 9.7: server 1 hosts virtual machine A and B and has only one free virtualization resource unit, server 2 hosts virtual machine C and has five free virtualization resource units, server 3 hosts virtual machine D and has seven free virtualization resource units. Virtual machines A, B, C, and D include respectively five, three, four, and two virtualization resource units. Server 1, 2, and 3 have respectively the 11.11%, 55.55%, and 77.77% of unused resources. As server 3 is turned on, even though 77.77% of its resources are unused, with a green computing strategy it is possible to reduce the energy consumption of the datacenter by migrating the virtual machine D from server 3 to server 2, so that it is possible to turn off server 3.

Such a scenario can be further improved by considering an Intercloud scenario. Let us suppose that the independent cloud of Figure 9.7 decides to federate itself with a foreign cloud, so that in Figure 9.8 we refer to such a cloud as the home cloud. In addition, let us assume that the foreign cloud lends

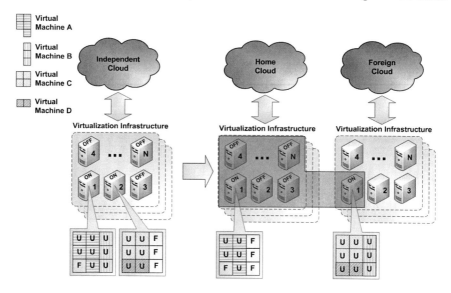

FIGURE 9.8
Example of Power Saving Scenario of Stage-3 (Intracloud/Intercloud).

resources to other clouds according to a given form of pay-per-use agreement. In addition, let us suppose that the costs to maintain turned-on servers 1, 2, and 3 of home cloud's virtualization infrastructure are different. As the cost due to keep turned-on server 2 is greater than the cost to host the C and D virtual machines in a foreign cloud, the home cloud decides to migrate the virtual machines C and D. Therefore, after the migrations, the home cloud's server 2 can be turned off, and the virtualization infrastructure of the home cloud logically includes server 1 of its own virtualization infrastructure and server 1 of the foreign cloud's virtualization infrastructure.

9.5 "High Cooperation Federation" Establishment

In the previous sections we debated some concepts related to both cloud federation and Intercloud, analyzing the evolutionary line of the cloud computing paradigm, also describing the involved advantages and concerns. We analyzed the state-of-the-art of the middlewares for cloud computing, evaluating their evolution status in the federation stages. In the following, we will try to understand which requirements a middleware for cloud computing should satisfy for enabling the ability of establishing stage-3 high cooperation federations.

We defined the Intercloud (stage-3) as an ephemeral interconnection of (homogeneous/heterogeneous) clouds, each laying out its own set of resources, managed using specific middleware, and probably accessible through different security mechanisms. Due to the high dynamism of the Intercloud, a flexible method for building dynamic high cooperation federation relationships, also enabling the coexistence of several different platforms and technologies, should be provided. A possible solution for building the Intercloud should answer all such issues, considering the requirements of interoperability, automatism and scalability.

In the Intercloud scenario, in fact, it may happen that new clouds, offering available resources (accessible through specific interfaces) with specific authentication mechanisms could appear, while others disappear. Taking into account such a dynamism, when a home cloud needs to "lease" external resources from a foreign cloud, the first phase it will perform refers to the *discovery* of the foreign cloud which properly *matches* (phase 2) its requirements (both in terms of available resources, technologies, policies, supported authentication mechanisms, etc.). Once these two phases have been performed, and the best foreign clouds have been identified, in order to establish a secure interaction between the home cloud and the selected foreign cloud, an *authentication* process (phase 3) will begin.

The accomplishment of the authentication process leads to the establishment of a secure and direct connection between the home cloud and the foreign cloud whereby the resource provisioning can be accomplished. As consequence, the home cloud will be able to instantiate (or migrate) VMs on the foreign cloud in a secure environment. The concept of migration can be seen as the opportunity to move the VMs not only in an intra-site domain but also to transfer them on federated inter-site domains. In this case the migration might occur across subnets, among hosts that do not share storage or even across administrative boundaries.

We can thus summarize the federation process as the accomplishment of a sequence of the above mentioned three different phases: *discovery*, *matchmaking* and *authentication* [169].

9.5.1 The Discovery Phase

During this first phase, all the available clouds within the dynamic environment have to be discovered. As this environment can not be known *a priori*, but it is pretty flexible and dynamic, the discovery process should be implemented in a totally distributed fashion: the middlewares of all the clouds which wish to attend the federation, must communicate exploiting a peer-to-peer (p2p) approach, based conveniently on the *presence* concept. This latter is indeed an enabling technology for p2p interactions, whose implementation follows the software design pattern known as publish-and-subscribe (pub-sub): an application publishes its own set of information to a centralized location (even though such a location is logically centralized, it is implemented in a distributed fashion, also granting fault-tolerance mechanisms), from which only a set of authorized (subscribed) entities are able to retrieve it.

The above mentioned approach, based on the "presence" concept, could represent a valid starting point to address the foreign cloud discovery problem in an federated Intercloud scenario. Each cloud middleware, according to the pubsub software pattern, broadcasts on-demand information about its supported features and resources state to other peers aiming to attend the Intercloud. When a home cloud needs to know the set of foreign clouds available to start the federation in a given moment, along with the related capabilities, its middleware will merely check the shared area on which this information is stored. To accomplish the previous mechanisms, a requirement the cloud middleware should have, consists of the integration of a specific component acting as a *presence daemon*. It may exploit one or more presence protocols, in order to distribute information about its features, state and availability as widely as possible. These protocols will be employed integrating features which regard service discovery and management of the capabilities information, for allowing clouds to gain knowledge about the resources made available from the other entities of the federation, providing real-time mechanisms for additional use of presence-enabled systems.

9.5.2 The Match-Making Phase

Once the discovery phase finishes, the more convenient foreign cloud(s) wherewith to establish a high cooperation federation has to be chosen. In order to accomplish this task, another specific component of the cloud middleware should be responsible to manage and enforce the policies defined by the clouds. In fact both the home cloud and the available (discovered) foreign clouds may be associated to a set of data containing several policies, each composed of a set of rules describing respectively which resources are required and which ones are offered according to some conditions. For example a home cloud could require a certain amount of CPU and storage, with a certain QoS, supporting particular security policies (as will be described later), from the 0.00 am to the 8.00 am. Instead, a foreign cloud could offer resources with a certain QoS, with a target security policy, at any time, denying the requests sent by certain home clouds. In order to enable the middleware to evaluate among all the available (discovered) clouds the ones that best "fit" the requirements of its home cloud, a policy matching should be performed.

Thus, each cloud middleware has to evaluate the applicable policies, returning authorization decisions, performing an access control by enforcing the stated authorization decisions, and converting the foreign cloud policies from the native format to its own supported policies format. More specifically, the cloud policies should be expressed by means of an extensible policy language able to integrate different policy formats using transformation algorithms.

9.5.3   The Authentication Phase

During the last phase, once the cloud to federate with has been selected, a mechanism for creating a security context among home and foreign clouds should be accomplished. When the authentication phase begins, the home cloud middleware will contact its "peer" on the foreign cloud: the authentication process will be lead exchanging authentication information in form of meta-data, involving trusted third parties in the process. In a distributed scenario composed of hundreds of clouds, the credential managements could be very hard: each home cloud should manage hundreds of credentials which can change over the time, each one needed for the authentication with a certain foreign cloud. In addition, each cloud could support different authentication mechanisms. These concerns raise a "cloud Single Sign On (SSO) authentication problem": a cloud should be able to authenticate itself with other homogeneous/heterogeneous clouds regardless of their security mechanisms, performing the log-in once, using unique credentials over the time, gaining the access to all the required, spread resources.

In order to accomplish these mechanisms, cloud middlewares could use the Identity Provider/Service Provider (IdP/SP) model. Such a model defines the exchange of authentication assertions between security domains, more specifically, between an IdP or asserting party (a producer of assertions) and a SP or relying party (a consumer of assertions). The model assumes that a subject, a person or a software/hardware entity (i.e., in our case the home cloud middleware) holds at least one digital identity on an IdP which supplies SSO authentication services. SSO is the property that allows an entity to perform its authentication only once to access the resources of different SPs (i.e., in our case one or more foreign clouds).

9.6   Technologies for Achieving the Intercloud: An Overview

In this section we present an overview of technological solutions which help to accomplish the three phases previously discussed.

9.6.1   Extensible Messaging and Presence Protocol (XMPP)

During the Discovery phase, it is necessary to identify which clouds are available to start the federation and to know the services they are able to offer: since the Intercloud is a dynamic scenario whose entities are not *a priori* established, some mechanisms for dynamically identifying the clouds and their status should be employed. We believe that the best way to accomplish this task is based on p2p mechanisms or communication protocols adopting the presence concept. A possible solution to the problem of finding the clouds during the discovery phase is XMPP. XMPP is a communication protocol based on the Extensible Markup Language (XML) whose original name was Jabber. It was developed by the Jabber open-source community in 1999 using open-standards for enabling extensible Instant Messaging (IM), presence information, and contact list maintenance for message-oriented middlewares. Since the protocol has been designed to be extensible, today it finds application also in VOIP, file transfer, signaling and so on.

The Internet Engineering Task Force (IETF) has formalized XMPP as an approved instant messaging and presence technology whose specifications have been published as [9] and [10]. No royalties are required to implement support of these specifications and their development is not tied to a single vendor. Unlike most instant messaging protocols, XMPP in fact uses an open systems approach of development and application, by which anyone may implement an XMPP service and interoperate with other organizations' implementations. Furthermore, the software implementation and many client applications are distributed as free and open source software.

The architecture of the XMPP network is quite similar to the approach used by the e-mail service: a strong decentralization is achieved because anyone can run his own XMPP server without the existence of a central master system. Moreover, XMPP servers may be isolated from the public XMPP network (e.g., on a company intranet), and robust security (via SASL and TLS) has been built into the core XMPP specifications. To encourage the use of channel encryption, the XMPP Standards Foundation currently runs an intermediate certification authority at StartSSL [11] offering free digital certificates to XMPP server administrators.

Last but not least, custom features can be built on top of XMPP; to maintain interoperability, common extensions are managed by the XMPP Software Foundation. XMPP applications beyond instant messaging include network management, content syndication, collaboration tools, file sharing, gaming, remote systems monitoring and even applications in cloud scenarios.

The XMPP network uses a client-server architecture because clients do not talk directly each other. However, this architecture, at the same time, can be considered as decentralized since there is no central authoritative server, unlike similar services (such as AOL Instant Messenger or Windows Live Messenger). Every user on the network is associated to a *Jabber ID* (JID). The employment of a central server which maintains a list of IDs, is avoided, structuring the JIDs like an e-mail address with a username and a domain name. The former identifies a single user, the latter refers to the server where that user resides; the two fields are separated by an at sign (@), such as username@exampledomain.com. Since users may wish to log in from multiple locations, they may specify a resource. A resource identifies a particular client belonging to the user (for example home, work, or mobile). This may be included in the JID by appending a slash followed by the name of the resource. For example, the full JID of a user's mobile account would be username@exampledomain.com/laptop. Each resource is also associated to a priority, expressed as a numerical value. Messages simply sent to username@exampledomain.com will go to the client with highest priority, but those sent to username@exampledomain.com/laptop will go only to the laptop client.

In order to better understand how messages are exchanged within an XMPP network, we suppose entity1@domain1.com wants to chat with entity2@domain2.net. Entity1 and Entity2 respectively have accounts on the domain1.com and domain2.net servers.

The following is the sequence of events happening when Entity1 sends a message to Entity2:

1. Entity1's client sends a message to the domain1.com server.

2. If domain2.net is blocked on domain1.com, the message will not be forwarded and will be dropped.

3. The domain1.com server opens a connection to the domain2.net server.

4. domain2.net checks if Entity2 is currently connected. If not, the message is stored for later delivery.

5. If Entity2 is online, the domain2.net server delivers the message to Entity2.

Another useful feature of the XMPP system is known as *transports* or *gateways*. It could be considered as a key feature for enabling interoperability and federation since it allows users to access networks using other different protocols. These latter can be other instant messaging, SMS or e-mail protocols. Unlike multi-protocol clients, XMPP provides this access at the server level by communicating via special gateway services running on a remote computer. Any user can "register" with one of these gateways by providing the information needed to log on to that network, and can then communicate with users of that network as though they were XMPP users. This means that any client which fully supports XMPP can be used to access any network for which a gateway exists, without the need for any extra code in the client and without the need for the client to have direct access to the Internet.

9.6.2 eXtensible Access Control Markup Language (XACML)

During the Match-Making phase, it is needed to pick out which of the discovered foreign cloud suits the home cloud. More specifically, the policies of both clouds must coexist without affecting each other. Nowadays, in large-scale distributed systems there is the need to enforce policies in many different *enforcement points*, representing critical points of a distributed system where the access to a target resource has to be controlled. Typically such policies govern which subjects can access to a target resource of a distributed system according to a policy matching task performed in order to check if the requirements of the subject match the requirements of the system. A typical example is a user who wants access to a secure web application using its own access credentials (e.g., username and password, digital certificate, biometrics parameters, etcetera). On the other hand, the web application might require to guarantee the access to its users in target period of time and/or to give a restrictive access to a target subset of users and/or to deny the access to its users according to some rules, etc. The current practice in major distributed systems is to independently manage the configuration of each enforcement point, in order to implement the security policies as accurately as possible. Consequently, this scenario is much too expensive and unreliable for complex systems.

One of the best solutions which is able to address the aforementioned scenarios is the XACML standard developed by OASIS, consisting of a declarative XML access control policy language and a processing model, describing how to interpret policies [1]. XACML permits to express policies by means of four major components.

I. Attributes and Functions. Attributes are characteristics of subjects, resources, actions, or environments that can be used to define a restriction. XACML does not predefine a list of attributes; instead it specifies a list of (normative) datatypes that subjects and/or communities can use to create the attributes which are valid for a given system. String, Time, Boolean, Double, and AnyURI are some examples of valid data-types. Similarly, functions are possible operators which can be used for normative data-types. String-Equal, Greater-Then, Date-Equal, String-Is-In, and AnyURI-One-And-Only are examples of valid functions.

II. Rule. A rule is the basic element of a policy. It identifies a complete and atomic authorization constraint which exists in isolation with respect to the policy in which it has been created. A rule is composed by a Target, to identify the set of requests the rule is intended to restrict, an Effect, which is either "Permit" or "Deny," and a set of Conditions, which represent predicates specifying when the rule applies to a request.

III. Policies. A policy is a combination of one or more rules. A policy contains a Target (specified using the same components as the rule Target), a set of rules, and a rule combination algorithm. The rule combination algorithm specifies the approach to be adopted to compute the decision result of a policy containing rules with conflicting Effects. The XACML normative rule combination algorithms are: Deny Override, Permit Override, First-one-applicable, Only-one-applicable.

IV. Policy Set. Like multiple rules compose a policy, multiple policies compose a policy set which represents the conditions to apply in case the authorization decision has to consider AC requirements of multiple parties. A policy set is defined by a Target, a set of policies (or other policy sets), and a policy combination algorithm which are the same of the rule combination algorithms. An XACML policy can also include Obligations, which represent functions to be executed in conjunction with the enforcement of an authorization decision (e.g., data can be accessed from 5 pm to 8 am, provided that the name of the requester is sent by email). XACML not only provides a language for policy specification; it also provides a method for evaluating a policy based on the values of the policy attributes associated with a request. The process for evaluating and enforcing AC policies is based on three main entities:

- A Policy decision point (PDP) – This is the entity in charge of evaluating applicable policies and returning an authorization decision.

- A Policy enforcement point (PEP) – This is the entity performing AC by enforcing the stated authorizations.

- A Context handler – This is the entity in charge of converting requests from the native format to the XACML canonical form and to convert the authorization decision from XACML to the native format supported by a PEP.

Commonly, an access request is evaluated as follows. The requester sends a request to a PEP which in turn contacts the context handler in charge of converting the attributes of the subject, resource, action, and environment describing the request from the PEP native format to the XACML request context. Such a context is then sent to the PDP, which evaluates the policy rules applicable to the request. A rule is applicable to a request if the information in the request context satisfies both the rule target and the rule condition predicates. If these predicates are not verified, the rule is "Not applicable" and its Effect is ignored. A policy is "Not applicable" if no policy rule applies to the request. The rule combination algorithm is used to resolve conflicts among applicable rules with different Effects, whereas the policy combination algorithm is used to resolve conflicts among applicable policies. The result of the policy evaluation ("Permit," "Deny," or "Not applicable") is returned to the context handler which first converts it in the native format supported by the PEP and then forwards it to the PEP. The PEP evaluates the policy evaluation result and takes the final authorization decision concerning the request.

Due to its nature, the XACML technology can be successfully used to achieve the match-making phase during the Intercloud relationship establishment. In order to understand if a foreign cloud is the right cloud for the home cloud for the establishment of an high cooperation federation, it is necessary that the home cloud's policies match the foreign cloud policies. Examples of clouds' policies are the following:

Home Cloud Policy: *I accept CPU X_1, RAM Y_1, with QoS Z_1 from any foreign cloud trusted with the Identity Provider (IdP) T (IdP will be better discussed in the following.).*

Foreign Cloud Policy: *I provide CPU X_2, RAM Y_2, with QoS Z_2 to all clouds except cloud A and I'm trusted with the IdPs T, H.*

The process of policy matching has to be accomplished for each discovered foreign cloud in order to do a sort of classification, from which the home cloud can select the foreign cloud candidate for the federation.

9.6.3 Security Assertion Markup Language (SAML)

During the Authentication phase it is necessary to establish a trust context between the home cloud and the foreign cloud(s) by means of SSO authentication mechanisms. In order to achieve this goal, one of the major technologies using the IdP/SP model is the Security Assertion Markup Language (SAML) [2]. SAML is an XML-based standard (a product of the OASIS Security Services Technical Committee) for exchanging authentication and authorization assertions between security domains between an IdP and a SP.

The aim of SAML is to enable a principal to perform SSO. This means a principal, by means of its IdP, must be able to authenticate itself once gaining the access to several trusted service providers which might use also different security technologies. Even though many single sign-on solutions exist in web application environments (using cookies, for example), the extension of these solutions has been problematic and has led to the proliferation of non-interoperable proprietary technologies. SAML assumes the principal (often a user) has enrolled (i.e., has an account) with at least one identity provider. This identity provider is expected to supply local authentication services to the principal. The way these local authentication mechanisms are implemented is not specified by SAML, which does not care how the services are implemented (although individual service providers most certainly will). A service provider relies on the identity provider to identify the principal. At the principal's request, the identity provider passes a SAML assertion to the service provider. On the basis of this assertion, the service provider makes an access control decision.

SAML combines four key concepts: assertion, binding, protocol and profile. Assertion consists of a package of information that supplies one or more statements (i.e., authentication, attribute, and authorization decision) made by the IdP. The authentication statement is perhaps the most important element, meaning the IdP has authenticated a subject at a certain time. A Protocol (i.e., Authentication Request, Assertion Query and Request, Artifact Resolution, etc.) defines how subject, service provider, and IdP might obtain assertions. More specifically, it describes how assertions and SAML elements are packaged within SAML request and response elements. A SAML binding (i.e., SAML SOAP, Reverse SOAP (PAOS), HTTP Redirect (GET), HTTP POST Binding, etc.) is a mapping of a SAML protocol message over standard messaging formats and/or communication protocols. For example, the SAML SOAP binding specifies how a SAML message is encapsulated in a SOAP envelope. A profile (i.e., Web Browser SSO, Enhanced Client or Proxy (ECP), Single Logout, Attribute Profiles, etc.) is a technical description of how a particular combination of assertions, protocols, and bindings defines how SAML can be used to address particular scenarios.

The authentication phase can be accomplished by means of the creation of specific SAML profiles, involving both the home cloud and the foreign middleware respectively representing the subject and the relying party, whereas the

IdP acts as the third party asserting to a foreign cloud the trustworthyness of the home cloud identity.

In an Intercloud scenario, a SAML profile should be accomplished by means of software integration within the involved cloud platforms. This allows a home cloud to perform SSO authentication on one or more foreign clouds, both having a trusted relationship with the a target IdP (e.g., the home cloud should have an account on an IdP acting as a provider of authentication services, and the foreign cloud has to trust the assertions of such an IdP).

9.7   Conclusions and Future Research Challenges

In this chapter, we provided an overview of the possible evolution of the cloud ecosystem in the coming years, also debating the impact that the Intercloud may have on clouds from the architectural and business point of view. More specifically, we analyzed several forms of cloud federation, focusing on the different involved factors for the establishment of High Cooperation Federation among heterogeneous cloud platforms. In addition, an overview of the possible new business advantages for federated clouds and IT societies, organizations, universities, desktop and mobile end-users has been discussed. The involved issues for the establishment of the High Cooperation Federation have been presented, also highlighting the technologies which could help cloud developers to design Intercloud federation utilities.

Nevertheless, many other research challenges have to be faced for the accomplishment of the Intercloud vision. In the following we present an overview of some of such research challenges.

I)Automatism and Scalability. Each home cloud, using discovery mechanisms, should be able to pick out the right foreign clouds which better satisfy its requirements, also reacting to changes in both each single cloud and the whole Intercloud.

II) Resource Management. Federated clouds need to keep the full control of their computing and storage resources, being aware of the amount of resources they require to arrange their own cloud-based services and the amount of resources they can share with other federated clouds. Foreign clouds can not decide to share their resources simply according to the instantaneous requests, in order to pick the best business benefits from the Intercloud, avoiding the run out of its capabilities. For these reasons, clouds need to take decisions according to different factors, for example by means of historical and statistical analysis.

III) Auto-Configuration and Resource Optimization. As both single cloud platforms and the whole Intercloud ecosystem are highly dynamic environments, changes can occur frequently. Often, resources may not be available any more due to hardware or software failures, or foreign clouds may not be able to provide resources any more according to certain SLAs, for example due to an overload of users' requests. In such situations home clouds need to react, auto-configuring themselves, in order to optimize their assets by means of migration of VMs and services, or by means of the establishment of a federation with other foreign clouds.

IV) Resource Migration. Clouds must be able to allocate and move VMs (e.g., by means of hot and cold migrations) into other federated clouds, independently from the virtualization format used in the destination place. In addition, as the current techniques for VMs and service migrations have been designed for LAN environments, it is needed to plan new migration techniques for WAN environments in order to reduce the costs due to the downtime.

V) Monitoring. As services can be based on resources spread over different federated clouds, the home clouds acting as the logical owner of *aaS(s) need monitoring mechanisms able to control the performance of the whole services.

VI) QoS and SLAs. Federated clouds need to be guaranteed regarding the fact that "lent" and "borrowed" computing and storage resources are provided according to *a priori* SLAs.

VII) Security. As the Intercloud scenario allows a home cloud to use the resources of other foreign clouds, privacy and trustworthiness mechanisms are required in order to allow the home clouds which borrow the resources of foreign clouds to gain the access only to the borrowed resources, isolating the other resources owned by the foreign clouds.

VIII) Policy Management. Home and foreign clouds need to choose a management model which suits as well as possible their business model. In order to achieve such a goal it is necessary to plan the policies and rules which will govern the use of their virtual and physical assets. Moreover, it is needed to integrate different security technologies, allowing clouds, to be able to join the Intercloud without changing their security policies.

IX) Naming and Information Retrieval. As cloud-based services can be distributed among different clouds, a mechanism is needed that is able to logically map cloud name spaces regardless of the underlying physical infrastructures. Moreover, it is necessary to manage and integrate heterogeneous and independent name spaces and naming systems, and to logically identify entities in a unique way, when they are physically moved.

In this chapter we provided a few possible research topics related to the cloud federation. We hope we succeed in stimulating your interest in further contributing in such a research area, as there is still a long way to go for a seamless worldwide Intercloud ecosystem.

Part II

Cloud Computing Functionalities and Provisioning

10

TS3: A Trust Enhanced Secure Cloud Storage Service

Surya Nepal

Information Engineering Lab, CSIRO ICT Centre, Australia

Shiping Chen

Information Engineering Lab, CSIRO ICT Centre, Australia

Jinhui Yao

Information Engineering Lab, CSIRO ICT Centre, Australia

CONTENTS

Cloud storage services have emerged as a way to provide effective utilization of storage capacity to address the explosive growth of personal and enterprise data. They allow clients to scale their storage space to meet expanding requirements while improving its utilization and manageability. Given its innovative nature compared to the standard model of service provision, cloud computing raises new questions in terms of *security*, *privacy* and *trust*. There is a need to understand the risks associated with the cloud as well as build technologies to address those risks. This chapter aims to discuss these issues, and provide a possible solution in the context of a framework called *TrustStore*. We first define a service provisioning architecture based on the *Virtual Network Operator* (VNO) model, and then realize it by developing a *Trust enhanced Secure cloud Storage Service* (TS3) using both private and public cloud infrastructures including Amazon S3. We also report on a developed prototype system for TS3 and the results of the evaluation of enabling technologies.

Keywords: cloud storage, trusted storage, storage architecture, secured storage

10.1   Introduction

Cloud computing is touted as the next generation of computing and considered to make significant advances in the next decade [301]. This is possibly due to the maturity and standardization of *Service-Oriented Architecture* (SOA) and *Web services* in the last decade [566]. The driving forces behind cloud computing are: (a) significantly reduced *Total Cost of Ownership* (TCO) of the required IT infrastructure and software including (but not limited to) purchasing, operating, maintaining and updating costs; (b) high *Quality of Service* (QoS) provided by cloud service providers such as availability, reliability and *Pay-As-You-Go* (PAYG) based low prices; and (c) easy access to organizational information and services any time, any where [155]. In a nutshell, cloud computing provides a new paradigm for delivering computing resources (e.g., infrastructure, platform, software, etc.) to customers such as utilities (e.g., water, electricity, gas, etc.) on demand. The current cloud computing architecture enables clients to interact with servers by providing three layers of services: *Software as a Service* (Saas), *Platform as a Service* (PaaS), and *Infrastructure as a Service* (IaaS) [463]. Our focus in this chapter will be on IaaS, and more specifically, on *cloud storage service*.

Cloud storage services have emerged as a way to address the effective utilization of storage space to meet explosive growth of personal and enterprise data. They allow clients to scale their storage space requirements to meet expanding needs while improving utilization and manageability. For example, the storage clouds such as *Amazon S3*, *Google documents*, and *RackSpace* demonstrate the feasibility of storage services in a new computing paradigm by offering (almost) unlimited storage for free or at very low prices yet with high availability (24X7 days). Since enterprises are attracted to cloud storage services due to these features and the potential savings in IT outlay and management, it is necessary to understand the risks involved in utilizing these services. Given its innovative nature compared to the standard model of service provision, cloud computing raises new questions in terms of security, privacy and trust [578]. Therefore, there are significant issues related to privacy, security, and trust for cloud storage services. There is a need to understand the risks associated with it as well as to build technologies to address those risks. Our focus in this book chapter is to analyze the security and trust aspects of cloud storage services, and propose a solution. We describe the issues in brief below.

Security: Despite what cloud service providers and vendors promise, cloud storage services are not secure by nature. The security challenges related to cloud storage services need deeper attention. In terms of security, cloud storage services must be managed and operated at equivalent security levels to enterprise storage systems. However, cloud users typically have no control over the cloud storage servers used. This means there is an inherent risk of data exposure to third parties on the cloud or by the cloud provider itself. The data must be properly encrypted both in motion (when transmitted) and at rest (when stored). The physical locations of the cloud must be secure, and the underlying business processes must be appropriate for the data storage and usage. One of the ways to satisfy these constraints is the segregation of data within the cloud infrastructure to ensure that each user can have full control over his/her data [368]. Similarly, encrypting data in transit has become common practice to protect secrecy and confidentiality of data in a hostile environment [189]. On the contrary, encrypting data at rest — while only end-users may hold the decryption keys — still poses some technical challenges, specifically on the issues of management of keys in the clouds [595].

Trust: Individual and enterprise consumers are generally excited about the potential benefits of cloud storage services, such as being able to store and manipulate high volume data in the cloud and capitalize on the promise of higher performance, more scalable and cheaper storage. These benefits of cloud storage services are very compelling. However, not all cloud storage services are created identically. If the wrong provider is chosen, all those benefits can vanish. The question is then how can you ensure the best possible storage services are selected for you and your business? Nowadays, many vendors/providers compete to offer cloud storage services; there are increasing numbers of solutions available. Not all of them are capable of or keen on providing an adequate level of Quality of Service (QoS) and delivering the promised qualities. When storing encrypted and segregated data, the data must be stored in a service that meets the required QoS. Therefore, it is essential to select cloud storage services with a proven track record on providing the promised QoS. Appropriate measures should be applied to address the issues in this aspect to gain trust from the (potential) customers in its capability to manage mission critical business data. Therefore, trust becomes an important and essential issue in selecting the appropriate cloud storage services, where trust is defined as the reflection of the QoS provided by the cloud storage services [536].

This chapter aims to discuss these issues and provide a possible solution in the context of a framework, called *TrustStore*, by developing TS3. The chapter is structured as follows. We next present a discussion on the proposed novel framework, called *TrustStore*. The proposed framework is based on VNO model and is built on the foundations of SOA and web services, through the use of service composition. We next describe an instance of the framework for providing TS3 in cloud infrastructures such as Amazon S3. We also define enabling technologies for TS3; this includes two categories of technologies: traditional security technologies based on cryptographic algorithms/protocols and the QoS driven trust management system. We then showcase the feasibility of the proposed framework and enabling technologies by developing a prototype in the cloud storage infrastructures such as Amazon S3. We analyze the costs and benefits of deploying our system in the cloud through a comprehensive evaluation of the developed prototype. We then present a section on related works and compare our solution with others. The final section draws the conclusions and outlines the possible future work.

10.2   The Framework — TrustStore

This section presents a cloud storage service architecture based on the concept of a *Virtual Storage Operator* (VSO) business model [180]. Our proposal for a VSO model for cloud storage is inspired by the VNO business model described in the Gartner report [400]. Similar to the tried and trusted VNO for networking, we offer the VSO as a new data storage business model for cloud storage infrastructures. Within our trusted cloud storage services architecture, individual cloud Storage Service Providers (SSPs) own and manage the different physical data storage infrastructures that provide basic cloud storage services such as reading/writing of raw data blocks. To construct a new set of value added services such as TS3 for their own customer base, a VSO will utilize the underlying services of one or multiple SSPs.

We also foresee that many VSOs will emerge to use cloud SSPs to deliver new value-added services that an individual SSP is either unable or unwilling to provide. For instance, a VSO can offer TS3 to back-up client data onto multiple cloud SSPs. Similarly, another VSO can offer a location-aware storage service to reduce the risk of data loss due to unpredictable incidents such as natural disasters in a specific area. This motivates our proposed virtual storage services architecture for cloud.

10.2.1   Architecture Overview

Figure 10.1 illustrates our virtual storage service architecture for a cloud, which consists of four layers: the Application Layer, the Virtual Storage Operator (VSO) Layer, the Storage Service Provider (SSP) Layer and the Storage Infrastructure Layer. Coordinating access to the services offered by the top three layers is done via a common service registry that spans these layers. Each of the main architectural entities is summarized as follows:

The Storage Infrastructure Layer is a heterogeneous mix of physical stor-

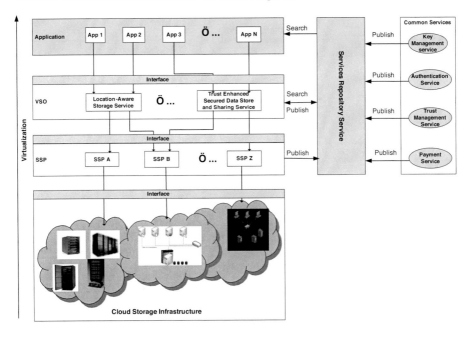

FIGURE 10.1
Cloud Storage Service Architecture.

age systems comprised of different technologies (e.g., RAID, SAN, NAS, DAS) and protocols (e.g., Fiber Channel, iSCSI). Each of these storage systems will be configured accordingly to meet its specific storage quality of service (e.g., latency, throughput, redundancy). For example, a storage area network (SAN) is a dedicated, high performance storage network that transfers data between storage servers and disks. In a SAN infrastructure, storage technologies such as NAS, DAS, RAID arrays or tape libraries are connected to servers using a high speed network protocol such as Fiber Channel.

The Storage Service Provider (SSP) Layer hides the complexity and heterogeneous nature of the underlying storage infrastructures from the higher layers by providing a location-independent cloud storage infrastructure. Therefore, this layer is also called the cloud vendor layer. In our architecture, a SSP provides an interface for the VSO to interact with it. That is, each SSP can interpret the operations supported by the interface into a format understood by its physical infrastructures. The SSP therefore abstracts the underlying Storage Infrastructure Layer allowing clients, e.g., a VSO, to put/get data to/from the storage infrastructure without needing to be concerned about storage technologies, topologies, location, etc. For example, if the VSO utilizes the Amazon S3 as a SSP, the simple put/get operations of the interface are translated into the equivalent protocol messages used by Amazon

S3, namely XML messages wrapped in HTTP carried by the BitTorrent P2P protocol.

The Virtual Storage Operator (VSO) Layer provides the capabilities that enable business entrepreneurs to meet storage requirements by using a combination of services offered by one or more cloud SSPs. The close relationship that exists between VSOs and their application needs enables them to offer "tailor-made" services to meet the specific demands of niche applications. The VSO model can therefore add value to the cloud storage by offering a more diverse range of storage services; for example, Trust enhanced Secure Storage Services, the focus of this chapter. We define a common interface for VSOs, called *Simple Distributed Storage Interface* (SDSI), as an enabling technology (which we describe in detail in a later section).

The Application Layer allows a variety of enterprise applications from different domains to transparently use, via VSOs, the underlying cloud storage infrastructures for their specific purposes. Note that the interface between the Application Layer and the VSO Layer in Figure 10.1 simply shows their logical relationship; i.e., depending on the choice of implementation VSO clients can either put/get data via a VSO or directly with the cloud SSPs at runtime. For the scenario where a client interacts with the SSPs directly, a VSO specific driver is downloaded by the client such that the put/get data operation is guided by predetermined VSO resource allocation and data distribution algorithms (one of such implementations will be explained in a later prototype implementation section).

The Services Registry (SR) provides functionalities such as service publishing, naming, searching, and addressing, using a standard interface across multiple layers. SRs functions are based on those defined in the Web Services UDDI standard. In a way, it is similar to how Amazon S3 is described using Web Services Description Language (WSDL) in a services directory.

Common Services are a set of supporting services that are required across the layers that help to deliver a specific storage service. We envisage that, as networks become faster and cheaper, more large-capacity cloud SSPs will emerge to meet the ever increasing data-intensive demands imposed on basic storage services. These services can be provided by a variety of dedicated storage technologies that constitute a heterogeneous mix of infrastructures hidden from SSP client applications. Since these storage infrastructures are operated by individual cloud SSPs (they are also often referred to as cloud vendors such as Amazon and Google), who may be competitors, they do not directly communicate with each other under normal circumstances. In order for the VSO to manage the distribution of client application data across multiple cloud SSPs, a set of VSO common service entities is required to provide a range of high level repository functions such as indexing, versioning, distribution, directory and authentication. As the aim of this chapter is to focus on trust and security aspects of cloud storage services, we focus on common services relevant to them, such as key management service and trust management service.

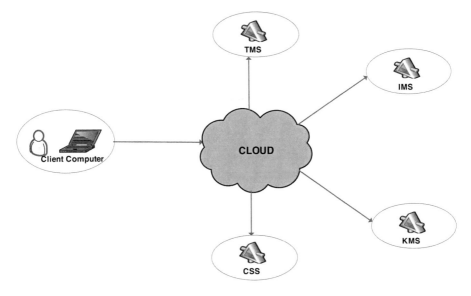

FIGURE 10.2
An Instance of Cloud Storage Service Architecture for Provisioning TS3.

10.3 Trust Enhanced Secure Cloud Storage Service (TS3)

This section presents the realization of the above proposed framework to deliver TS3.

10.3.1 Architecture Instance

Figure 10.2 illustrates an instance of our cloud storage services architecture for provisioning TS3. There are five entities involved in our architecture: *client application, cloud Storage Service Provider* (SSP), *Key Management Service Provider* (KMSP), *Virtual Service Operator* (VSO) and *Trust Management Service Provider* (TMSP). The client application is responsible for invoking VSO for storing data. VSO is composed by utilizing functionalities provided by SSP, TMSP and KMSP. It also provides an interface to the client application. The encrypted data is stored in the SSP. The KMSP acts as a key registry that issues, stores, and manages the encryption keys for the client. It has no knowledge about the use of the stored keys. This means KMSP and SSP are autonomous services provided by two independent entities. The TMSP manages the quality of service based reputation for SSPs. In general, the working of our architecture can be explained as follows.

The client application first sends the confidential data to the VSO by invoking

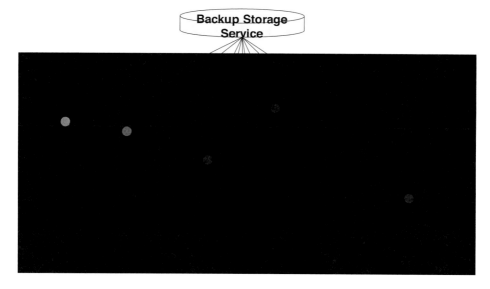

FIGURE 10.3
VSO for Backup Storage Service.

operations supported by its interface. The confidential data is then processed with an encryption mechanism to transform the data into two parts: *cipher text* and *encryption key*. The VSO then finds a suitable SSP that meets its QoS requirements using a reputation-based trust provided by TMSP. The cipher text is uploaded to SSP. Without the key, the SSP can not reverse the process and access the confidential data. The key is stored with KMSP. It has the capability to reverse the encryption process, but it does not have the cipher text. We next explain a motivating example followed by protocols and algorithms involved in our architecture.

10.3.2   Motivating Example

We consider a running example of a trusted storage backup service of high volumes of confidential data for a personal computer to explain our architecture, algorithms, and protocols. In addition to privacy, security, and trust, a particular issue to this example is that the backed-up data must be recoverable even when large scale disaster scenarios occur (e.g., system failures, natural catastrophes, terrorist attacks, and security breaches). One way of satisfying this requirement is to replicate the data in various data centers at different global locations. An efficient and cost-effective way of achieving this goal is to utilize our virtual storage services architecture where multiple SSPs are selected by the VSO to meet the requirements of globally replicating the data as illustrated in Figure 10.3.

10.3.3 Encryption Algorithm

TS3 provides security based on the following assumptions. We assume that a client computer is entirely trusted and secured for sensitive data operation and computation. VSO provides a trusted code, called a *VSO driver*, which can be downloaded and executed in the trusted environment within the client computer. Both SSP and KMSP are semi-trusted and there is no collusion between them. They will manage to provide services they claim. However, they are subject to both internal and external attacks. Our aim in the following is to protect data from such attacks. The data in a client computer is stored as a file. Let a file (Ψ) be modeled as:

$$\Psi = \{M, \chi, \Sigma_{i=0}^{n}\Psi_i\} \tag{10.1}$$

where M represents meta-data including filename, size, type and modified time. In our model, a file can be either a concrete file with content χ or a directory that contains n child files. The meta-data and directory structure of a file may reveal private and confidential information about the data itself. Therefore, they need to be treated at the same level to that of file content. To encrypt a concrete file, it is thus important to minimize the disclosure of its meta-data in the generated cipher text. To achieve this, we propose to partition the file into a set of fragments, which is defined as:

$$F = \{fID, m, x\} \quad where \quad m = \{kID, filename, order\} \tag{10.2}$$

where fID is a fragment ID, which is a randomly generated number and used as a *filename* for that fragment; m represents the fragment meta-data including the ID of the key used to encrypt this fragment, the *filename* of the original file and the fragmentation order; x is the content of the fragment. We can redefine a concrete file ($\overline{\Psi}$) as a set of fragments as follows:

$$\overline{\Psi} = \Sigma_{i=1}^{p}f_i = \{\Sigma_{i=1}^{p}fID_i, \Sigma_{i=1}^{p}m_i, \Sigma_{i=1}^{p}x_i\} = \{\omega, \Sigma_{i=1}^{p}x_i\} \tag{10.3}$$

where p is the number of fragments for a file and represents all meta information (meta-info) of these fragments. Note that not only p needs to be generated randomly on the fly, but also the size of each fragment needs to be different. In this way, we can have the meta-data (M in (1)) completely hidden in the fragments. As illustrated in (1), the hierarchical structure in folders also contains some information about the files. We need a model to represent such structure so that file related information can be stripped off. To achieve this, we define *File Object* (ϕ):

$$\phi = \{ID, \omega, \Sigma_{i=1}^{n}\phi_i\} \tag{10.4}$$

Each file has an associated *File Object* with a unique ID that is randomly assigned to the file, and a meta-info of the file's fragments in case of a concrete file. In case of a directory, the *File Object* contains the File Objects of all its underlying child files. With this model, the meta-info and file structure of a whole collection of files can be encapsulated into one single *File Object*, which we call the *Root File Object* (ϕ_R). Therefore, the encryption of a collection of files can be expressed as below:

$$Enc(\Sigma_{i=1}^{n}\Psi_i) = \{\Sigma_{i=1}^{q}x_i \otimes k_i, \phi_R \otimes k_{q+1}\} \qquad (10.5)$$

where $\otimes$ represents the encryption operation; $q+1$ keys are involved in which one is used for the *Root File Object* and the rest are used for fragments generated from n files in the collection. The file meta-data and structure are indeed transformed into the cipher text similar to the content. The keys used in (5) are issued from KMSP upon request, which is described as follows:

$$Keys = \{\Sigma_{i=1}^{q+1}(k_i, kID_i), R_{ID}, R_{kID}\} \qquad (10.6)$$

The Keys part of the data includes a collection of $q+1$ encryption keys with their key IDs. It also includes the pair of *Root File Object* (R_{ID}) *ID* and its associated key ID. Since the fragment IDs and the associated fragment key IDs are all contained in the *Root File Object*, this is the first file that needs to be fetched and decrypted so that other fragments can be processed. Therefore, its file name (R_{ID}) and key ID (R_{kID}) are required for the key part of the data. The key part is stored at the KMSP, but neither the cipher text nor the original content is disclosed to it. Hence, the purpose of separation of encryption key from the cipher text can be achieved.

Another important aspect of security is preserving data integrity. Users do not have a sense of security if their data can be altered by others without them being aware of it. In our model, we approach the data integrity from the viewpoint of *soundness* and *freshness*. *Soundness* for integrity refers to the awareness of the data owner for any tampering happened to the data [387]. To incorporate this property, we utilize a collision-resistant hash function [288] to calculate the hash of all fragments, and put the hashes into their meta-data (m in (2)) :

$$m = \{kID, filename, order, hash(x)\} \qquad (10.7)$$

where x refers to the content of this fragment. Every time a fragment is fetched from a SSP, the hash contained in the *Root File Object* is used to validate its *soundness*. If the hash of the fetched data is different, the data must have been tampered with. *Freshness* requires the fetched data to be in its most recent version. Any attempts of SSP/KMSP to revoke the change done by the user must be caught. In our model, this is achieved as follows. The cipher text in SSP must have a correct match with the key in KMSP. If either of the two service providers rolls back the stored data to an earlier state, this is noticed by users when they cannot conduct a successful decryption.

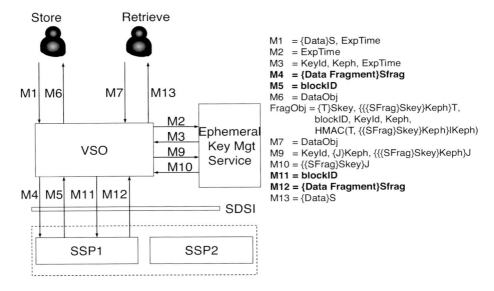

FIGURE 10.4
Key Management Service Protocol.

10.3.4 Key Management Service

One of the most common, and yet significant problems in providing secure storage service in the cloud is the key management (i.e., generation and maintenance of cryptographic keys described earlier). In our case study, we utilize a key manager known as the *Ephemerizer* [577]. This service enables users to "keep" data for a finite time and then makes them unrecoverable after that, which can be one of the desirable policies related to data to maintain privacy. The Ephemerizer service creates keys, makes them available for encryption, aids in decryption, and destroys the keys when they expire. We can make the data persistent by defining the expiration time for the keys as infinity. This service also keeps a database of the mapping information that links data with ephemeral keys. The service may routinely obtain a list of expired ephemeral keys which will be translated into corresponding data objects. Depending on the policy, these objects can be used to instruct the SSPs to reclaim storage space occupied by expired (disappeared) data. Figure 10.4 defines a set of messages and the message exchange protocol implemented for our backup storage service.

Storing data: In our scenario, a user first encrypts a data object with a per-data secret key S and sends it to the VSO along with expiration time (message M1). The VSO then fragments the data as per equation (5). For each fragment, the VSO sends the expiration time of the data to the Ephemeral key management service (message M2) which returns a set of parameters required for time-limited data management (message M3). These parameters are the unique key identifier for the ephemeral key, the ephemeral key itself and its expiration time. The VSO then encrypts each fragment with a symmetric per-fragment key (Sfrag) and stores the fragment (message M4) via the SDSI. The hosting SSP returns an identifier (BlockID) required for subsequent retrieval of the stored fragment (message M5). This process is repeated for all data fragments. The VSO combines the information from M3 and M5 for all fragments, and creates a data object using identifiers of all related fragments and forwards it (message M6) to the user.

Retrieving data: The stored data can be securely retrieved by using the data object along with the encrypted per-data secret S (message M7). The data object also includes an ephemeral session key T that is used to check the integrity of the data linking per-fragment secret Sfrag with Keph. It also includes the ephemeral key and its identity. The triple encryption for per-fragment secret is necessary to deal with dishonest Ephemeral key management services (see details in R. Perlman, [576]). If the ephemeral key has not expired, the Ephemeral key management service returns the per fragment secret (message M8) to the VSO. The VSO then extracts BlockIDs from the data object and sends to the hosting SSP (message M11) via the SDSI and receives the encrypted fragment record (message M12). The VSO combines the information from M10 and M12 (message M13). A user is now able to decrypt the data as the owner of per-data secret key.

10.3.5   Reputation Assessment Algorithm

In our proposed architecture, there are many SSPs providing a variety of persistent storage services. Each service provides a different quality of service. Thus, individual users or businesses face the difficulty of selecting the best service that meets their quality requirements. The quality requirements of a user are expressed through a Service Level Agreement (SLA). Not all service providers meet the promised qualities all the time. Some may exceed on certain aspects of qualities some times and fail to meet the promised qualities at other times. Therefore, there is a need to monitor these SSPs by a trusted third party that can monitor the delivered qualities against promised qualities as specified in SLA. We define a Trust Management Service Provider (TMSP) as a trusted third party for monitoring and assessing SSPs against their SLAs.

Trust reflects the overall performance of the service. It can be measured in different ways. In our model, we are advocating the use of a reputation-based trust model. There are two ways of computing reputation-based trust: computation and experience. The experience based reputation model has been very successful in Internet marketplaces such as Amazon and eBay, and social networks [748]. However, we believe that a computation based reputation model is suitable for infrastructure services such as storage services. We next describe a reputation-based trust model we have proposed, TS3

Let QoS_p be the quality of service promised by a SSP and N be the number of quality of service parameters specified in the SLA, such as latency and throughput. The TMSP monitors the SLA as a trusted third party for VSO and records the delivered quality, denoted by QoS_d . Let M be the number of invocations to SSP. The reputation of a SSP R_{ssp} is given by

$$R_{ssp} = \frac{\Sigma_{i=1}^{M}\Sigma_{j=1}^{N}(QoS_d^{ij} - QoS_p^{ij})}{M \times N} \qquad (10.8)$$

The positive value of Equation (10.8) indicates that a SSP has on average exceeded the promised quality and has a positive reputation. The negative value indicates the opposite, i.e., a SSP has failed to deliver the promised QoS and acquired negative reputation. Not all QoS parameters are equally important. For example, the latency may be more important than throughput in certain scenarios and less important in others. To represent the importance of QoS parameters, we restate the reputation of a SSP R_{ssp} as follows:

$$R_{ssp} = \frac{\Sigma_{i=1}^{M}\Sigma_{j=1}^{N}w_j(QoS_d^{ij} - QoS_p^{ij})}{M \times N}, \quad where \quad \Sigma w_j = 1 \qquad (10.9)$$

Another important aspect in reputation assessment is the temporal sensitivity. Recently monitored data are more valuable than old data in terms of currency. This means the value of the old data should decay with time. The decaying function could be different for different QoS parameters. To represent the temporal sensitivity of monitored data, we restate the reputation of a SSP R_{ssp} as follows:

$$R_{ssp} = \frac{\Sigma_{i=1}^{M}\Sigma_{j=1}^{N}\alpha_t^{ij} w_j(QoS_d^{ij} - QoS_p^{ij})}{M \times N}, \quad where \quad \Sigma w_j = 1 \qquad (10.10)$$

where α_t^{ij} represents the temporal decay function of QoS parameter j for the invocation i that occurred at time t.

10.3.6 Trust Management Service

The functionality of the trust management service can be explained in three phases as follows:

Agreement: The agreement phase involves defining SLAs and agreeing on the terms of SLAs between the VSO and SSPs. SLAs specify QoS properties that must be maintained by SSP during service invocation. WS-Agreement provides a specification for defining SLA [218]. This step in this phase is not shown in Figure 10.5.

Monitoring: Monitoring plays an important role in determining the delivered QoS and calculates the reputation of a service using agreed SLAs. The VSO invokes the service operation such as storing a data fragment to a SSP (step 1). The SSP finishes the result and returns the result of the operation (step 2). At the same time, the SSP logs the SLA relevant data to a trusted third party, trust management service (step 3). Once the result is received by the VSO, it also logs the SLA relevant data to the trusted third party (step 4). We have

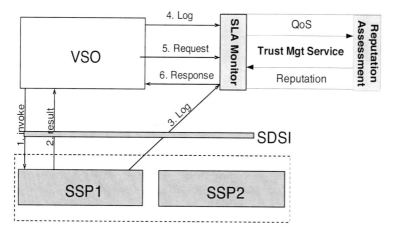

FIGURE 10.5
Trust Management Service Protocol.

used a very simple, but effective way of monitoring the SLAs. A number of variants of it have been proposed in the literature. We refer readers to [599] for details of one of such approaches.

Evaluation: Before using a particular SSP, the VSO can request its reputation from a trust management service (step 5). The logged data is then used to evaluate the reputation of the service using the equation (10). The reputation value is then returned to the VSO (step 6). The VSO then selects the highly reputed SSPs for storing the data.

10.3.7 Simple Distributed Storage Interface (SDSI)

The VSO uses the trust management service to select SSPs that meet its requirements. The VSO must interact with multiple SSPs. Each SSP has its own interface. We need to provide a single interface to the VSO. To achieve this objective, we propose a Simple Distributed Storage Interface (SDSI). The purpose of the SDSI is to facilitate the VSO business model in the sense that it enables the VSO to use the services of multiple SSPs without needing to understand the technical complexities of the storage infrastructures underlying the SSPs. As such, the VSO does not need to understand each interface presented by the SSPs. The SDSI sits between the VSO and these SSP interfaces to present simple generic put/get operations, similar to existing approaches used by Amazon S3 and the Google File System. To make the VSO business model viable, the SDSI, apart from simple put/get operations, needs to reveal sufficient information about the SSPs in use to the client, where a client can be a VSO, another SSP, or even an application. As such, each client can negotiate and purchase storage services, store and access data, launch enquiries and obtain status information about their accounts. To cover all these aspects of

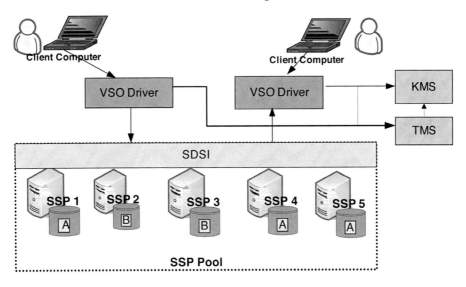

FIGURE 10.6
Prototype Implementation Cloud Storage Service Architecture for Provisioning TS3.

interaction between a client and an SSP, the SDSI API provides *functional* and *non-functional* components. The functional component includes the minimal set of requirements for a client in its ability to put data into, and get data from, the abstracted storage infrastructure. Our SDSI supports two types of non-functional API: *Service Level Agreement* (SLA) and *Account Management*. A VSO must enter into a Service Level Agreement (SLA) contract with SSPs in order to use the storage facilities provided by the SSPs. The contract structure records information about the service provider, the service consumer, and quality of service parameters. The quality of service parameters are monitored and used to assess reputation by a trust management service. The Account Management API includes activities related to managing user accounts such as disk space, billing, etc. We refer readers to [179] for details on SDSI.

10.4 Prototype Implementation

In order to illustrate our framework, we have implemented a prototype system for TS3. The prototype system has a pool of SSPs, two remote services for key and trust management, a VSO and a client application, as shown in Figure 10.6. We next describe these components in brief.

SSPs: In our prototype implementation, we have used five SSPs. One of the

SSPs was Amazon S3, whereas the rest of the four SSPs were provided in a private cloud. Amazon S3 allows standard storage operations such as: PUT, GET, LIST, and DELETE. The client application directly executes these operations through sending SOAP messages to Amazon S3. We have used a Microsoft SQL Server 2000 as the back-end storage system to host and manage data blocks for other four SSPs that supports SDSI as an interface. Therefore, these SSPs do not need to translate SDSI calls.

KMS: A key management Web service was implemented and deployed in EC2. It generates the encryption keys upon request and stores the keys, represented by Equation (10.6).

TMS: A trust management Web service was implemented and deployed in EC2. It logs the values against the agreed SLAs and evaluates the reputation of SSPs based on their past performance, as represented by Equation (10.10).

VSO Driver: The client computer runs applications via a specific VSO. Depending on the choice of implementation, VSO clients can either put/get data via a VSO or directly with the SSPs at runtime. In our implementation, the VSO features are embedded in a specific driver that is downloaded by the client and run along with their application. The put/get data operation is guided by predetermined VSO resource allocation and data distribution algorithms. For example, we have configured five different SSPs in the driver for our prototype implementation and selected them based on their reputation value.

Client Application: It is implemented as a GUI-based Java application. Figure 10.7 shows a screen-shot of the client application. Its appearance is very similar to a normal split-view file explorer in modern operating systems. The application is to be used as a tool to take care of tasks of data processing and service coordinating at the background. Users can utilize the application to achieve TS3 by simple standard operations such as drag and drop. The detail functionalities are illustrated through the following standard operations:

Log in: A user is authenticated by both SSP and KMSP. A Root File Object is first fetched. Upon its successful decryption, the file collection will be displayed. An empty one is created for the first time.

PUT: The user can upload a file by simply drag-n-drop it into the GUI. The target file is then transformed according to the model and uploaded to SSP and KMSP by using TMSP.

GET: If the user double clicks on a file when its status is uploaded (see Figure 10.7), its respective cipher text and key data will be fetched. After the decryption, its hash will be computed to be validated with the stored hash value.

Structural operation: The changes in the file structure cause changes to the Root FileObject. The structure information is used for display and does not affect the stored fragments.

Delete and Update: Delete can be done by pressing the Del key to the target file. Related data is deleted from both storage and key management services. Update is achieved through a Delete followed by a Put.

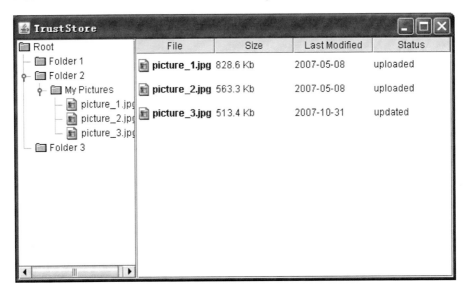

FIGURE 10.7
A Screen-Shot of Client Application.

Log off: When the user logs off the application, the updated Root File Object is encrypted and uploaded to replace the old version. Changes to the file structure and meta-data (e.g., file name) do not suffer delay because they are done locally and transmitted at the end of every user session.

10.4.1 Performance Evaluation

This subsection presents the performance evaluation of our prototype system TS3. The report on our evaluation focuses on two aspects of the proposed framework: (a) the cost of trust and security protocols and algorithms, and (b) the cost of involving multiple SSPs from a variety of sources within a unified interface (SDSI) to the client. We next describe evaluation environments and results for these two aspects of our proposal.

To demonstrate the cost related to the implementing security and trust protocols, we have conducted baseline testing to observe the latency of our system as compared to the ordinary use of Amazon S3. As described in a previous section, the PUT/GET operations in our system involve several extra processes apart from the transmission, including key requesting, trust evaluation, content hashing, fragmentation and cryptography. We first show the testing results of each of these extra processes, and then demonstrate our approach to minimize the overall costs through multi-threading. The tests are carried out on a computer with an Intel Pentium 1.72Ghz processor and 512 MB RAM. The computer is located in Sydney Australia as the KMS, while the Amazon

TABLE 10.1
Specification of the Test Environment.

4 SSP Server	Hardware	Dell 1x3.0GHz Hyper-Thread P4 CPU
4 SSP Server	Software	1.0GB ROM, 1Bbps dedicated LAN Microsoft Windows 2003 SP1 IIS,
1 SSP Client VSO	Hardware	ASP.NET 1.1, SQL Server 2000 SP3 Dell 4x 1.6GHz Xcon CPU,
1 SSP Client VSO	Software	3.5GB ROM, 1Bbps dedicated LAN Microsoft Windows 2003 SP1, .NET 1.1

S3 and EC2 servers are likely located in US as the SSP. The Internet access was ADSL2+ with theoretical capacity of 24,000kbps (about 2.4MB/sec) download speed and 1000kbps (about 0.1MB/sec) upload speed. We observed the following latency for the individual processes: Key Requesting −1.08 sec, Trust Requesting − 0.80 sec, Hashing (SHA-1) -0.03 sec/MB, Fragmentation − negligible, AES 256 bit key Encryption -0.16 sec/MB, Decryption-0.12 sec/MB. The key requesting test is conducted by transmitting 50 keys of 256 bit.

We noticed that the quantity of keys in the request does not much affect the latency because the size of keys are relatively small in comparison to messages. The time is mainly determined by the distance to the KMS. The same is true for requesting reputation values of SSPs. The other processes are all internal computations whose latency grows linearly with the amount of data to process. The extra latency to perform a GET operation on a file of 10MB was around 2.98 sec. The direct transmission of the same file took around 14.7 sec. This latency is noticeable, but can be substantially reduced by applying multi-threading techniques described below, as shown in Figure 10.8. In the second part, we wanted to test the cost of implementing SDSI. We have defined and implemented a SOAP based protocol for SDSI, which we refer to as SDSI/SOAP. For performance evaluation of the SDSI/SOAP, four SSP storage services that use the SDSI were deployed on four identical Dell machines using SOAP and SQL databases. A tester, which ran on a separate Dell machine to read and write data to the SSP storage servers, was utilized to emulate a VSO client. Table 1 shows the specification of the test environment. To benchmark the system, we used the following performance metrics, Wall-Time and Throughput. The WallTime presents the overall time for writing a certain amount of data into the cloud storage (see Equations (10.11) and (10.12)).

$$Latency = \frac{WallTime}{DataAmount} \tag{10.11}$$

$$Throughput = \frac{1}{Latency} \tag{10.12}$$

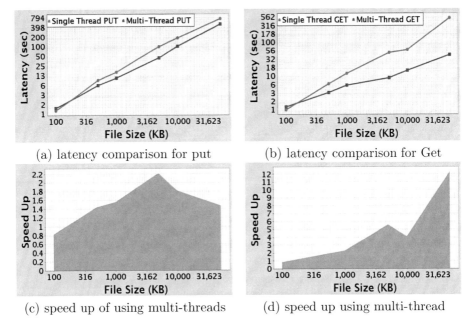

(a) latency comparison for put   (b) latency comparison for Get

(c) speed up of using multi-threads   (d) speed up using multi-thread

FIGURE 10.8
Performance Improvement: Single Thread vs. Multi-Thread SSS.

We uploaded data blocks in various sizes (100kB, 1MB, 10MB, 100MB) via the SDSI/SOAP to a single SSP and measured the time taken to complete the operation. A parallelization technique for uploading the data was used in order to optimize the system performance by segmenting the data (i.e., SOAP messages) into blocks of 4kB and by setting the number of simultaneous threads to 20. These parameters were arrived at empirically. The test results for uploading data to an SSP via the SDSI/SOAP are shown in Figures 10.9 (a) and (b). As expected, for smaller data blocks (i.e., $< 4kB 20 threads = 80kB$) the benefits of parallelization are not fully realized since the processing capacity of the system is under-utilized, i.e., the system is still ramping up. Furthermore, as shown in Figure 10.9 (b), for a data size $> 1MB$ the system throughput saturates at about 1.5MB/s.

A single SSP in our architecture is comparable to Amazon S3 in terms of its put/get functionalities. However, we can ramp up the performance by scaling a single SSP to multiple SSPs in our architecture. This is possible due to the existence of a VSO that can operate over multiple SSPs using standard SDSI/SOAP protocol. We next describe the benchmarking of multiple SSPs in our architecture.

When evaluating the scalability of the SDSI/SOAP across multiple SSPs, we selected an operating region that was neutral to the data amount being stored (i.e., linear section of Figure 10.9 (a)) while the data throughput had

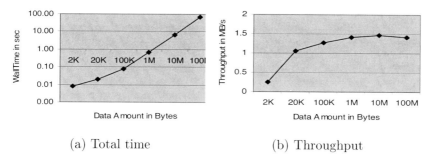

(a) Total time   (b) Throughput

FIGURE 10.9
SDSI Benchmarking for a Single SSP

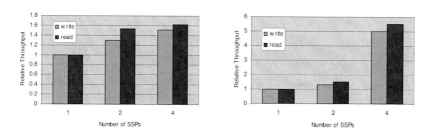

(a) Using Single SSP Parameters   (b) Using Multiple SSPs Parameters

FIGURE 10.10
SDSI Benchmarking for Multiple SSPs.

reached saturation. Under these conditions the scalability of the SDSI/SOAP was evaluated based on the following parameter set, number of SSPs, data block size and number of threads. We evaluated the performance of the SDSI for 2 SSPs and then 4 SSPs. From empirical results, we found that the optimal data block size and number of threads for 2 SSPs was 4kB and 20, respectively. Similarly, for 4 SSPs, the optimal data block size and the number of threads was 128kB and 10, respectively.

Referring to Figure 10.10 (a), we used the single SSP benchmark parameters (normalized), namely data block size = 4kB and number of threads = 20, and performed write/read operations for 2 and then 4 SSPs. The first observation is that there is a marginal difference in the throughput for write/read operations. Further, the much more interesting observation is that as the number of SSPs doubles each time, there is no significant increase in throughput as perhaps would be expected. This result motivated further investigations on fine tuning of data block size and thread number for 2 and 4 SSPs to see if throughput could be increased. Referring to Figure 10.10 (b), for 4 SSPs it can be seen that by optimizing the data block size to 128kB and the thread number to 10 the, throughput is at least 5 times greater than for a single optimized SSP.

However, for 2 SSPs, the optimized values are the same as for a single SSP, and as already observed in Figure 10.9 (a), the increase in throughput for 2 SSPs is only 1.3 to 1.5 times greater than for a single SSP. We believe these empirical results indicate that there are a number of complex interactions at work which require much deeper investigation which are beyond the scope of this chapter.

10.5   Related Work

While cloud storage is new in terms of the concept, considerable technologies have been developed from both research communities and industries to directly and/or indirectly contribute to it. As such, we review these related works from the following different aspects:

Architecture: FARSITE [48] is Microsoft's initiative to use a large amount of desktop PCs as data storage to replace dedicated storage devices to reduce costs. As a result, FARSITE is more likely to be used within an organization for general application storage, rather than serious enterprize storage. OceanStore [446], and PAST [618] are large-scale distributed storage systems utilizing Peer-to-Peer (P2P) technologies to automatically maintain data replication [797]. However, these architectures may not be suitable for cloud storage. It is unlikely for the cloud storage providers to open their services to be connected using P2P technologies. Three popular storage infrastructures related to our work are: GFS [302]; Akamai [697] and Amazon S3 [561]. However, their approaches are different from ours in terms of scope and business model. While GFS is for Google's internal use for storing and managing almost-read-only data, Akamai is a content service provider used to globally deliver web content hosting services. Although Amazon S3 provides a similar web services interface as our SDSI does, Amazon S3 is a SSP in our architecture and thus it is unable to support virtual storage services by itself. In addition, our SDSI provides non-functional API, that Amazon S3 does not have, making it easy to facilitate the integration between virtual storage services and SSPs.

Middleware: The Storage Resource Broker (SRB) [597] is the middleware deployed by the Grid Computing community to provide a uniform interface for heterogeneous data storage resources. While SRB strives to provide an integrated solution for online storage, the SDSI is a flexible component that can be readily composed into a Services Oriented Architecture (SOA). MetaCDN [144, 570] provides a Web portal for customers to utilize different cloud storage services, where a set of storage clouds can be combined intelligently to meet the customer's needs. On the other hand, approaches like WuKong [491] aim to overcome the storage constraints of mobile devices by extending its file system to closely interact with the cloud. Users can manipulate the data in the cloud transparently, as if they are stored in the device.

Protocols: Although the SDSI/SOAP shares the same storage goals as iSCSI (RFC3720), FCIP (RFC3821) and iFCP (RFC 4172) for storing data via the WAN, the motivation is somewhat different from these protocols in the sense that the SDSI/SOAP is a simple web services-based abstraction which facilitates the viability of the VSO business model. The SDSI therefore operates at a higher level than the aforementioned protocols and in fact may utilize these protocols for the actual storage of data depending on the underlying technology.

Security: There is considerable research work conducted to enhance the security and privacy of data storage. Jammalamadaka et al. [387] deployed a master password to generate encryption keys, which are used to encrypt the data stored remotely. Jammalamadaka et al. [386] use a similar method; instead, a random binary key is used as the master key. SHAROES [656] issues each user a private/public key pair. The private key is used to encrypt both the content and the metadata of the data stored. The aforementioned approaches used the concept similar to ours. However, all keys are derived from the same master key. This makes the master key very important in the system and it is the only key that needs to be brute forced. Sirius [308] and Plutus [405] allow clients to use different keys generated on-the-fly without any master key. Clients are then responsible for the key distribution and management. Fu, Kaashoek and Mazieres [288] and Kher and Kim [429] use asymmetric cryptography to maintain undeniable records to protect the data integrity. They do not focus on the privacy of the data.

Some studies also proposed to compose multiple service entities for storage outsourcing. Aggarwal et al. [51] breaks down the relational information in the sensitive data (e.g., user-name and password) and stores them in different service providers. Heitzmann et al. [340] computes the hashes of all files stored in one service provider and stores them in another. Popa et al. [585] proposed to use cloud proof, which is a signed attestation about the state and property of the stored data object. In this approach, the client can always prove if the object is corrupted and the cloud can always defend it from a false accusation. Cepheus [403] introduces a key service to contain the keys while the other data service contains the cipher. Each client manages a private/public key pair to encrypt the keys before they are stored. Again the private key is the only secret in the entire protection. Their concept of separation is greatly refined in our approach. We rely on the authentication credentials in two different service providers to separate the cipher form and the key form data. This allows us to improve the strength of the encryption by the quantity of keys used without introducing a single master key.

Trust and Reputation: Reputation management has been studied in a variety of disciplines including economics, computer science, marketing, politics, sociology, and psychology [230]. In computer science, major applications where reputation has been effectively used include e-business, peer-to-peer (P2P) networks, grid computing systems, multi-agent systems, web search engines, and ad-hoc network routing. Reputation systems have benefited electronic commerce in recent years. Amazon, eBay and Yahoo! Auction are examples of businesses that have deployed reputation systems successfully. These reputation systems use feedbacks from the consumers as the reputation measure, and have received considerable attention in the literature [175, 774]. Examples of reputation systems include SPORAS [789], PeerTrust [773], Eigen-Trust [408], PowerTrust [799], PRIDE [236], XRep [217], P-Grid [45], RE-GRET [627], FIRE [357], and TRAVOS [696]. The evaluation of the reputation value in these approaches is classified into two categories: feedback-based and monitoring-based. In the feedback-based approach [261], the reputation value is calculated using users' feedback. In the monitoring based approach [290], the reputation value is evaluated using the delivered QoS. We have defined and used the later model in our approach.

10.6   Conclusions and Future Work

Cloud computing has emerged as a way to provide a cost effective computing infrastructure for enterprises. Given its innovative nature compared to the standard model of service provision, cloud computing introduces new challenges in terms of security, privacy and trust. We have discussed these issues and presented a solution in the context of a framework, called *TrustStore*. We have presented a novel architectural solution for provisioning trusted cloud storage service in the public cloud and the relevant technologies. Our focus in the current architecture is on the security and trust aspects of cloud storage solutions. The data privacy in the cloud, including authorization, is a challenging problem. We plan to extend our model and address this issue in future.

As more and more enterprises start moving their data and processes into the cloud, the cloud environment will emerge as a hybrid environment where the public cloud has to work seamlessly with the private clouds. This poses interesting research and business challenges such as interoperability between clouds, selective data and process migration into the cloud. Furthermore, the issues of dealing with transactional data as well as management of data in the cloud need to be addressed. In a nutshell, we believe that the success of the cloud as a new computing paradigm depends on the success of overcoming the psychological barriers and technical challenges in the area of security, privacy, and trust. It is important to emphasize here that the technological solutions alone may not be enough to address security, privacy, and trust concerns associated with the cloud.

11

High Performance Computing Clouds

Andrzej Goscinski

School of Information Technology, Deakin University, Geelong, Australia

Michael Brock

School of Information Technology, Deakin University, Geelong, Australia

Philip Church

School of Information Technology, Deakin University, Geelong, Australia

CONTENTS

In recent times, cloud computing has gained the interest of High Performance Computing (HPC) for providing HPC environments on-demand, at low cost and in an easy to use manner. While the potential for HPC clouds exists, there are challenges (such as performance and communication issues) and some experts deem cloud computing inappropriate for HPC.

This chapter investigates what challenges exist when attempting to use clouds for HPC and demonstrates the effectiveness of HPC cloud computing through broad benchmarking and through a new prototype cloud called HPCynergy - a cloud built solely to satisfy the needs of HPC clients. This chapter demonstrates, via benchmarking comparisons between clouds such as EC2, HPCynergy, virtual and physical clusters that HPC clouds have a bright future and that most claims against the use of cloud computing for HPC are exaggerations.

11.1   Introduction

A recent trend in computing research is the use of clouds to support and/or provide complete environments for high performance computing (HPC) applications. The motivation for this trend is to support HPC applications without the excessive cost and complexity overheads. HPC applications are software used to carry out complex algorithms on large data set (often terabytes in size) in the following fields:

- Bioinformatics – to understand diseases and biological systems through DNA analysis

- Physics – to discover particles often using the Large Hadron Collider

- Engineering – to design better cars and planes and make environments green(er)

Traditionally, clusters have been used to run HPC applications: rather than run the whole application on a single system, subsets of the application are run concurrently across multiple cluster nodes so that results are returned within reasonable timeframes. However, clusters are not cheap to purchase and maintain and are usable only by computing specialists. Hence the attention of cloud computing for HPC is to reduce costs, offer on-demand services and simplify use.

Currently specialized High Performance Compute Clouds (HPC Clouds) are only offered by a few providers such as Amazon [60] and SGI [373]. These specialized cloud instances usually make use of nodes with HPC-specific hardware (Infiniband or Gigabit Ethernet interconnects between nodes, special processors, etc.), HPC middleware and some form of hardware monitoring.

While there is the promise of cheap, on-demand HPC, cloud computing is a completely different paradigm thus exhibits unexpected side-effects. Cost wise, there is no guarantee that paying more for additional cloud nodes improves performance [636]. As most clouds use virtualization, HPC software applications may execute abnormally and there is no guarantee that all cluster nodes in a cloud are within close proximity to each other.

Furthermore, cloud interoperability does not exist: whole computer systems created on one cloud are not transferable to another cloud. This issue is so significant that even Intel proposed it as a requirement in their vision of cloud computing [343]. Finally, HPC clouds still lack a number of important software features: (i) current cloud management tools focus on single servers, not whole clusters, (ii) tools for publication and discovery only focus on service functionality and ignore other parameters such as quality and current activity. In general, while the potential of on-demand HPC via clouds exists, there is no solution that combines cloud computing and HPC together. The aim of this chapter is to present a review of publications on combining clouds and HPC, highlight existing challenges faced when using clouds for HPC, present a new HPC cloud, and demonstrate, via experimentation, the applicability of HPC clouds.

The rest of this chapter is as follows. Section 11.2 presents arguments by leading experts both against and in favor of using cloud computing for HPC. Section 11.3 provides a definition of HPC clouds and makes a comparison with current physical and virtualized HPC cloud offerings. Section 11.4 details various challenges faced (based on the definition presented in Section 11.3) when using clouds for HPC. In response the design and implementation of a new HPC cloud called HPCynergy is presented in Section 11.5. The feasibility of running bioinformatics and physics applications on HPCynergy (when compared to other cloud computers and dedicated HPC platforms) is investigated in 11.6. Conclusions and future trends are presented in Section 11.7.

11.2 High Performance Computing (HPC) vs. Cloud Computing

Recently, vendors and researchers have turned to cloud computing as a way to access large amounts of computational resources, storage, and software cheaply. In terms of clouds for HPC environments, experts are both for and against the idea.

Dr. Iordache rejects the possibility claiming that HPC is about performance while virtualization in clouds is about adding more latency (overhead) [344]. This claim is not entirely correct as virtualization provides a required platform independent of an existing physical platform. While latency exists, it is not such a significant problem now as it was when it first emerged. Another opponent to HPC clouds is Dr. Kranzmuller, who stated *"there are only very limited uses for the public cloud in HPC ... For applications with larger storage requirements or closely coupled parallel processes with high I/O requirements, clouds are often useless."* [345].

Dr. Yelick, director of NERSC, noted about current developments in the Magellan cloud, *"there's a part of the workload in scientific computing that's well-suited to the cloud, but it's not the HPC end, it's really the bulk aggregate serial workload that often comes up in scientific computing, but that is not really the traditional arena of high-performance computing."* [345].

Not all experts are against the idea. An initial study of the effectiveness of clouds for HPC shows that the opportunity is there and is compelling but *"the delivery of HPC performance with commercial cloud computing is not yet mature"* [736]. However, the computational environments used in the study (Amazon's EC2 and NCSA cluster) have inconsistent hardware and software specifications, and the experiment details are not specified precisely.

Professor Ian Foster stated in his blog when addressing these results that *"based on the QBETS predictions, if I had to put money on which system my application would finish first, I would have to go for EC2."* He also added that when HPC clouds are to be assessed, optimization for response time should be considered rather than the utilization maximization used on supercomputers [278].

There are also other strong voices supporting HPC in clouds. In response to the increasing costs of IT infrastructure, use and maintenance, Microsoft's Dan Reed stated, *"fortunately, the emergence of cloud computing, coupled with powerful software on clients ... offers a solution to this conundrum."* He adds that *"the net effect will be the democratization of research capabilities that are now available only to the most elite scientists."* [346].

An analysis of HPC and clouds shows that (i) the concept of HPC clouds has been rejected by some researchers because clouds use virtualization; (ii) not all problems require powerful and expensive clusters; (iii) organizations, although they depend on storing and processing large data, they are not prepared to invest in private HPC infrastructures; (iv) HPC on clouds offer cost and scalability advantages to clients (users).

In support to Item (i), vendors such as SGI [373] and Penguin [201] do not offer virtualization in their HPC offerings; they provide customized servers.[1] In this context it is worth presenting the thought of Ms. Hemsoth, *"It's not really the cloud that everyone recognizes if you remove the virtualization, is it? At least not by some definitions. But then again, getting into complicated definitions-based discussions isn't really useful since this space is still evolving (and defining it ala the grid days) will only serve to stifle development."* [343].

Overall, it is possible to offer HPC as a service: if there are many clusters that provide HPC services, cloud management software offers services transparently, and clients can discover and select a service that satisfies their requirements, then there is a satisfactory condition to say that there is a HPC cloud. While virtualization is often a common element for cloud computing, it should be noted that not all expert definitions of cloud computing [298, 300] consider virtualization as a vital element.

In support to Item (ii), there are very many bioinformatics, engineering, social sciences problems that could be successfully solved on less powerful clusters and on demand (without a need for waiting in a queue[2] for execution). Furthermore, many clients do not wish to purchase, deploy, run and maintain even smaller clusters.

In support to Items (iii) and (iv), organizations prefer to concentrate on their mission statement activities, and clients prefer to concentrate on their problems, research, and development. Clouds allow on demand resource usage: the allocation and release is proportional to current client consumption, thus clients only pay for what they use. This implies the availability of cloud resources is more important than performance. However, there is no guarantee

---

[1] We will show in this chapter the influence of virtualization on the execution performance of high performance applications on different clouds.

[2] Queues are used with clusters to hold jobs that require more nodes than currently available.

that paying for (renting) extra nodes on a cloud will guarantee significant performance improvement [636]. Finally, even when purchase, installation and deployment of a cluster is complete, the cluster still incurs ongoing costs (mostly electrical power for cooling) long after research projects are finished.

Besides the low cost and simplified access to specialized, high throughput systems, there is also the opportunity to improve collaboration between research projects. Instead of only making large datasets available to various research teams, it is now possible to expose whole research environments as on-demand services. Thus, it is clear that HPC clouds have a bright future.

11.3   Taxonomy of HPC Clouds

This section presents an accurate cloud computing definition based on current literature. Based on the presented definition, a taxonomy of current cloud and HPC solutions is presented. Problems are identified based on what properties from the supplied definition in Section 11.3.1 are not satisfied by existing clouds. If the means by which a property has been exposed is deemed inadequate, the poorly exhibited property is also considered a problem.

11.3.1   HPC Cloud Definitions

Neither cloud computing nor HPC cloud computing have a concrete and commonly accepted definition. According to Gartner [298], clouds are characterized by of five properties:

1. Service-based – all resources within the cloud system are to be made available via services that have well defined interfaces.

2. Scalable and Elastic – clients only use resources (and amounts thereof) they need.

3. Shared – all resources are used at their full utilization and not dedicated to any client.

4. Metered by Use – clients make use of clouds on a pay-as-you-go basis.

5. Uses Internet Technology – cloud services are built and accessed using Internet technologies, often the World Wide Web.

The NIST definition of cloud computing [508] furthers Gartner's definition in that clients have broad network access (the cloud is accessible via any network enabled device) and require clouds to monitor and reconfigure resources based on service usage.

In another study, 21 experts were asked for their definitions of cloud computing [300]. Each expert used at least two of Gartner's properties and some consider virtualization as a property of cloud computing. Some experts even considered that cloud computing offerings fall under one of three broad categories:

- Infrastructure as a Service (IaaS) – virtual servers are provided where clients are able to create their required platforms from the (virtual) hardware upwards.

- Platform as a Service (PaaS) – complete hardware and software configurations (platforms) are offered to clients and maintained by cloud providers.

- Software as a Service (SaaS) – clients are offered software applications/services and only have to focus their use.

In this chapter, cloud computing (based on the description by Gartner and current cloud offerings) is made possible through the combination of virtualization, Web services, and scalable data centers [145]. With the recent emergence of HPC in cloud computing (thus giving rise to the idea of HPC clouds [201, 373] and/or HPC as a service [60]) once again there is the question of what the term HPC cloud stands for and what properties need to be exhibited. Thus HPC clouds are viewed as having the following properties:

1. Web Service-Based – all resources from data storage to cluster job management are done via self-describing Web services with Web APIs for software processes and Web forms for human operators.

2. Use the Pay as You Go Model – all HPC clients are billed for the resources they use and amounts thereof.

3. Are Elastic – if a client has a task that consists of multiple jobs in a required sequence (workflow) HPC clouds are expected to allocate and release required services/resources in response to changes in the workflow.

4. Clusters Are Provided On-Demand – clients should be able to specify requirements and then (i) discover an existing cluster for immediate use or (ii) have an existing cluster reconfigured to satisfy client requirements.

5. Guaranteed Performance – typically, if cluster nodes are allocated to clients, all nodes are expected to be within close proximity to each other.

6. Virtualization – for flexibility, cloud computing will require the use of virtualization. However, virtualization should be made an optional feature to HPC clients: they may have applications that do not work in virtual environments due to high latency sensitivity.

7. Are Easy to Use – clusters (physical or virtual) can be complex thus need to be simplified. First, unintuitive user interfaces (such as command lines) can frustrate human users into finding work around. Second, some researchers (even experts) may wish to spend more

time conducting research rather than on HPC application and application configuration/modification.

These properties of the provided HPC cloud definition are used to propose a simple HPC cloud taxonomy that considered both major types of HPC cloud solutions: virtualized HPC clouds and physical HPC clouds.

11.3.2   Virtualized HPC Offerings in the Cloud

While cloud computing has much to offer HPC (such as low cost, availability and on-demand service provision), few cloud based HPC solutions exist. There are even vendors that claim to provide HPC cloud solutions but are simply resellers for other cloud computing vendors. CloudCycle [214] is such a vendor that sits on top of Amazon and sells HPC products to clients. This section shows what HPC product offerings currently exist from cloud computing providers and examines what properties they exhibit from the HPC cloud definition presented in Section 11.3.1.

11.3.2.1   Amazon Cluster Compute Instances

An IaaS cloud, Amazon's Elastic Compute Cloud (EC2) [59] offers virtual machines (instances) to clients who then install and execute all required software. EC2 offers a catalog of instance types that detail the (virtual) hardware specifications and range from small 32 bit systems to large multicore 64 bit systems [61]. There have even been successful attempts to create clusters within EC2 manually [259, 349] yet others have deemed EC2 unviable solely on the network communication performance [736].

Recently, Amazon has offered the Cluster Compute Instance for HPC [62] which offers a (combined) 23 gigabytes of RAM, 16 Terabytes of storage and use 10 Gigabit Ethernet. In terms of the HPC cloud definition, Cluster Compute Instances only exhibit the properties of being service based, clients paying for what they use, and the use of Gigabit Ethernet for performance. However, Cluster Compute Instances are still not adequate, as the property of ease of use is not exhibited. Clients are required to install cluster management software.

11.3.2.2   Microsoft Azure

A PaaS cloud, Microsoft Azure [515] provides complete hardware and software configurations on-demand. Thus, clients focus on the development and management of their own services and can make the services discoverable via the .NET Service Bus [661]. However, nothing more than a unique URI of each service is publishable.

In terms of support for HPC, Azure only acts as an "auxiliary" to clusters running Microsoft's HPC Server 2008 [516]. On site clusters lacking available resources can connect and off load tasks[3] to Azure. Overall, the only properties exhibited by Azure are the use of Web services, the pay-as-you-go model, and (through the use of clusters extending to Azure) elasticity.

11.3.2.3   Adaptive Systems' Moab

Adaptive Systems is a provider of HPC cluster management software and a provider who helped in the genesis of Amazon's Cluster Compute Instances [341]. While Adaptive does not provide a cloud perse, it is worth describing, as clusters are comprised of hardware and software. Of interest to this chapter are the Moab Adaptive HPC Suite [603] and the Cluster Management Suite which contains a Cluster Manager [604], Workload Manager [605] and Access Portal [602].

The Moab Adaptive HPC Suite allows each cluster node to automatically change its installed operating systems on-demand. For example, a cluster comprised of Linux nodes is able to change a required number of nodes to Windows, thus allowing the job to run. This feature, based on the definition in Section 11.3.1, exhibits the properties of clusters on-demand and ease of use.

The Cluster Manager offers the ability to partition clusters into virtual clusters, thus giving clients the illusion they are operating whole clusters dedicated to them. While Quality of Service functionality is offered, it is limited to functional aspects of cluster jobs such as the maximum amount of time the client will tolerate the job being in a queue. In general, this offering exhibits the cluster on demand property (as with the Adaptive HPC Suite).

In terms of workflows, orchestration of multiple services/executables, there is the Moab Workload Manager. It is a scheduler with a few extra features: it is possible for jobs to change their resource allocations during execution (for example, the first five minutes use a single node, the next five use two nodes, etc.) and integration with existing grids.

Finally, Access Portal provides job submission, monitoring and control to clients with a graphical view of the cluster job load based on the individual nodes. The problem with the portal is that getting jobs on the cluster is user unfriendly; the Web page is mostly a wrapper for a command line. While Adaptive Systems does make reference to HPC on cloud computing, they do not provide a tangible solution. All they offer is architecture for resource consolidation [200].

11.3.3   Physical HPC Offerings in the Cloud

With cloud computing's use of virtualization, performance overheads are often exaggerated by some experts. Irrespective, there are clients that simply prefer to use physical servers over virtual servers. The purpose of this section is

---

[3]Given Azure is a .NET based environment, it is assumed only .NET tasks can be offloaded.

to examine HPC cloud offerings that use physical servers instead of virtual servers.

11.3.3.1 Silicon Graphics Cyclone

Cyclone [373] is a cloud that does not use virtualization and offers itself to clients as both a SaaS and an IaaS cloud. As a SaaS cloud, Cyclone provides a wide array of applications for numerous scientific fields [374]: from BLAST for biology experiments to SemanticMiner for Web ontologies.

In terms of an IaaS cloud, all that can be learned is that clients are able to install their own applications as in EC2 — clients are provided with a server with an installed operating system as a starting point. However, the hardware specifications are extremely vague with the only detail published that the machines use Intel Xeon processors. As with previous cloud offerings, the pay-as-you-go billing model is exhibited. However, that is the only property exhibited.

11.3.3.2 Penguin Computing on Demand (POD)

A similar solution to Cyclone, Penguin Computing's POD [201] offers physical cluster nodes to clients as utilities. Physical servers are used to carry out application execution while clients access POD via allocated (virtual) login servers. While Penguin Computing claims security on its POD offering, an examination of the documentation shows that the security relies on best-practices (trust) [202, 203]. Furthermore, when a client submits a job to a compute node in POD, the nodes performing the execution become dedicated to the client. While advertised as a cloud, no HPC cloud property is exhibited by POD. In general, POD is a remotely accessible cluster — not a cloud by definition.

11.3.4 Summary

In general, the idea of hosting HPC applications on public clouds such as EC2 is possible. However, there is yet to be a solution that combines both the convenience and low cost of clouds with high performance. Some HPC cloud solutions are clouds only in name and only focus on making clusters accessible over the Internet. Thus, clients are significantly involved with the installation and management of software required by their applications.

Of all the properties presented in Section 11.3.1, only the properties of clouds being Web service based, using the pay-as-you-go model, elasticity and ease of use are exhibited. Other important properties such as transparent cloud management, and guaranteed performance, are not exhibited. The latter property is of significant importance — HPC is about running large software applications within reasonable time frames, thus performance is a must.

11.4 HPC Cloud Challenges

While the term HPC cloud is in use, current solutions only exhibit a minimal set of required properties. Besides requiring clients to (in the worst case) be involved in the infrastructure establishment of HPC environments (from finding and configuring data storage to building and configuring HPC applications for execution on clusters), literature shows [342, 636] that clients may also encounter issues with placing HPC software on cloud hosted clusters, interoperability if resources are spread across multiple clouds (or they wish to migrate), discovering required resources and even ensuring they get acceptable cost-to-performance value.

Based on the literature presented in Section 11.3, this section presents the challenges faced when using clouds to support HPC. While the focus is on HPC clouds specifically, many of the issues presented in sections exist in a same or similar form for cloud computing [733] — specifically the infrastructure aspects of clouds.

11.4.1 Interface Challenges

Most cloud and cluster solutions offer command line interfaces [201, 373] but rarely (if at all) present graphical interfaces. Even when provided, graphical interfaces tend to act as command line wrappers (as seen with Moab [602]). In some cases, the only Web interface is an API for Web-enabled software processes. Besides frustrating and confusing to general users, users who are computing experts may prefer to spend less time getting applications to work on clusters and more time conducting research.

The best cloud interface so far is that offered by CloudSigma [196]. Using a desktop-like interface, CloudSigma is easy and best suited for human operators as it uses a common interface approach. What is lacking is a facility to manage multiple servers at once. While the need for interfaces is mentioned in [733], it examined the need for having a private cloud communicate with a public cloud in the event of resource shortages. This supports the claims that most interfaces offered are programming/infrastructure oriented and often overlook the need for human oriented interfaces.

11.4.2 Performance Challenges

At the very least, a cluster (physical or virtual) hosted in a cloud should run applications as fast (or even faster) than existing physical clusters. Currently, there is no mechanism in place to evaluate performance factors such as current CPU utilization and network IO. Furthermore, there is no service that allows the evaluated factors to be published.

Immediately, this makes it impossible to discover servers and clusters based on quality of service (QoS) parameters. Also, it makes it impossible for the cloud to be intelligent (Section 11.4.4) — the cloud is not able to see that (for example) a virtual cluster is under performing due to high network latency thus will not make attempts to transparently reconfigure the cluster to use a different (idle) network.

Currently, the only parameter guaranteed with clouds is high availability (as seen with EC2 [62]). Similarly, this issue appears as a resource management and energy efficiency issue in [733]. While it is true that allocation of physical CPU cores to virtual machines can significantly improve performance, the approach can increase network latency as virtual machines are placed farther and farther apart so they have effective use of physical CPU cores.

11.4.3 Communication Challenges

A common issue with clouds is their network performance, e.g., EC2 [736]. Network performance cannot be solved by using an Infiniband network alone. There is still the issue of network topologies within clouds. Even if a cloud is hosted within a single location (organization), the virtualized nature of clouds may cause a single (virtual) cluster to be hosted across multiple systems (separate datacenters).

At the time of writing, no cloud could be found that details how or where client virtual machines are placed. As stated at the end of Section 11.4.2, communication overheads and performance are at odds with each other — spreading out virtual machines may improve (CPU) performance, but the overall performance may remain unchanged due to increased network latency caused by VM placement.

11.4.4 Intelligence Challenges

While cloud computing offers resources as a utility, they are not always intelligent when allocating resources to clients. For example, if a client needs additional (virtual) servers, the client has to invoke functionality in the cloud itself to create the new servers.

The other issue is the placement of (virtual) servers in relation to each other. While some HPC cloud offerings claim to use Infiniband, they do not give any guarantee that client nodes are kept in close proximity of each other.

11.4.5 Configuration Challenges

As well as providing clusters in clouds, it has to be possible to reconfigure the specifications of existing clusters for different types of cluster applications. The configuration of a cluster cannot become rigid, otherwise application developers will be forced to build the applications specifically for the cloud cluster and not the problem they intend to solve [636]. So far, the only such solution that allows a cluster to be dynamically reconfigured is the Adaptive

HPC Suite [603]. However, the solution is too course grained: it allows for the complete change of an installed operating system, not individual software libraries.

11.4.6 Publication and Discovery Challenges

If a cluster is deployed in a cloud, current information about the cluster has to be published in a well-known location. Otherwise, clients cannot find and later make use of the cluster. Currently, there is no cloud registry for clusters or other HPC resources. This is also an issue for other cloud and HPC related resources such as individual servers and data services.

Furthermore, clients should not be subjected to information overload. For example, if a client only wants a 10 node Linux cluster with 2 Gigabytes of RAM per node, the client should only be informed of clusters that satisfy those requirements and how. If the client does not care if the cluster is using Infiniband or Gigabit Ethernet, then the client should not have to be told about it. Overall, publication in clouds only varies between having static catalogs [61, 374, 632] to only tracking unique addresses of services [661].

11.4.7 Legacy Support Challenges

Often, the term legacy is used to describe software written years ago but still in use. However, it is possible for the legacy unit itself to be data generated from experiments years ago and stored using software systems current at the time. The challenge with such data is ensuring that it can be maintained in its original form (thus avoiding loss) but also making the data accessible to newer software (compatibility). In some cases, the data is too big to be converted to a newer format.

A similar challenge was faced with the CloudMiner [254]: the goal was to use cloud computing to provide data mining in an effective manner across heterogeneous data sources. The solution proposed was to have all data and data mining algorithms exposed via services that are published to well-known brokers. The end result is an environment where providers could publish their data and data mining techniques in a single, cloud based environment thus allowing clients to discover required data and run required algorithms over the data, irrespective of how the data is stored or where it is located.

11.4.8 SLA Challenges

One of the most difficult challenges is forming and maintaining agreements between clients and services. Computer systems (all types from standalone systems to clouds) are built using technological objectives. Even if the underlying computer system is the most effective at getting a given task done, it is ignored if it does not satisfy client (personal) requirements. For example, a database system may offer data storage using RSA encryption for strong security but the client (for whatever reason) may require Triple-DES. RSA is better but it is not what the client requires hence is not acceptable.

Besides satisfying client preference, mechanisms need to be in place to ensure that the terms outlined in the SLA at met for the life of the agreement and that violations of the agreement are properly addressed. For example, if the SLA states a job will be completed in two week but the job takes four, that is a violation, for which the provider must bear a penalty. Based on the current cloud offerings, SLAs only cover against loss of availability and state damages awarded (as seen with EC2 [62]). No other issues, such as security violations, are effectively covered.

While security and availability are presented in [733], possible resolutions to them are unconvincing at best. In terms of security, it is suggested that virtual VLANs (virtual, overlay, LANs that exist independent of physical LANs) could help improve security via traffic isolation. There is still the open issue of what international law applies when there is a security breach. However, when it comes to availability, [661] proposes that SLAs go beyond availability, need to consider QoS parameters and agreements properly formed between service providers and clients so that violations can be settled properly.

11.4.9 Workflow

The final challenge when creating HPC clouds is supporting client workflows. Not all HPC problems are processed using a single application; some problems require a chain of multiple different applications. The challenge with workflows is ensuring that each application in the workflow has all of its required resources before execution. Furthermore, it is unrealistic to assume that all resources will exist on a single cloud; the resources may be spread across separate clouds.

11.5 HPC Cloud Solution: Proposal

In response to the challenges presented in Section 11.4, this section proposes a new cloud solution called HPCynergy (pronounced HP-Synergy). The goal of HPCynergy is to provide an HPC oriented environment where HPC resources (from data storage to complete cluster environments) are accessible to clients as on-demand, easy to use services. While the HPCynergy solution aims to satisfy all properties presented in Section 11.3.1 the virtualization property is made optional for clients who have HPC applications that are sensitive to network/CPU latencies.

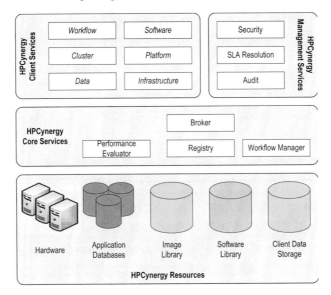

FIGURE 11.1
HPCynergy Structure.

11.5.1 HPCynergy Design

HPCynergy is designed as a collection of services, each service belonging to one of four broad groups. To address the Interface Challenge (Section 11.4.1) all services presented in this section have a Web API for software process clients and intuitive Web pages for human operators using Web browsers. Figure 11.1 shows the structure of HPCynergy with numerous resources at its base ranging from large hardware systems to various database services.

On top of the HPCynergy resources are the Core Services (Section 11.5.1.2): provided in response to the challenges of Publication and Discovery (Section 11.4.6), Configuration (Section 11.4.5) and Performance (Section 11.4.2). In relation to [733], the Cores Services are a realization of client on-demand resource provisioning and negotiation. Based on supplied requirements, clients learn what resources are available and decide for themselves which ones they will use. If matches cannot be found, clients retry with updated requirements.

On top of the Core Services are the HPCynergy Management Services (Section 11.5.1.3): provided in response to the Communication (Section 11.4.3) and SLA (Section 11.4.8) challenges. Security is also offered so that all clients only use services (and resources) intended for them.

Finally there are the HPCynergy Client Services (Section 11.5.1.4). The services provided come from either HPCynergy itself (such as cluster discovery) or services that have been provided by other HPCynergy clients. For example, when a HPC application completes and generates new results, the results can be kept in place in the Application Database thus supporting further projects.

11.5.1.1 HPCynergy Resources

HPCynergy provides a number of resources to users, which are described below:

Hardware: Hardware can exist as physical servers or virtual servers to support two scenarios for clients. First, a client may require a special purpose environment, hence will wish to start with raw (virtual) hardware and construct a suitable system from there. The other scenario is the client's requirements of an environment may be flexible, thus only a reconfiguration of existing services and resources (such as a pre-made cluster) are needed.

Application Databases: To support HPC applications, application databases (such as genome databases) are offered to any client. By providing the database the client can focus on the application to be run and avoid lengthy transmissions from remote locations to the cloud.

This element is similar to the Storage Virtualization feature in [733]. In that work, the focus was on providing a uniform means of data storage to support virtual machines — most likely how the virtual machines could be stored and how virtual hard drives could be provided to them. The HPCynergy realization is different — as well as providing a store of many application databases, there is a plan to offer the databases in a format clients require. E.g., if a client prefers SQL, all databases (regardless of type or format) are presented as SQL databases to the client. In effect, the aim is to provide the same uniform environment as CloudMinder [254].

Image Library: The Image Library contains numerous operating systems in both initial states (just the operating system installed and nothing else) and in pre-packaged states (have various combinations of installed software applications and libraries).

Software Library: The Software Library is provided so that HPC clients can create virtual machines and then either make use of or install required software to run their applications. In general, the Software Library keeps software as operating system specific installable (such as executable for Windows and packages for Linux). By keeping a library, clients will be able to install required software (along with any required dependent software).

Client Data Storage: The final element of the HPCynergy resources group is data storage for individual clients. When a client application is complete, the virtual machine needs to be disposed of so that the hardware resources can be released and used for new virtual machines. However, if a virtual machine is disposed, any data in it may be lost.

Thus, the Client Data Storage is provided so that clients can keep their data (results and log files) separate from their virtual machines. Furthermore, providing data storage to clients allows easy sharing and makes it possible for clients to share the results of the experiments with other clients. As with the

Application Databases, the Client Data Storage is a virtualization of data storage [733] which aims to provide all data in a required client format [254].

11.5.1.2 HPCynergy Core Services

Clients to HPCynergy are not limited to computing specialists needing to run an application on a (virtual) cluster. There is also the possibility where the client is a research team that intends to realize a new form of research environment. Thus such clients can be considered Service Providers and require facilities to make their new services known to other clients.

The purpose of the Core Services is to provide a set of generalized and reusable services to support the operation of HPCynergy and to help support clients who wish to provide their own HPC services to other clients. The goal of the Core Services is similar to the objectives behind CloudMiner [254]. CloudMiner allows any form of data mining algorithm to be carried out on any form of data source without any dependency between each other. HPCynergy allows any form of HPC application to be carried out using any HPC resource necessary without any dependencies between the two either.

Registry: The Registry allows for extensive publication of all services and resources (from databases to clusters) within HPCynergy. Having a registry system in a cloud is a significant feature. At the time this was written, all other clouds used catalogs for their services/resources instead of actual discovery systems ([61, 374, 632] to name a few).

Broker: Working with the registry, the Broker is a service that allows clients (human operators or other services within the HPCynergy) to find and select required services and resources [145]. Like the registry, such a service does not exist as yet in cloud computing hence its presence is a unique feature.

It could be considered that a Broker and Registry could be combined into a single service. While that may be true, the two services are separate to keep client request load low — high demand for the registry service will not affect attempts to discover, locate and arrange access to required services and resources.

Performance Evaluator: The Performance Evaluator service performs routine inspections on all services and resources within HPCynergy. By keeping current information about performance factors (considered quality of service parameters — QoS), it is possible to allocate high quality (thus high performance) resources and services to HPCynergy clients. Furthermore, other management services such the Workflow Management service can use the same information to avoid poor quality resources.

Workflow Management: Clients do not always require just a single service or resource. Clients may require a set of resources and services and (within that set) a required workflow. Like all resources and services, workflows also have requirements of their own: specifically, they all require a series of tasks to be carried out on specific services. Furthermore, each task within the workflow will have requirements on what data is passed in and what information comes out.

Thus, the Workflow Management service is a high level version of the Broker: the requirements going in will be the description of an overall workflow and the outcome is to be an orchestration that can execute the workflow within client specified limitations (for example, do not take longer than two weeks and do not cost the client more than $200).

At the time of writing, no adequate services for workflow management in clouds could be found. The closest solution found is the Moab Workload Manager [605] and even then it is ineffective, because jobs must be prepared before being ordered into a workflow.

11.5.1.3 HPCynergy Management Services

As with any service based environment, access to the services (and the resources offered by them) has to be monitored. Thus, a means of security has to be in place not just to prevent unauthorized client access to services, resources and data, but so that HPCynergy can comply with any regional laws regarding information security.

While good detail of management services/features are presented in [733], this chapter considers them in the Core Services group — they are fundamental to getting HPCynergy working. Thus in this group HPCynergy offers services that address difficult (and often nation spanning) issues.

Security: The role of the security service is to provide the three core elements of security: confidentiality — clients only access and use services, resources and data meant for them; integrity — data and communications between clients and services are not modified in transit; and authentication — to ensure that all entities (clients, services and resources) can have their identities verified.

SLA Resolution: If services and resources are to be presented to a client based on their requirements, then the parameters of the services when they were discovered need to be maintained. Thus, when services (and the resources behind them) are discovered and selected, a service level agreement (SLA) needs to be formed and enforced. That is the role of the SLA Resolution service. Once a SLA has been formed between a client and a service, any violation that occurs goes to the SLA Resolution service. This service seeks to settle disputes and carry out any penalty/compensation agreements outlined in the SLA.

This service is especially important when supporting client workflows. Workflows have very long lifetimes and involve numerous services and resources directly (may utilize them without services). Thus a single workflow will have numerous SLAs, one between each service for a given task and one for the complete workflow as a whole.

Audit: A counterpart to the SLA Resolver, the Audit service monitors all interactions with all services and ensures that clients pay for what they use. Another scenario addressed by the Audit service is where clients make use of services but fail to pay any fees. Thus, another role of the Audit service is to check clients wishing to make use of services in HPCynergy. Even if the client has security clearance, if the Audit service detects that the client has failed to pay for service use earlier, the Audit service blocks the client's access.

11.5.1.4 HPCynergy Client Services

In this group, the services offered are intended primarily for client usage and in turn make use of services in HPCynergy. As such the services in this category will vary from services provided by HPCynergy itself to services created and later offered by clients (often called reselling in business worlds).

In general, services in this collection are intended to provide high level abstraction for otherwise complex specialized tasks. For example, to ease clients in their discovery of a required cluster (or reconfiguration of an existing cluster), infrastructure-like services are placed in this category so that clients only focus on what they require in a cluster and all other resources and services are arranged for them.

Infrastructure: Infrastructure services offer functionality to clients so that systems from standalone virtual machines to complete cluster arrays (without software) can be created. For example, if the client requires a 20 node cluster, each with a single core, a service can exist to find physical servers on the cloud to host 20 virtual machines and run an appropriate hypervisor.

Platform: Like Infrastructure services, platform services offer complete hardware and software environments. For example, a client may require a 20 node Linux cluster but does not care how it is built. Thus a platform service can be presented that offers prebuilt clusters and/or stand-alone (virtual) machines.

Software: Software services are focused on offering required HPC applications and other support services to clients. For example, if a client only needs access to mpiBLAST and the delivery is irrelevant, mpiBLAST can be exposed as a service and offered in this area of HPCynergy.

Another example is a software compiler: instead of burdening clients with the task of installing and configuring all required software libraries, the task can be encapsulated in a service instead. All that is required of the client is the source files and the compiler input parameters.

Data: Data services are provided so that data sources in HPCynergy can be seen by clients as huge, homogeneous databases. For example, Client A will see the HPCynergy will all data as a relational SQL Database while Client B will see the data as XML files. Furthermore, if a version of CloudMiner [254] existed for HPCynergy, this would be the most likely location to place it.

Cluster: Cluster services are provided to ease the discovery, selection, and use of clusters. If a cluster cannot be found, the cluster services can attempt to make a required cluster using the Management Services (Section 11.5.1.3).

Workflow: Similar to the software category, if a research team finds an effective means of using existing services to quickly solve a problem, the workflow

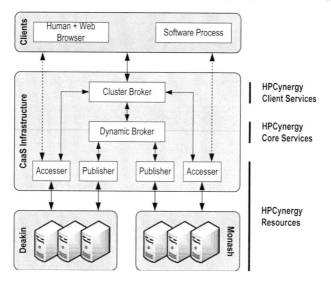

FIGURE 11.2
HPCynergy Implementation Stack.

used itself is a valuable contribution thus should be made accessible as a service.

11.5.2 HPCynergy Implementation

HPCynergy is implemented using two clusters and new Web service-based middleware called Cluster as a Service (CaaS) Infrastructure. Figure 11.2 shows the implementation stack of HPCynergy in relation to clients and the two clusters. Annotations on the right indicate where the current services relate to the HPCynergy design in Section 11.5.1.

As this is the first implementation of HPCynergy, the focus was mostly on easing the publication and access to existing clusters. The reason for this is that if clients cannot learn of existing resources (in this case, a cluster) then it does not matter how well managed or secure or reliable the resources are as the client will not know of then thus will not make use of them.

With respect to the features shown in [733], the current implementation aimed to have the features of virtualization support, providing services and resources on demand, dynamic allocation and reservation/negotiation.

The Deakin cluster is comprised of both physical and virtualized nodes. In terms of hardware, the cluster has been constructed using 20 servers, each with two quad-core Xeon processors. Of these 20 servers, 10 have VMware vSphere [730] installed thus allowing them to run virtual machines instead of a single operating system. Subsequently, 20 virtual machines (each with 6 virtual cores) have been created thus (theoretically) forming a 30 node cluster. All nodes (virtual and physical) run the CentOS operating system and have Sun Grid Engine and Ganglia installed.

The Monash cluster is a physical cluster of 36 nodes — 20 nodes are the same as the Deakin Cluster in hardware and software, and the remaining 16 nodes have the same software but instead have two dual core processors each.

In terms of virtualization support [733], the Deakin cluster supports this feature. However, the focus is on using virtualization to gain more efficiency from the cluster — virtual machines can be migrated/rearranged so that busy virtual machines can be dedicated whole physical systems, while idle virtual machines can be consolidated to single physical nodes. There was no focus on clients being able to create and maintain their own virtual machines. Also, while there is mention of supporting multiple hypervisors, that was not considered this in the HPCynergy implementation. First, a single type of hypervisor to simplify testing was kept. Second, there is yet to be the invention of a cloud that uses more than one form of hypervisor.

In terms of dynamic resource allocation, the best that is in the current implementation is that offered by individual cluster schedulers — cluster jobs start off in a queue and are executed when enough nodes become available to be allocated to them.

To make the clusters accessible to clients, the Cluster as a Service (CaaS) Infrastructure solution is used. An evolution of the previous CaaS Technology [146], the CaaS Infrastructure provides an environment where numerous clusters can be easily published by their providers and later easily discovered and used by clients.

Clusters offered via the CaaS Infrastructure are coupled with a pair of services: a Publisher Service that makes current and detailed information about clusters known and an Accesser Service that allows clients to use the related cluster in an abstract manner. In relation to the HPCynergy design, the clusters along with their Publisher and Accesser Services belong to the Resources group (Section 11.5.1.1).

To make information about the cluster known to clients, the Publisher Services forward all information to a discovery system called the Dynamic Broker [145]. In relation to the HPCynergy design, the Dynamic Broker is an implementation of both the Registry and Broker services (Core Services, See Section 11.5.1.2).

Finally, clients discover and select required clusters through a high level service called the Cluster Broker service — a simplified form of the Dynamic Broker that specializes in taking client requests, encoding them into attributes for use with the Dynamic Broker and then displaying result information in an easy to read manner. After learning and selecting a required cluster via the Cluster Broker service, clients approach the related Accesser service and make use of the cluster. In terms of the HPCynergy design, the Cluster Broker service belongs to the Client Services group (Section 11.5.1.4).

In relation to [733], the Cluster Broker service exhibits the Reservation and

Negotiation Mechanism — albeit manual. In terms of negotiation, clients are presented with search results and decide for themselves which cluster is suitable for their needs. If a required cluster cannot be found, the only negotiation possible is where clients update their requirements and run the search again. In future work, it is planned to create additional services to complete the HPCynergy Core Services group and implement the Management Services group. The Client Services group will evolve as more applications and ideas to HPCynergy are applied.

11.6   Cloud Benchmark of HPC Applications

As seen in Sections 11.3 and 11.4, cloud computing is an emerging technology and there are many concerns about the security, performance, reliability and usability. Despite this, the advantage of on-demand computing is beneficial when solving research problems that are data heavy and/or computationally intensive. It is this adoption of cloud computing by the scientific community that has paved the way for HPC clouds (such as HPCynergy described in Section 11.5). Performance of HPC clouds is important due to the nature of scientific problems. This section uses benchmarking to compare the usability and performance of cloud computers and dedicated HPC machines.

11.6.1   Tested Applications

To investigate the feasibility of running HPC applications on HPCynergy, and other clouds, two practical applications were tested. The first application is an embarrassingly parallel system biology pipeline developed at Deakin University (Section 11.6.1.1). The second application is called GADGET; a communication bound N-body physics simulation (Section 11.6.1.2).

11.6.1.1   System Biology Pipeline

Bioinformatics studies complex biological systems by applying statistics and computing to genomic data. While the first human genome was sequenced at a cost of $2.7 billion dollars improvements to sequencing chemistry has reduced the price to $10,000 [256]. As a result bioinformatics has seen a large increase in available data. It has been estimated that in 2 years the cost of sequencing will be $1000 per human genome, thus making it economically feasible to sequence large amounts of human and other complex organisms.

Collecting large amounts of genomic data has a number of important implications in cancer treatment and medicine, in particular through personalized cancer treatments. These treatments rely on first identifying cancer subtypes which can be found or diagnosed by building system models [255], which show

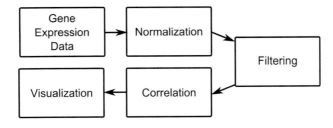

FIGURE 11.3
A Common System Network Workflow.

Gene 1	Gene 2	Correlation
SGT20c1_H07	SGT20c1_H07	1.0
SGT20c1_H07	SGT20c1_H04	-1.0
SGT20c1_H07	SGT20c1_F04	-0.8
...	...	...
SGT20c1_H04	SGT20i4_E07	0.0

FIGURE 11.4
An Example of the Simple Interaction Format.

the interaction of genes in a biological system. Building system models is a multi-step process (see Figure 11.3), consisting of; normalization and filtering of data, statistically correlating genes, and then visualizing results. Of these steps, correlating genes can be the most time consuming; a list of N genes requiring N correlations to be made for each gene.

This system biology workflow has been implemented using a combination of R [372] and C++. In this implementation cDNA array gene expression data is first normalized using a cross channel, quantile-quantile approach. Filtering is then used to remove noise and find significant genes. After filtering, a distributed implementation of the Pearson's correlation algorithm [351] is used to find relationships between genes. Data is outputted in Simple Interaction Format (SIF) [170], a tab delimited text format used to store network data (see Figure 11.4). Correlated gene data is then visualized as a network using Cytoscape [257].

The following benchmark utilized gene expression data collected before and during the Tammar Wallaby lactation period. Eight observations taken over this lactation period were first normalized and then filtered. Remaining genes were then correlated using Pearson's R. Genes which were strongly correlated were visualized (Figure 11.5).

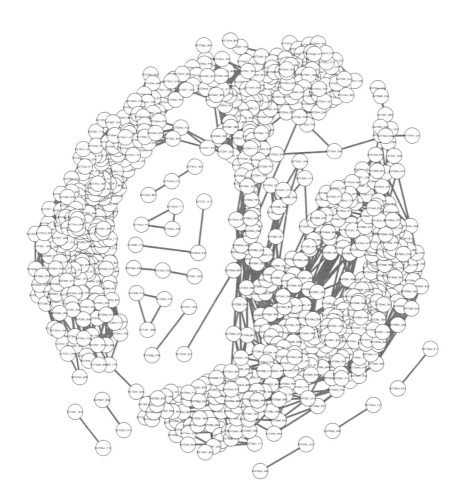

FIGURE 11.5
Network of Genes Expressed during Tammar Wallaby Lactation.

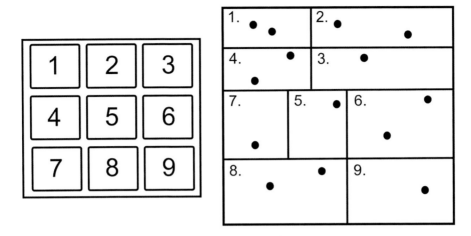

FIGURE 11.6
Example of Space to Processor Mapping during an N-Body Simulation.

11.6.1.2 GADGET: Communication Bound N-body Simulation

Like bioinformatics, physics also contains many computational intensive problems. Particle accelerators such as synchrotrons and the Large Hadron Collider (LHC) generate large amounts of data. The LHC smashes bundles of Quarks (the smallest known particles at this time) together close to the speed of light in the hopes of finding evidence for the existence of several theoretical particles, such as the Higgs boson and those predicted by Super Symmetry [576]. During these experiments terabytes of data are collected using multiple sensors which then must be verified by comparisons to particle simulations based on current theories [652].

The implementation of particle simulations, also known as N-body simulations, is diverse. A common implementation method, called physical mesh, maps compute nodes to physical space [110]. As particles move and collide in simulated space, particle data is also distributed to the node which simulates that space. This transfer of particles between nodes is facilitated by message passing. Because communication is a key requirement it is common for logical space to be ordered in a similar way to physical space (Figure 11.6). By ensuring communication is between physically close nodes, communication speed is made as fast as possible. How this mapping is accomplished depends on the hardware and N-body algorithm used.

GADGET [672] is a well-known astrophysics application designed to simulate collision-less simulations and smoothed particle hydrodynamics on massively parallel computers. This program was used to run a simulation of two colliding disk galaxies for a total of 10 simulation time-steps. To ensure a standard level of accuracy, particle positions were calculated at 0.025 current Hubble time

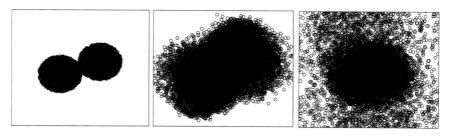

FIGURE 11.7
Visualization of Two Disk Galaxies Colliding.

intervals. Snapshots of particles were written out as binary files every 0.1 simulation time-steps, some of which have been visualized using R (Figure 11.7).

11.6.2 Benchmarked Platforms

Two physical clusters and three clouds were used during the benchmark study. Naming conventions of the machines are as follows: each cluster is referred to by network interface (InfiniBand Cluster, Ethernet Cluster) and each cloud is referred to by the cloud management interface (vSphere [730], Amazon, HPCynergy). In terms of hardware, these computer platforms were chosen to be as similar as possible to each other; even when possible utilizing the same hardware. Of the five systems described below, vSphere and the InfiniBand cluster run on subsets of the Deakin Cluster[4](described in Section 11.5.2), while the Ethernet and Amazon machines use their own dedicated hardware. Despite the large effort taken to minimize hardware differences, the Ethernet and some Amazon instances differ in the amount of cores per processor. Because of this variation, each process was mapped to a single core and when possible a single node. To validate the mapping process, CPU usage was monitored during data collection, for example a dual core system with a single process would be using 50% capacity. This methodology was chosen because it is similar to that used by the clouds, in that virtual machines are mapped to physical hardware.

The InfiniBand cluster used in this benchmark consists of the physical nodes on the Deakin Cluster. This cluster is a bare-metal system consisting of 10 nodes, each with an Intel Quad Core Duo processor running at 2.33 GHz. Each node utilizes 8 GB of RAM and runs a 64-bit version of CentOS. As a cluster dedicated to HPC, nodes are connected using 10 GB InfiniBand, and a

---

[4]Testing has been on the Deakin cluster directly and (later) through HPCynergy. This is to assess the claim that using middleware with high levels of abstraction results in clusters being unusable due to poor performance.

mounted network drive allows users to easily setup MPI applications. In terms of CPU speed and RAM, this cluster is equivalent to the documented specification of the large Amazon Instance but differs by having a faster network interconnect.

The Ethernet cluster used in this benchmark is also devoid of virtualization and is equivalent to the documented specification of the small Amazon instance. This four node cluster was constructed using Intel Dual Core computers running at 1.6 GHz each with 2 GB of RAM. A 32-bit version of Ubuntu 9.10 was used as the OS. Compute nodes were connected by a low I/O Ethernet network (1 Gb/sec).

Three Amazon instance types were tested; small, large and cluster. It has been documented that Amazon uses a modified version of Xen as the hypervisor [59]. In each case the Amazon Elastic Block Store (an Amazon service which provides persistent storage of virtual hard-drives) was used to store the state of the deployed virtual machines. Each instance type differed in CPU, RAM and network I/O. Amazon measures the performance of CPUs in Amazon Compute Units (ACUs), this is equivalent to an Intel Xeon chip.

Each Amazon small compute instance contains 1 ACU and 1.7 GB ram. Connection between Amazon small instances is documented as low I/O [61]. The large instances contain a dual core CPU (each with 2 ACU of power) and 7.5 GB of RAM. Connection between Amazon large instances is documented as high I/O [61]. The Amazon Cluster Compute Instances is the best defined, these machines contain two Intel "Nehalem" quad-core CPU running at 2.98 GHz and 26 GB of RAM [61]. Connection between cluster instances uses a 10 GB Ethernet connection. The small and large instance types were used to setup a 17 node cluster; however, the allocation of the Cluster Compute Instance was capped at 8 nodes.

The second cloud used in this benchmark utilized VMware vSphere. This private cloud runs on the physical nodes of the Deakin Cluster. A ten node virtual cluster was deployed through this VMware cloud, each with dual core processors running at 2.33Ghz. A 10 GB InfiniBand network was used to provide inter-node communication.

The final cloud used in this benchmark is HPCynergy (see Section 11.5). This cloud platform exposed the whole Deakin Cluster through the underlying CaaS Infrastructure. A total of seventeen compute nodes were utilized through HPCynergy, each node containing a hexa-core processor running at 2.33 GHz[5]. A 10 GB InfiniBand network provided inter-node communication.

Specifications of all platforms used in the benchmark tests are summarized in Table 11.1.

---

[5]Only virtual nodes were utilized during the benchmark to enable comparisons between HPCynergy and the vSphere cloud.

TABLE 11.1

List of Computer Platforms Broken Down by Specifications.

Name	Nodes	Hyper-visor	Plat-form	Hard Disk	CPU	RAM	Network
Amazon Cluster	8	Modified Xen: HVM	64-bit CentOS	Elastic Block Store	2 x Intel quad-core Nehalem (2.93GHz)	23GB	10Gb Ethernet
Amazon Large	17	Modified Xen: Para-virtual	64-bit Ubuntu 9.10	Elastic Block Store	2 x 2007 Xeon equivalent (2.2GHz)	7.5GB	High I/O
Amazon Small	17	Modified Xen: Para-virtual	64-bit Ubuntu 9.10	Elastic Block Store	2007 Xeon equivalent (1.1Ghz to 1.6GHz)	1.7GB	Low I/O
vSphere Cloud	10	VMware	64-bit Ubuntu 9.10	Separate Drives	2.33Ghz Intel Dual Core	2GB	10Gb InfiniBand
Infini-Band Cluster	10	None	64-bit CentOS	Shared Drive	2.33GHz Intel Quad Core Duo	8GB	10Gb InfiniBand
Ethernet Cluster	4	None	64-bit Ubuntu 9.10	Separate Drives	1.6GHz Intel Dual Core	2GB	1Gb Ethernet
HPC-ynergy	30: 10 physical 20 virtual	VMware	64-bit CentOS	Shared Drive	Virtual-Hexa-cores (2.33GHz) Physical-Dual Quad Cores	8GB	10Gb InfiniBand

11.6.3   Setting up the Clouds: Methodology

Setting up computer resources for HPC is a time consuming task and often serves to interrupt research. While the Ethernet and InfiniBand clusters used in these benchmarks could be used once code had been compiled, the Amazon and vSphere clouds required modification to enable HPC. The HPCynergy cloud solution aims to reduce setup time by exposing systems which have middleware already setup. The modification scope has been defined by both the application and cloud infrastructure.

11.6.3.1   General Setup

Before benchmarking could occur, each cloud required a number of steps including: (i) transferring data and source code, (ii) configuring the dynamic linker, (iii) compiling the source code and any dependencies, (iv) configuring the sshd client, (v) generating public and private keys, (vi) passing public keys to all nodes and (vii) creating a machineFile for MPI.

In addition to the above steps, each cloud had limitations which required additional setup time. The vSphere system did not contain any VM templates thus installation of the Ubuntu OS was required before operation. While all Amazon E2C instances used in these benchmarks did not have common utilities such as the g++ compiler, the g77 compiler, vim or zip, setting up these applications required additional time. Once each system was setup, transfer of benchmark specific data and compilation of necessary software was required.

11.6.3.2   System Biology Pipeline

A number of programs were used during the bioinformatics benchmark; filtering and normalization was performed using C++ and correlation utilized the R runtime environment (see Section 11.6.1.1). The data and source code was zipped and transferred through the scp utility; once on the target machines, the source code was then configured and compiled. The InifiniBand cluster used a shared drive and therefore compilation was only required once. Set up of both the Amazon and vSphere cloud systems was simplified through use of virtualization; a template containing the necessary software and data was created and then cloned. The Ethernet cluster made use of separate drives but because each node was homogeneous to each other (in hardware architecture and software versions) each could use the same compiled binary.

Figure 11.8 shows the necessary setup times for each platform; these times are based on best case scenarios and do not take into account problems occurring during compilation. It should be noted that this is not usually the case. Common problems include missing configuration augments, missing library dependencies, and compiler specific code. If any of these problems occur compilation has to start over.

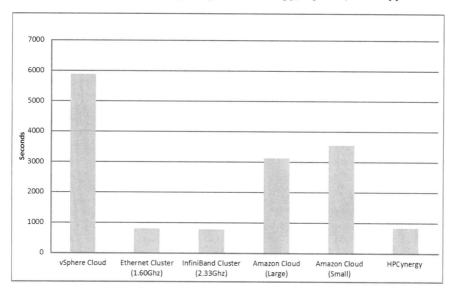

FIGURE 11.8
Total Setup Time of the 4-Node System Biology Benchmark.

11.6.3.3 GADGET: Communication Bound N-body Simulation

The physics application GADGET (Section 11.6.1.2) required a large number of dependencies. These dependencies are as follows:

- OpenMPI for distribution of GADGET across nodes,
- GSL - a general scientific library for C++,
- FFTW - an MPI implementation of the Fourier transformation, and
- HD5F - a binary output format.

As in the bioinformatics setup, data and source code was zipped and transferred through the SecureCopy (scp) utility from a local machine. Dependencies were compiled from source code on each target machine and GADGET was setup for HD5F output and FFTW double floating point support. Reported setup time (see Figure 11.9) is based on best case scenario and was further minimized by use of cloning or shared drives. Despite these advantages each cloud system required on average 7 hours to setup, while the cluster setup time was less than an hour.

11.6.3.4 HPCynergy Features and Setup

When setting up HPCynergy for benchmarking, setup time was minimized due to its unique interface. Like other clouds, HPCynergy monitors and acts as a broker to linked (physical and virtual) hardware. However, instead of hiding

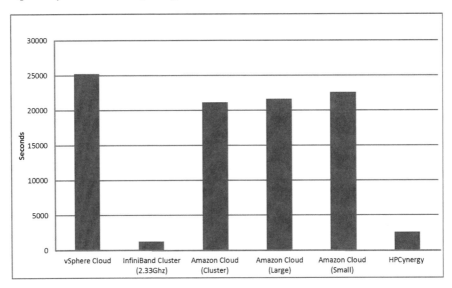

FIGURE 11.9
Total Setup Time of the 17-Node Physics Benchmark.

the state and specification of hardware from the users, the opposite approach is taken. Users are informed of the software and underling (virtual) hardware specifications of each machine. This allows jobs to be optimized to the CPU architecture as well as minimizing the need to install specific libraries.

When compared to other clouds (see Figure 11.10) the HPCynergy solution has a reduced setup time. This is due to a Web interface which allows users to search for a cluster that contains enough resources to support their job. Once a user has selected a cluster, minimal user requirements are used to configure an underlying scheduler — such as the number of required nodes. On submission of a job, a user provides only a bash script which oversees the whole cluster job (workflow) and any data files they wish to upload. While compilation of application level software is sometimes necessary, cluster middleware is already setup (for example, schedulers and libraries such as OpenMPI). The advantage of this method is that a user has only to configure and start the applications.

11.6.4 Benchmarking

Comparisons made between collected results highlight the advantages and weaknesses of utilizing specific cloud platforms for high performance scientific computing. HPCynergy addresses many of these weaknesses (see Section 11.4) through a combination of Web services and easy to use Web forms, but benchmarking is necessary to prove that HPCynergy is feasible with regard to performance. It is often claimed that providing ease of use or higher levels of

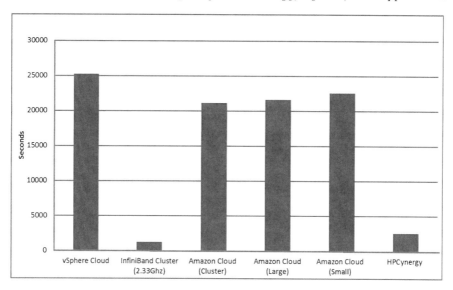

FIGURE 11.10
Total Setup Time of the HPCynergy System.

abstraction results in poor performance — despite the fact that performance and ease of use are two unrelated fields. To test performance, the system biology pipeline (Section 11.6.4.1) and GADGET application (Section 11.6.4.2) were run on a number of commercial cloud solutions, dedicated clusters, as well as (virtual) nodes discovered and used via HPCynergy.

11.6.4.1   Bioinformatics Benchmark

Performance of the system biology pipeline (described in Section 11.6.1.1) was recorded from five machines, the small and large Amazon virtual clusters, the private vSphere cloud, the Ethernet cluster and the InfiniBand cluster. As stated in Section 11.6.2, the Ethernet cluster is equivalent to the Amazon small instance and the InfiniBand cluster comparable to the Amazon large cluster and vSphere cloud.

Once set up according to the methodology specified in Section 11.6.3, the state of the virtual machine was saved as an Amazon Machine Image (AMI). Creation of this AMI allowed easy scaling of the bioinformatics application through virtual machine cloning. Access to each machine used in this benchmark was achieved remotely through an SSH client. Only one process was run on each node in order to minimize overhead of any other operating system processes. Performance results for each machine were measured up to four nodes; each test was run three times in order to ensure the validity of results. As seen in Figure 11.11, the results show a nearly linear increase of perfor-

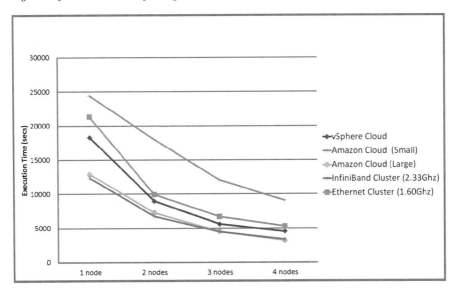

FIGURE 11.11
Building a System Network on Different Computer Architectures.

mance to available resources; this is expected, as most of the system network workflow is embarrassingly parallel. When compared to physical hardware, the vSphere cloud shows a noticeable increase in required computational time. It is likely that this increase is due to virtualization overhead, in which part of the CPU is constantly being delegated to simulate the specified environment. This virtualization overhead leads to an interesting relationship where the smaller a job is, the closer cloud performance will match physical hardware performance. These results also highlight that different hypervisors and cloud service implementations also affect performance. Performance of Amazon which uses a modified Xen hypervisor is very close to physical hardware, while the vSphere cloud which makes use of VMware virtualization suffered the most overhead.

From a user view, setting up a cloud for HPC is time consuming. If nonpersistent storage is utilized, setup must occur every time a job is run. In embarrassingly parallel applications, data and code must first be transferred over the internet to the cloud. Depending on the size of problem or result data, a considerable delay is required before a job can be started. In the case of this system biology workflow, the delay of retrieving results from the cloud is an even bigger problem, the correlated data becoming many times larger than the original gene set.

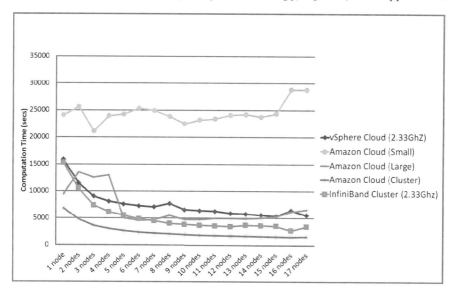

FIGURE 11.12
Simulating Particles over a Range of Computer Architectures.

11.6.4.2   Physics Benchmark

As in the bioinformatics example, many cloud architectures were compared, each running similar virtualized hardware. The platforms utilized in this application study were the small, large and cluster Amazon E2C clouds, the private vSphere cloud and an InifiniBand cluster. Benchmarking made use of full machine capacity, tests running on up to 17 nodes. Each Amazon machine was based on template AMI created during the setup process and started via the Elastic Block Store. Access to each machine used in this benchmark was achieved through an SSH client.

To view the effect of network speed, only one GADGET process was run on each node. Each point was run three times in order to ensure the validity of results. The results from this GADGET benchmark study can be seen in Figure 11.12.

The physical hardware results represent the ideal performance of this study, a near constant computational decrease as more compute nodes are added. The vSphere cloud, which runs on the same hardware, shows this relationship but with a similar offset as seen in bioinformatics study (Section 11.6.1.1).

Performance of the Amazon EC2 cloud varies depending on the instance type chosen. The small instance shows a sharp computational increase at 2 nodes before performance becomes optimal at 3 nodes. The large instance with higher I/O shows a similar early computational spike before optimizing at 5 nodes. Both the small and large EC2 cloud instances show an increase in computation time as more nodes are added past this optimal performance threshold. This relationship is an indication of a communication bottleneck, where each node is spending more time communicating than processing.

Amazon recently added a Cluster Compute Instance which has been optimized for running computation heavy applications. The performance of this instance shows a decrease in execution time mirroring other high speed clusters. This optimal performance is dependent on allocating cluster instances at the same time. Because of this requirement the user loses one of the biggest draws to the cloud, the ability to elastically scale their applications.

Unlike the system biology problem presented in Section 11.6.1.1, this N-body algorithm requires communication between nodes. Collected results from Amazon show that performance is not necessarily linked to the amount of machines used. When running communication based applications, it is important that load is balanced between nodes and that communication is minimized. If each node is communicating more than it is processing, the computation time increases as resources are added. Cloud computing resources are highly distributed and performance of communication heavy applications can vary depending on the network architecture and the location of machines that have been allocated to the user.

11.6.4.3   Bioinformatics-Based HPCynergy Benchmark

Performance assessment of the execution of bioinformatics application on HPCynergy was carried out when building the lactation system network visualized in Figure 11.5. In this test jobs were submitted to the virtual nodes of the Deakin Cluster (as described in Section 11.6.2). Each job submitted to this system utilized only a single core of each node. Through this methodology it was hoped that overhead of the operating system and CPU memory caching would be minimized. As in the previous tests, the bioinformatics benchmark was run on up to four virtual nodes.

Results shown in Figure 11.13 indicate that HPCynergy scales nearly linearly like all other computers tested in this benchmark. While not performing as well as physical hardware or the Amazon cloud, HPCynergy incurs a smaller overhead than the vSphere interface. On average HPCynergy performance results were 5% faster than vSphere. This saving can be significant when running large-scale distributed applications. In this scenario, utilizing the HPCynergy solution simplifies submission of HPC applications while reducing performance overhead (when compared to other clouds).

11.6.4.4   Physics-Based HPCynergy Benchmark

Results were also taken while running the N-body application on HPCynergy. Like the previous benchmark, jobs were submitted to the virtual nodes of the Deakin Cluster (as described in Section 11.6.2). Up to seventeen nodes were utilized during this physics benchmark

Collected results (see Figure 11.14) show a similar decreasing trend to the

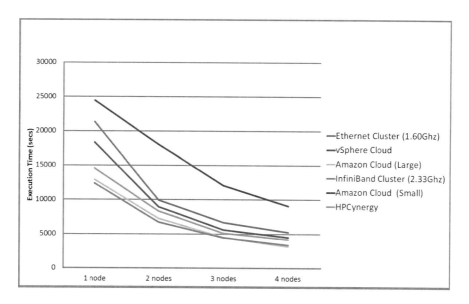

FIGURE 11.13
Performance of System Biology Workflows on HPCynergy.

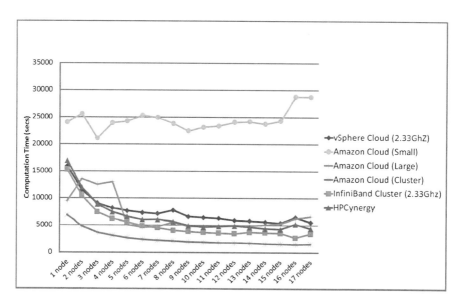

FIGURE 11.14
Performance of N-Body Simulations on HPCynergy.

other cluster utilizing high speed networks. When compared to the vSphere cloud, an average performance improvement of 16% is observed. The simple interface of HPCynergy allows for this improved performance, but it is not streamlined enough match the performance of the physical hardware. In conclusion, HPCynergy is a viable alternative to other cloud platforms providing both improved usability and performance

11.6.4.5   Conclusion

The results in this section show that even standard clouds can achieve performance similar to that of dedicated HPC clusters, depending on the class of problem. When running embarrassingly parallel applications a near linear speed up is achievable. Collected results show clearly that the effects of virtualization vary with the type of hypervisor used. The use of Xen seems to have minimal performance effect on computation while VMware is noticeable. This overhead can be reduced through the addition of multiple nodes, becoming insignificant once computation is reduced to 3 hours. When running communication bound applications performance results vary. On the clouds with slow network speeds the n-body application achieved maximum performance early on, past this point the required compute time steadily increased due to communication overhead. The three clouds with HPC hardware (Amazon Cluster Compute Instance, HPCynergy and VMware vSphere) showed the same decreasing performance trend as the InfiniBand cluster. These performance results indicate it is feasible to run communication bound applications only when cloud providers make use of HPC dedicated hardware.

The usability and setup time of cloud computers is another major issue. Setup time can be greatly reduced through use of pre-existing templates. Templates can be difficult to utilize as they are often missing common dependencies (compilers, text editors, etc.) and may have a range of security access setups. Amazon makes use of a Machine Image repository sortable only by operating system, architecture and image name. Using this search interface it is difficult to know what software each image contains until an instance of the template is launched. Ideally when launching an Amazon cluster instance a message passing interface such as MPI and common compilers should be pre-compiled; when using the recommended cluster compute image this was not the case.

HPCynergy addresses issues in cloud performance and usability, providing an easy to use job submission interface and resource monitor. In both benchmarks the HPCynergy cloud platform showed a scaling trend similar to that of dedicated HPC clusters. When compared to the vSphere VMware cloud, HPCynergy showed an average performance improvement of 10% but did not match the physical hardware. It is hypothesized that the remaining performance lag is due to the hypervisor, and further improvement could be made if a Xen hypervisor was utilized instead. In conclusion, the HPCynergy platform addresses a number of HPC cloud challenges while improving computational performance through reduction of management overhead.

Along with the benchmark results, the HPCynergy design is a significant contribution: it is the first (and possibly, only) existing design that encompasses many issues faced when using cloud computing to supplement or fully conduct HPC research. All issues from required resources to publication and discovery to management to (finally) high level simplifications have been presented and full details given on how all elements relate to each other.

The current HPCynergy implementation is the first step to realizing a promising cloud environment for HPC: where clients can quickly learn of and make use of existing clusters with minimal effort and do not have to spend significant amounts of time installing additional software. It has even been found through benchmarking that the CaaS Infrastructure, employed by the HPCynergy, has little to no performance penalty when running HPC applications. In effect, HPC clients are able to get the turnaround time of application results without the installation and setup overhead.

11.7 Conclusions and Future Trends

The main advantage of utilizing cloud computing is that it allows users to scale resources to their problem without investment in hardware. A key example of this is using clouds to handle increased traffic of web sites and web applications during peak times. When running high performance applications on clouds, hardware requirements are different (high speed network interconnect, RAID drive for parallel read and writes, etc.).

Furthermore, the use of virtualization incurs an overhead and makes it difficult to optimize code at the hardware level. With these weaknesses, running HPC applications on clouds has thought to have been unfeasible [345].

This chapter demonstrates that HPC clouds have a bright and promising feature. The combination of cloud computing elasticity, pay-as-you-go payment, and flexibility through virtualization offer significant opportunities in accessing HPC infrastructure and furthering collaboration with other research teams/projects.

This chapter focused on what challenges exist when using cloud computing to support HPC and in response proposed and demonstrated via HPCynergy the possibility of accessing HPC clusters as on demand utilities. Despite what many experts have claimed, no significant overhead when using cloud for HPC was found. Through the tests with Amazon, various configurations of cluster, and HPCynergy, it was found that (given a high speed network interconnect) the performance of HPC application in clouds is comparable to that of physical HPC clusters.

There are, however, other significant issues that still restrict the use of cloud for HPC. First, performance - there is no means to measure the quality of service (or whole clusters) within clouds nor can such information be made known.

There is also the issue of security - just as how some organizations may have obligations to the privacy of result data, so too do research project. Besides obligations, security is vital so that malicious users cannot access and destroy research results. While cloud computing can make HPC more reachable to teams that cannot afford the costs of clusters, it also makes it more accessible to malicious users who have access to the same cloud.

The last problem of interest is interoperability between HPC clouds. With the volume of result data generated by research projects, there is a strong need to provide data in a uniform manner so that clients can access and contribute data irrespective of the storage format. Using cloud computing for HPC data storage will have a significant impact as research teams will no longer need to wait excessively long periods of time to transfer huge volumes of data between locations. Result data itself will become an on demand utility, thus allowing research teams to focus more time on data processing.

12

Multitenancy: A New Architecture for Clouds

Enrique Jiménez-Domingo
Universidad Carlos III de Madrid, Spain

Ángel Lagares-Lemos
Universidad Carlos III de Madrid, Spain

Juan Miguel Gómez-Berbís
Universidad Carlos III de Madrid, Spain

CONTENTS

12.1 Abstract

Multitenancy is one of the basic principles of Cloud Computing. Sharing resources is one of the main advantages of the Cloud Computing paradigm and multitenancy technology is one of the most effective cornerstones from a Software Architecture perspective. Cloud Computing has provided a reliable, cost-efficient, cutting-edge infrastructure for advanced software delivery models such as Software-as-a-Service (SaaS). Most SaaS platforms are built on Multi-Tenant Architectures (MTA), which allows modeling and exploiting efficiently huge amounts of computing resources and large data management structures, such as databases. As a result of the Multi-Tenancy model, the Cloud can be used in a more efficient way, changing the increase of the profits for the sellers and the decrease of the prices for the customers due to the economy of scale. Nevertheless, from a software engineering standpoint, the need for flexible, context-aware and dynamic Multi-Tenant Architectures is gaining momentum, particularly because when dealing with Business Information Systems (BIS) scenario requirements such as Customer Relationship Management (CRM) or Enterprise Resource Planning (ERP) systems, which require the capability for extending Data Models inside the MTA and also provide more knowledge-oriented metadata-driven approaches such as inference, where data relationships can be entailed.

Keywords: Multitenancy, Cloud Computing, SaaS, Sharing Resources.

12.2   Introduction: Concepts and Features

Salesforce.com CEO, Marc Benioff, stated at a particular time: "I don't buy that there's going to be all these single-tenant databases everywhere, that everybody's going to have their own server, which everybody's going to have their own stack of SAP code that everybody's going to have their own version of NetWeaver customized with their schema. I believe that the world in the future is multitenancy." In this chapter we are going to describe all the features that make Multi-Tenancy the future, or rather, the near future.

Getting the gist of the definitions by [332] [382] we can conclude in the following definition of multitenancy: "The technology that allows the customers (tenants) to share the resources (mainly servers) maintaining the customization of every particular tenant, isolating traffic, data and configurations by means of a uniform framework that supports scale up and scale out and can theoretically maintain an infinite number of users providing an infinite amount of systems and resources. All these features converge in a higher profit margin by leveraging the economy of scale."

This definition could not fit every current Multi-Tenant system, and it does not fit a lot of previous released Multi-Tenant systems. The main objective is trying to make clear how a Multi-Tenant system should be in the Cloud. For example, scaling out is not absolutely necessary in Multi-Tenant systems, but without the scaling out we can consider them incomplete, especially if we are in the Cloud/SaaS field where one of the main goals is to take advantage of the economy of scale and this goal would not be perfectly achieved ruling out the scale out feature.

A Multi-Tenant architecture allows managing different kinds of users of a system in a very flexible way due to the great amounts of configurations that can be adopted easily. For this reason, it is necessary to establish the difference between tenant and user. We consider it essential to include this in the introduction because it is going to be mentioned widely in the rest of the chapter. A tenant is considered the owner or the supplier of a SaaS application. He is responsible for the management, maintenance and update of the application and should ensure its availability and security. On the other hand, it is possible to find the users that properly use the applications provided by the tenants. The wider idea is that the users are subscribed to the tenant's applications and establish a pay-per-use method, more accepted and extended each day. This conceptual differentiation makes possible increasing the flexibility of these systems, thanks to the fact that a tenant can have many users subscribed to its applications and a user can be subscribed to different tenants' applications, paying only for the ones that he really needs, representing a cost advantage compared with previous models.

As a summary of this section we would like to emphasize that the given definition can be discussed as it is happening nowadays in forums, conferences and research exchanges. Despite this lack of exactness or academic consensus in the definition, we find that it covers the most important features of multitenancy and represents the concept in a complete and precise way. Bearing this in mind, we can consider that there are different levels of multitenancy, and the one given in the definition is the highest one; it depends on the IT companies, tenants and customers that opt for real multitenancy systems or others that just for using approaches (lower level multi-tenancy systems).

Consequently, the customers have to be careful about the kind of multitenancy that is being provided, considering the number of instances, how the versions are maintained, the security of the system, the management of the customizations' ensuring that the chosen solution is the most appropriate one. On the other side, when designing the architecture for a SaaS system to be provided, the IT company has to consider the level of multitenancy depending on the customer requirements, cost of rewriting or updating the software, cost of technology, cost of maintenance, support, etc. [757].

The remainder of this chapter is as follows. Section 3 provides a thorough state-of-the-art and brief introduction about the more relevant research lines in this scope. In Section 4 a more detailed explanation about the features, advantages and problems of multitenancy is provided. What concepts have to be taken into account when developing multi-tenant architectures or applications is detailed in Section 5 and an example of how to apply the previous information is provided in Section 6. Finally, Section 7 outlines future work and research lines in this topic and Section 8 concludes the chapter.

12.3   Background

Multi-Tenant Architecture (MTA) maximizes sharing between users and increases the total revenue and economy of scale for the client; however, it increases development costs [416]. A Multi-Tenant system should be able to support scale-up (consolidating multiple customers onto the same server) and scale-out (spanning a farm of servers which consolidates customers) as we already mentioned in the previous definition. Not in vain, multi-tenant operations require an exceedingly complex cloud operating system [416]. As it is long detailed in [538], to create a configurable and customizable SaaS system able to fulfil the needs of every user is not a trivial issue, and it requires an extended analysis and, in most of the cases, a very complex architecture and implementation. This makes sense, keeping in mind that the source code has to be the same for every customer, therefore there is just one version of the software; it is obvious that creating a tailored system for every tenant will destroy the SaaS philosophy, and the company responsible of the cloud will incur in extra costs that probably will be unacceptable for the tenants. In conclusion, the target is to create a Multi-Tenant environment configurable and customizable for every kind of tenant, involving no modifications in the source code (extra development) or any extra costs in time or money. There are two fields to consider when we are designing a Multi-Tenant system, the application and the database systems. The selected way to configure them will affect the performance of the multitenancy in the whole system.

According to [735] there are four approaches to deploy the application servers:

- Fully Isolated Application Server. Each tenant accesses an application server running on a dedicated server. In this scenario the multitenancy is not very efficient, because most of the resources are not shared and idle resources that could be used by a different user are not, because of the architecture of this model. The performance of the system is the best, but it is the most expensive solution. At the same time, the management of this solution is the easiest one.

- Virtualized Application Server. Each tenant accesses a dedicated application running on a separate virtual machine. Still, this solution is not sharing most of the resources, but it can have more than one tenant per server; that means decreasing the cost, but the performance of the system usually will not be as good as the previous one.

- Shared Virtual Server. Each tenant accesses a dedicated application server running on a shared virtual machine. In this model, the shared resources are more than in the "Virtualized Application Server" therefore the collisions trying to use the same resource will occur more often, but the efficiency will be increased and it will be cheaper.

- Shared Application Server. The tenants shared the application server and access application resources through separate sessions or threads. This is the solution that best fits the philosophy of the SaaS systems and multitenancy trend. The resources are sharing as much as possible and the efficiency is the best. On the other side the performance is supposed to be the worst (although with proper implementation there should not be a big difference between this one and the two previous ones). The cost of this model is the cheapest regarding the costs of the hardware, but the implementation is the most complex one.

The other issue to take into account is the database systems. The main challenge for creating and implementing a Multi-Tenant oriented Data Model precisely stems from building a secure and robust system, providing optimized scalability and over-performance [186]. Hence, this model also points to several research problems for the Database community, including security, contention for shared resources, and extensibility [355]. According to [104], there are three major approaches to Multi-Tenant DBMS design, namely: Tagging, Integrated schema and Database space. According to [161] the main limitation of the first two approaches is that they cannot effectively support SLAs for reliability and performance, provide data isolation, nor enable differential user classes. The database space approach provides better security than the other two since its security is implemented within the infrastructure, while the security of the other two is implemented within the users' applications. Given the importance of Multi-Tenancy for SaaS companies and users, many works have been devoted to this research field [355], [327], [658], [466], [777]. The most famous company providing SaaS and multitenancy is Salesforce.com. They have a metadata-driven architecture that delivers scalability, and customization for on-demand, Multi-Tenant applications. They work with a Multi-tenant programming language named Apex which uses Java-like syntax. Their Multi-Tenant software infrastructure is based on a software stack composed of both commodity RDBMS servers (Oracle) and specialty toolkits (the Apex language and its interpreter). They can be considered as pioneers in the use of multitenancy [633]. Another example that is worth mentioning is that the SAP Company has reconsidered its point of view about multitenancy. They were not very enthusiastic supporters of multitenancy but since January 2010 they seem to have turned around and they have released a new version of their product Business ByDesign that comes in single-tenant and multi-tenant flavors [92]. At this time, more companies are using multitenancy in the provided services and they are really influential; some examples are [537], [641], [526].

12.4   Features, Advantages and Problems

The adoption of Software as a Service and the Cloud Computing business model (and the associated technology infrastructure) have important benefits for both software vendors and their customers. The pay-per-use method eliminates the need for costly investments to purchase perpetual software licenses. It is paid solely for the use that is given to the product. It also eliminated the costs associated with installing the product (and the need for dedicated machines for this) and maintenance. Furthermore, the only requirement necessary to make use of the application will be to have Internet access to get into it. The SaaS model remains in the same way, the feature of multi-property (Multi-Tenant) advocated by Cloud Computing. Another benefit of implementing this business model is the existence of central, single point updates that allow all customers to have the latest updates made on the application in the instant it becomes available [538]. Therefore, reducing customer costs is considerable, and in terms of accessibility and availability, significantly improves customer experience.

Multitenancy architecture in Saas allows a single implementation of a service application to multiple customers. The importance of this programming technique is that by allowing a single service application to be used by a large number of customers, it generates scale economies derived from the efficient use of resources (both hardware and human) and this translates into a low priced software [187]. This is often associated with SaaS, but in reality and for the same reason, the multitenancy is used at all levels of Cloud Computing (IaaS, PaaS or SaaS).

- Multi-user environment. Multitenancy provides this environment to make the most of the possibility for sharing resources and costs among a large number of users. This leads to improved business process as companies share data and applications, focusing on business processeses rather than the infrastructure and technologies that support them. At the same time, users of all kinds have available the latest versions of systems and services they use. Multi-tenant model providers make it much easier to offer instant access to new features to all users, compared with traditional models.

- Multi-property. It facilitates the sharing of resources and costs among a large number of tenants. This allows, for example, centralization of infrastructure in areas with low costs (in properties, electricity, etc.), and improving the use and efficiency of computer systems (usually underutilized).

- Scalability. Providing resources or services on demand and using coupled architectures can improve system scalability.

- Cost for the clients. Logically it is not the same to acquire or maintain N infrastructures than just one, and if the maintenance cost is higher, what will happen to the price for the client?

- Higher probability of fault maintenance. Likewise, the larger the infrastructure to maintain, the greater probability of failure, especially if each infrastructure is custom tailored to each client.

- Increasingly competitive prices. A growing number of customers who use SAAS solutions will optimize the use of infrastructure. This fact combined with a competitive market make SaaS prices lower. This means that an ASP can not compete even though they use virtualization to provide the service.

- Ability to access the application from the very beginning. It may not be a requirement for all customers, but multitenancy architecture offers the flexibility to be able to access the application from the first minute.

Probably the main advantage of a Multi-Tenant application is the operational benefit. This is because all the code is in one place and it is much easier to maintain, update and back up (and consequently, cheaper). Another advantage is the lower system requirements needed due to the sharing of application and databases by multiple clients, a fact that makes unnecessary the use of one server for every client, which means a clear improvement in the use of resources. Dealing with this, when sharing servers between multiple clients, scalability can appear to be a problem. In single-tenant applications, all clients have their resources and when a new client wants to use his application the necessary resources are added to the system. When talking about Multi-Tenant applications all the clients share the same resources. For this reason, it is possible that these resources become overloaded with derived problems carried. A decision that can influence in this is the implementation and configuration of the database, a key concept for achieving multitenancy that will be explained in the next section.

12.5 Modeling Multitenancy

As has been mentioned before, Cloud Computing is evolving from a mere "storage" technology to a new vehicle for Business Information Systems (BIS) to manage, organize, and provide added-value strategies to current business models. However, the underlying infrastructure for Software-as-a-Service (SaaS) to become a new platform for trading partners and transactions must rely on intelligent, flexible, context-aware and, of course, Multi-Tenant Architectures that provide the necessary features to make the most of this technology.

One of the basic topics that have to be covered is the configuration and design of the database model which will support the power of MTAs. An appropriate model for the database can take an important increase in the scalability of the system. For achieving that, it is possible to find different configurations, each of them with its domain of applications, powers and limitations [297]:

- Separate databases

- Shared database, separate schema

- Shared database, shared schema

As it is easy to see, the possibilities are quite different and offer several perspectives. To choose the best option in each case it is necessary to study the case in which the configurations are going to be used and try to make the most of them in terms of efficiency and cost. Now a brief description of each one will be provided in order to get a better understanding of the issue that it has been treating.

- The separate databases configurations is the simpler way of managing tenant information. Each tenant has his own database and all the information is stored there, avoiding management and configuration problems. The great problem is that the space needed grows rapidly and sometimes it is wasted, fact that increases the costs.

- The second approach involves housing multiple tenants in the same database, with each tenant having its own set of tables that are grouped into a schema created specifically for the tenant. This is also relatively easy to implement and tenants can extend the data model with no hard difficulties. The main disadvantage is that in case of failure it is more difficult to recover, and it still consumes more resources than necessary.

- Finally, the shared database, shared schema present an approach that involves using the same database and the same set of tables to host multiple tenants' data. This allows using only the stored space that is necessary in each case and avoiding the problems explained before, with lower costs.

One way to avoid limitations of "pre-allocated fields" is to allow customers to extend the data model arbitrarily, storing custom data in a separate table and using metadata to define labels and data types for each tenants custom fields. Dealing with this, the metadata table stores important information about every custom field defined by each tenant. When an end user saves a record with a custom field, two things happen: first, the record itself is created or updated in the primary data table; values are saved for all of the predefined fields, but not the custom field. Instead, the application creates a unique extension table that contains the following pieces of information. First, the particular ID data of the associated record in the primary data table is vital for the approach. Second, the extension ID associated with the correct custom field definition and finally, the value of the custom field in the record which is being saved and eventually transformed into a string.

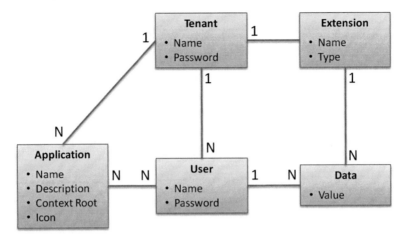

FIGURE 12.1
Example of An Extensible Database Model.

In Figure 12.1 the following tables are showed:

1. Tenant. Stores tenant information: username, password and any necessary administrative data.

2. User. Basic information of users: user name, password and tenant to which they belong.

3. Application. It contains the necessary data from an application, like name and description, the path to the application and its icon, for example. It has a relationship of belonging with a tenant (which may have several applications) and availability to users (a user can have available several applications and an application may be available to multiple users).

4. Extension. Description of the custom fields of a tenant for its users: field name and data type (integer, real number, string, date, etc.).

5. Data. Content (values) of the virtual table made up of users and extensions of a tenant. It stores the value of each pair user-extension that belongs to the same tenant, if it is not empty.

This approach allows each tenant to create as many custom fields as necessary to meet its business needs [323]. When the application retrieves a customer record, it performs a lookup in the extension table, selects all rows corresponding to the record ID, and returns a value for each custom field used. To associate these values with the correct custom fields and cast them to the correct data types, the application looks up the custom field information in metadata using the extension IDs associated with each value from the extension table.

Obviously, this is the configuration that satisfied the majority of the requirements due to its flexibility and adaptability to different situations and contexts. Thanks to this, the tenants that use the system where this configuration is developed do not have to be worried about technical details, which allows centering their efforts in the really important thing: their applications. There are some requirements that must be fulfilled for the development of this kind of extensible database [679]:

1. The database has to be able to store the information of different tenants in the same database.

2. The objects in the database (users, extensions, applications) are property of one single tenant, who is the one that has access to this data.

3. Each user must have a list of applications or similar items to which he has access, which are previously determined by the tenants.

4. The tenant should define some default fields for the users subscribed to his applications. These fields can be customized for the user after that.

12.6   An Original Example

The architecture of this system has been designed to offer the best flexibility and usability for the multitenancy database. The major contribution of this work is the two levels of multitenancy that this architecture provides (the architecture is reflected graphically in the Figure 12.2). The first level of multitenancy is due to different tenants being assigned to different databases (different instances), and the second one is invoked for the ability of the tenants to share the tables of the database (shared schema). It has been made possible by adding two innovative elements to the typical architecture: the "JDBC Driver Manager" and the "Multi-Tenant DB Index," which together make the transition from Single-Tenant to Multi-Tenant almost automatic in a very flexible way. These two artifacts make this architecture different from a regular Multi-Tenant application working in Amazon or another cloud infrastructure.

The SaaS application which is in the top of the architecture is a J2EE application that needs to store information in a database. Actually, it does not need to be a "SaaS" application but in this context usually is.

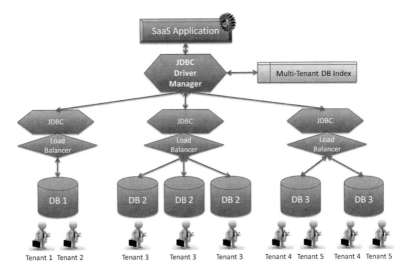

FIGURE 12.2
System Architecture.

The JDBC Driver Manager is just above the predefined JDBC driver and allows, as it was explained, a Multi-Tenant database to be created automatically with no effort on the tenant side. In the same layer there is the Multi-Tenant DB Index that contains the data of which database is the correspondent for each tenant, therefore it is implemented as a hash map of key; value pairs, being the key the tenant ID, and the value the connection parameters and the subjacent JDBC driver to use.

Under the JDBC Driver Manager is located the JDBC Driver that really connects to the database; any regular driver works. The cloud infrastructure is in charge of balancing the amount of work of every database, to keep them working in a high-performance status.

At the bottom are the databases. The data of a tenant exists only in one database, although that database can be replicated as many times as needed. This means that the queries to the database dependent on the Multi-Tenant DB Index will consult just one database. A database can be held by many tenants; in consequence every row of the database has to have a Tenant ID field in order to identify the owner of the data. More information about the system can be found in [394].

12.6.1 Implementation

The implementation consists of two parts: the necessary modifications to the database schema and the driver implementation.

12.6.1.1   Database Modification

For the driver to be able to discern the owner of each bit of data, some new data must be added to the database model. First, a new table should be created to store tenant data, using only a single field, the tenant ID. This table is not strictly needed, but it allows the user to associate data with tenants (e.g., tenant name) and to maintain data consistency in the database, forcing all data to be owned by one of the tenants on this table. The other bit of data that is added to the database is a new column, storing the identifier of the data owner, in every existing table. This new column allows to yielding loosely decoupled virtual tables of each tenant, so they can only access the data rows that are really owned by them. By adding a reference from these columns to the tenants table, we achieve consistency, in the sense that invalid tenants cannot be specified as owners of database rows. The Figure 12.3 shows the configuration of the database.

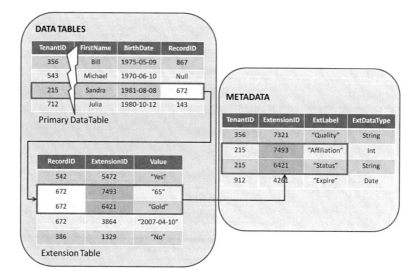

FIGURE 12.3
Shared Database, Shared Schema.

These modifications to the database schema can be automated with a program (in this case a Java program) that connects to the specified database, creates a tenant table with a single tenant_id field, and then lists all the original tables and adds the owner column to each of them. These columns are initialized with a default value, for which a row is inserted in the tenant's table. Finally, references are created from all those tables to the new one, linking the owner field with the tenant identifiers in the tenant's table. No database indexes are created during this process and they are even left to the database administrator to implement. The main reason for this is because in several cases, indexing the data by the tenant identifier can actually decrease dramatically the performance due to the ever-growing index size and, eventually, the database administrator can do a better job than an automated program.

12.6.1.2   Driver Implementation

The current driver is implemented as a JDBC driver, so any Java application can use it without significant modifications. When a Java program requests a new connection, the driver starts a real connection to the database using a second database driver. The back-end driver to use actually depends on the tenant that is starting the connection, which is passed as a custom parameter to the front-end JDBC driver. This approach allows the application developer to use any method he sees appropriate to determine the tenant that is connecting to the database. The tenant identifier is looked up in a hash table, which associates it with a parameter set that allows the opening of a JDBC connection. The selection of a hash table allows unevenly distributed tenant identifiers, while maintaining a short look-up time. This division of tenants in different databases implements the first level of multitenancy, distributing their data in different database instances, which in fact could be using different database engines.

Once a connection is established, the application can begin sending queries. Each query will then be analyzed and then rewritten to include the extra data needed to handle the multitenancy schema. In case of an UPDATE or SELECT query, an extra WHERE parameter is added filtering by the owner of the data to ensure no reads or writes fall outside the tenant virtual table. Similarly, for INSERT, a new value is added to the query to set the owner to the correct tenant identifier.

This re-implementation of the queries implements the second level of multitenancy, where the data is separated into virtual tables while sharing the same tables of the same database instance. This query is then sent to the database through the back-end driver, and the results are returned verbatim to the application, which can use them as normally.

12.6.1.3   Security Limitations

As this Multi-Tenant Architecture is intended for application developers, it does not enforce data separation between tenants. Any tenant can access the data of other tenants just by modifying the identifier passed to the front-end driver. Thus, the application developer must ensure that the correct tenant is sent to the driver each time. Also, the queries sent to the drivers are not checked before being rewritten, which means that SQL injection-like attacks can be used to ignore the data separation between tenants. Queries should always be checked with standard techniques before being sent to the driver to ensure this cannot happen.

12.7   Future Research Directions

Leveraging the potential of Cloud Computing technologies in the Business Information Systems (BIS) application domain has been deemed as a critical challenge for a particular number of disciplines that interrelate intensively in optimizing and harnessing synergies in both domains. Nevertheless, a number of research initiatives for future directions are being fostered by R&D stakeholders, both in the public and private sector.

Regarding multitenancy, the particular topic of this chapter, a combination of data-intensive and more efficient data management strategies can be followed. One of those strategies is based on data mining techniques, namely Clustering. Clustering can be considered one of the most important unsupervised learning problems. A common definition of clustering can be "the process of organizing objects into groups whose members are similar in some way." Another way to say it is that a cluster is a group of objects which are "similar" and are "dissimilar" to the objects that belong to other clusters [336]. Particularly in this case, the aim of clustering algorithms is to classify the data into groups that share some features between them. The problem merges when we try to establish a unique criterion to determine the best solution independently of the aim of the clustering process because this criterion does not exist. Therefore, it must be the user who determines this criterion in order to satisfy his needs. At this point the goal of the system converges with clustering algorithms properties. The user decides the direction in which the algorithms should group the information provided, in order to obtain the desired data.

A second approach would include Database Segmentation, which follows up the work presented in the Implementation section of this chapter, where partitioning a set of database tables into "virtual tables" allows a more powerful data management procedure. Those "virtual tables" could follow a number of optimization criteria, but fundamentally could solve some of SQL and Relational Databases problems in terms of scalability.

Finally, the research directions follow-up of the work presented in this chapter will also rely on a twofold basis. On the one hand, testing and verifying on a real-world industry-related multitenancy requirement scenario, where the proficiency of multitenancy is measured and quantified.

Potential additional research issues could be the assessment of those systems under particular "on the edge" conditions, such as Very Large Dataset Management Systems.

12.8 Conclusions

Integration of Enterprise Information Systems is a growing and recognized challenge. With the emergence of Cloud Computing, new strategies and approaches for Enterprise Application Integration have gained momentum. In such a context, enterprise data integration comprises problems like optimizing the data model with schema integration, combining multiple data sources queries and answers and, finally, transforming and integrating the different Information Systems to enable knowledge-based interoperability.

Cloud Computing systems are a particular type of data-intensive and data-oriented information systems aiming at storing and providing a huge amount of corporate data.

In this chapter, we have presented multitenancy as one of the major tenets of Cloud Computing and Software-as-a-Service, since it provides an optimized and efficient technological means for enhancing Large Datasets Management among others. Multitenancy appears as a new way to manage Information Systems due to it flexibility and its intrinsic capacity of dealing with multiple users avoiding the limitations of previous models and architectures and obtaining good measures of performance. These features allow providers to offer more accurate work environments for users which only have to be worried about their activities, leaving all the customization and configuration processes to the company that provides the services. Besides, this improved situation has the consequent impact on the costs associated with these systems that, especially for companies, mean an important cost saving also related to the pay-per-use method commented on in the chapter.

In addition, in this work, we have presented a novel approach to achieve Database Tables integration and interoperability by means of using those particular technologies. The use of the extensible database models explained better management of users and applications data for Cloud Computing systems, which are able to save space and time only by using this approach.

Finally, we have proved that Service-Oriented systems such as Service-Oriented Architecture (SOA) or Enterprise Resource Planning systems (ERPs) could largely benefit from including a knowledge-intensive mechanism such as multitenancy to harness Intelligent Data Management inside Relational Databases.

13

SOA and QoS Management for Cloud Computing

Vincent C. Emeakaroha, Michael Maurer, Ivan Breskovic
Vienna University of Technology, Austria

Ivona Brandic, Schahram Dustdar
Vienna University of Technology, Austria

CONTENTS

13.1   Introduction

In the recent years, Cloud computing has become a key IT megatrend that will take root, although it is at infancy in terms of market adoption. Cloud computing is a promising technology that evolved out of several concepts such as virtualization, distributed application design, Grid, and enterprise IT management to enable a more flexible approach for deploying and scaling applications at low cost [150].

Service provisioning in the Cloud is based on a Service Level Agreement (SLA), which is a set of non-functional properties specified and negotiated between the customer and the service provider. It states the terms of the service including the quality of service (QoS), obligations, service pricing, and penalties in case of agreement violations.

Flexible and reliable management of SLAs is of paramount importance for both Cloud providers and consumers. On the one hand, the prevention of SLA violations avoids penalties that are costly to providers. On the other hand, based on flexible and timely reactions to possible SLA violation threats, user interaction with the system can be minimized, enabling Cloud computing to take root as a flexible and reliable form of on-demand computing.

In order to guarantee an agreed SLA, the Cloud provider must be capable of monitoring its infrastructure (host) resource metrics to enforce the agreed service level objectives. Traditional monitoring technologies for single machines or Clusters are restricted to locality and homogeneity of monitored objects and, therefore, cannot be applied in the Cloud in an appropriate manner. Moreover, in traditional systems there is a gap between monitored metrics, which are usually low-level entities, and SLA agreements, which are high-level user guarantee parameters.

In this chapter we present a novel framework for the mapping of Low-level resource Metrics to High-level SLA parameters named a LoM2HiS framework, which is also capable of evaluating application SLAs at runtime and detecting SLA violation situations in order to ensure the application QoS. Furthermore, we present a knowledge management technique based on Case-Based Reasoning (CBR) that is responsible for proposing reactive actions to prevent or correct detected violation situations.

The LoM2HiS framework is embedded into a FoSII infrastructure aiming at developing an infrastructure for autonomic SLA management and enforcement. Thus, LoM2HiS represents the first building block of the FoSII [277] infrastructure. We present the conceptual design of the framework including the run-time and host monitors, and the SLA mapping database. We discuss our novel communication model based on queuing networks ensuring the scalability of the LoM2HiS framework. Moreover, we demonstrate sample mappings from the low-level resource metrics to the high-level SLA parameters. Thereafter, we discuss some details of the CBR knowledge management technique thereby showing how the cases are formulated and their similarities. Furthermore, we describe a utility function for calculating the quality of a proposed action.

The main contributions of this chapter are: (i) the design of the low-level resource monitoring and communication mechanisms; (ii) the definition of mapping rules using domain specific languages; (iii) the mapping of the low-level metrics to high-level SLA objectives; (iv) the evaluation of the SLA at run-time to detect violation threats or real violation situation; (v) the design of the knowledge management technique; (vi) the evaluation of the LoM2HiS framework in a real Cloud testbed with a use-case scenario consisting of an image rendering application such as POV-Ray [305].

The rest of the chapter is organized as follows. Section 13.2 presents the related work. In Section 13.3 we present the background and motivation for this research work. The conceptual design and implementation issues of the LoM2HiS framework are presented in Section 13.4. In Section 13.5 we give the details of the case-based reasoning knowledge management technique. Section 13.6 deals with the framework evaluation based on a real Cloud testbed with POV-Ray applications and the discussion of the achieved results. Section 13.7 presents the conclusion of the book chapter and our future research work.

13.2 Related Work

We classify related work on SLA management and enforcement of Cloud based services into (i) Cloud resource monitoring [289,322,768] (ii) SLA management including QoS management [134,204,242,287,439,703] and (iii) mapping techniques of monitored metrics to SLA parameters and attributes [137,216,616]. Since there is very little work on monitoring, SLA management, and metrics mapping in Cloud systems we look particularly into related areas such as Grid and SOA based systems.

Fu et al. [289] propose GridEye, a service-oriented monitoring system with flexible architecture that is further equipped with an algorithm for prediction of the overall resource performance characteristics. The authors discuss how resources are monitored with their approach in Grid environment but they consider neither SLA management nor low-level metric mapping. Gunter et al. [322] present NetLogger, a distributed monitoring system, which monitors and collects information from networks. Applications can invoke NetLogger's API to survey the overload before and after some request or operation. However, it monitors only network resources. Wood et al. [768] developed a system, called Sandpiper, which automates the process of monitoring and detecting hotspots and remapping/reconfiguring VMs whenever necessary. Their monitoring system reminds ours in terms of goal: avoid SLA violation. Similar to our approach, Sandpiper uses thresholds to check whether SLAs can be violated. However, it differs from our system by not allowing the mapping of low-level metrics, such as CPU and memory, to high-level SLA parameters, such as response time.

Boniface et al. [134] discuss dynamic service provisioning using GRIA SLAs. The authors describe provisioning of services based on agreed SLAs and the management of the SLAs to avoid violations. Their approach is limited to Grid environments. Moreover, they do not detail how the low-level metric are monitored and mapped to high-level SLAs. Theilman et al. [703] discuss an approach for multi-level SLA management, where SLAs are consistently specified and managed within a service-oriented infrastructure (SOI). They present the run-time functional view of the conceptual architecture and discuss different case studies including Enterprise Resource Planning (ERP) or financial services. But they do not address low-level resource monitoring and SLA mappings. Koller et al. [439] discuss autonomous QoS management using a proxy-like approach. The implementation is based on WS-Agreement. Thereby, SLAs can be exploited to define certain QoS parameters that a service has to maintain during its interaction with a specific customer. However, their approach is limited to web services and does not consider requirements of Cloud Computing infrastructures like scalability. Frutos et al. [287] discuss the main approach of the EU project BREIN [140] to develop a framework, which extends the characteristics of computational Grids by driving their usage inside new target areas in the business domain for advanced SLA management. However, BREIN applies SLA management to Grids, whereas we target SLA management in Clouds. Dobson et al. [242] present a unified quality of service (QoS) ontology applicable to the main scenarios identified such as QoS-based web services selection, QoS monitoring and QoS adaptation. However, they do not consider low-level resource monitoring. Comuzzi et al. [204] define the process for SLA establishment adopted within the EU project SLA@SOI framework. The authors propose the architecture for monitoring of SLAs considering two requirements introduced by SLA establishment: the availability of historical data for evaluating SLA offers and the assessment of the capability to monitor the terms in an SLA offer. But they did not consider monitoring of low-level metrics and mapping them to high-level SLA parameters for ensuring the SLA objectives.

Brandic et al. [137] present an approach for adaptive generation of SLA templates. Thereby, SLA users can define mappings from their local SLA templates to the remote templates in order to facilitate communication with numerous Cloud service providers. However, they do not investigate mapping of monitored metrics to agreed SLAs. Rosenberg et al. [616] deal with QoS attributes for web services. They identified important QoS attributes and their composition from resource metrics. They presented some mapping techniques for composing QoS attributes from resource metrics to form SLA parameters for a specific domain. However, they did not deal with monitoring of resource metrics. Bocciarelli et al. [216] introduce a model-driven approach for integrating performance prediction into service composition processes carried out using BPEL. In their approach, they compose service SLA parameters from resource metrics using some mapping techniques. But they neither consider resource metrics Đ nor SLA monitoring.

To the best of our knowledge, none of the discussed approaches deal with mapping of low-level monitored metrics to high-level SLA guarantees as those necessary in Cloud-like environments.

13.3 Background and Motivations

The processes of service provisioning based on SLA and efficient management of resources in an autonomic manner are major research challenges in Cloud-like environments [150, 425]. We are currently developing an infrastructure called FoSII (Foundations of Self-governing Infrastructures), which proposes models and concepts for autonomic SLA management and enforcement in the Cloud. The FoSII infrastructure is capable of managing the whole life cycle of self-adaptable Cloud services [136].

The essence of using SLA in Cloud business is to guarantee customers a certain level of quality for their services. In a situation where this level of quality is not met, the provider pays penalties for the breach of contract. In order to save Cloud providers from paying costly penalties, and increase their profit, we devised the Low Level Metrics to High Level SLA—*LoM2HiS framework* [248], which is a core component of the FoSII infrastructure for monitoring Cloud resources, mapping the low-level resource metrics to high-level SLA parameter objectives, and detecting SLA violations as well as future SLA violation threats so as to react before actual SLA violations occur.

13.3.1 FoSII Infrastructure Overview

Figure 13.1 presents an overview of the FoSII infrastructure. Each FoSII service implements three interfaces: (i) negotiation interface necessary for the establishment of SLA agreements, (ii) application management interface necessary to start the application, upload data, and perform similar management actions, and (iii) self-management interface necessary to devise actions in order to prevent SLA violations.

The self-management interface shown in Figure 13.1 is implemented by each Cloud service and specifies operations for sensing changes of the desired state and for reacting to those changes [136]. The host monitor sensors continuously monitor the infrastructure resource metrics (input sensor values arrow a in Figure 13.1) and provide the autonomic manager with the current resource status. The run-time monitor sensors sense future SLA violation threats (input sensor values arrow b in Figure 13.1) based on resource usage experiences and predefined threat thresholds.

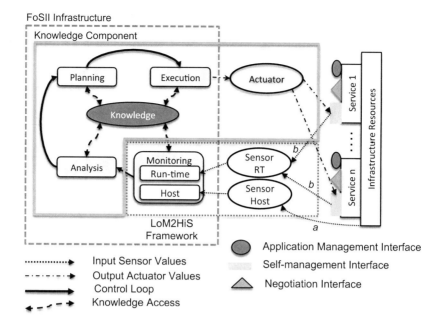

FIGURE 13.1
FoSII Infrastructure Overview.

Logically, FoSII infrastructure consists of multiple components working together to achieve a common goal. In this chapter, we focus on the LoM2HiS framework and give some details of the knowledge management technique, since they are responsible for system monitoring, detection of SLA violations, and proposing of reactive actions to prevent or correct the violation situation.

13.4   Design of the LoM2HiS Framework

The LoM2HiS framework is the first step toward achieving the goals of the FoSII infrastructure. In this section, we give the details of the LoM2HiS framework and in the following sections we describe the components and their implementations.

In this framework, we assumed that the SLA negotiation process is completed and the agreed SLAs are stored in the repository for service provisioning. Beside the SLAs, the predefined threat thresholds for guiding the SLA objectives are also stored in a repository. The concept of detecting future SLA violation threats is designed by defining more restrictive thresholds known as threat thresholds, that are stricter than the normal SLA objective violation thresholds. In this chapter, we assume predefined threat thresholds because the autonomic generation of threat thresholds is far from trivial and is part of our ongoing research work.

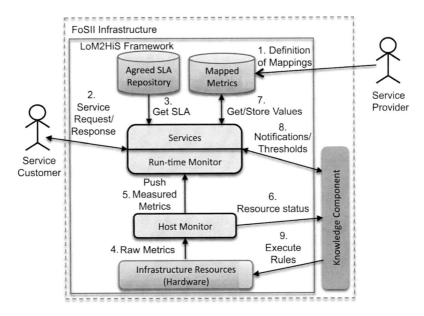

FIGURE 13.2
LoM2HiS Framework Architecture.

Figure 13.2 presents the architecture of our LoM2HiS framework. The service component including the run-time monitor represents the application layer where services are deployed using a Web Service container e.g., Apache Axis. The run-time monitor is designed to monitor the services based on the negotiated and agreed SLAs. After agreeing on SLA terms, the service provider creates mapping rules for the low-level to high-level SLA mappings (step 1 in Figure 13.2) using Domain Specific Languages (DSLs). DSLs are small languages that can be tailored to a specific problem domain. Once the customer requests the provisioning of an agreed service (Step 2), the run-time monitor loads the service SLA from the agreed SLA repository (Step 3). Service provisioning is based on the infrastructure resources, which represent the physical-, virtual machines, and network resources in a data center for hosting Cloud services. The resource metrics are measured by monitoring agents, and the measured raw metrics are accessed by the host monitor (step 4). The host monitor extracts metric-value pairs from the raw metrics and transmits them periodically to the run-time monitor (step 5) and to the knowledge component (step 6) using our designed communication mechanism.

Upon receiving the measured metrics, the run-time monitor maps the low-level metrics based on predefined mapping rules to form an equivalent of the agreed

SLA objectives. The mapping results are stored in the mapped metric repository (step 7), which also contains the predefined mapping rules. The run-time monitor uses the mapped values and the predefined thresholds to monitor the status of the deployed services. In case future SLA violation threats occur, it notifies (step 8) the knowledge component for preventive actions. The knowledge component also receives the predefined threat thresholds (step 8) for possible adjustments due to environmental changes at run-time. This component works out an appropriate preventive action to avert future SLA violation threats based on the resource status (step 6) and defined rules [501]. The knowledge component's decisions (e.g., assign more CPU to a virtual host) are executed on the infrastructure resources (step 9).

13.4.1   Host Monitor

This section describes the host monitor component, which is located at the Cloud infrastructure resource level. We first explain its design and later present the implementation details.

13.4.1.1   Host Monitor Design

The host monitor is responsible for processing monitored values delivered by the monitoring agents embedded in the infrastructure resources. The monitoring agents are capable of measuring both hardware and network resources. Figure 13.3 presents the host monitoring system.

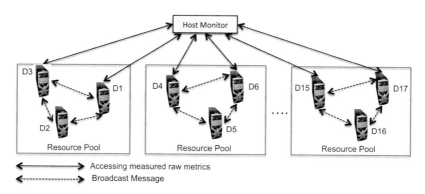

FIGURE 13.3
Host Monitoring System.

As shown in Figure 13.3, the monitoring agent embedded in Device 1 (D1) measures its resource metrics and broadcasts them to D2 and D3. Equally, D2 measures and broadcasts its measured metrics to D1 and D3. Thus, we achieve a replica management system in the sense that each device has a complete result of the monitored infrastructure. The host monitor can access these results from any device. It can be configured to access different devices at the same time for monitored values. In case one fails, the result will be accessed from the other. This eradicates the problem of a bottleneck system and offers fault-tolerant capabilities. Note that a device can be a physical machine, a virtual machine, a storage device, or a network device. It should also be further noted that the above described broadcasting mechanism is configurable and can be deactivated in a Cloud environment where there are lots of devices within resource pools to avoid communication overheads, which may consequently lead to degraded overall system performance.

13.4.1.2   Implementation of Host Monitor Component

The host monitor implementation uses the GMOND module from the GAN-GLIA open source project [500] as the monitoring agent. The GMOND module is a standalone component of the GANGLIA project. We use it to monitor the infrastructure resource metrics. The monitored results are presented in an XML file and written to a predefined network socket. We implemented a Java routine to listen to this network socket where the GMOND writes the XML file containing the monitored metrics to access the file for processing. Further-more, we implemented an XML parser using the well-known open source SAX API [634] to parse the XML file in order to extract the metric-value pairs. The measured metric-value pairs are sent to the run-time monitor using our implemented communication mechanism. These processes can be done once or repeated periodically depending on the monitoring strategy being used.

13.4.2   Communication Mechanism

The components of our FoSII infrastructure exchange large numbers of mes-sages with each other and within the components. Thus, there is a need for a reliable and scalable means of communication. Figure 13.4 presents an exam-ple scenario expressing the usage of the communication mechanism.

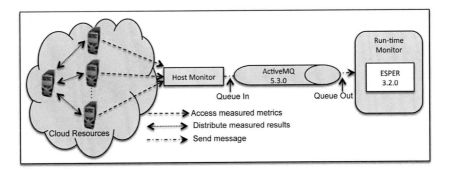

FIGURE 13.4
Communication Mechanism Scenario.

The scenario of Figure 13.4 depicts the processes of extracting the low-level metrics from the monitoring agents embedded in the Cloud resources and passing them to the run-time monitor for mapping and further processing.

To satisfy this need of communication means, we designed and implemented a communication mechanism based on the Java Messaging Service (JMS) API, which is a Java Message Oriented Middleware (MOM) API for sending messages between two or more clients [397]. In order for us to use JMS, we need a JMS provider that manages the sessions and queues. Thus, we use the well-established open source Apache ActiveMQ [47] for this purpose.

The implemented communication model is a sort of queuing mechanism. It realizes an inter-process communication for passing messages within FoSII infrastructure and between components of the LoM2HiS framework, due to the fact that the components can run on different machines at different locations. This queue makes the communication mechanism highly efficient and scalable.

13.4.3   Run-Time Monitor

The run-time monitor component, which is located at the application level in a Cloud environment is presented in this section. We first describe the design of the component and later explain its implementation details.

13.4.3.1   Run-Time Monitor Design

The run-time monitor performs the mappings and based on the mapped values, the SLA objectives, and the predefined thresholds; it continuously monitors the customer application status and performance. Its operations are based on three information sources: (i) the resource metric-value pairs received from the host monitor; (ii) the SLA parameter objective values stored in the agreed SLA repository; and (iii) the predefined threat threshold values. The metric-value pairs are low-level entities and the SLA objective values are high-level entities, so for the run-time monitor to work with these two values, they must be mapped into common values.

Mapping of low-level metrics to high-level SLAs: As already discussed in Section 13.4, the run-time monitor chooses the mapping rules to apply based on the service being provisioned. That is, for each service type there is a set of defined rules for performing their SLA parameter mappings. These rules are used to compose, aggregate, or convert the low-level metrics to form the high-level SLA parameter. We distinguish between simple and complex mapping rules. A simple mapping rule maps one-to-one from low-level to high-level, as for example mapping low-level metric *Òdisk spaceÓ* to high-level SLA parameter *ÒstorageÓ*. In this case only the units of the quantities are considered in the mapping rule. Complex mapping rules consist of predefined formulae for the calculation of specific SLA parameters using the resource metrics. Table 13.1 presents some complex mapping rules.

In the mapping rules presented in Table 13.1, the downtime variable represents

TABLE 13.1
Complex Mapping Rules.

Resource Metrics	SLA Parameter	Mapping Rule
downtime, uptime	Availability (A)	$A = 1 - \frac{downtime}{uptime}$
inbyte, outbytes, packetsize, *avail.bandwidthin,* *avail.bandwidthout*	Response Time (R_{total})	$R_{total} = R_{in} + R_{out}$ *(ms)*

the *mean time to repair (MTTR)*, which denotes the time it takes to bring a system back online after a failure situation, and the uptime represents the *mean time between failure (MTBF)*, which denotes the time the system was operational between the last system failure to the next. R_{in} is the response time for a service request and is calculated as $\frac{packetsize}{availablebandwidthin - inbytes}$ in milliseconds. R_{out} is the response time for a service response and is calculated as $\frac{packetsize}{availablebandwidthout - outbytes}$ in milliseconds. The mapped SLAs are stored in the mapped metric repository for usage during the monitoring phase.

Monitoring SLA objectives and notifying the knowledge component: In this phase the run-time monitor accesses the mapped metrics' repository to get the mapped SLA parameter values that are equivalent to the agreed SLA objectives, which it uses together with the predefined thresholds in the monitoring process to detect future SLA violation threats or real SLA violation situations. This is achieved by comparing the mapped SLA values against the threat thresholds to detect future violation threats and against SLA objective thresholds to detect real violation situations. In case of detection, it dispatches notification messages to the knowledge component to avert the threats or correct the violation situation. An example of an SLA violation threat is something like an indication that the system is running out of storage. In such a case the knowledge component acts to increase the system storage. Real violations probably occur if the system is unable to resolve the cause of a violation threat notification.

13.4.3.2 Implementation of Run-Time Monitor Component

The run-time monitor receives the measured metric-value pairs and passes them into the Esper engine [253] for further processing. Esper is a component for CEP and ESP applications, available for Java as Esper, and for .NET as NEsper. Complex Event Processing (CEP) is a technology to process events and discover complex patterns among multiple streams of event data. Event Stream Processing (ESP) deals with the task of processing multiple streams of event data with the goal of identifying the meaningful events within those streams, and deriving meaningful information from them.

We use this technology because the JMS system used in our communication model is stateles and as such makes it hard to deal with temporal data and real-time queries. From the Esper engine the metric-value pairs are delivered as events each time their values change between measurements. This strategy drastically reduces the number of events/messages processed in the run-time monitor. We use an XML parser to extract the SLA parameters and their corresponding objective values from the SLA document and store them in a database. The LoM2HiS mappings are realized in Java methods and the returned mapped SLA objectives are stored in the mapped metrics database.

13.5   Knowledge Management

In this section we give some details about Case-Based Reasoning (CBR), which is the knowledge management technique we are currently investigating for proposing reactive actions to SLA violation threats or real violation situations. CBR was first built on top of FreeCBR [286], but is now a completely independent Java framework taking into account, however, basic ideas of FreeCBR. We first explain the ideas behind CBR, describe how to infer the similarity of two cases, and finally derive a utility function to estimate the "goodness" of a reactive action in a specific situation.

13.5.1   Case-Based Reasoning Overview

Case-Based Reasoning is the process of solving problems based on past experiences [43]. In more detail, it tries to solve a *case*, which is a formatted instance of a problem, by looking for similar cases from the past and reusing the solutions of these cases to solve the current one.

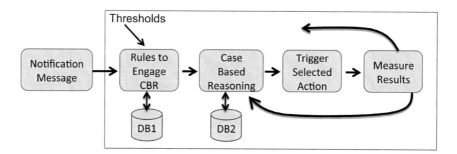

FIGURE 13.5
Case-Based Reasoning Process Overview.

As shown in Figure 13.5 the idea of using CBR in SLA management is to have rules stored in a database that engage the CBR system once a threshold value has been reached for a specific SLA parameter. The notification information is fed into the CBR system as new cases by the monitoring component. Then, CBR, prepared with some initial meaningful cases stored in database 2 (Figure 13.5), chooses the set of cases, which are most similar to the new case by various means as described in Section 13.5.2. From these cases, we select the one with the highest utility measured previously and trigger its corresponding action as the proposed action to solve the new case. Finally, we measure in a later time interval the result of this action in comparison to the initial case and store it with its calculated utilities as a new case in the CBR. Doing this, we can constantly learn new cases and evaluate the usefulness of our triggered actions.

In general, a typical CBR cycle consists of the following phases assuming that a new case was just received:

1. Retrieve the most similar case or cases to the new one.

2. Reuse the information and knowledge in the similar case(s) to solve the problem.

3. Revise the proposed solution.

4. Retain the parts of this experience likely to be useful for future problem solving. (Store new case and corresponding solution into knowledge database.)

In order to relate the cases to SLA agreement, we formalize the language elements used in the remaining of the chapter. Each SLA has a unique identifier *id* and a collection of Service Level Objectives (SLOs), which are predicates of the form

$$SLO_{id}(x_i, comp, \pi_i) \text{ with } comp \in \{<, \leq, >, \geq, =\}, \tag{13.1}$$

where $x_i \in P$ represents the parameter name for $i = 1, \ldots, n_{id}$, π_i the parameter goal, and *comp* the appropriate comparison operator. Additionally, action guarantees that state the amount of penalty that has to be paid in case of a violation can be added to SLOs, which is out of scope in this book chapter. Furthermore, a case *c* is defined as

$$c = (id, m_1, p_1, m_2, p_2, \ldots, m_{n_{id}}, p_{n_{id}}), \tag{13.2}$$

where *id* represents the SLA id, and m_i and p_i the measured (m) and provided (p) value of the SLA parameter x_i, respectively. The measured value (m) indicates the current amount of this specific Cloud resource used by the running application and the provided value (p) shows the amount of this specific Cloud resource allocated to this application. These two parameters are paramount for efficient Cloud resource management in proposing reactive actions to prevent or correct SLA violation situation.

A typical use case for the evaluation might be: SLA id $= 1$ with SLO_1("Storage", $\geq$, 1000, ag_1) and SLO_1 ("Bandwidth", $\geq$, 50.0, ag_1), where ag_1 stands for the appropriate preventive action to execute after an SLO

violation. A simple case that can be notified by the measurement component would therefore look like $c = (1, 500, 700, 20.0, 30.0)$. A result case
$rc = (c^-, ac, c^+, utility)$ includes the initial case c^-, the executed action ac,
the resulting case c^+ measured after some time interval later and the calculated *utility* as described in Section 13.5.3.

13.5.2   Inferring Similarity of Two Cases

To retrieve similar cases already stored in the database in order to propose
an action for a new case, the similarity of the two cases has to be calculated.
However, there are many metrics that can be considered in this process.
We approach this problem using a strategy similar to Euclidean distance.
However, the problem with Euclidean distance, for instance, is due to its symmetric nature and therefore cannot correctly fetch whether a case is in a state
of over- or under-provisioning. Additionally, the metric has to treat parameters in a normalized way so that parameters that have a larger distance range
are not over-proportionally taken into account compared to parameters with a
smaller difference range. For example, if the difference between measured and
provided values of parameter A always lies between 0 and 100 and of parameter B between 0 and 1000, the difference between an old and a new case can
only be within the same ranges, respectively. Thus, just adding the differences
of the parameters would yield an unproportional impact on parameter B.
This leads to the following equation whose summation part follows the principle of semantic similarity [339]:

$$d(c^-, c^+) = \min(w_{id}, |id^- - id^+|) + \sum_{x \in P} w_x \left| \frac{(p_x^- - m_x^-) - (p_x^+ - m_x^+)}{max_x - min_x} \right|,$$

(13.3)

where $w = (w_{id}, w_{x_1}, \ldots, w_{x_n})$ is the weight vector; w_{id} is the weight for non-
identical SLAs; w_x is the weight, and max_x and min_x the maximum and
minimum values of the provided and measured resource differences $p_x - m_x$
for parameter x. As it can be easily checked, this indeed is a metric also in
the mathematical sense.
Furthermore, the match percentage mp of two cases c^- and c^+ is then calculated as

$$mp(c^-, c^+) = \left(1 - \frac{d(c^-, c^+)}{w_{id} + \sum_x w_x}\right) \cdot 100.$$

(13.4)

This is done because the algorithm does not only consider the case with the
highest match, but also cases in a certain percentage neighborhood (initially
set to 3%) of the case with the highest match. From these cases the algorithm
then chooses the one with the highest utility. By calculating the match percentage, the cases are distributed on a fixed line between 0 and 100, where
100 is an identical match, whereas 0 is the complete opposite.

13.5.3 Utility Function and Resource Utilization

To calculate the utility of an action, we have to compare the initial case c^- vs. the resulting final case c^+. The *utility function* presented in Equation 13.5 is composed by a SLA violation and a resource utilization term weighed by the factor $0 \leq \alpha \leq 1$:

$$utility = \sum_{x \in P} violation(x) + \alpha \cdot utilization(x) \qquad (13.5)$$

Higher values for α indicate desire to achieve high resource utilization, whereas lower values implies the wish of non-violation of SLA parameters. Thus, to achieve a balance, a trade-off must be found. We further note that $c(x)$ describes a case only with respect to parameter x. E.g., we say that a violation has occurred in $c(x)$, when in case c the parameter x was violated.

The function *violation* for every parameter x is defined as follows:

$$violation(x) = \begin{cases} 1, & \text{No violation occurred in } c^+(x), \text{ but in } c^-(x) \\ 1/2, & \text{No violation occurred in } c^+(x) \text{ and } c^-(x) \\ -1/2 & \text{Violation occurred in } c^+(x) \text{ and } c^-(x) \\ -1 & \text{Violation occurred in } c^+(x), \text{ but not in } c^-(x) \end{cases} \qquad (13.6)$$

For the *utilization* function we calculate the utility from the used resources in comparison to the provided ones. We define the distance $\delta(x, y) = |x - y|$, and utilization for every parameter as

$$utilization(x) = \begin{cases} 1, & \delta(p_x^-, m_x^-) > \delta(p_x^+, u_x^+) \\ -1, & \delta(p_x^-, m_x^-) < \delta(p_x^+, u_x^+) \\ 0, & \text{otherwise.} \end{cases} \qquad (13.7)$$

We get a utilization utility of 1 if we experience less over-provisioning of resources in the final case than in the initial one, and a utilization utility of -1 if we experience more over-provisioning of resources in the final case than in the initial one.

We map utilization, u, and the number of SLA violations, v, into a scalar called *Resource Allocation Efficiency* (RAE) using the following equaltion

$$RAE = \begin{cases} \frac{u}{v}, & v \neq 0 \\ u, & v = 0, \end{cases} \qquad (13.8)$$

which represents an important goal for the knowledge management. High utilization leads to high RAE, whereas a high number of SLA violations leads to a low RAE, even if utilization is in normal range. This can be explained by the fact that having utilization at a maximum — thus being very resource efficient in the first place — does not pay if the SLA is not fulfilled at all.

13.6   Evaluations

In this section we present an image rendering use-case scenario to evaluate our approach in this chapter. The knowledge management technique is developed as an independent work and has not yet been integrated with the LoM2HiS framework. Thus our evaluations here covers only the monitoring framework. The goal of our evaluations is to determine the optimal measurement interval for monitoring agreed SLA objectives for applications at runtime. We first present our real Cloud experimental environment, after which we discuss in detail the use-case scenario.

13.6.1   Experimental Environment

The resource capacities of our Cloud experimental testbed is shown in Table 13.2. The table shows the resource compositions of the physical and the virtual machines being used in our experimental setups. We use Xen virtualization technology in our testbed; precisely, we run Xen 3.4.0 on top of Oracle Virtual Machine (OVM) server.

TABLE 13.2
Cloud Environment Resource Setup Composed of 10 Virtual Machines.

Machine Type = Physical Machine				
OS	CPU	Cores	Memory	Storage
OVM Server	Pentium 4 2.8 GHz	2	2.5 GB	100 GB

Machine Type = Virtual Machine				
OS	CPU	Cores	Memory	Storage
Linux/Ubuntu	Pentium 4 2.8 GHz	1	1 GB	5 GB

We have in total five physical machines and, based on their resource capacities as presented in Table 13.2, we host two VMs on each physical machine. We use an Automated Emulation Framework (AEF) [158] to deploy the VMs onto the physical hosts, thus creating a virtualized Cloud environment with up to ten computing nodes capable of provisioning resources to applications and one front-end node responsible for management activities.

The front-end node serves as the control entity. It runs the automated emulation framework, the application deployer [249] responsible for moving the application data to the virtual machines, and the LoM2HiS framework to monitor and detect SLA violation situations. We use this Cloud environment to evaluate the use case scenario presented in the next section.

13.6.2   Image Rendering Application Use-Case Scenario

We developed an image rendering application based on the Persistence of Vision Raytracer (POV-Ray)[1], which is a ray tracing program available for several computing platforms [305]. In order to achieve a heterogeneous load in this use-case scenario, we experiment with three POV-Ray workloads, each one with a different characteristic of time for rendering frames, as described below and illustrated in Figures 13.6 and 13.7:

- **Fish:** rotation of a fish on water. Time for rendering frames is variable.

- **Box:** approximation of a camera to an open box with objects inside. Time for rendering frames increases during execution.

- **Vase:** rotation of a vase with mirrors around. Time for processing different frames is constant.

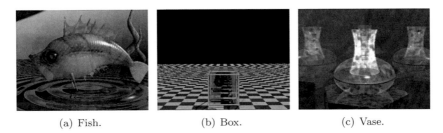

    (a) Fish.   (b) Box.   (c) Vase.

FIGURE 13.6
Example of Images for Each of the Three Animations.

Three SLA documents are negotiated for the three POV-Ray applications. The SLA documents specify the level of Quality of Service (QoS) that should be guaranteed for each application during its execution. Table 13.3 presents the SLA objectives for each of the applications. These SLA objective thresholds are defined based on test runs and experiences with these applications in terms of resource consumption. With the test runs, the Cloud provider can determine the amount and type of resources the application requires. Thus, the provider can make a better resource provisioning plan for the applications. Based on these SLA objectives, the applications are monitored to detect SLA violations. These violations may happen because SLAs are negotiated per application and not per allocated VM, considering the fact that the service provider may provision different application requests on the same VM.

Figure 16.4 presents the evaluation configurations for the POV-Ray applications. We instantiate 10 virtual machines that execute POV-Ray frames submitted via Application Deployer. The virtual machines are continuously

---

[1]www.povray.org

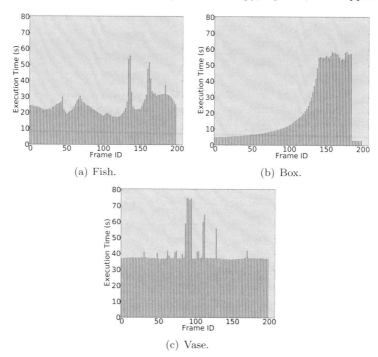

(a) Fish.   (b) Box.

(c) Vase.

FIGURE 13.7
Behavior of Execution Time for Each POV-Ray Application.

monitored by Gmond. Thus, LoM2HiS has access to resource utilization during execution of the applications. Similarly, information about the time taken to render each frame in each virtual machine is also available to LoM2HiS. This information is generated by the application itself and is sent to a location where LoM2HiS can read it. As described in Figure 13.8, users supply the QoS requirements in terms of SLOs (Step 1 in Figure 13.8). At the same time the images with the POV-Ray applications and input data (frames) can be uploaded to the front-end node. Based on the current system status, a SLA negotiator establishes an SLA with the user. Description of the negotiation process and components is out of scope of this chapter and is discussed by Brandic et al. [136]. Thereafter, a VM deployer starts configuration and allocation of the required VMs whereas an application deployer maps the tasks to the appropriate VMs (step 3). In step 4 the application execution is triggered.

13.6.3   Achieved Results and Analysis

We defined and used six measurement intervals to monitor the POV-Ray applications during their executions. Table 13.4 shows the measurement intervals

TABLE 13.3
POV-Ray Applications SLA Objective Thresholds

SLA Parameter	Fish	Box	Vase
CPU	20%	15%	10%
Memory	297MB	297MB	297MB
Storage	2.7GB	2.6GB	2.5GB

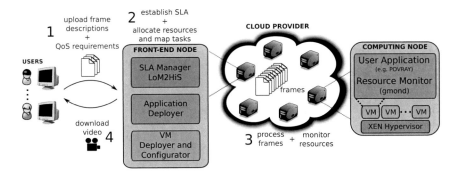

FIGURE 13.8
Pov-Ray Evaluation Configuration.

and the number of measurements made in each interval. The applications run for about 20 minutes for each measurement interval.

TABLE 13.4
Measurement Intervals.

Intervals	5s	10s	20s	30s	60s	120s
Nr. of Measurements	240	120	60	40	20	10

The 5 second measurement interval is a reference interval, meaning the current interval used by the provider to monitor application executions on the Cloud resources. Its results show the present situation of the Cloud provider.

Figure 13.9 presents the achieved results of the three POV-Ray applications with varying characteristics in terms of frame rendering as explained in Section 13.6.2. We use the ten virtual machines in our testbed to simultaneously execute the POV-Ray frames. There is a load-balancer integrated in the application deployer, which ensures that the frame executions are balanced among the virtual machines.

The *LoM2HiS* framework monitors the resource usage of each virtual machine to determine if the SLA objectives are ensured, and reports violations otherwise. Since the load-balancer balances the execution of frames among the virtual machines, we plot in Figure 13.9 the average numbers of violations encountered in the testbed for each application with each measurement interval.

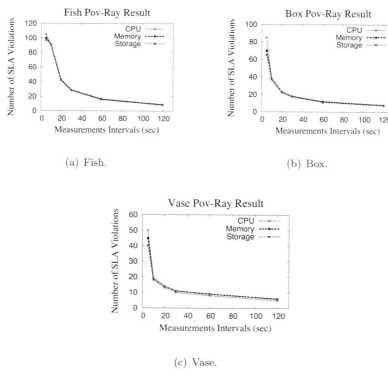

(a) Fish.

(b) Box.

(c) Vase.

FIGURE 13.9
POV-Ray Experimental Results.

To find the optimal measurement interval for detecting applications' SLA objectives violations at runtime, we discuss the following two determining factors i) the cost of making measurements; and ii) the cost of missing SLA violations. The acceptable trade-off between these two factors defines the optimal measurement interval.

Using these two factors and other parameters we define a cost function (C) based on which we can derive an optimal measurement interval. The ideas of defining this cost function are derived from utility functions discussed by Lee et al. [458]. Equation 13.9 presents the cost function.

$$C = \mu * C_m + \sum_{\psi \epsilon \{cpu, memory, storage\}} \alpha\left(\psi\right) * C_v \qquad (13.9)$$

where μ is the number of measurements, C_m is the cost of measurement, $\alpha\left(\psi\right)$ is the number of undetected SLA violations, and C_v is the cost of missing an SLA violation. The number of undetected SLA violations is determined based on the results of the 5 second reference measurement interval, which is assumed to be an interval capturing all the violations of applications' SLA objectives. This cost function now forms the basis for analyzing the achieved results of our use-case scenario.

The cost of making measurements in our testbed is defined by considering the intrusiveness of the measurements on the overall performance of the system. Based on our testbed architecture and the intrusiveness test performed, we observed that measurements have minimal effects on the computing nodes. This is because measurements and their processing take place in the front-end node while the services are hosted in the computing node. The monitoring agents running on computing nodes have minimal impact on resource consumption. This means a low cost of making measurements in the Cloud environment.

The cost of missing SLA violation detection is an economic factor, which depends on the SLA penalty cost agreed to for the specific application and the effects the violation will have on the provider, for example in terms of reputation or trust issues.

Applying the cost function on the achieved results of Figure 13.9, with a measurement cost of 0.5 dollar and an aggregated missing violation cost of 1.5 dollar, we achieve the monitoring costs presented in Table 13.5. These cost values are example values for our experimental setup. They neither represent nor suggest any standard values. The approach used here is derived from the cost function approaches presented in literature [457, 780].

TABLE 13.5
Monitoring Cost.

Intervals / SLA Parameter	Fish POV-Ray Application				
	Reference 10s	20s	30s	1min	2min
CPU	0	22.5	94.5	115.5	133.5	145.5
Memory	0	13.5	85.5	106.5	126	136.5
Storage	0	9	81	102	120	132
Cost of Measurements	120	60	30	20	15	5
Total Cost	120	105	291	344	394.5	419
Box POV-Ray Application						
CPU	0	19.5	42	49.9	58.5	64.5
Memory	0	10.5	33	40.5	49.5	55.5
Storage	0	9	81	102	120	132
Cost of Measurements	120	60	30	20	15	5
Total Cost	120	97.5	135	147.5	169.5	177.5
Vase POV-Ray Application						
CPU	0	10.5	18	22.5	25.5	30
Memory	0	6	13.5	18	21	25.5
Storage	0	7.5	15	19.5	22.5	27
Cost of Measurements	120	60	30	20	15	5
Total Cost	120	84	76.5	80	84	87.5

The monitoring cost presented in Table 13.5 represents the cost of measurement for each interval and for failing to detect SLA violation situations for each application. The reference measurement captures all SLA violations for each application, thus it only incurs measurement cost. Taking a closer look at Table 13.5, it is clear that the values of the shorter measurement interval are closer to the reference measurement than those of the longer measurement interval. This is attributed to our novel architecture design, which separates management activities from computing activities in the Cloud testbed.

The relations of the measurement cost and the cost of missing SLA violation detection is graphically depicted in Figure 13.10 for the three POV-Ray applications. From the figures, it can be noticed in terms of measurement cost that the longer the measurement interval, the smaller the measurement cost and in terms of detection cost, the higher the number of missed SLA violations detected, the higher the detection cost rises. This implies that to keep the detection cost low, the number of missed SLA violations must be low.

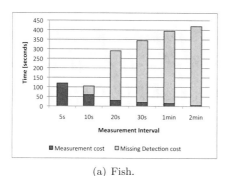

(a) Fish.

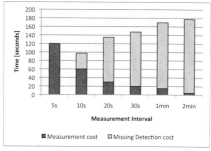

(b) Box.

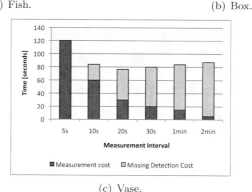

(c) Vase.

FIGURE 13.10
POV-Ray Application Cost Relations.

Considering the total cost of monitoring the fish POV-Ray application in Table 13.5 and Figure 13.10(a), it can be seen that the reference measurement is not the cheapest although it does not incur any cost of missing SLA violation detection. In this case the 10-second interval is the cheapest, and in our opinion, the most suited measurement interval for this application. In the case of box POV-Ray application, the total cost of monitoring, as shown in Table 13.5 and depicted graphically in Figure 13.10(b), indicates that the lowest cost is incurred with the 10-second measurement interval. Thus we conclude that this interval is best suited for this application. Also from Table 13.5 and Figure

13.10(c), it is clear that the reference measurement by far does not represent the optimal measurement interval for the vase POV-Ray application. Based on the application behavior, longer measurement intervals are better fitted than shorter ones. Therefore, in this case the 20-second measurement interval is best suited for the considered scenario.

Based on our experiments, it is observed that there is no best suited measurement interval for all applications. Depending on how steady the resource consumption is, the monitoring infrastructure requires different measurement intervals. Note that the architecture can be configured to work with different intervals. In this case, specification of the measurement frequencies depends on policies agreed to by users and providers.

13.7 Conclusion and Future Work

Flexible and reliable management of SLA agreements represents an open research issue in Cloud computing infrastructures. Advantages of flexible and reliable Cloud infrastructures are manifold. For example, prevention of SLA violations avoids unnecessary penalties the provider has to pay in case of violations, thereby maximizing the profit of the provider. Moreover, based on flexible and timely reactions to possible SLA violations, interactions with users can be minimized. In this chapter we presented the *LoM2HiS framework* — a novel framework for monitoring low-level Cloud resource metrics, mapping them to high-level SLA parameters, using the mapped values and predefined thresholds to monitor the SLA objectives at runtime, detecting and reporting SLA violation threats or real violation situations. We also presented the knowledge management technique based on Case-Based Reasoning for managing the SLA violation situation and proposing preventive or corrective actions.

We evaluated our system using a use-case scenario consisting of image rendering applications based on POV-Ray with heterogeneous workloads. The evaluation is focused on the goal of finding an optimal measurement interval for monitoring application SLA objectives at runtime. From our experiments we observed that there is no particular well suited measurement interval for all applications. It is easier to identify optimal intervals for applications with steady resource consumption, such as the "vase" POV-Ray animation. However, applications with variable resource consumption require dynamic measurement intervals. Our framework can be extended to tackle such applications but this will be in the scope of our future work.

Currently on the FoSII project, we are working toward integrating the knowledge management with the LoM2HiS framework in order to achieve a complete solution pack for the SLA management. Furthermore, we will design and implement an actuator component for applying the proposed actions from the knowledge database on the Cloud resources.

Thus, in the future, besides our investigation on dynamic measurement intervals, we will evaluate the influence of such intervals on the quality of the reactive actions proposed by the knowledge database. If the effects of measurement intervals are known, the best reactive actions may be taken, contributing to our vision of flexible and reliable on-demand computing via fully autonomic Cloud infrastructures.

Acknowledgments

This work is supported by the Vienna Science and Technology Fund (WWTF) under grant agreement ICT08-018 Foundations of Self-governing ICT Infrastructures (FoSII). We would like to thank Marco A. S. Netto for his support in carrying out the evaluations. The experiments were performed in the High Performance Computing Lab at Catholic University of Rio Grande do Sul (LAD-PUCRS), Brazil.

14

Auto-Scaling, Load Balancing and Monitoring in Commercial and Open-Source Clouds

Eddy Caron

University of Lyon - CNRS - ENS Lyon - UCB Lyon 1 - INRIA
LIP Laboratory, Lyon, France

Frédéric Desprez

University of Lyon - CNRS - ENS Lyon - UCB Lyon 1 - INRIA
LIP Laboratory, Lyon, France

Luis Rodero-Merino

University of Lyon - CNRS - ENS Lyon - UCB Lyon 1 - INRIA
LIP Laboratory, Lyon, France

Adrian Muresan

University of Lyon - CNRS - ENS Lyon - UCB Lyon 1 - INRIA
LIP Laboratory, Lyon, France

CONTENTS

Abstract

Over the past years, the Cloud phenomenon had an impressive increase in popularity in both the software industry and research worlds. The most interesting feature that Cloud Computing brings, from a Cloud client's point of view, is the on-demand resource provisioning model. This allows Cloud client platforms to be scaled up in order to accommodate more incoming clients and to scale down when the platform has unused resources, and this can all be done while the platform is running. As a result, the physical resources are used more efficiently and the Cloud client saves expenses. The importance of this feature is unquestionable to Cloud client users.

Achieving the above mentioned is not trivial and is done by leveraging the more direct functionalities that Clouds provide. Three of these key functionalities are automatic scaling, load balancing and monitoring. They represent the focus of the current work.

This chapter is divided into three main parts, one for each of the three main topics of interest: auto-scaling, load balancing and monitoring. We detail each of the three topics and present details on their presence and implementation in the commercial, open-source and research worlds.

Among the commercial Cloud platform providers we have focused on Amazon EC2, Microsoft Azure, GoGrid and RackSpace. From the open-source initiatives we have focused our attention on Nimbus, Eucalyptus and OpenNebula. We have also detailed endeavors in the research world that are relevant to the topic of interest and are not connected to any of the Cloud providers. By studying this chapter, the reader will be able to understand the role that each of the three Cloud features plays for a Cloud client, will understand their inner workings and will be aware of the state-of-the-art available in current commercial and open-source Cloud platforms.

14.1 Introduction

The increasing relevance of Cloud computing in the IT world is undeniable. Cloud providers have focused a lot of attention on providing facilities for Cloud clients, that make using the Cloud an easy task. These facilities range from automatic and configurable platform scaling and load balancing services to platform monitoring at different levels of granularity and configurable alert services. Given that there are no formal standards related to this topic, each Cloud provider has their own interpretation of the problem and their own way of addressing it.

In what follows, we will detail the topics of auto-scaling, load-balancing and monitoring. We will explore both Cloud provide and Cloud client points of view and examine what approaches are taken by the commercial providers and their open-source counterparts. Where an implementation is not available for one of the discussed Clouds, we will present alternative solutions by turning to available commercial services and open-source software. We will also present research work that has been done around the topic of interest.

The rest of this chapter is organized as follows: in Section 14.2 we present the notion of auto-scaling and detail its implementation in the current commercial and open-source Cloud providers, as well as in third party software. Section 14.3 presents the concept of load balancing related to the topic of Cloud platforms, its usefulness and its presence in the current available Cloud platforms and possible alternatives. The last of the three topics of interest, platform monitoring, is detailed in Section 14.4 and we conclude the current chapter in Section 14.5.

14.2 Cloud Auto-Scaling

Elasticity is regarded as one of the differentiating features of clouds. In fact, for some authors, it is considered the characteristic that makes clouds something other than "*an outsourced service with a prettier face*" [558]. Cloud users can quickly deploy or release resources as they need them, thus taking benefit of the typical pay-per-use billing model. They avoid potential over-provisioning of resources which implies investment in resources that are not needed. Also, increases on demand can be quickly attended to by asking the cloud for more resources, thus preventing a possible degradation of the perceived service quality.

However, to benefit from elasticity in typical Infrastructure-as-a-Service (IaaS) settings, the cloud user is forced to constantly control the state of the deployed system. This must be done in order to check whether any resource scaling action has to be performed. To avoid this, several auto-scaling solutions have been proposed by academia [496, 613] and by different cloud vendors. All these solutions allow users to define a set of scaling rules regarding the service hosted in the clouds. Each rule is composed by one or more conditions and a set of actions to be performed when those conditions are met. Conditions are typically defined using a set of metrics (which must be monitored by the cloud platform) like for example CPU usage, and some threshold. When the threshold is traversed then the condition is met.

Although cloud auto-scaling proposals all are based on this conditions + actions approach, they vary substantially in several aspects: which metrics are monitored (and so included in the rules definition); expressiveness of the conditions defining mechanism; and which actions can be taken. Many of them focus on horizontal scaling, i.e., deploying or releasing VMs, while vertical scaling (like for example increasing physical resources of an overloaded server) is not considered, possibly due to the impossibility of changing the available CPU, memory, etc., on-the-fly in general purpose OSs.

Here we analyze the different auto-scaling solutions used by several cloud proposals, commercial ones such as Amazon EC2 and open source solutions such as Open Nebula. Also, we take a look at solutions developed by academia.

14.2.1 Auto-Scaling in Commercial Clouds

14.2.1.1 Amazon EC2

Amazon provides auto-scaling as part of the service offered by their IaaS EC2 public cloud. This service can be accessed by a web services API or through the command line. Auto-scaling in EC2 is based on the concept of *Auto Scaling Group* (ASG). A group is defined by:

- The configuration of the VMs that will be part of the group, where the configuration is given by the virtual image (that contains the OS and software stack of the VM) and hardware characteristics. As there it can only be an unique configuration per group, then all machines must by force have the same configuration. As a matter of fact each ASG represents a cluster, so this limitation can be assumed in many scenarios as real world clusters are often built by similar machines. But the lack of heterogeneous groups impedes certain useful configurations. For example, some users could benefit by replacing several "small" machines with one single "powerful" machine for cost reasons. Such replacement cannot be done automatically by EC2 auto-scaling service.

- Certain parameters such as the zone where VMs of the group will be deployed (among EC2's available regions, i.e., EU, US East...) or the minimum and maximum amount of VM instances allowed for the group. When setting a minimum size on the group, the user implicitly configures EC2 to automatically create a new VM whenever some of the running instances are shut down (e.g., because of a failure) and the minimum limit is exceeded.

Finally, the user can define a set of rules for each ASG. In EC2 jargon, the possible actions to be run are denoted *policies*. Each policy defines the amount of capacity (in absolute or relative values) to be deployed or released in a certain group. The platform will create or shut down VM instances in the ASG to meet that capacity demand. Triggers are denoted *metric alarms* and are based on the metrics served by EC2's monitoring service *CloudWatch* (see Section 14.4). Each metric alarm is defined by the metric and related statistic to be observed (like the average value), the evaluation period, and the threshold that will trigger the alarm. When the alarm is triggered, the action defined by the corresponding policy is run. Load balancers are automatically notified to start/stop sending requests to the created/stopped VMs.

14.2.1.2 Microsoft Azure

Microsoft Azure is considered a Platform-as-a-Service (PaaS) cloud as the Google App Engine or salesforce.com. PaaS clouds offer a runtime environment system (e.g., a servlets container) where users' components can be deployed and executed in a straightforward manner. Thus, PaaS clouds are said to offer an *additional abstraction level* when compared to IaaS clouds [722], so users do not have to handle virtual resources such as machines or networks to start running their systems. In such settings is the cloud system who must scale resources as needed by the container platform in a transparent manner, without any user intervention. Users are not required to monitor its service to scale resources, nor to define scalability rules.

Azure, nonetheless, is an exception. In Azure is the user who must configure the scalability settings, i.e., the platform does not handle resources scaling on behalf of the user. As part of the Azure cloud, there is a "Compute" service where users can request VMs instances, so Azure does not isolate users from resources, in contrast with other PaaS platforms. There is also available a *diagnostics mechanism* to obtain several metrics, and a *management API* to request certain actions on the deployed servers. But it is the user who must code the auto-scaling functionality she may require by using these APIs, i.e., Azure does not implement any embedded auto-scaling solution ready to be used by only configuring it, as for example EC2 does.

Not surprisingly, some commercial offers such as Paraleap [35] have emerged that try to address this severe limitation. Paraleap automatically scales resources in Azure to respond to changes on demand.

14.2.1.3 GoGrid

GoGrid does not implement any auto-scaling functionality. Similarly to Azure, it does provide an API to remotely command the addition or removal of VMs (servers), but it is up to the user to call on this API method when required. RightScale [37] is a *cloud management platform* that offers control functionality over the VMs deployed in different clouds. It provides auto-scaling functionality on top of GoGrid and EC2, based on *alerts* and associated *actions* to be run (one or more) each time an alarm is triggered. This is similar to the auto-scaling services of EC2. But there are some differences. First, alerts can be defined based not only on hardware metrics. Metrics regarding the state of software applications such as the Apache web server and MySQL engine are also available. Also, several actions can be associated to the alert, like for example sending emails to administrators. Besides, these actions can be run periodically for defined intervals of time, not just once. Finally, alerts and actions are usually defined at the server level. But scaling actions are an exception: they are performed only if a certain user-defined percentage of servers "vote" for the action to be run.

14.2.1.4 RackSpace

As in the case of GoGrid, RackSpace has not built in auto-scaling capabilities, although it does provide an API for remote control of the hosted VMs. Thus, the user is responsible for monitoring the service and taking the scaling decisions. The creation and removal of resources is done through calls to the remote API.

Enstratus [21] is a *cloud management platform* that offers control functionality over the VMs deployed in different clouds, as RightScale does. It provides auto-scaling functionality on top of all those clouds, including RackSpace, again very similar to the one provided by the auto-scaling services of EC2. RightScale [37] has recently added support for VMs deployed on RackSpace. Scalr [38] is an open source project that handles scaling of cloud applications on EC2 and RackSpace. It only manages web applications based on Apache and MySQL, and the scaling actions are decided by a built-in logic. Users cannot configure how their applications must scale in response to changes on load.

14.2.2 Implementations of Auto-Scaling in Open-Source Clouds

14.2.2.1 Nimbus

Nimbus does not have any embedded auto-scaling functionality, it is a tool to create clouds from datacenter resources. In [496] the authors introduce a *resource manager* able to ask for resources from third-party clouds when the local resources are under high demand. The decisions about when to scale the deployment by adding new VMs is taken by *policies*, implemented as Python modules. These policies are the equivalent to the elasticity rules present in other systems.

14.2.2.2 Eucalyptus

In Eucalyptus there is no out-of-the-box auto-scaling functionality. Eucalyptus is focused on the management at virtual resource level, and does not control the services running on it. Hence it cannot be aware of their state and so it cannot decide when to add or release resources to the service.

Makara [25] offers a "wrapper" service for the deployment and control of services running in the cloud, an approach similar to RightScale, but aimed at Java and PHP-based applications. Makara can manage services running on top of one or many cloud providers, including EC2 and private clouds based on Eucalyptus. The auto-scaling functionality in Makara allows to define when to create or release VMs using thresholds over two metrics: CPU, and requests to node.

Scalr can also be used for applications running on Eucalyptus clouds.

14.2.2.3 OpenNebula

OpenNebula does not provide auto-scaling. It promotes a clear distinction of the roles at each layer of the cloud stack. Open Nebula, which would be at the bottom of this stack, aims to ease the management of the virtual infrastructure demanded, and it assumes that scaling actions should be controlled and directed at a higher level. I.e., OpenNebula is not "aware" of the services it hosts, or of their state. Hence, OpenNebula is not in charge of supervising the deployed systems.

The work in the "*OpenNebula Service Management Project*" [34] tries to develop a component to be run on top of OpenNebula that handles services (understood as a cluster of related VMs) instead of raw hardware resources. Support for auto-scaling could be added in the future. Most mature seems Claudia, discussed in Section 14.2.3.

14.2.3 Claudia: A Proposal for Auto-Scaling

Auto-scaling is getting some attention from the research community, as the limitations of the present solutions become more and more clear. One of such limitations is the lack of flexibility. The auto-scaling configuration is set using a GUI with limited choices regarding the metrics to be monitored and the conditions that will trigger the corresponding scaling actions.

In this regard, Claudia [613] proposes a more flexible solution. Claudia is a Cloud tool designed to handle not just VMs, but whole services which in turn are usually composed of a set of connected VMs. Claudia's declared goal is to offer cloud users an abstraction level closer to the services' lifecycle, that does not force users to deal with single VMs. As part of the service definition the user will describe the VMs that must be instantiated and the dependencies among them. Claudia will deploy the service following the instructions given by the user, including the instantiation order of VMs. Also, it will make available to the VMs the configuration information they require, including the one that will be only available at deployment time. For example, a webserver could need the IP address to connect to a database server which is part of the same service, which is only known once the database has been started. Such dependencies are managed by Claudia (such a configuration process is also denoted *contextualization*).

Also, as part of this service-level handling, Claudia is in charge of scaling the resources associated to the service. As part of the service definition (format based on an extension to the *Open Virtualization Format*) users can include *elasticity rules* based on arbitrary service level metrics (such as the number of transactions). The service will forward these values to Claudia through the monitoring system. Claudia uses a rule engine to constantly check the conditions defined for each rule and trigger the corresponding actions when the conditions are met. Claudia was built and tested as a set of components running on top of OpenNebula, but is not tied to any particular cloud platform. The Claudia proposal goes a step further than other auto-scaling systems. First, scaling decisions can be based on *any* metric, while other auto-scaling proposals restrict the metrics that can be used to define the scalability rules that will govern the service. Therefore, scaling can be applied to any service regardless of the metrics that represent its state. Also, rules are defined using the extensions that Claudia proposes to the Open Virtualization Format [210] (format to define the deployment of collections of VMs, promoted by the DMTF), that have a richer expressiveness than typical rule definition schemes in present cloud offers. Finally, each rule can have many actions associated (not just one).

14.2.4   Implementing and Auto-Scaling Cloud System

To conclude this section, we briefly enumerate the three main functionalities to be implemented in a cloud system to enable it to scale the software it hosts:

- *Monitoring* (see Section 14.4) is required to get information about the state of the running services and the resources they use. Two challenges are involved: first, being able to handle metrics from up to thousands of services, each one running on several VMs; second, adding arbitrary service-level metrics that represent the state of the deployed software beyond simple hardware-level metrics.

- *Load-balancing* (see Section14.3) mechanisms should be ready to update the set of available server replicas, as it can change at any moment due to the creation and removal of VM instances.

- *Rule checking* is in charge of reading the metrics values to verify when the conditions that trigger the scaling actions are met.

14.3 Cloud Client Load Balancing

This concept of load balancing is not typical to Cloud platforms and has been around for a long time in the field of distributed systems. In its most abstract form, the problem of load balancing is defined by considering a number of parallel machines and a number of independent tasks, each having its own load and duration [188]. The goal is to assign the tasks to the machines, therefore increasing their load, in such a way as to optimize an objective function. Traditionally, this function is the maximum of the machine loads and the goal is to minimize it. Depending on the source of the tasks, the load balancing problem can be classified as: *offline load balancing* where the set of tasks is known in advance and cannot be modified and *online load balancing* in the situation that the task set is not known in advance and tasks arrive in the system at arbitrary moments of time.

In the case of Cloud computing we can consider load balancing at two different levels: Cloud provider level and Cloud client level. From the point of view of the Cloud provider, the load balancing problem is of type online and is mapped in the following way:

- The parallel machines are represented by the physical machines of the Cloud's clusters

- The tasks are represented by client requests for virtual resources

- Cloud client requests can arrive at arbitrary moments of time

In the case of Cloud client's virtual platform, the load balancing problem is mapped in the following way:

- The parallel machines translate into the virtual resources that the Cloud client has currently running

- The tasks translate into client requests to the Cloud client's platform

- End user requests can arrive at arbitrary moments of time

As a consequence, in Cloud platforms, load balancing is an online problem where end user requests that enter the Cloud client's application need to be distributed across the Cloud client's instantiated virtual resources with the goal of balancing virtual machine load or minimizing the number of used virtual machines.

Although load balancing is not a unique feature to Cloud platforms, it should not be regarded as independent from auto-scaling. In fact, the two need to work together in order to get the most efficient platform usage and save expenses.

The end goal of load balancing from the Cloud client's point of view is to have a more efficient use of the virtual resources that he has running and thus reduce cost. Since most Cloud providers charge closest whole hour per virtual resource, then the only way that cost saving is achieved is by reducing the number of running virtual resources, while still being able to service client requests. It follows that load balancing should be used in conjunction with auto-scaling in order to reduce cost. As a result we have the following usage scenarios:

1. **high platform load** when the Cloud client's overall platform load is high, as defined by the Cloud client. The platform needs to scale up by adding more virtual resources. The load balancing element automatically distributes load to the new resources once they are registered as elements of the platform and therefore reduces platform load.

2. **low platform load** when the Cloud client's overall platform load is low, as defined by the Cloud client. In this situation, the platform needs to scale down by terminating virtual resources. This is not as trivial as the previous scenario, because the load balancing element typically assigns tasks to all resources and therefore prevents resources from reaching a state where their load is zero and can be terminated. In this situation, the load balancing element needs to stop distributing load to the part of the platform that will be released and, even more, the currently-running tasks of this part of the platform need to be migrated to ensure that a part of the platform will have zero load and therefore can be released.

Load balancing also brings some issues as side effects along with it. One of these is session affinity. Because load balancers distribute load evenly among available nodes, there is no guarantee that all the requests coming from one user will be handled by the same node from the pooled resources. This has the implication that all context related to the client session is lost from one request to another. This is usually an undesired effect. In the great majority of situations, it is desired that requests from the same client be handled by the same node throughout the duration of the client's session. In modern clouds this is referred to as session stickiness.

Mapping of virtual resources to physical resources also has an impact on Cloud clients. There is usually a compromise between the following two opposite use cases:

- The Cloud provider achieves a more efficient resource usage by trying to minimize the number of physical hosts that are running the virtual resources. The downside for the Cloud client is the fact that his platform is at a greater risk in case of hardware failure because the user's virtual resources are deployed on a small number of physical machines.

- The virtual resources are distributed across the physical resources. Thus the risk of failure is less for Cloud clients in case of hardware failure. On the downside, there is a greater number of physical machines running and thus more power usage.

14.3.1 Load Balancing in Commercial Clouds

The problem of load balancing in all Cloud platforms may be the same, but each Cloud provider has its own approach to it, which is reflected in the services they offer and their differences with respect to other providers.

14.3.1.1 Amazon EC2

Amazon EC2 offers load balancing through their *Amazon Elastic Load Balancing* service [63]. The Cloud client can create any number of *LoadBalancers*. Each *LoadBalancer* will distribute all incoming traffic that it receives for its configured protocol to the EC2 instances that are sitting behind the *LoadBalancer*. One single *LoadBalancer* can be used to distribute traffic for multiple applications and across multiple *Availability Zones*, but limited to the same Amazon EC2 *Region*.

If an instance that is behind the *LoadBalancer* reaches an unhealthy state, as defined by the *LoadBalancer*'s health check, then it will not receive any new load, until its state is restored so that it passes the health check. This feature increases the fault tolerance of Cloud client applications by isolating unhealthy components and giving the platform notice to react.

Amazon's *Elastic Load Balancing* service has two ways of achieving stickiness:

1. A duration-based sticky session in which case the load balancers themselves emit a cookie of configurable lifespan, which determines the duration of the sticky session.

2. An application-controlled sticky session in which case the load balancers are configured to use an existing session cookie that is completely controlled by the Cloud client's application.

Related to sticky sessions, it is worth noting that Amazon EC2 does not support load balancing for HTTPS traffic. This is due to the fact that the cookies are stored in the HTTP header of the request and all HTTPS traffic is encrypted. So to perform session stickiness when load balancing HTTPS traffic, the balancers would need to have the application's SSL certificate, which cannot be done in the current version of the load balancers.

As for pricing, the Cloud user is charged for the running time of each *Load-Balancer*, rounded up to an integer number of hours, and also for the traffic that goes through the *LoadBalancer*. Pricing for load balancers is calculated identically to pricing for any other instance type, given that the balancers are not hardware, but regular instances configured to work as load balancers.

14.3.1.2   Microsoft Azure

In Windows Azure, Microsoft has taken an automatic approach to the load balancing problem. Load is automatically distributed among available work resources by using a round robin algorithm in a way transparent to the platform's users [28].

Microsoft's Windows Azure offers a Cloud middleware platform for managing applications. This middleware is known under the name of *AppFabric* [27]. Essentially, this is a managed PaaS Cloud service. For Cloud clients, the AppFabric service abstracts away deployment and service management. The control unit for the AppFabric service is the *AppFabric Controller*.

Load balancing for applications running under the AppFabric service is achieved by using hardware load balancers. The load balancers have redundant copies to reduce failure.

The load balancing mechanism is aware of the health state of the nodes that it balances load to (in the Azure platform, these nodes are called roles and can be either of type *Web role* or *Worker role*). The *AppFabric Controller* maintains an updated list of health states for the nodes in the platform. This is done in two ways: the node itself communicates a bad health state to the controller, or the controller queries the node for its health (usually by pinging it). When a node becomes unhealthy, traffic will cease to be distributed to it by the load balancers, until its health state is restored to normal.

The *AppFabric Controller* has fault tolerance mechanisms when it comes to virtual resource mapping to physical resources. This is done by introducing the concept of *fault domains*. A fault domain represents a point of hardware failure and in Azure it is actually a collection of physical network racks. When a physical failure happens inside a fault domain, it is very probable that all the virtual resources running in that fault domain will be affected.

A Cloud client application has a number of fault domains associated to it. They are controlled by the *AppFabric Controller* and cannot be changed by the Cloud client. The default number of fault domains per application is 2.

14.3.1.3   GoGrid

With respect to load balancing, GoGrid uses redundant f5 hardware load balancers [24]. Each account has free usage of the load balancers.

The load balancers can be configured in terms of a load balancing algorithm to use. The user has a choice between two available approaches:

1. Round robin: with this configuration, traffic is balanced evenly among available pooled nodes.

2. Least connect: this configuration makes the load balancers send new traffic to the pooled node with the least number of currently active concurrent sessions.

Load balancing can disturb client sessions if traffic for the same session is not routed to the same server node that initiated the session throughout the whole duration of the session. To prevent this, load balancers can be configured with a persistency option. The user can choose one of the following three:

1. None: in which situation, traffic is distributed as according to the balancing algorithm selected, ignoring possible session problems.

2. SSL Sticky: in which situation, all SSL traffic is routed to the same destination host that initiated the session. When a new SSL session is initiated, the destination node for handling the first request is chosen based on the balancing algorithm selected for the load balancer.

3. Source address: in which situation, all traffic from a source address is routed to the same destination node after the initial connection has been made. The destination node for handling the first connection of a new source is chosen based on the algorithm that the load balancer is configured to use.

The load balancers also check the availability of nodes in the balancing pool. If one node becomes unavailable, the load balancer removes it from the pool automatically.

14.3.1.4   Rackspace

Rackspace Cloud offers two types of Cloud services: *Cloud Servers* and *Cloud Sites*. The *Cloud Servers* service is of type IaaS in which all auto-scaling, load balancing and backup related issues are left in the hands of the Cloud client. On the other hand, the *Cloud Sites* service targets automated scaling, load balancing and daily backups [39]. Thus it leverages the benefits of Clouds without a need for too much interaction from the clients. In fact, the Cloud clients do not see the virtual resources that they are currently using. The Cloud platform is presented in a very abstract manner, the Cloud client deploys his application to the Cloud without having intimate knowledge of the Cloud's underlying scaling, balancing, and backup platforms.

The algorithm used for distributing the load is round robin.

Pricing is done based on the client's platform usage in terms of disk space, bandwidth and compute cycle, which is a unit that enables Rackspace to quantify their platform's computational usage. Users are not charged for use of load balancing service.

14.3.2 Implementations of Load Balancing in Open-Source Clouds

14.3.2.1 Nimbus

From the Cloud provider's point of view, there is a feature still under development in Nimbus that allows back-filling of partially used physical nodes. This will also allow preemptable virtual machines, an identical concept to Amazon EC2's spot instances.

From the virtual platform level, there is ongoing work for a high-level tool that monitors virtual machine deployment and allows for compensation of stressed workloads based on policies and sensor information.

14.3.2.2 Eucalyptus

Eucalyptus does not contain an implicit load balancing service for low-level virtual machine load balancing or high-level end-user request load balancing. Nor does Eucalyptus have a partnership program similar to RackSpace's Cloud Tools or GoGrid's Exchange programs.

As alternatives, one can opt for a complete managed load balancing solution offered by third party providers. Given that Eucalyptus implements the same management interface as Amazon EC2 does, it is relatively easy to find such commercial services.

14.3.2.3 OpenNebula

OpenNebula is service agnostic. This means that the service being deployed on OpenNebula needs to take care of load balancing on its own.

From a virtual resource balancing point of view, OpenNebula's virtual resource manager [33] is highly configurable. Each virtual machine has its own placement policy and the virtual resource manager places a pending virtual machine into the physical machine that best fits the policy. This is done through the following configuration groups:

- The *Requirements* group is a set of boolean expressions that provide filtering of physical machines based on their characteristics.

- The *Rank expression* group is a set of arithmetic statements that use characteristics of the physical machines and evaluate to an integer value that is used for discriminate between the physical machines that have not been filtered out. The physical host with the highest rank is the one that is chosen for deploying the virtual machine.

To choose the best physical machine is done by first filtering based on the requirements of the virtual machine and then choosing the physical machine with the highest rank for deployment.

It is trivial to obtain a policy that minimizes the number of used physical resources. It is also possible to obtain a policy that achieves a good distribution of virtual machines among the physical machines with the goal of minimizing the impact that a hardware failure would have on a Cloud client's platform. The virtual machine placement policies can be configured per virtual machine instance; however, when using a Cloud interface, this cannot be specified by the user and so they are defined by the Cloud administrator per virtual machine type.

14.4   Cloud Client Resource Monitoring

Keeping track of the platform health is crucial for both the platform provider and the platform user. This can be achieved by using platform monitoring systems. Monitoring can be done on two different levels, depending on the beneficiary of the monitoring information:

1. Low-level platform monitoring is interesting from the point of view of the platform provider. Its purpose is to retrieve information that reflects the physical infrastructure of the whole Cloud platform. This is relevant to the Cloud provider and is typically hidden from the Cloud clients, as their communication to the underlying hardware goes through a layer of virtualization. In general, it is the responsibility of the Cloud provider to ensure that the underlying hardware causes no visible problems to the Cloud clients. For commercial Cloud providers, the low-level monitoring service is usually kept confidential.

2. High-level monitoring information is typically interesting for Cloud clients. This information is focused on the health of the virtual platform that each individual Cloud client has deployed. It follows that the Cloud providers have little interest in this information, as it is the up to the client to manage his own virtual platform as he sees fit. Due to privacy constraints, platform monitoring information is only available to the virtual platform owner and is hidden from the other Cloud clients.

Although this separation is intuitive, there is no clear separation between the two. Each Cloud provider comes with its own interpretation and implementation of resource monitoring.

14.4.1   Monitoring in Commercial Clouds

14.4.1.1   Amazon EC2

As a commercial Cloud, the low-level monitoring system that Amazon uses for acquiring information on its physical clusters is kept confidential.

The approach that Amazon EC2 has taken with respect to high-level resource monitoring is to provide a service called *CloudWatch* [19] that allows monitoring of other Amazon services like EC2, Elastic Load Balancing and Amazon's Relational Database Service. The monitoring information provided to a Cloud client by the *CloudWatch* service is strictly related to the Cloud client's virtual platform.

The *CloudWatch* service collects the values of different configurable measurement types from its targets and stores them implicitly for a period of two weeks. This period of two weeks represents the expiration period for all available measures and is, in essence, a history of the measure that allows viewing of the evolution in the measurements. *CloudWatch* is actually a generic mechanism for measurement, aggregation and querying of historic data. All the measurements are aggregated over a period of one minute.

In association with the Elastic Load Balancer service and the Auto-scaling feature, *CloudWatch* can be configured to automatically replace platform instances that have been considered unhealthy, in an automatic manner.

CloudWatch comes with an alarm feature. An alarm has a number of actions that are triggered when a measure acquired by the monitoring service increases over a threshold or decreases under a threshold. The measures are configurable and the thresholds correspond to configurable limits for these measures. The possible actions are either a platform scaling action or a notification action. In the case of notification, there are a number of possible channels for doing this. They include Amazon's SNS and SQS services, HTTP, HTTPS or email. The actions are executed only when a measure transitions from one state to another and will not be continuously triggered if a measure persists on being outside the normal specified working interval.

Related to pricing, the *CloudWatch* service is charged separately with a single price per hour, regardless of the resource that is being monitored. Recently, Amazon changed the basic monitoring plan to be free of charge. This includes collection of values every five minutes and storage of these values for a period of two weeks. A detailed monitoring plan is also available that offers value collection at a rate of once per minute and is charged per hour of instance whose resource values are collected.

14.4.1.2   Microsoft Azure

Information about the monitoring system used for low-level platform monitoring of the whole Azure platform has not been given. However, the approaches that the Azure Cloud client has to application monitoring have been documented [29].

For monitoring applications deployed on Microsoft Windows Azure, the application developer is given a software library that facilitates application diagnostics and monitoring for Azure applications. This library is integrated into the Azure SDK. It features performance counters, logging, and log monitoring. Performance counters are user-defined and can be any value related to the Cloud application that is quantifiable.

The logging facilities of the library allow tapping into:

- Application logs dumped by the application. This can be anything that the application developer wants to log.

- Diagnostics and running logs

- Windows event logs that are generated on the machine that is running a worker role

- IIS logs and failed request traces that are generated on the machine that is running a web role

- Application crash dumps that are automatically generated upon an application crash

The storage location for the log files is configurable. Usually one of two storage environments is used: local storage or Azure storage service. The former is a volatile storage that is included in the virtual machine's configuration, while the latter is a storage service offered by Azure and has no connection to the virtual machine's storage. Usually the latter is preferred for what the Cloud user considers to be permanent logs while the former is used as a volatile storage.

There is no automatic monitoring mechanism for web roles and worker roles running on Microsoft Azure.

The cost implications of using the diagnostics and monitoring libraries are only indirect. There is no fee associated to using them, but there is a fee for storing information in a non-volatile persistence storage service and also in querying that storage service.

14.4.1.3 GoGrid

There is currently no public information related to how GoGrid achieves low-level monitoring on their platform.

GoGrid features a collaboration program. This program runs under the name of *GoGrid Exchange* [23] and presents third-party services that can prove useful to Cloud clients. These services include third-party packages that target monitoring features ranging from platform security monitoring to resource usage monitoring and database monitoring. These services also include the possibility of configurable alerts based on the values of the monitored measures.

14.4.1.4   RackSpace

As in the case of the other commercial Cloud providers, the approach used by RackSpace for low-level platform monitoring is not public. In what follows we will detail how Cloud clients can monitor their platform on RackSpace.

The Rackspace *Cloud Sites* service offers monitoring capabilities at the whole application level for fixed parameters that include used compute cycle count, used bandwidth and storage. This fits well into the usage scenario that *Cloud Sites* offer: that of a PaaS service; but lack of finer-grained sub-application level monitoring can be a downside for some Cloud clients. Again at an application level, logging for applications deployed on *Cloud Sites* is offered, but in a per-request manner.

On the other hand, the *Cloud Servers* service, which is an IaaS-type of service, does also have monitoring capabilities through the use of third-party partner software, especially tailored for Rackspace's *Cloud Servers* service. These partner solutions are aggregated by Rackspace under the name of *Cloud Tools* [36]. Among these partner services, one can find complete monitoring solutions ranging from general virtual machine monitoring to specialized database monitoring. The services that are specialized on monitoring also feature configurable alert systems.

Recently, RackSpace has acquired CloudKick [20], a multi-cloud virtual platform management tool. CloudKick has a broad range of monitoring features for virtual machines. These include different monitoring metrics from low-level metrics like CPU / RAM / disk utilization to high-level metrics like database statistics, HTTP / HTTPS and others. The monitoring metrics can be extended by custom plugins that are able to monitor anything that the user defines. Measured data can be presented in raw form or aggregated by user-defined means. For data visualization, a real-time performance visualization tool is also provided.

CloudKick also features alerts that have a configurable trigger and repeat interval. The alert prompt can be sent by SMS, email or HTTP.

14.4.2   Implementations of Monitoring in Open-Source Clouds

14.4.2.1   Nimbus

Nimbus features a system of Nagios [31] plugins that can give information on the status and availability of the Nimbus head node and worker nodes, including changes of the virtual machines running on the worker node.

Also, there is active work being done around building a higher level tool that is able to monitor deployed virtual machines and compensate for stress points by using monitor information from sensors and configurable policies.

14.4.2.2 Eucalyptus

Since version 2.0, Eucalyptus has introduced monitoring capabilities [22] for the running components, instantiated virtual machines and storage service. This is done by integrating Eucalyptus monitoring into an existing and running monitoring service. Currently, monitoring has been integrated with Ganglia [500] and Nagios. In Eucalyptus this is done by means of scripts that update the configuration of the running monitoring service to also monitor Eucalyptus components and virtual machines.

As alternative solutions to achieving monitoring at a hardware level, one can employ one of the monitoring systems that have been designed and used in grid environments. Some such systems have been detailed in Section 14.4.3.

We can also opt for a completely managed monitoring solution offered by third party providers. Given that Eucalyptus implements the same management interface as Amazon EC2 does, it is relatively easy to find such commercial services. Among the commercial solutions available, we can enumerate the following as compatible with Eucalyptus (but the list is not exhaustive):

RightScale offers a Cloud management environment for Cloud clients. This management environment also includes a real-time platform monitoring sub-service.

enSTRATUS provides a set of tools for the provisioning, managing and monitoring of Cloud infrastructures with applicability to private and public Cloud platforms.

Makara offers a PaaS service on top of an IaaS Cloud. They focus on the deployment, management, scaling and monitoring of PHP and Java applications deployed on Clouds. With respect to monitoring, Makara supports real-time and historical performance monitoring.

14.4.2.3 OpenNebula

The built-in monitoring capabilities of OpenNebula focus on the Cloud provider's interest — the physical resources. This functionality is found in the OpenNebula module called the *Information Manager* [32].

The Information Manager works by using *probes* to retrieve information from the cluster's nodes. The probes are actually custom scripts that are executed on the physical nodes and output pairs of Attribute=Value on their standard output. The pairs are collected and centralized. As a requirement, the physical nodes should be reachable by SSH without a password.

Currently, the probes are focused on retrieving only information that underlines the state of the physical nodes and not its running virtual machines (CPU load, memory usage, host name, hypervisor information, etc.). It is advised that this information not be mixed with information of interest to the Cloud client. For such a task, the OpenNebula community recommends using a service manager tool that is a separate entity from OpenNebula. As possible solutions, we can consider commercial services that are specialized in Cloud platform management, including monitoring. Such solutions have been described in the previous sections. Alternatively, we can also turn to cluster monitoring solutions that come from the open-source world, some of which are the result of long research endeavors and have been described in Section 14.4.3.

While still under development, the next version of the Information Manager is based on the Ganglia multi-cluster monitoring tool.

14.4.3   Other Research Endeavors That Target Monitoring in Large-Scale Distributed Systems

Over the years as grid computing evolved, so did the need for monitoring large-scale distributed platforms that are built on top of grids. There have been many fruitful research efforts for designing and implementing monitoring systems for large-scale platforms. In the following, we will highlight some of these efforts. The list of research projects that we present is not exhaustive for the field of large-scale platform monitoring.

14.4.3.1   The Network Weather Service - NWS

NWS [766] has the goal of providing short-term performance forecasts based on historic performance measurements by means of a distributed system. To achieve this, NWS has a distributed architecture with four different types of component processes:

Name Server process – responsible for binding process and data names with low level information necessary when contacting a process

Sensor process – responsible for monitoring a specified resource. The first implementation contained sensors for CPU and network usage. Sensors can be added dynamically to the platform.

Persistent state process - responsible for storing and retrieving monitoring data. By using this type of process, the process of measuring is disconnected from the place where measurements are stored.

Forecaster process – responsible for estimating future values for a measured resource based on the past measure values. The forecaster applies its available forecasting models and chooses the value of the forecaster with the most accurate prediction over the recent set of measurements. This way, the forecaster insures that the accuracy of its outputs is at least as good as the accuracy of the best forecasting model that it implements.

To increase fault tolerance, NWS uses an adaptive and replicated control strategy by an adaptive time-out discovery and a distributed leader election protocol. Sensors are grouped into hierarchical sets called *cliques*. A Sensor can only perform intra-clique measurements, thus limiting contention and increasing scalability of the system.

The implementation uses TCP/IP sockets because they are suited for both local area and wide area networks and they provide robustness and portability.

14.4.3.2 Ganglia

Ganglia [500] addresses the problem of wide-area multi-cluster monitoring. To achieve this it uses a hierarchy of arbitrary number of levels with components of two types:

Gmon component responsible for local-area monitoring. To gather information from cluster nodes, Gmon uses multicast over UDP, which has proved to be an efficient approach in practice, and it also makes Ganglia immune to cluster node joins and parts.

Gmeta component is responsible for gathering information from one or more clusters that run the Gmon component or from a Gmeta component running in a lower level of the tree hierarchy. Communication between the two components is done by using XML streams over TCP.

In order to achieve almost linear scalability with the total number of nodes in the clusters, the root Gmeta component should not be overwhelmed with monitoring data. To do this, an 1-level monitoring hierarchy should be avoided. Instead, an N-level monitoring tree hierarchy should be deployed, where it is dependent on the number of nodes in the cluster. There is a limitation here in the sense that although nodes can be dynamically added and removed from a cluster without needing to manually update the Ganglia hierarchy, the same cannot be said for Gmon and Gmeta components. The Gmeta needs to have *a priori* knowledge of the of its underlying child nodes.

14.4.3.3 Supermon

Supermon [668] aims at providing a high-speed cluster monitoring tool, focusing on a fine-grained sampling of measurements. To achieve this, Supermon uses three types of components:

Kernel module for monitoring provides measurements at a high sampling rate. Values are represented in the form of s-expressions.

Single node data server (mon) is installed for each monitoring kernel module. It parses the s-expressions provided by the kernel module. For each client connected to this server, it presents measurement data filtered by the client's interest. Data is sent by means of TCP connections.

Data concentrator (Supermon) gathers data from one or several mon or Supermon servers. They also implement the same per client filtering capability that mon servers have. Hierarchies of Supermon servers are useful to avoid overloading a single Supermon, especially in situations where a large number of samples is required or there is a large number of nodes that are monitored.

14.4.3.4 RVision

RVision (Remote Vision) [267] is an open tool for cluster monitoring. It has two basic concepts that make it highly configurable:

Monitoring Sessions are actually self-contained monitoring environments. They have information on what resource to monitor and what acquisition mechanism to use for the monitoring process associated to the resource.

Monitoring Libraries are actually collections of routines that are used for resource measurement information acquisition. These routines are dynamically linked at runtime and thus information acquisition, which is occasionally intimately tied to the resource that is being measured, is disconnected from the general mechanisms of monitoring.

The architecture of RVision is a classical master-slave architecture. The master node is represented by the RVCore. It is responsible for managing all the active sessions and distributing the monitoring information to the clients. The communication layer is implemented by using TCP and UDP sockets. The slave part corresponds to the RVSpy component. There is one such component running on each node of the cluster that is to be monitored. The monitoring libraries that are needed for information acquisition are dynamically linked to the RVSpy. The slave component communicates its acquired information to the master by means of UDP, as this is a low-overhead protocol, and cluster nodes are usually connected by means of a LAN, where package loss is usually small.

14.5 Conclusions

The current work aims to familiarize the reader with the importance of auto-scaling, load balancing and monitoring for a Cloud client platform. The elasticity of Cloud platforms is reflected in their auto-scaling feature. This allows Cloud client platforms to scale up and down depending on their usage. Achieving automatic client platform scaling is done in different ways, depending on the Cloud platform itself. One can opt for the Cloud's built-in auto-scaling feature if present — in Amazon EC2, Microsoft Azure, or use a third party Cloud client management platform that offers this functionality: RightScale, Enstratus, Scalr, that are usable in most Cloud platforms. GoGrid and RackSpace have partner programs — the GoGrid Exchange and the RackSpace Cloud

Tools — that provide custom-made tools that work on top of the hosting service that they offer.

Load balancing has the goal of uniformly distributing load to all the worker nodes of the Cloud client's platform. This is done by means of an entity called a load balancer. This can be either a dedicated piece of hardware that is able to distribute HTTP/HTTPS requests between a pool of machines — the case of Microsoft Azure, GoGrid, or with a Virtual Machine instance that is configured by the platform provider or by the client himself to do the same job — the case of Amazon EC2, Rackspace, Nimbus, Eucalyptus and OpenNebula. In order to make sure that the platform is in a healthy working state, platform monitoring is used. This is done by means of a service that periodically collects state information from working virtual machines. The monitoring service can either be built as a part of the Cloud platform, as is the case of Amazon's CloudWatch, Microsoft Azure's monitoring package, Nimbus' Nagios plugins, Eucalyptus 2.0's monitoring service and OpenNebula's Information Manager. GoGrid and RackSpace offer this feature by means of their partner programs. Alternatively, Cloud clients can always choose a third party monitoring solution. As examples, we can enumerate CloudKick, Makara or any of the third party Cloud client platform management tools presented above.

Ultimately, all the three presented features are designed to work hand-in-hand and have the high-level goal of ensuring that the Cloud client's platform reaches a desired QoS level in terms of response time and serviced requests, while keeping the cost of running the platform as low as possible.

The readers of the current chapter should now have a clear understanding of the importance of the three main topics discussed here with respect to Cloud client platforms. We have presented the working principles of the three topics, along with a survey on the current available commercial and open-source Clouds. We hope that the current work can help current and future Cloud clients in making informed decisions.

15

Monitoring: A Fundamental Process to Provide QoS Guarantees in Cloud-Based Platforms

Gregory Katsaros

High Performance Computing Center Stuttgart, Germany

Roland Kübert

High Performance Computing Center Stuttgart, Germany

Georgina Gallizo

High Performance Computing Center Stuttgart, Germany

Tinghe Wang

High Performance Computing Center Stuttgart, Germany

CONTENTS

15.1   Introduction

The adoption of Cloud computing in the last years has increased notably [487] [153]. Cloud computing has had a major commercial success over recent years and will play a large part in the ICT domain over the next 10 years or more [487]. However, there are still major limitations which current Cloud computing solutions are not yet able to address adequately. Among others, the difficulties to provide guaranteed levels of service (also known as Quality of Service, QoS) for all kind of applications is one of the major barriers for the adoption of this technology by a wider range of users. One of the main reasons is the lack of a complete monitoring solution which provides information at all levels of the environment, from application, to platform, down to the physical infrastructure. The existence of an integrated monitoring infrastructure within the Cloud environment would bring benefits to both users and providers of Cloud services in their different delivery models (Infrastructure-as-a-Service, Platform-as-a-Service and Software-as-a-Service) [54]. A well designed monitoring infrastructure is crucial to provide and guarantee most of the essential features of Cloud computing, such as accounting and billing (which allows the pay-per-use model), access control (contribution towards trust and security mechanisms), resource planning and optimization (crucial to fulfill the Cloud provider business interests), elasticity (allowing to scale resources as needed by the user application) and Service Level Agreements (SLAs) (the more complete the monitoring solution, the more fine-grained the SLA can be, which gives the user of the Cloud solution the chance to run more demanding applications on the Cloud, in terms of QoS requirements). This chapter analyzes the requirements of an efficient monitoring mechanism for Cloud environments. Furthermore, it presents an analysis of different existing approaches for collecting monitoring information at different levels of Cloud based platforms, and compares different monitoring tools and solutions available (both commercial and open source). Finally, it elaborates a proposal on how a full monitoring solution should be designed and implemented, in order to serve the needs and business interests of different Cloud computing stakeholders (users and providers of different delivery models).

15.2   Monitoring in the Cloud

The emergence of the Cloud computing paradigms of Infrastructure as a Service, Platform as a Service and Software as a Service (IaaS, PaaS, SaaS) along with the actual appearance of providers for such services (Amazon, Google, VMware etc.), lifted the topic of monitoring resources, services and applications onto a new level [643] [680]. With the introduction of new architectural approaches, new technologies and different business models a brand new market place has been created [313]. In order for this market to be prosperous and receive the acceptance of the consumers, the QoS of the Cloud infrastructure as well as of the provided service must be ensured and effectively monitored. The use of virtualization technologies, the distribution of the infrastructure, the scalability of the applications and the energy consumption metrics are just a few challenges to be overcome toward the development of an effective monitoring system in the Cloud. Defining and designing a monitoring mechanism for Cloud computing is a complex problem while we must take into consideration the different levels of abstraction that exist into the Cloud service model (IaaS, PaaS, SaaS), the various roles participating in every level, as well as the different use cases applicable. In the following paragraphs we elaborate on the requirements identified for each level of monitoring within a Cloud environment.

15.2.1 IaaS - Virtual and Physical Infrastructure Monitoring

When it comes to providing IaaS, the two most important things are:

- ensuring the QoS agreed with the customer regarding the Virtual Environment delivered and

- managing the physical resources in the most efficient way.

Achieving those two goals is a demanding and sometimes complex process. The requirements that we can identify when designing a monitoring mechanism for the Cloud infrastructure are listed below:

- Support various types of resources: the Infrastructure monitoring mechanism should be able to retrieve performance parameters from different kinds of resources. Information regarding the storage, network utilization, CPU speed and temperature are only some examples related to the physical infrastructure. In addition, it is of major importance to keep track of the status of the virtual infrastructure (e.g., virtual machines running, virtual network deployment, mapping between physical and virtual infrastructure).

- Managing big amount of data: when monitoring many parameters on heterogeneous and distributed systems, the management of that information becomes a challenging job. The produced data should be managed and aggregated in an effective way in order to be useful for evaluation.

- Minimum impact on the resources: when realizing a business model like IaaS, any Cloud Provider would want to maximize the capacity of his infrastructure and eventually use the least resources possible for the internal monitoring mechanism. As resources we consider computing nodes, network and storage. In this context, the Infrastructure monitoring should be lightweight and efficient without adding substantial overload. As such, the implementation of the data collection process with either pushing or pulling or even both techniques combined should be done very carefully, taking also into consideration the time interval (granularity) of the polling mechanism.

- Configurable and Scalable solution: the monitoring mechanism of the Cloud Infrastructure should be dynamic and configurable. It should be easily extensible to add more resources and scale horizontally. Finally, the monitoring information should be provided by the resources automatically on boot up.

15.2.2   PaaS - Management and Collection in the Platform Level

Regarding the Platform level of the Cloud service model, the responsibilities of the monitoring infrastructure are to gather and aggregate all the information coming from the Infrastructure as well as the Application and provide that data to any consumer (end-user or administrator) through different interfaces (programmatic APIs, Web GUIs, reports etc). While the Cloud Platform is the means of a developer or a user towards deploying a service or an application in a Cloud, there are several requirements raised regarding the monitoring process. Information deriving from the Cloud infrastructure should be returned to the consumer of the platform. The user should be able to process that data, compare them with older historical information and combine them with monitoring data coming from his own application. Another objective of the monitoring functionality is to maximize the usage of the data collected toward SLA evaluation, resource usage optimization, energy efficiency, auditing and accounting. In this sense, the framework should offer different features through services, as part of the PaaS toolkit, such as a stream interface allowing to obtain monitoring information in "real-time", generation of monitoring statistics based on previously collected monitoring information, GUIs for the visualization of both "real-time" and historical values, etc. The existence of a powerful platform through which a user could efficiently develop, deploy, control and monitor his services over a private, a public, or hybrid Cloud would further uplift the acceptance and utilization of Cloud technologies. In this context, we foresee that the core of such a platform would be the monitoring and management framework.

15.2.3   SaaS – Monitoring the Cloud-Hosted Application

The main challenge and change caused by the realization of a SaaS business model through Cloud computing, was the introduction of the virtualization layer. On the one hand, this was the solution for offering Infrastructure and Software as a Service but, on the other hand, it is a fact that this complicated the application development and execution monitoring. The physical infrastructure is transparent for the end user and the capacity in a virtualized environment is shared and dynamic most of the times. As a result, the performance of the application cannot be directly linked with the CPU utilization or other resource level parameters. In that context, the metrics that should be monitored and could indicate the performance of an application deployed in a Cloud based environment must be revised. While resource utilization is no longer the absolute and reliable indicator, we should start focusing on different values, such as the response time of the application. This evolution is based on the fact that the consumer of the Cloud environment (end user) is not interested in the resources, which are transparent to him, but whether the application is responding as expected within the Cloud. Another side effect of the virtualization technology used in Cloud computing is the limited access (or even no access) to the underlying infrastructure. To this end, the monitoring mechanism should be a part of the application deployed and consequently provided and installed by the end user. Hybrid Cloud[1] scenarios should also be considered when monitoring an application deployed in such architecture. The security restrictions arose by the virtualization layer and the isolation that is offered makes the communication between the monitoring agents of the different Clouds a challenging process. To this end, the monitoring mechanism of the application should be designed in a smart way and utilize Internet friendly protocols and ports, but at the same time secure and encrypt any sensitive data.

15.3 Available Monitoring Tools/Solution

In this section, we will present and analyze a selection of available tools and APIs used widely for resource monitoring. Most of those have been developed to serve clusters and Grid infrastructure but with proper configuration and additions can be used as the baseline of the monitoring infrastructure for Cloud environments.

15.3.1 Ganglia

Ganglia [628] [293] is an open source monitoring solution designed for high performance computing systems, such as clusters and grid environments. It is based on a hierarchical architecture, which provides high scalability and high performance, being able to support clusters with up to 2000 computer nodes. Management of data is done by means of XML, for data provision, and XDR for data transport, while monitoring results are stored and provided in a graph format, by means of RRDtool. Visualization of performance data is

---

[1]A part of the application, deployed in one Cloud provider, and another part in a different, private Cloud.

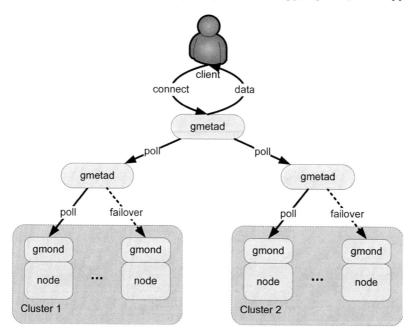

FIGURE 15.1
Ganglia's Architecture.

supported, as well as the realization of diagnostics tuning and history trend. However, it does not support automated alert responses. Furthermore, Ganglia is only aware of metrics available from its internal monitoring module, not being able to support active service level monitoring external from the node. A notable characteristic of Ganglia is that it introduces low overhead over the monitored environment. The metrics supported by this monitoring system are classified in two main groups: built-in metrics and user-defined metrics. The first are capable of capturing computing node information, while the second may represent application-specific states, allowing an adaption and extension of the monitored data set. Every built-in metric has a default threshold value as a baseline to decide whether the local node data would be collected and sent to the multicast traffic. That baseline could be modified at compile time to accommodate different environments. Ganglia is a BSD-licensed open-source project that grew out of the University of California, Berkeley Millennium Project [124].

In Figure 15.1 we present Ganglia's hierarchical tree-like architecture design. It consists of four main components:

- Data collector daemon gmond
- Data aggregator daemon gmetad

- RRDtool (round-robin database tool)

- PHP Web Frontend

In a compute cluster *gmond* gathers monitoring metrics for each node and forwards the collected metrics to the aggregator *gmetad*. *Gmetad* aggregates the node level metrics in an XML-based data format by polling computer nodes at periodic intervals. In the same way, in a Grid environment, the state of a federated cluster node is achieved by *gmetad* aggregating the states of their cluster child nodes by polling. The monitored performance data is stored in RRDtool. The monitored performance information can be presented through a PHP script and visualized via a web based GUI generating graphs dynamically.

15.3.2   Nagios

Nagios [530] [531] is another open source monitoring system which allows monitoring of resource infrastructures through status checks. It also provides an automated alerting mechanism to warn about identified problems. One of the key features of Nagios is the plug-in concept which it relies on in order to perform various kinds of monitoring tasks and provide local and remote service status. Furthermore, the plug-in architecture enables the extensibility and flexibility of Nagios' capabilities. From a functional point of view, Nagios' architecture is split in two layers: a monitoring logic layer and a monitoring operation layer. The Nagios daemon resides on the monitoring logic layer and the plug-ins reside on the operation layer. Nagios itself does not have internal mechanisms to perform monitoring tasks, but the external plug-ins are the ones which provide the monitoring functionalities. Every plug-in is in charge of a specific task, performing actual checks and sending back the result to Nagios' monitoring logic layer. Therefore, if suitable plug-ins are available, Nagios can, in principle, monitor any kind of resources. The core API already includes some plug-ins for the monitoring of basic resources, while custom ones can be developed according to the plug-in guideline and API [532]. Regarding data evaluation and alerting, the Nagios daemon performs evaluation of the monitoring data provided by plug-ins, by comparing them against defined thresholds. Based on the result (alert level OK, Warning, Critical or Unknown), it takes necessary actions, such as triggering an action or sending a notification to maintain the health of the IT infrastructure. The notification can be sent in various ways, including email, instant message or cellular text messaging, in order to alert technical staff to the problem. The Nagios core is licensed under GPL but built on the open source version there are commercial versions available, such as Nagios XI, which provides extra capabilities, including a powerful web interface and advanced configuration management, in order to make problematic IT monitoring tasks simple.

In more detail, Nagios is built on a client/server architecture to monitor a distributed system. It is composed of three main parts:

- Nagios daemon

- Web-based GUI

- Plug-ins

In a distributed environment, the Nagios daemon runs on the server host, which is responsible of starting the monitoring and taking necessary action based on the monitoring results. Nagios plug-ins run on all the remote hosts that need to be monitored. The plug-ins communicate with the elements to be monitored and return the results to the Nagios Daemon.

15.3.3   Hyperic HQ

Hyperic HQ [362] is an application monitoring software designed to manage and monitor web applications and infrastructures in heterogeneous environments. It is based on server/agent architecture, where an agent performs monitoring tasks and a central server is used to process the results. In contrast to many monitoring solutions, which cannot easily monitor applications, Hyperic HQ supports the detailed monitoring of not only operating systems but of various web servers, relational databases, application servers, mail servers etc. Because of its extensible framework, additional metrics can easily be defined. HQ management capability is based on the inventory model [363], which specifies the relationships among the managed resources. It is a key concept to enable HQ to manage resources and present information about a large amount of IT resources. HQ will automatically detect the software resources running on a machine and stored in the HQ Database according to a hierarchical inventory model. Moreover, HQ supports auto-discovery and remote control of resources, which helps to simplify tasks for administrative management. There are three kinds of resource plug-ins, which build the monitoring capability: pre-existing built-in resource plug-ins, community-contributed plug-ins and customer defined resource plug-ins (also called extension plug-ins). The built-in plug-ins can therefore not only be extended by one's self but also through resource plug-ins contributed by the Hyperic community [361]. Agents depend on plug-ins in order to collect performance and track performance and event data, create alerts and report performance. Hyperic HQ provides a portal component, HQ Portal, for the presentation of monitoring data through a graphical user interface. It can be customized in order to provide specific reports for monitoring, performance and availability data. Currently there are two editions of Hyperic HQ available, an open source edition and an enterprise edition, which is available under a commercial license. Compared to the open source version; the enterprise edition adds various functions, for example deployment automation, advanced alerting options, and scheduling of service actions, for example restarts, etc.

The architecture of Hyperic HQ is based on an agent/server principle, where the monitoring task itself is executed by agents, who report back to the main

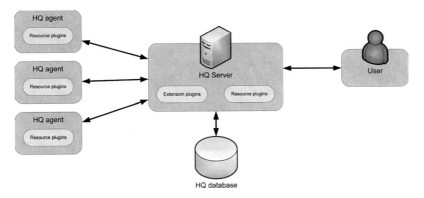

FIGURE 15.2
Hyperic HQ's Architecture.

monitoring server. An agent is running on each host that is monitored and employs various plug-ins in order to monitor the host itself and the applications detected on the host. The server stores the data reported by agents in a database and provides a user-friendly presentation via a portal (see Figure 15.2).

15.3.4 Lattice Monitoring Framework

Lattice monitoring framework [453] is an open source API for building monitoring systems for physical, virtualized environments, virtual networks as well as services. Through the Java based API that is provided someone can develop an overlay management system for monitoring resources and services in virtualized environments [193]. Like a toolbox, the Lattice framework provides various components and functionalities to build up a customer-specific monitoring system. In that context, Lattice is not exactly offered as a ready-to-use monitoring solution, but mainly as an API including all the necessary libraries for implementing a custom monitoring system. The use of Lattice allows the user to build a management infrastructure that collects, processes, and disseminates network and system information from/to network entities in real-time. In practice, the Lattice framework has been successfully used in the AutoI project [106] for monitoring virtual networks as well as in the RESERVOIR project [601] where it has been utilized in building a system to allow the monitoring of service Clouds. The Lattice Monitoring Framework, while it provides the functionality for monitoring physical hosts (e.g., CPU, hard disk etc.), is primarily utilized to gather measurement data on virtual machines and provide the result for the management overlay. Its functionalities are focused on the collection and distribution of monitoring data through either multicast or UDP protocol. Therefore, Lattice doesn't provide functionalities for visualization, evaluation and automated alerting. The monitoring of

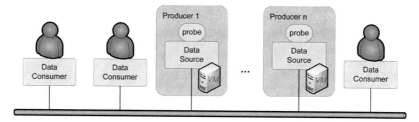

FIGURE 15.3
Lattice's Operation.

resources in virtualized environments is realized by interacting with a virtual machine hypervisor, and specifically with the Libvirt API [472].

As presented in Figure 15.3, Lattice design is based on the concept of producers and consumers. Therefore, the framework is consists of the following elements: Consumers, Producers, Data Sources, Probes and the Distribution Framework. In the system Producers, Data Sources and Probes are collecting data and Consumers are reading and utilizing those. The communication between the Producers and the Consumers occurs with the support of the Distribution Framework, which can distribute the measurements collected by the monitoring system. On the side of data collection it is the Probes that define the measurements and collect the result into a Data Source, which acts as a control point and a container for one or more Probes.

15.3.5   Zenoss

Zenoss Core [794] is an open source IT management and monitoring software based on the Zope Application Server. In order to organize and manage resources in large IT environments efficiently, Zenoss uses a concept called Zen-Model to specify resources and their relationships. Based on the ZenModel, Zenoss enables automatic monitoring and alerting after auto-discovery of devices; through an auto-discovery process, IT infrastructure is detected and is used to populate the model. To deal with complex monitoring information efficiently, Zenoss utilizes three separate databases to store monitoring related data: it stores the monitoring performance data in an RRDtool file, event data in a MySQL database and uses an object database, which is built into the Zope Application Server [803], to store configuration model-related device data. It is capable of availability monitoring as well as performance monitoring and has several different ways to monitor performance metrics of devices and components, by using SNMP, ZenCommands or XML-RPC; configuration information and rules for performing monitoring are specified in performance templates. Zenoss makes uses of a command process (Zen-Command) in order to run scripts, and uses plug-ins to collect performance data locally and remotely. The solution for automated alerting is dependent

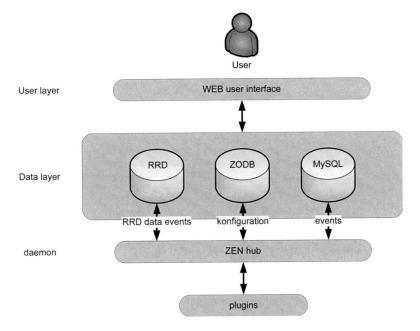

FIGURE 15.4
Zenoss' Architecture.

on Zenoss' Event Management System which, can report monitoring results with embedded graphs as well as in a customizable dashboard. There are two types of available products, Zenoss Core and Zenoss Enterprise. Zenoss Core is licensed under GPL while Zenoss Enterprise is a commercial product with additional features built on a Zenoss Core.

The Zenoss System can be divided into three distinct layers (Figure 15.4):

- User layer

- Data layer

- Collection and Control Service layer.

The User Layer consists of web-based graphical user interfaces that allow user access to monitoring data. The Data layer is home to the three separate databases, as previously mentioned. Various kinds of collection daemons, which are responsible for collecting specific types of data, are associated with the Collection and Control Service layer. The main component of this layer is the ZenHub, a component that brokers information between the Data layer and the collection daemons.

15.3.6   Summary

In Table 15.1 below we have summarized the features and characteristics of each one of the selected monitoring tools for a better one-to-one comparison. As mentioned in the beginning, we selected and presented the monitoring mechanisms and APIs that we consider the most popular. Besides those, there are various other solutions proposed in research, like MonaLisa [584] and service oriented solutions based on Globus MDS4 [415] as well as other complete tools and applications such as Zabbix [788] and Monit [521]. Each one of which has specific capabilities and might also be a good choice, depending on the requirements and needs each system has. In some cases, a couple of tools can be combined in order to create a hybrid solution, thereby making use of the benefits of both products. GroundWork (GWOS) [324], for example, offers a hybrid solution combining Ganglia and Nagios.

15.4   Monitoring Infrastructure: A Generic Approach

Taking into consideration the analysis above and the investigation of some of the most popular monitoring solutions provided, we have identified and present the characteristics and capabilities of a Cloud-enabled monitoring infrastructure. In that context, we propose that an efficient monitoring mechanism for a Cloud should follow the Service Oriented Architecture design principles. It should operate on at least two levels: collection of information and management. The aggregation of the information is a step that is usually underestimated. This task is of major importance when having a computer system that provides many types of different information. A monitoring infrastructure for a Cloud system could gather information from the physical infrastructure, the virtual environment, the application execution, or even energy efficiency related data. That information must be sorted and stored in such a way that any consumer of the information could effectively utilize and process that data. A third operational level is the assessment layer that could be situated on top of the management for post-processing of the information retrieved. In addition, a very important characteristic of a modern monitoring infrastructure is to be compatible or even reuse available technologies and tools. This will extend the acceptance and effectiveness of the system. The use of open source APIs will contribute a lot toward that direction while it is common habit to use such utilities on Grid and Cloud computing systems and offers many advantages when monitoring distributed and scalable systems (easy to set up, zero license costs etc.). Furthermore, the monitoring infrastructure should be dynamic and scalable, being able to follow the elasticity of the Cloud paradigm. Finally, when designing such a monitoring infrastructure

TABLE 15.1

Feature Comparison

	Ganglia	Nagios Core	Hyperic HQ (OS)	Zenoss Core	Lattice
Generic Info					
Current version	3.17	3.2.3	4.4.0	2.5.2	0.6.4
Platforms	Linux, Solaris, BSD	Linux, Unix	Windows, Linux, Mac, Solaris	RHEL, CentOS, Novell, SuSE	Platform independent
License	BSD	GPLv2	GPLv2	GPLv2	LGPL
Architecture					
Frontend	Yes	Yes	Yes	Web GUI	Yes
Server/Agent	No	Yes	Yes	Yes	No
Plug-in structure	No	Yes	Yes	Yes	No
Configuration					
GUI	No	No	Partial	Yes	No
Permission Management	No	Yes	Yes	Yes	No
Auto devices	In cluster	No	Yes	Yes	No
Monitoring					
Metrics	Yes	Yes	Yes	Yes	Yes
Availability	No	Yes	Yes	Yes	No
Track performance, configuration	Yes	Yes	Yes	Yes	
Physical environment	Yes	Yes	Yes	Yes	Yes
Virtual environment	No	Yes	Yes	Yes	Yes
Evaluation, Alerting, Notification					
Alerting	No	Yes	Yes	Yes	No
Recovery alerts / corrective actions	No	Yes	Yes	Yes	No
Visualization					
Diagrams and Charts	Yes	Yes	Yes	Yes	No
Customizable Dashboard	No	Yes	Yes	Yes	No
Extensibility					
Web Service API	Yes		Yes	Yes	
Nagios plug-in integration	Yes	Yes	Yes	Yes	No
Plug-in Development Kit	No	Yes	Yes	Yes	No
Support Service					
Document	Yes	Yes	Yes	Yes	Yes
Forum	Yes	Yes	Yes	Yes	No

we should consider the operation of the monitoring mechanism in a private Cloud, as well as hybrid architectures.

15.4.1 Architecture

In the context of the previous requirement analysis we have defined and proposed an architectural model (Figure 15.5) that addresses the aforementioned challenges while at the same time is utilizing open source available APIs. The presented monitoring infrastructure consists of three different layers, with several components spread across those layers implementing different functionalities.

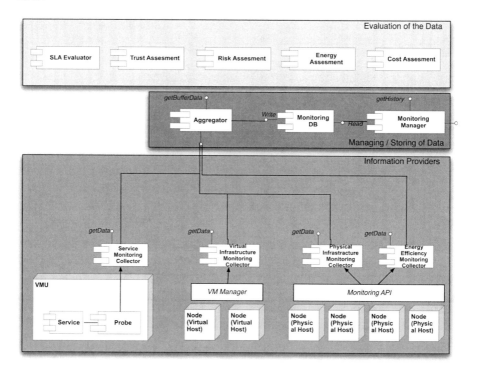

FIGURE 15.5
Multi-Layered Monitoring Infrastructure.

In detail we have defined the following hierarchical layers:

- Information Providers: it comprises the different sources where monitoring data is collected from, as well as the components in charge of collecting them (known as "Collectors"). The Monitoring Infrastructure is designed in a way that this layer is scalable, allowing the incorporation of additional sources, through the corresponding collectors. In the presented infrastructure we have defined four types of Monitoring Collectors: the physical infrastructure Collector extracts data such as CPU load, disk space and memory, while the Energy Efficiency Monitoring Collector retrieves from the same nodes (physical infrastructure) information related with the energy efficiency of the system (e.g., Processor Performance in MHz/Watt, Power Usage Effectiveness (PUE)). Both of them interact with a monitoring API installed in the physical host. In our case we have tested and propose the Nagios API which is widely used and can cover to a great extent our needs in that specific part of the system. The Virtual Monitoring Collector keeps track of the Virtual Machines (VMs) deployment by interacting with the VM Manager. The latter is able to collect such information through the Libvirt API. The fourth Monitoring Collector is related to the Service execution within the virtual environment. Using certain mechanisms and available APIs we can extract data from the service being executed and push them to a higher hierarchical level. In our case we have tested and propose the Lattice API, which uses customized probes in order to get the monitoring information from the service. Overall, all the collectors will push the monitored data to the Management and Storing layer but also expose an interface in order for the latter to be able to pull information on demand.

- Managing/Storing of the data: this layer includes those components in charge of managing the data collected from the monitoring information providers and storing them in an aggregated way into a database. Apart from the Aggregator and the database components there will be a Monitoring Manager Web Service that will act as an orchestrator of the monitoring infrastructure (start, stop, configure operations) and also expose interfaces the rest of the internal components (assessment and evaluation tools) as well as any external component (GUI or other monitoring infrastructures). Finally, a specific functionality is being proposed at that point in relation to the data providing. Two different interfaces are being implemented: the first (getBufferData) will offer the latest data set retrieved and aggregated through a buffer mechanism while the second (getHistory) will provide data sets collected in the past through queries onto the database.

- Evaluation of the Data: we have identified a number of evaluators starting from Service Level Agreement (SLA) and including Trust, Risk, and Cost as well as Energy Assessment tools. Each one of those components will interact with the Monitoring Manager WS in order to acquire the historical data from the database but also will communicate with the Aggregator directly in order to get the last data set from the buffer and proceed to direct assessing, such as SLA violation detection. Overall, the post-processing of the data and the assessment of the monitoring information is very important, and will allow us to optimize the resource utilization, identify the relation of the high-level service deployment with the low-level energy consumption, reduce the related costs, and diminish risks and uncertainties when managing Cloud environments. The presented monitoring infrastructure is applicable to private Cloud architectures but the overlay management framework as well as the Evaluation and assessment tools can be utilized in public as well as hybrid Clouds.

15.4.2 Data Model

Apart from an effective architecture for the monitoring infrastructure of a private cloud, a data model that would manage the information, entities and relationship between them in a well-defined and efficient way is of major importance. In that context, we have designed the following, generic but yet consistent with the proposed architecture data model (Figure 15.6).

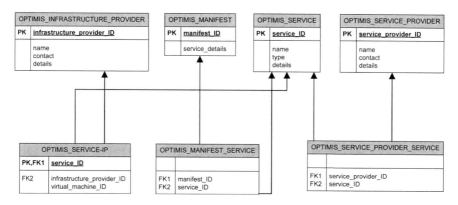

FIGURE 15.6
Data Model Relationship Diagram.

The major entities within our proposed model are those of the Infrastructure Provider (IP) and the Service Provider (SP). The first is the one who offers IaaS through his virtual environment over physical infrastructure. The second is interacting with the IP by deploying his own services on the virtual machines in order to offer SaaS to customers. In that scenario we could consider the IP as the cloud provider and the SP as the consumer. Of course the role of the actual customer/end-user exists but while it is an external entity is not depicted in our model. The Manifest entity presented in Figure 15.6 is the definition of a service while the Service entity is an instance generated from the respective Manifest. While the Manifest serves as a template of a service that can be deployed and executed within the virtual infrastructure, there is a one-to-many relationship between the Manifest and the Service entities. In addition, each SP owns several Manifests from which he can instantiate and deploy different services. Finally, while any IP can host several service instances we have defined the SERVICE-IP relationship in order to keep track of the deployments. The definition of such data models when deploying and executing services into cloud environments is critical for the effectiveness of the data collected throughout the monitoring mechanisms. A clear description of the relationships of the involved actors will make feasible the correlation of the reports coming from the different information providers. At last, this very model can be transformed to a database schema that can be used for storing the historical data collected by the monitoring mechanism.

15.5 Conclusions

The evolution of Cloud computing into an architectural paradigm with important advantages and significant capabilities increased its acceptance on the part of the industrial community. In that context, setting up a private Cloud infrastructure using our in-house resources becomes a realistic, affordable and reliable solution. To this end, the monitoring mechanism of a private Cloud constitutes a necessary tool for achieving infrastructure maintenance, optimization of resource usage, cost and energy efficiency etc. Indeed there are several toolkits and APIs available that have been mainly designed and used for resource monitoring on clusters and Grids and these very solutions may well serve as the baseline technology for designing a monitoring system placed within a Cloud environment. The service oriented approach, the introduction of virtualization technology and the application of alternative business models are some of the requirements for developing such monitoring infrastructure. In sum, we defined a generic architecture of a monitoring framework for a private Cloud. This approach adds two hierarchical layers on top of the monitoring collection level (i.e Information Providers): management/storing and data evaluation. We believe that the definition of a powerful monitoring infrastructure, compatible with several available APIs will add value to the Cloud paradigm and further strengthen this technology. Even though the topic of computing resources monitoring is fairly old, there are several steps to be taken towards the production of a Cloud-enabled monitoring mechanism.

16

Cloud Bursting: Managing Peak Loads by Leasing Public Cloud Services

Michael Mattess

The University of Melbourne, Australia

Christian Vecchiola

The University of Melbourne, Australia

Saurabh Kumar Garg

The University of Melbourne, Australia

Rajkumar Buyya

The University of Melbourne, Australia

CONTENTS

In general, small and medium-scale enterprises (SMEs) face problems of unpredictable IT service demand and infrastructure cost. Thus, the enterprises strive towards an IT delivery model which is both dynamic and flexible, and able to be easily aligned with their constantly changing business needs. In this context, Cloud computing has emerged as a new approach allowing anyone to quickly provision a large IT infrastructure that can be completely customized to the user's needs on a pay-per-use basis. This paradigm opens new perspectives on the way in which enterprises' IT needs are managed. Thus, a growing number of enterprises are outsourcing a significant percentage of their infrastructure to Clouds.

However, from the SMEs perspective, there is still a barrier to Cloud adoption, being the need of integrating current internal infrastructure with Clouds. They need strategies for growing IT infrastructure from inside and selectively migrating IT services to external Clouds in a way that enterprizes benefit from both Cloud infrastructure's flexibility and agility as well as lower costs.

In this chapter, we present how to profitably use Cloud computing technology by using Aneka, which is a middleware platform for deploying and managing the executions of applications on different Clouds. This chapter also presents public resource provisioning policies for dynamically extending the enterprise IT infrastructure to evaluate the benefit of using public Cloud services. This technique of extending capabilities of enterprise resources by leasing public Cloud capabilities is also known as Cloud bursting. The policies rely on a dynamic pool of external resources hired from commercial IaaS providers in order to meet peak demand requirements. To save on hiring costs, hired resources are released when they are no longer required. The described policies vary in the threshold used to decide when to hire and when to release; the threshold metrics investigated are queue length and queue time as well as a combination of these. Results demonstrate that simple thresholds can provide an improvement to the peak queue times, hence keeping the system performance acceptable during peaks in demand.

16.1 Introduction

Recently there has been a growing interest in moving infrastructure, software applications and hosting of services from in-house server rooms to external providers. This way of making IT resources available, known as Cloud computing, opens new opportunities to small, medium, and large sized companies. It is not necessary any more to bear considerable costs for maintaining the IT infrastructures or to plan for peak demand, but infrastructure and applications can scale elastically according to the business needs, at a reasonable price. The possibility of instantly reacting to the demand of customers without long term infrastructure planning is one of the most appealing features of Cloud computing and it has been a key factor in making this trend popular among technology and business practitioners. As a result of this growing interest, the major players in the IT playground such as Google, Amazon, Microsoft, Oracle, and Yahoo, have started offering Cloud computing based solutions that cover the entire IT computing stack, from hardware to applications and services. These offerings have quickly become popular and led to the establishment of the concept of the "Public Cloud," which represents a publicly accessible distributed system hosting the execution of applications and providing services billed on a pay-per-use basis.

Because Cloud computing is built on a massively scalable shared infrastructure, Cloud suppliers can in theory quickly provide the capacity required for very large applications without long lead times. Purchasers of Infrastructure as a Service (IaaS) capacity can run applications on a variety of virtual machines (VMs), with flexibility in how the VMs are configured. Some Cloud computing service providers have developed their own ecosystem of services and service providers that can make the development and deployment of services easier and faster. Adding SaaS capacity can be as easy as getting an account on a supplier's website. Cloud computing is also appealing when we need to quickly add computing capacity to handle a temporary surge in requirements. Rather than building additional infrastructure, Cloud computing could in principle be used to provide on-demand capacity when needed. Thus, the relatively low upfront cost of IaaS and PaaS services, including VMs, storage, and data transmission, can be attractive. Especially for addressing tactical, transient requirements such as unanticipated workload spikes. An additional advantage is that businesses pay only for the resources reserved; there is no need for capital expenditure on servers or other hardware.

However, despite these benefits, it has become evident that a solution built on outsourcing the entire IT infrastructure to third parties would not be applicable in many cases. On the one hand, enterprise applications are often faced with stringent requirements in terms of performance, delay, and service uptime. On the other hand, little is known about the performance of applications in the Cloud, the response time variation induced by network latency, and the scale of applications suited for deployment; especially when there are mission critical operations to be performed and security concerns to consider. Moreover, with the public Cloud distributed anywhere on the planet, legal issues arise and they simply make it difficult to rely on a virtual public infrastructure for some IT operation. In addition, enterprises already have their own IT infrastructures, which they have been using so far.

In spite of this, the distinctive feature of Cloud computing still remains appealing to host part of applications on in-house infrastructure and others can be outsourced. In other words, external Clouds will play a significant role in delivering conventional enterprise compute needs, but the internal Cloud is expected to remain a critical part of the IT infrastructure for the foreseeable future. Key differentiating applications may never move completely out of the enterprise because of their mission-critical or business-sensitive nature. Such infrastructure is also called a hybrid Cloud, which is formed by combining both private and public Clouds whenever private/local resources are overloaded. The notion of hybrid Clouds that extend the capabilities of enterprise infrastructures by leasing extra capabilities from public Clouds is also known as Cloud bursting.

For this vision to be achieved, however, both software middleware and scheduling policies supporting provisioning of resources from both local infrastructures and public Clouds are required, so that applications can automatically and transparently expand into public virtual infrastructures. The software middleware should offer enterprises flexibility in decision making that can enable them to find the right balance between privacy considerations, performance and cost savings. Thus, in this chapter, we first describe the architecture of a Cloud middleware called Aneka, which is a software platform for building and managing a wide range of distributed systems. It allows applications to use resources provisioned from different sources, such as private and public IaaS Clouds. Such a hybrid system, built with resources from a variety of sources, is managed transparently by Aneka. Therefore, this platform is able to address the requirements to enable execution of compute intensive applications in hybrid Clouds.

In summary, in this chapter, we will present the architecture of the Aneka middleware for building hybrid Clouds by dynamically growing the number of resources during periods of peak demand. We will then propose and evaluate three scheduling approaches to manage the external pool of resources and achieve cost benefits for enterprises.

16.2   Aneka

Aneka [724] is a software platform and a framework for developing distributed applications in the Cloud harnessing the computing resources of a heterogeneous network. It can utilize a mix of resources such as workstations, clusters, grids and servers in an on demand manner. Aneka implements a Platform as a Service model, providing developers with APIs for transparently exploiting multiple physical and virtual resources in parallel. In Aneka, application logic is expressed with a variety of programming abstractions and a runtime environment on top of which applications are deployed and executed. System administrators leverage a collection of tools to monitor and control the Aneka Cloud, which can consist of both company internal (virtual) machines as well as resources from external IaaS providers.

The core feature of the framework is its service oriented architecture that allows customization of each Aneka Cloud according to the requirements of users and applications. Services are also the extension point of the infrastructure: by means of services it is possible to integrate new functionalities and to replace existing ones with different implementations. In this section, we briefly

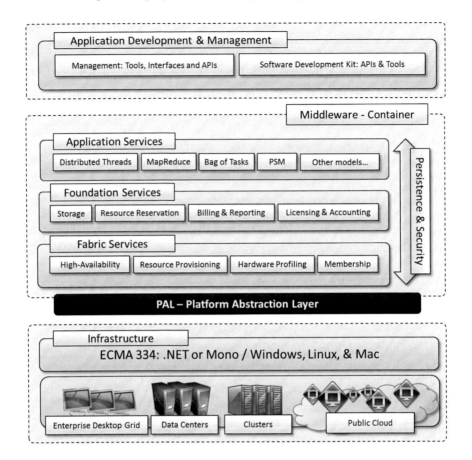

FIGURE 16.1
The Aneka Framework.

describe the architecture and categorize the fundamental services that build the infrastructure.

Figure 16.1 provides a layered view of the framework. Aneka provides a runtime environment for executing applications by leveraging the underlying infrastructure of the Cloud. Developers express distributed applications by using the APIs contained in the Software Development Kit (SDK). Such applications are executed on the Aneka Cloud, consisting of a collection of worker nodes hosting the Aneka Container. The Container is the core building block of the middleware and can host a number of services. These include the runtime environments for the different programming models as well as middleware services. Administrators can configure which services are present in the Aneka Cloud. Using such services for the core functionality provides an extendible and customizable environment. There are three classes of services that characterize the Container:

Execution Services: These services are responsible for scheduling and executing applications. Each programming model supported by Aneka defines specialized implementations of these services for managing the execution of a work unit defined in the model.

Foundation Services: These services are the core management services of the Aneka Container. They are in charge of metering applications, allocating resources for execution, managing the collection of available nodes, and keeping the services registry updated.

Fabric Services: These services constitute the lowest level of the services stack of Aneka and provide access to the resources managed by the Cloud. An important service in this layer is the Resource Provisioning Service, which enables horizontal scaling (e.g, increase and decrease in the number of VMs) in the Cloud. Resource provisioning makes Aneka elastic and allows it to grow and shrink dynamically to meet the QoS requirements of applications

The Container relies on a Platform Abstraction Layer that interfaces with the underlying host, whether this is a physical or a virtualized resource. This makes the Container portable over different platforms that feature an implementation of the ECMA 335 (Common Language Infrastructure) specification. Two well known environments that implement such a standard are the Microsoft .NET framework and the Mono open source .NET framework.

16.3   Hybrid Cloud Deployment Using Aneka

Hybrid deployments constitute one of the most common deployment scenarios of Aneka [726]. In many cases, there is an existing computing infrastructure that can be leveraged to address the computing needs of applications. This infrastructure will constitute the static deployment of Aneka that can be elastically scaled on demand when additional resources are required. An overview of such a deployment is presented in Figure 16.2.

This scenario constitutes the most complete deployment for Aneka which is able to leverage all the capabilities of the framework:

- Dynamic Resource Provisioning.

- Resource Reservation.

- Workload Partitioning.

- Accounting, Monitoring, and Reporting.

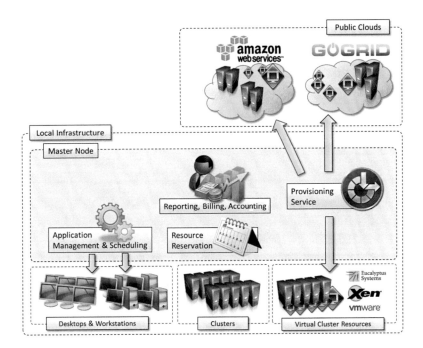

FIGURE 16.2
Aneka Hybrid Cloud.

In a hybrid scenario heterogeneous resources can be used for different purposes. For example, a computational cluster can be extended with desktop machines, which are reserved for low priority jobs outside the common working hours. The majority of the applications will be executed on these local resources, which are constantly connected to the Aneka Cloud. Any additional demand for computing capability can be leased from external IaaS providers. The decision to acquire such external resources is made by provisioning policies plugged into the scheduling component, as this component is aware of the current system state. The provisioning service then communicates with the IaaS provider to initiate additional resources, which will join the pool of worker nodes.

16.4   Motivation: Case Study Example

Enterprises run a number of HPC applications to support their day-to-day operation. This is evident from the recent Top500 supercomputer applications [710]. Many supercomputers are now used for industrial HPC applications, such as 9.2% of them are used for Finance and 6.2% for Logistic services. Thus, it is desirable for IT industries to have access to a flexible HPC infrastructure which is available on demand with minimum investment. In this section, we explain the requirements of one of such enterprise called the GoFront Group, and how they satisfied their needs by using the Aneka Cloud Platform, which allowed them to integrate private and public Cloud resources.

The GoFront Group in China is a leader in research and manufacture of electric rail equipment. Its products include high speed electric locomotives, metro cars, urban transportation vehicles, and motor train sets. The IT department of the group is responsible for providing support for the design and prototyping of these products. The raw designs of the prototypes are required to be rendered in high quality 3D images using the Autodesk rendering software called Maya. By examining the 3D images, engineers are able to identify any potential problems from the original design and make the appropriate changes. The creation of a design suitable for mass production can take many months or even years. The rendering of the three dimensional models is one of the phases that absorbs a significant amount of time, since the 3D model of the train has to be rendered from different points of view and for many frames. A single frame with one camera angle defined can take up to 2 minutes to render. To render one completes set of images from one design takes over 3 days. Moreover, this process has to be repeated every time a change is applied to the model. It is then fundamental for GoFront to reduce the rendering times, in order to be competitive and speed up the design process.

To solve the company's problem, an Aneka Cloud can be set up by using

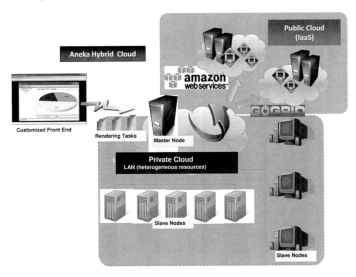

FIGURE 16.3
Aneka Hybrid Cloud Managing GoFront Application.

the company's desktop computers and servers. The Aneka Cloud also allows dynamic leasing of resources from a public Cloud in the peak hours or when desktop computers are in use. Figure 16.3 provides an overall view of the installed system. The setup is constituted by a classic master slave configuration in which the master node concentrates the scheduling and storage facilities and thirty slave nodes are configured with execution services. The task programming model has been used to design the specific solution implemented at GoFront. A specific software tool that distributes the rendering of frames in the Aneka Cloud and composes the final rendering has been implemented to help the engineers at GoFront. By using the software, they can select the parameters required for rendering and leverage the computation on the private Cloud.

The rendering phase can be sped up by leveraging the Aneka hybrid Cloud, compared to only using internal resources. Aneka provides a fully integrated infrastructure of the internal and external resources, creating, from the users' perspective, a single entity. It is, however, important to manage the usage of external resources as a direct hourly cost is incurred. Hence policies are needed that can be configured to respond to the demand on the system by acquiring resources only when it is warranted by the situation. Therefore, we have designed some scheduling policies which consider the queue of waiting tasks when deciding to lease resources. These policies are further described in detail with the analysis of their cost saving benefits in subsequent sections.

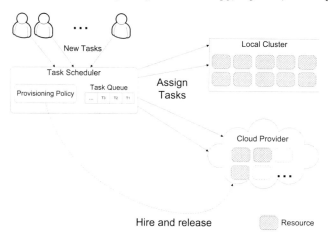

FIGURE 16.4
The Task Scheduling and Resource Provisioning Scenario.

16.5   Resource Provisioning Policies

In Aneka, as discussed in the previous section, there are two main components which handle the scheduling of tasks on the hybrid Cloud. They are the scheduling and provisioning components. The Provisioning component is responsible for the interactions between Aneka and the external resource providers, be they commercial IaaS Cloud or private clusters with a Cloud software stack like Eucalyptus. As each IaaS provider and software stack have their own API, a pluggable driver model is needed. The API specifics are implemented in what are termed resource pools. A number of these pools have been implemented to support commercial providers such as Amazon EC2 and GoGrid. Private Cloud stacks such as Eucalyptus and VMware products are also supported. The scheduling component is responsible for assigning tasks to resources; it keeps track of the available resources and maintains a queue of the tasks waiting to execute. Within this component there is also a pluggable scheduling and provisioning interface, so that Aneka can be configured to use different scheduling policies. The decision to request more resources or release resources (that are no longer needed) is also made by the selected scheduling policy. A number of such provisioning policies are implemented and can be selected for use.

These provisioning policies are the core feature of Aneka to enable the dynamic acquisition and release of external Cloud resources. As such resources have a cost attached, it is important to maximize their utilization and at the same time, request them only when they will be of benefit for the system. In the most common scenario, Cloud resources are billed according to time blocks that have a fixed size. In case of peak demands new resources can be provisioned to address the current need, but then used only for a fraction of the block for which they will be billed. Terminating these resources just after their usage could potentially lead to a waste of money, whereas they could be possibly reused in the short term and before the time block expires. In order to address this issue, we introduced a resource pool that keeps resources active and ready to be used until the time block expires.

Figure 16.4 describes the interaction between the scheduling algorithm, the provisioning policy, and the resource pool. Tasks in the queue are processed in order of arrival. A set of different events, such as the arrival of a new task, its completion, or a timer, can trigger the provisioning policy that, according to its specific implementation, will decide whether to request additional resources from the pool or release them. The pool will serve these requests, first with already active resources, and if needed, by provisioning new ones from the external provider.

The policies described below rely on the common principle of grow and shrink thresholds to determine when additional resources will be of benefit. That is, once a metric passes the defined growth threshold a request for additional resources is triggered. Conversely, the shrink threshold is used by the policy to determine if a resource is no longer required.

16.5.1   Queue Length

The *Queue Length* based policy uses the number of tasks that are waiting in the queue as the metric for the grow and shrink thresholds. When a task arrives the number of tasks in the queue is compared to the growth threshold; if the queue length exceeds the threshold an additional resource is requested. When a resource in the external Cloud becomes free, the queue length is compared to the shrink threshold. If the number of waiting tasks is less than the shrink threshold the resource is released back to the pool, otherwise it will be used to execute the next task.

Algorithm 1 Queue Length: Task Arrival

---

  if $Queue.Length \geq Growth\_Threshold$ **then**
    return *RequestResource*
  else
    return *NoAction*
  end if

16.5.2   Queue Time

The *Queue Time* based policy uses the time individual tasks have spent in the queue to determine when additional resources are requested. This policy periodically checks for how long the task at the head of the queue has been

---

Algorithm 2 Queue Length: Task Finished

---

  if $Queue.Length \leq Shrink\_Threshold$ **then**
    return $ReleseResource$
  else
    return $KeepResource$
  end if

---

waiting. If this time exceeds the growth threshold, additional resources are requested. The number of requested resources is the same as the number of tasks that have exceeded the growth threshold. When a resource in the Cloud becomes free, the amount of time the task at the head of the queue has been waiting is compared to the shrink threshold; if it is less than the threshold the resource will be released, otherwise it will be used to execute the next task.

Algorithm 3 Queue Time: Periodic Check

  $requestSize \leftarrow 0$
  $index \leftarrow 0$
  while $Queue[index].QueueTime \geq Growth\_Threshold$ **do**
    $requestSize \leftarrow requestSize + 1$
    $index \leftarrow index + 1$
  end while
  $requestSize \leftarrow requestSize - OutstandingResourceCount$
  if $requestSize > 0$ **then**
    $RequestResources(requestSize)$
  end if

Algorithm 4 Queue Time: Task Finished

---

  if $Queue[head].QueueTime \leq Shrink\_Threshold$ **then**
    return $ReleseResource$
  else
    return $KeepResource$
  end if

16.5.3 Total Queue Time

The *Total Queue Time* policy sums the queue time of each of the tasks in the queue, starting from the tail of the queue. When the sum exceeds the growth threshold, before reaching the head of the queue, a resource is requested for each of the remaining tasks. The releasing of a resource occurs when a task on a external Cloud resource has finished. At that time the total queue time is established and compared to the shrink threshold. If it is less than the threshold the resource is released, otherwise it will be used for the next task.

Algorithm 5 Total Queue Time: Periodic Check

$requestSize \leftarrow 0$
$totalQueueTime \leftarrow 0$
for $i = 0$ to Queue.Length **do**
   $totalQueueTime \leftarrow totalQueueTime + Queue[i].QueueTime$
   if $totalQueueTime \geq Growth\_Threshold$ **then**
     $requestSize \leftarrow requestSize + 1$
   end if
end for
$requestSize \leftarrow requestSize - OutstandingResourceCount$
if $requestSize > 0$ **then**
   $RequestResources(requestSize)$
end if

Algorithm 6 Total Queue Time: Task Finished

$totalQueueTime \leftarrow 0$
for $i = 0$ to Queue.Length **do**
   $totalQueueTime \leftarrow totalQueueTime + Queue[i].QueueTime$
end for
if $totalQueueTime < Shrink_T hreshold$ **then**
   return $ReleseResource$
else
   return $KeepResource$
end if

16.5.4 Clairvoyant Variant

In our scenario we consider the execution time of a task an unknown until the task finishes. Hence these policies can not know if a particular task will complete before a time block finishes or if charges for additional time blocks will be incurred. This makes it difficult to utilize the remaining time of external resources, which are being released by the policy. However, to explore the impact of better utilizing this otherwise wasted time we created clairvoyant variants of these policies. These variants are able to predict the run time of each task in the queue. As the policy decides to release a resource, the clairvoyant variants perform a pass over all the tasks in the queue, searching for tasks which will complete within the remaining time. The best fit task is then assigned to the resource, which remains in use. If no task is found which will complete in time, the resource is released.

Algorithm 7 Clairvoyant Variants: Task Finished

$remainingTime \leftarrow ResourceToBeReleased.RemainingTimeInBlock$
$bestFitRunTime \leftarrow 0$
$bestFitTask \leftarrow NULL$
for $i = 0$ to Queue.Length **do**
  $task \leftarrow Queue[i]$
  $runTime \leftarrow task.RunTime$
  if $runTime \leq remainingTimeANDrunTime > bestFitRunTime$
  then
    $bestFitTask \leftarrow task$
    $bestFitRunTime \leftarrow runTime$
  end if
end for
if $bestFitTask \neq NULL$ **then**
  $AssignTask(bestFitTask, ResourceToBeReleased)$
  return $KeepResource$
else
  return $ReleseResource$
end if

16.6 Performance Analysis

The policies described in the previous section have been implemented in Aneka and have been used to observe the behavior of the system for small and real workloads. In order to extensively test and analyze these policies we built a discrete event simulator that mimics the Aneka scheduling interfaces, so that policies can be easily ported between the simulator and Aneka. We also identified a performance metric that will drive the discussion of the experimental results presented in this section.

16.6.1 Simulator

To analyze the characteristics of these provisioning policies it is necessary to construct a simulator. Using a simulator in this case allows policies to be compared with exactly the same workload. It also means that any interruptions to production Aneka environments are avoided. But most importantly the simulations can be run much faster than real-time, so that a whole month or year can be simulated in minutes.

The simulator used here is especially constructed to reflect the internal interfaces of Aneka, so that implementation of scheduling and provisioning policies could be easily ported from the simulator to the Scheduling module in Aneka. In the simulator design, the fundamental goal was to gain an insight into these policies while keeping the design as simple as possible. Hence the simulator does not attempt to capture all the complexities of a distributed environment, instead it focuses on exposing the behavior of the policies.

16.6.2 Performance Metric

To examine and compare the performance characteristics of the policies we use a metric which we termed the *Top Queue Time Ratio*. This metric comprises the average of the 5000 longest queue times, which is then divided by the average execution time of all the tasks.

We consider the average of the largest queue times rather than an overall average as our focus is on managing the spikes in demand rather than improving the average case. We decided to consider as a reference the top 5000 largest queue times. The reason for this choice is that this number represents about 15% of the workload used for the analysis. In our opinion, the acceptability of a waiting time is to some extent relative to the execution time of the task. Hence, we use the ratio between the largest queue times and the average task duration. Using a metric which is relative to the average duration of the tasks also allows us to scale time in the workload without affecting the metric, thus making direct comparisons possible.

16.6.3 Workload

To drive the simulation we use a trace from the Auver Grid [107]. It was chosen as it only consists of independent tasks, which reflects the scenario described in the motivation section (Section 16.4). This trace covers the whole of 2006 and has been analyzed in [376]. The Auver Grid is a production research grid located in France including 475 Pentium 4 era CPUs distributed over six sites. We divided the trace into calendar months to compare periods with high and low demand on the system. As it is clear from Table 16.1, October experienced

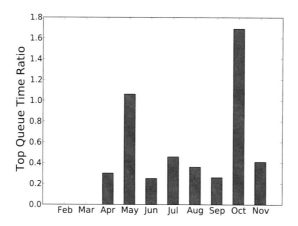

FIGURE 16.5
Top Queue Time Ratio of Each Month.

the greatest load on the system. This is also reflected in the *Top Queue Times* of the months shown in Figure 16.5. Hence, we used the trace of October to carry out the experiments.

	Av.RunTime(min)	Task Count	Total RunTime
Jan	318	10,739	3,415,002
Feb	255	15,870	4,046,850
Mar	242	39,588	9,580,296
Apr	484	27,188	13,158,992
May	493	32,194	15,871,642
Jun	361	35,127	12,680,847
Jul	311	46,535	14,472,385
Aug	605	28,934	17,505,070
Sep	552	30,002	16,561,104
Oct	592	33,839	20,032,688
Nov	625	14,734	9,208,750
Dec	382	24,564	9,383,448

TABLE 16.1
Characteristics of Each Month.

The average duration of a task in this workload during the month of October is almost 10 hours. This is a significantly longer time period than the one hour blocks, which form the *de facto* atomic unit for buying time in the Cloud. We felt that the relationship between the average task duration and the size of the blocks in which time is bought is worth exploring. Hence, we used a

Scaling Factor	1	10	20	40
Av. Task Duration (minutes)	592	59	29	15

TABLE 16.2
The Average Task Duration, during October, for Scaling Factors Used.

scaling factor to divide the task duration and arrival interval times, essentially compressing time. This reduced the average task duration (see Table 16.2) while maintaining the same load on the system. The Top Queue Time Ratio is also left unaffected by the scaling, thus allowing it to be used to compare the results with different scaling factors. The cost, on the other hand, is affected by the scaling as the absolute total amount of task time is scaled down. As less time is required to process tasks, less time ends up being bought from the Cloud provider. To allow cost comparisons of different scaling factors we multiply the cost by the scaling factor such that the cost is representative of the entire month rather than the scaled down time.

16.6.4   Experimental Setup

Although our simulations do not intend to replicate the Auver Grid (see Workload Section 16.6.3) we referred to it when configuring the simulated environment. In particular, the number of permanent local resources in the simulation, is set to 475, reflecting the number of CPUs in the Auver Grid. Also, when considering the relative performance between the local and Cloud resources the Pentium 4 era CPUs of the Auver Grid were compared to Amazon's definition of their EC2 Compute Unit [641] and it was concluded that they are approximately comparable.

Hence, a Small EC2 Instance was used as the reference point, which has one EC2 compute unit. Thus for all the simulations the relative performance was configured such that a particular task would take that same amount of time to run locally or in the Cloud. To calculate a cost for the use of Cloud resources we again use a Small EC2 Instance as a reference point with an hourly cost of USD $0.10 [1]. Although the true cost of using Cloud resources includes other costs such as network activity both at the provider's end and the local end, as well as data storage in the Cloud, we have not attempted to establish these costs as they are dependant on the individual case. Hence, the cost for using Cloud resources in this work is solely based on the amount of billable time.

When requesting resources from a Cloud provider, there is generally a *time block* which represents an atomic billable unit of usage. In the case of Amazon's EC2 this time block is one hour. Meaning that if a resource is used for 1.5 hours the cost will be that of two time blocks. This behavior is reflected in the simulations where the size of the time block is also set to one hour. As requested resources take time to come on line, we have set a three minute

---

[1] Amazon has since reduced its prices.

delay between a resource being requested and it becoming available to execute tasks. Our own usage of EC2 has shown this to be a reasonable amount of time for a Linux instance to become available. Although these times can vary considerably on EC2 we do not anticipate getting a better insight into the policies by introducing random delays to the simulation.

16.6.5   Experimental Results

In this section we explore the effects of the parameters on the performance of the policies. First we examine the effects of the settings associated with the policies. Later we look at the impact the task size has on the efficiency in terms of the unused time of resources in the Cloud.

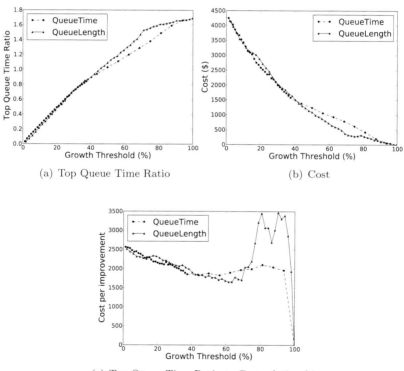

(a) Top Queue Time Ratio   (b) Cost

(c) Top Queue Time Ratio to Cost relationship

FIGURE 16.6
The Performance of the Queue Time and Queue Length–Based Policies as the Growth Threshold Increases.

First of all we take a look at the Growth Threshold's impact. A low Growth Threshold means that the policy will request additional resources sooner. As the Growth Threshold increases, the burden on the system needed to trigger a request for additional resources also increases. Hence, as the Growth Threshold increases, the time some tasks spend waiting in the queue also increases. The graph in Figure 16.6(a) demonstrated this trend showing the Top Queue Time as the Growth Threshold increases. In this figure the Growth Threshold is expressed as a percentage of the maximum Growth Threshold, which is a threshold so high that the workload we are using does not trigger a request for additional resources. Such a threshold has been determined by repeatedly increasing the value set until no more additional resources are requested.

As one would expect, the Growth Threshold to cost relationship (see Figure 16.6(b)) is inversely proportional to that of the Top Queue Time Ratio (see Figure 16.6(a)). A high Top Queue Time Ratio means that less external resources were requested resulting in a lower cost. The graph in Figure 16.6(c) shows the cost per unit reduction of the Top Queue Time Ratio. The most striking features of this graph are the two spikes of the Queue Length based policy. When the growth threshold reaches the size where these spikes occur there is only one instance of high demand in the workload, which is large enough to cause additional resources to be requested. The Queue Length based policy in this case requests more resources than would be ideal, hence incurring costs, which are not reflected in a reduction of the Top Queue Time Ratio. The Queue Time based policy, on the other hand, triggers fewer additional resources to be requested, which means that the cost is in proportion to the improvement in the Top Queue Time Ratio. As at that point (around 80%) the cost and improvement in the Top Queue Time Ratio are quite small; the graph comparing these two values becomes more sensitive to small changes, resulting the large size of these spikes. Apart from the spikes, Figure 16.6(c) shows that the inverse relationship between the Top Queue Time Ratio and the Cost has a slight downward trend until around the middle of the growth threshold range, meaning that the best value is with a growth threshold around the middle of the range of values.

Figure 16.6 also demonstrates that the Queue Length and Queue Time based policies behave very similarly when the task size is large compared to the size of the time blocks purchased from the Cloud provider. Comparing the lines of scale factor one in Graph 16.7(a) and 16.7(b) further confirm this by indicating that the cost of achieving a particular Top Queue Time Ratio is virtually identical.

However, as the scale factor is increased, the cost of reaching a particular Top Queue Time Ratio also increases in Figure 16.7. In this figure the cost has been multiplied by the scaling factor, as scaling reduces the total amount of work that needs to be done, which we have to normalize against to be able to compare the results using different scaling factors (see Section 16.6.3). Hence a smaller task size increases the amount of time that needs to be purchased from the Cloud to reach a particular Top Queue Time Ratio. Comparing the graphs in Figure 16.7 also shows that the Queue Time based policy performs slightly better at lower Top Queue Time Ratios and that the Queue Length

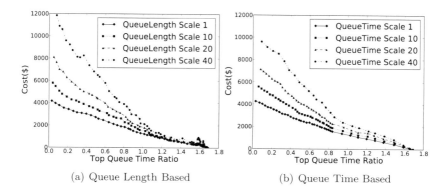

(a) Queue Length Based   (b) Queue Time Based

FIGURE 16.7
Demonstrating the Relationship between the Top Queue Time and the Cost.
The growth threshold is varied between the points on a line. The scaling factor
is changed between lines.

based policy has a slight edge at higher Top Queue Time Ratios. But these
slight differences only become apparent as the tasks become smaller.

To examine the increased cost with the smaller task size we first look at
the utilization of the time that is bought from the Cloud provider in Figure
16.8. We see that with smaller tasks a significant proportion of the available
resource time in the Cloud goes unused. Comparing Figures 16.7 and 16.8 it
can be noticed that some features correlate between the cost and utilization.
In particular for the Queue Length based policy the lines move closer for larger
Top Queue Time Ratios. The Queue Time based policy, on the other hand,
maintains a more consistent separation in both the cost and utilization graphs.
Hence, we attribute the increased cost with smaller tasks to the decreased
efficiency with which resources in the Cloud are used.

In Figure 16.9(a) we compare the utilization of the Queue Time based policy
to its clairvoyant variant. This graph is essentially Figure 16.8(b) with the
clairvoyant results superimposed. We can see that at very low Top Queue
Time Ratios the clairvoyant variant performs only marginally better. At higher
queue time ratios the clairvoyant variant is able to significantly improve the
utilization of the time bought in the Cloud. However, in Figure 16.9(b), which
is derived from Figure 16.7(b) we see that the improvement in utilization has
not lead to a comparable reduction of cost for achieving a particular Top
Queue Time Ratio. This is because the clairvoyant policy uses backfilling
to utilize otherwise wasted time of resources being released. Hence, it does
not directly change the growing of resources in the Cloud. Its effect on cost
is only indirect by removing some tasks from the queue, which will affect
when the thresholds are crossed. Also, our metric focuses on the longest queue
times, whereas the clairvoyant variant allows more tasks to be processed when
resources are being released. Hence the benefit of the clairvoyant variant comes

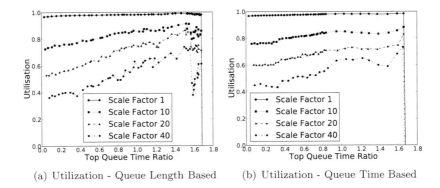

(a) Utilization - Queue Length Based   (b) Utilization - Queue Time Based

FIGURE 16.8

Shows the Utilization of the Time Bought on the Cloud. The utilization is calculated using only time available for executing tasks and not startup times for resources in the Cloud. The growth threshold is varied between the points on a line. The scaling factor is changed between lines.

at a time when the queue has already reached an acceptable size and will only have a limited effect on the largest queue times.

Hence, to impact cost we need to be looking at when additional resources are requested. The Total Queue Time policy is in essence a combination of the simpler Queue Length and Queue Time based policies. In Figure 16.10 the cost is compared to that of the Queue Time based policy discussed previously. We see that for lower Top Queue Time Ratios and bigger tasks, a clear improvement is possible. But once the average task duration becomes less than the time blocks being purchased in the Cloud the advantage disappears. For small average task durations (scaling factor 40) this policy can perform significantly worse when aiming at higher Top Queue Time Ratios.

Up to now our investigation has focused on the month of October in the 2006 Auver Grid trace. We will now take a look at how both the Queue Time and Queue Length based policies behave in the other months of the trace. For this comparison we configured both policies with a growth threshold value that for October resulted in a Top Queue Time Ratio close to 0.7 and did not apply any scaling of the trace. Figure 16.11 presents the Top Queue Time Ratio and corresponding cost for each month. Clearly during October and May the system is under the greatest load, which is considerably higher than any other months. Hence as expected these policies have their greatest impact during these months. There is, however, an interesting observation to make. During November, both policies trigger requests for additional resources, however during July, which has a slightly higher Top Queue Time Ratio, the policies do not trigger any requests.

To understand this behavior we need to look at the average task duration and

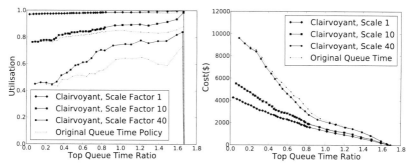

(a) Utilization comparison to original
data from Figure 16.8(b)

(b) Cost comparison to original data from
Figure 16.7(b)

FIGURE 16.9
Contrasts the Queue Time–Based Policy with its Clairvoyant Variant. Scale
factor 20 is omitted in the interest of clarity.

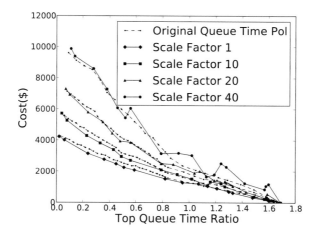

FIGURE 16.10
Contrasts the Queue Time Total Policy with the Queue Time Policy.

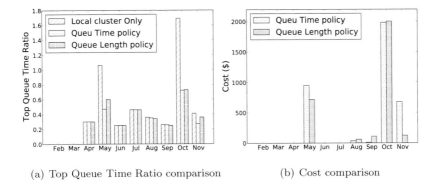

(a) Top Queue Time Ratio comparison   (b) Cost comparison

FIGURE 16.11
Month by Month Performance.

task count for each month, which are presented in Table 16.1. First we note that the average task duration of November is almost exactly double that of July. The Top Queue Time Ratio is relative to the average task size. Hence as both July and November have similar Top Queue Time Ratios, the actual top 5000 queue times in November are almost double the length of those in July. The Queue Time based policy however, uses a fixed growth threshold, which leads it to trigger requests for additional resources. The Queue Length based policy is affected to a lesser extent.

16.7 Related Work

This work considers the case where an enterprise wants to provision resources from an external Cloud provider to meet its peak compute demand in the form of Bag-of-Tasks applications. To keep the queue waiting time of HPC tasks at an acceptable level, the enterprise hires resources from public Cloud (IaaS) providers.

A key provider of on-demand public Cloud infrastructure is Amazon Inc. with its Elastic Compute Cloud (EC2) [641]. EC2 allows users to deploy VMs on Amazon's infrastructure, which is composed of several data centers located around the world. To use Amazon's infrastructure, users deploy instances of pre-submitted VM images or upload their own VM images to EC2. The EC2 service utilizes the Amazon Simple Storage Service (S3), which aims at providing users with a globally accessible storage system. S3 stores the users' VM images and, as EC2, applies fees based on the size of the data and the storage time.

Previous work has shown how commercial providers can be used for scientific applications. Deelman et al. [229] evaluated the cost of using Amazon EC2 and S3 services to serve the resource requirements of a scientific application. Palankar et al. [562] highlighted that users can benefit from mixing Cloud and Grid infrastructure by performing costly data operations on the grid resources while utilizing the data availability by the Cloud.

The leading IaaS provider, Amazon, charges based on one hour time block usages. In our scenario we consider unbounded tasks, which can produce a small amount of wasted computation time with the hired resources; detailed evaluation is given in Section 16.6. For example, the policy hires a resource for one hour and the scheduled task on it takes 40 minutes, as the demand is lower and the scheduler does not schedule any other task on it. Therefore, in the next section, we describe a *clairvoyant* policy that relies on an adaptation of backfilling techniques to give an indication of the impact of wasting this time. We consider the conservative backfilling [528]. In that case the scheduler knows the execution time of tasks beforehand, the scheduler allows a reservation for each task when it arrives in the system, and tasks are allowed to jump ahead in the queue if they do not delay the execution of other tasks. Hence, the scheduler can execute smaller tasks on hired resources that have unused time remaining before being terminated.

Our work is based on a similar previous work [223] that evaluates different strategies for extending the capacity of local clusters using commercial providers. Their strategies aim to schedule reservations for resource requests. A request has a given ready time, deadline, walltime, and the number of resources needed. Here, we consider unbounded tasks that require a single container to be executed. Tasks are also executed in a First-Come-First-Served manner.

Several load sharing mechanisms have been investigated in the distributed systems realm. Iosup et al. [377] proposed a matchmaking mechanism for enabling resource sharing across computational Grids. Wang and Morris [749] investigated different strategies for load sharing across computers in a local area network. Surana et al. [682] addressed the load balancing in DHT-based P2P networks. Balazinska et al. [112] proposed a mechanism for migrating stream processing operators in a federated system.

Market-based resource allocation mechanisms for large-scale distributed systems have been investigated [765]. In this work, we do not explore a market-based mechanism because we rely on utilizing resources from a Cloud provider that has cost structures in place. We evaluate the cost effectiveness of provisioning policies in alleviating the increased waiting times of tasks during periods of high demand on the local infrastructure.

16.8 Conclusions

In this chapter, we discussed how an Aneka based hybrid Cloud can handle the sporadic demand on IT infrastructure in enterprises. We also discussed some resource provisioning policies that are used in Aneka to extend the capacity of local resources by leveraging external resource providers. Once requested, these resources become part of a pool until they are no longer in use and their purchased time block expires. The acquisition and release of resources is driven by specific system load indicators, which differentiate the policies explored.

Using a case study example of Go-Front, we explained how Aneka can help in transparent integration of public Clouds with local infrastructure. We then introduced two policies based on queue length and queue time. We analyzed their behavior from the perspective of the growth threshold, which is in the end responsible of the acquisition of new resources from external providers. The experiments conducted by using the workload trace of October 2006 from Auver Grid show that the best results in term of top queue time and cost are obtained when the growth threshold ranges from 40% to 60% of the maximal value. The behavior is almost the same for both of the two policies. Since the average task duration in the October trace is almost 10 hours, we decided to scale down the trace by different factors and explore the behavior of the policies as the average task size changes. We concluded that smaller task size leads to more wastage when trying to maintain the queue time comparable to average task duration. We then compared the performance of the Queue Time policy with its clairvoyant variant that performs backfilling by relying on the task duration. The experiment showed that the improvement in utilization obtained does not reflect a corresponding cost improvement. We also introduced a Total Queue Time policy that, in essence, combines the original two policies, and studied its performance at different scale factors of the workload. There is only a marginal improvement at the original scale of the trace, whereas for smaller task sizes the behavior is similar.

A comparison with the execution of the same trace without any provisioning capability, clearly demonstrates that the policies introduced are effective in bringing down the Top Queue Time Ratio during peak demands. The cost spent on buying computation time on EC2 nodes to obtain such a reduction in one year can roughly accommodate the purchase of a single new server for the local cluster that definitely does not guarantee the same performance during peak.

17

Energy-Efficiency Models for Resource Provisioning and Application Migration in Clouds

Young Choon Lee

Centre for Distributed and High Performance Computing, School of Information Technologies, University of Sydney, Australia

Dilkushan T. M. Karunaratne

School of Information Technologies, University of Sydney, Australia

Chen Wang

CSIRO ICT Center, Epping, Australia

Albert Y. Zomaya

Centre for Distributed and High Performance Computing, School of Information Technologies, University of Sydney, Australia

CONTENTS

17.1   Introduction

In the recent past, we have witnessed that performance of distributed computing systems particularly at large scale (i.e., large-scale distributed computing systems or LDCSs) has been significantly constrained by their excessive power requirements. This ever increasing energy consumption for powering and cooling is a major limiting factor in the running and the expansion of data centers. Besides, the energy consumption issue in these systems raises both business and environmental concerns. A recent study on power consumption by servers (the type used in data centers and clouds) shows that electricity use for servers (compute nodes) and their associated cooling and auxiliary equipment worldwide—not including storage devices and network equipment—in 2005 cost 7.2 billion US dollars [440]. The study also indicates that the electricity consumption in that year had doubled compared with consumption in 2000.

Cloud computing systems typically consist of many power-hungry components including processors, memory modules, and network devices. Energy efficiency in these systems is investigated at different levels, such as hardware level, operating system (OS) level, virtualization technique level, and operations management level. Recent developments and advancements in low-power microprocessors and virtualization technologies have made a significant impact on the improvement of energy efficiency. However, most current energy-efficiency solutions tend to be specific to certain computing components like processors; and at the same time they are very general in that a set of common techniques and technologies is applied to their target levels irrespective of different working circumstances. Typical example energy-efficiency solutions include dynamic voltage/frequency scaling (DVFS), solid state drives (SSD), server virtualization, workload offloading/outsourcing to public clouds like Amazon Elastic Compute Cloud (EC2), and Intel Atom servers. While the use of new and more energy-efficient hardware devices can surely lead to energy savings to a certain degree, the effectiveness of these "hardware" devices in energy reduction may not be fully realized due to their inefficient usage. For example, applications running on LDCSs (with those new devices) typically have performance requirements that might hinder the effective exploitation of energy reduction.

Since large-scale distributed computing systems like clouds deal with dynamic and heterogeneous resources and applications, scheduling and resource allocation play a crucial role in optimization of resource utilization; hence, better energy efficiency. Efficient resource utilization is in fact a major driving force in cloud computing. Two main factors affecting resource utilization are clearly workloads (applications running on clouds) and resources provisioned. Typically, resources are overly provisioned to meet application performance requirements (or service level agreements, SLAs) even in the case of sudden workload surges; this results in poor resource utilization. The characterization and profiling of applications and its incorporation into scheduling and runtime monitoring/control decisions greatly help alleviate the resource over-provisioning issue. In other words, clouds with virtual machine techniques in particular make good use of such application information for effective resource management. Figure 17.1 illustrates two different resource provisioning scenarios, i.e., static and dynamic provisioning.

In this chapter, we investigate the energy efficiency in LDCSs in terms particularly of resource allocation, review current practices with respect to energy models for applications, and discuss issues in this regard with a practical example.

The remainder of the chapter is organized as follows: Section 17.2 overviews energy efficiency in LDCSs with issues; Section 17.3 discusses the current approaches that incorporate energy models derived with the consideration of application characteristics; Section 17.4 presents results of our preliminary study on server (VM) consolidation with a hope to guide the development of energy efficiency solutions; Section 17.5 summarizes and concludes the chapter.

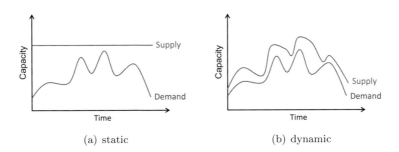

(a) static   (b) dynamic

FIGURE 17.1
Resource Provisioning.

17.2   Energy Efficiency in LDCSs

In this section, we overview energy-efficient computing (or more broadly, sustainable computing), and describe current computing practices and issues in LDCSs in terms of energy efficiency.

17.2.1   Sustainable Computing

Sustainable/green computing is the application of ecological-friendliness to the practice of computing as an essential principle. In other words, the impact of computing facilities and their use is much widened as the modern society is increasingly dependent on computers and their services, and the scale of computer systems is at an unprecedented level. Efficient resource management (or the optimization of resource utilization) is clearly one of the core aspects of sustainable computing. The notion of efficiency in the current computing environment is expanded, explicitly involving energy consumption (or energy efficiency). The scope of sustainable computing is not limited to main computing components (e.g., processors, storage devices and visualization facilities), but it can expand into a much larger range of resources associated with computing facilities including auxiliary equipments, water used for cooling and even the physical/floor space that these resources occupy. Energy consumption in computing facilities raises various monetary, environmental and system performance concerns. According to a recent IDC (International Data Corporation) report [17], energy cost will overtake IT equipment cost in the near future if current data center power consumption continues to grow at the current rate. In addition, some $26.1 billion was spent in 2005 to power and cool the worldwide installed base of servers; this is more than double the cost from 10 years ago of $10.3 billion.

17.2.2   Power Usage Reality of LDCSs

For decades, we have vigorously strived for (high) performance with computer systems often neglecting energy efficiency — more specifically, the utilization of resources. One "surprising" (in fact not very surprising, rather obvious) fact about LDCSs like data centers and clouds, is low resource utilization (10–20% on average). What's more, even at a very low load, such as 10% CPU utilization, the power consumed is over 50% of the peak power [119], i.e., a large portion of energy is dissipated when servers run at idle or low utilization. Similarly, if the disk, network, or any such resource is the performance bottleneck, the idle power wastage in other resources goes up. Clearly, the energy proportionality of LDCSs should be much improved. In principle, power drawn when servers idle should be minimal, i.e., close to zero. Figure 17.2(a) shows the typical power usage and energy efficiency in LDCSs. In the meantime, ideal and theoretical energy efficiency curves are shown in Figure 17.2(b). Since there is some transition cost between different power states, the theoretical energy efficiency (perfect energy proportionality) might not be possible to realize in practice.

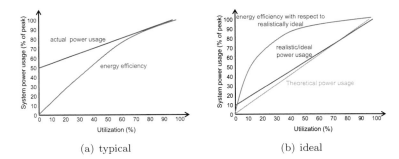

(a) typical   (b) ideal

FIGURE 17.2
Energy Proportional Computing.

Another interesting fact with respect to data center energy efficiency is the actual power usage of "computing" resources. In other words, the total power supplied to a data center is typically (far) greater than the amount of energy consumed by IT equipment (i.e., servers) due to power consumed by other infrastructural components including cooling and power distribution.

Since the amount of power consumed by these components is directly related to resources provisioned, efficient resource utilization has an inherent and significant impact on the reduction of such power consumption [333]. The relationship between the total power consumption and IT equipment power consumption is a good indicator of data center energy efficiency. The Green Grid (http://www.thegreengrid.org/)—a global consortium of various members mostly from IT industry for improving data center energy efficiency—has defined power usage effectiveness (PUE) and also its reciprocal (Data Center Infrastructure Effectiveness, DCiE) to quantify that relationship. More formally, these two energy efficiency metrics are defined as:

$$PUE = \frac{Total\ data\ center\ power}{IT\ equipment\ power} \tag{17.1}$$

$$DCiE = \frac{IT\ equipment\ power}{Total\ data\ center\ power} \cdot 100\% \tag{17.2}$$

According to the US Environmental Protection Agency (EPA) report to the U.S. Congress [600], typical enterprise data centers in 2006 have a PUE of 2.0 or higher. Data center efficiency in terms of PUE in particular is expected to be continually improved. The report suggests that a PUE of 1.7 by 2011 is anticipated for typical data centers, and 1.5 and 1.4 for data centers with best practices and state-of-the-art measures, respectively. A data center with a PUE of 2.0 indicates that only 50% of its energy consumption is used to actually run applications on servers; and the other 50% is consumed by power distribution and cooling facilities.

17.3   Energy Efficiency and Applications

In this section, we review energy efficiency in three different levels of LDCSs and how each of these relates to applications.

17.3.1   Platform Level

Typical LDCSs are equipped with a large number of compute nodes in each of which there are various components including processors, memory modules and disks. For example, Amazon Web Services (AWS) Compute Cluster consists of 1760 Intel Xeon Quad-core processors (i.e., 7040 cores).[1] Each of these processors consumes as much as 95W [207]. More importantly, in general, this power consumption accounts for more than a third of server power draw [262]. Table 17.1 shows a more detailed breakdown of component peak power draw for a typical server [262].

Chip makers, such as Intel and AMD, have put a lot of effort into the energy efficiency of their products. These efforts include more fine-grained power states using DVFS, multi-core technologies, and more recently, Intel Turbo Boost Technology.[2] However, irrespective of all these efforts, the power consumption of processors still remains high, i.e., they are energy inefficient and account for more than a third of total server power consumption, as shown in Table 17.1. Therefore, the reduction in energy consumption of processors is critical to improve energy efficiency in computer systems in general.

Since the execution of an application involves a number of different tasks including computation and I/O, the utilization of processors in particular is far from its full utilization, resulting in the frequent occurrence of idle time slots. Probably, the most common approach to processor energy efficiency is *slack reclamation* using the DVFS technique, e.g., [800], [460]. Specifically, this technique temporarily decreases voltage supply level at the expense of lowering processing speed (frequency). Slack reclamation is made possible primarily by recent DVFS-enabled processors (with technologies like Enhanced Intel Speed-

---

[1]The   November   2010   update   of   Top   500   Supercomputer   Sites   list (http://www.top500.org) includes AWS Compute Cluster as the 231st most powerful super computer site in the world.

[2]http://www.intel.com/technology/turboboost/

Step,[3] and AMD Cool 'n' Quiet [4] and PowerNow! [5]) and the parallel nature of the deployed tasks. For example, when the execution of a task is dependent on two predecessor tasks and these two tasks have different completion times, the predecessor task with an earlier completion time can afford additional run-time (slack); this slack can be then exploited using under-volting for energy saving. Since most DVFS-based energy-aware scheduling and resource allocation techniques are static (offline) algorithms with an assumption of tight coupling between tasks and resources (i.e., local tasks and dedicated resources), their application to our cloud scenario is not apparent, if not possible. Another recent solution to processor energy efficiency is servers with low-power processors like SeaMicro's Intel Atom servers [6] and the system in the FAWN project [589], [83]. This type of system aims specifically for I/O-intensive applications in which a large amount of application execution time is in waiting for the completion of I/O tasks; and thus, those low-performance processors are sufficient to meet the performance requirement of those I/O-intensive applications.

Storage area network (SAN) and network-attached storage (NAS) are two common networked storage systems that can unburden storage energy issues in LDCSs to a certain degree. Solid state drives (SSDs) can be considered as a good alternative (in terms of energy efficiency) to traditional storage media, such as hard disk drives (HDDs) and tape drives. Data de-duplication might be of a great practical importance since the avoidance of redundant data improves disk utilization; hence, storage energy efficiency is improved.

Although the power consumption of memory is rather significant, its optimizations are less mature due mainly to memory technologies not supporting enough ACPI (Advanced Configuration and Power Interface) [7]states and not enough control in memory controllers.

TABLE 17.1
Power Breakdown of Typical Server Components

Component	Count	Peak power	Total	Relative power
CPU	2	40W	80W	37%
PCI slots	2	25W	50W	23%
Memory	4	9W	36W	17%
Motherboard	1	25W	25W	12%
Disk	1	12W	12W	6%
Fan	1	10W	10W	5%

---

[3]http://www.intel.com/support/processors/sb/CS-028855.htm?wapkw=(eist)
[4]http://www.amd.com/us/products/technologies/cool-n-quiet/Pages/cool-n-quiet.aspx
[5]http://www.amd.com/us/products/technologies/amd-powernow-technology/Pages/amd-powernow-technology.aspx
[6]http://www.seamicro.com/
[7]http://www.acpi.info/

17.3.2   Virtual Machine Monitor Level

Server virtualization is one of the current trends in energy efficient computing. Virtualization enables resources in physical machines, such as processors, network devices and memory, to be shared among multiple isolated platforms, also known as virtual machines (VMs). This resource sharing much helps efficient resource management leading to the improvement in resource utilization. A hypervisor (virtual machine monitor or VMM) enables multiple servers (guests) to run on a single machine (host) and each guest server could run with a different operating system [310]. Para-virtualization and full-virtualization are two types of virtualization available. In para-virtualization, the guest servers run a customized kernel (virtualization aware) while the standard operating systems without any customizations could run as guest operating systems in a full-virtualized environment. VMMs have incorporated various techniques to optimize energy consumption. For example, Dynamic Resource Scheduler (DRS) with Distributed Power Management (DPM) from VMware dynamically scales resources as applications' needs shrink or grow.[8]

In principle, VMs can migrate from one physical server to another without causing a significant impact on the application running within them. However, in practice, much attention should be paid to VM migration due to possible resource contention between VMs on the same physical machine; that is, certain resources cannot be completely isolated between VMs and/or simultaneously shared by multiple VMs. Therefore, accurate application profiling and modeling is of great practical importance in efficient VM migration. While there have been extensive studies on energy reduction using (live) VM migration (or server consolidation), e.g., [192], [674], [402], the understanding of the effectiveness of VM migration in terms particularly of VM-level energy efficiency is relatively limited. It is only recently that some noticeable studies have been conducted, e.g., Joule meter [401], [409].

In the server (VM) consolidation approach, virtualization technologies play a key role. These technologies with the prevalence of many-core processors have greatly eased and boosted parallel processing; that is, those many-core processors can effectively provide resource isolation to VMs. Server consolidation in [674] is approached using the traditional bin-packing problem with two main characteristics, i.e., CPU and disk usage. The algorithm proposed in [674] attempts to consolidate tasks balancing energy consumption and performance on the basis of the Pareto frontier (optimal points). Two steps incorporated into the algorithms are: (1) the determination of optimal points from profiling data, and (2) energy-aware resource allocation using the Euclidean distance between the current allocation and the optimal point at each server.

In [409], a utility analytic model for Internet-oriented server consolidation is proposed. The model considers servers being requested for services like e-books database or e-commerce web services. The main performance goal is the maximization of resource utilization to reduce energy consumption with the

---

[8]https://www.vmware.com/products/drs/

same quality of service guarantee as in the use of dedicated servers. The model introduces the impact factor metric to reflect the performance degradation of consolidated tasks.

Server consolidation mechanisms developed in [711], [533] deal with energy reduction using different techniques, especially [711]. Unlike typical server consolidation strategies, the approach used in [711] adopts two interesting techniques, memory compression and request discrimination. The former enables the conversion of CPU power into extra memory capacity to allow more (memory intensive) servers to be consolidated, whereas the latter blocks useless/unfavorable requests (coming from web crawlers) to eliminate unnecessary resource usage. The VirtualPower approach proposed in [533] incorporates server consolidation into its power management combining "soft" and "hard" scaling methods. These two methods are based on power management facilities (e.g., resource usage control method and DVFS) equipped with VMs and physical processors, respectively.

A recent study on VM consolidation [402] has revealed (or confirmed) that live VM migration may even lead to significant performance degradation; and thus, VM migration should be modeled explicitly taking into account the current workloads (i.e., application characteristics) and their changes.

17.3.3 Management Level

In the management level, the resource management system may direct applications (or their components) to the servers most suitable to run them, as long as the quality of service requirements can be satisfied.

In the context of energy efficiency in clouds, load imbalance/flux is a serious cause of inefficient resource use, resulting in poor energy proportionality to performance. Energy efficiency in this level can be addressed using the existing scheduling and resource allocation strategies since energy efficiency can possibly be considered as an additional objective of those strategies. Energy efficiency may be translated and incorporated into part of the utility function of those strategies. However, the dynamic and heterogeneous nature of clouds hinders the direct application of the existing strategies to the cloud. Also, users in the cloud pay for services they request; this further complicates such a direct application. Therefore, in the (public) cloud environment energy efficiency solutions may be incorporated into scheduling and resource allocation; and additional SLAs metrics (e.g., response time) other than the current "availability" metric adopted by most cloud service providers, if not all should be explicitly taken into account.

Alternatively, some more flexible scheduling and resource allocation strategies (e.g., Amazon Spot instances) can be adopted to more effectively manage workload flux. In other words, applications/workloads can be better profiled and scheduling decisions are made based on such a profiling, e.g., alternating workload surges shown in Figure 17.3.

In [461], a three-tier cloud structure—consisting of service consumer, service provider and infrastructure provider—is adopted to more practically model dynamics of the cloud. Although the scheduling algorithms in [461] do not explicitly consider energy efficiency, resource utilization obtained using them can be a good indicator of their capability to achieve good energy efficiency. What's more, those algorithms essentially exploit application characteristics in order to optimize their performance metrics.

In the management level, service level agreements may play a key role in resource allocation. In other words, resource allocation decisions can be made in an energy-efficient manner as long as performance requirements in SLAs can be met.

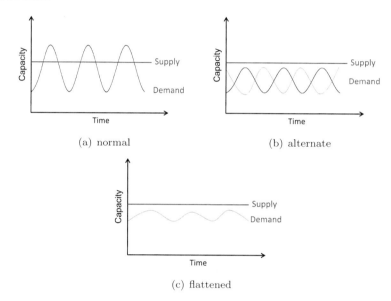

FIGURE 17.3
Workload Alternation.

17.4   Energy Efficient VM Consolidation

In this section, we investigate VM consolidation using the Xen hypervisor [118], [117] and its impact on performance and power consumption. Specifically, the power management (PM) facility in the Xen hypervisor and the energy efficiency of its built-in power management (PM) governors—i.e., power-save, ondemand, performance and userspace—are scrutinized and preliminary results obtained from our study are presented.

17.4.1 Xen Hypervisor

Xen is a VMM or hypervisor to enable multiple VMs to run on a single hardware system of the hosting physical machine (Figure 17.4). The Xen hypervisor acts as a software layer positioned between hardware and virtual machines. In Xen terminology, virtual machines are called domains. There are two types of domain guests: domain0 (dom0) and domainU (domU). Dom0 is a special domain, also known as privileged domain, which is the fundamental element of Xen's functionality; the hypervisor only operates on dom0. domU refers to a production or usable virtual machine with a guest operating system managed by dom0. dom0 hosts the network and I/O drivers for all domUs. The administration and management of the entire environment is also controlled via dom0. domUs have no direct access to hardware (disks, memory and CPU); hence, they are unprivileged domains.

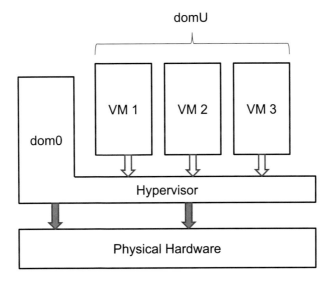

FIGURE 17.4
Abstract Xen Architecture.

17.4.2 Xen Power Management (xenpm)

Many recent microprocessors support multiple performance and idle states for efficient power management. Xenpm as the PM subsystem of Xen enables hypervisor-level frequency scaling using cpufreq and cpuidle. Xenpm controls the CPU frequency and idle states via the Xen hypervisor as a measure of energy saving. The (Linux) kernel level PM has been in existence for the last five years. What is special about Xenpm is the integration of existing kernel-level PM features with Xen; that is, PM control can be performed via the hypervisor.

17.4.2.1   Power States

There are three power state categories of microprocessors: G, C and P states. The first two control sleep states while the last one controls performance states.

- Global (G-states) or System (S-states): ACPI (an open standard for OS-directed device configuration and power management) defines these states in its specifications. States in this category define whether the system is on, soft off, or mechanically off [117].

- Processor States (C-states): G-states at the core processor level are managed by C-states. The C states are defined as C1, C2, etc. The larger the C-state index, the more energy efficient; for example, C4 saves more energy compared with C3. When the system is in active state usually it is on C0.

- Performance States (P-states): processor frequency and supply voltage may change depending on P-states. There are one or more P-states available on a processor (e.g., P0, P1, P2, etc). Similar to C-states, the lowest P-state (P0) runs the highest frequency while the highest P-state (e.g., P11 in Intel Xeon X3440) runs the lowest frequency. Processor frequency is directly proportional to the supply of voltage and vice versa.

17.4.2.2   Xempm Features

Xenpm as a power management facility was included in XEN hypervisor version 3.4 and later releases. C-states are maintained via the cpuidle driver while cpufreq controls P-states. It has the following features [117].

- Better support to deep C-states with ACPI timer stop

- More efficient cpuidle "menu" governor

- More cpufreq governors (performance, userspace, powersave and ondemand) and drivers supported

- Enhanced xenpm tool to monitor and control Xen power management activities

- MSI-based HPET (High Precision Event Timer) delivery, with less broadcast traffic when CPUs are in deep C-states

- Power aware option for credit scheduler (`sched_smt_power_savings`)

- Timer optimization for reduced break events (range timer, vpt align)

17.4.2.3 Cpufreq and Cpuidle

The cpufreq driver is in charge of dynamically scaling processor frequency. The decision on a frequency change is made via the cpufreq governors. Specfically, the cpufreq driver contains four default power management governors. The Xenpm utility could change the governors on real-time. The governors also could set per physical core.

- Performance – Selects the highest frequency and operates on the highest performance level. (When an operation arrives on the queue it selects the highest frequency, completes the operation, and drops back to the lowest frequency).

- Ondemand – Selects the best frequency to operate (it increases the frequency as the load on CPU increases and vice-versa).

- Powersave – Operates on the lowest frequency.

- Userspace – Allows the user to change processor frequency.

These governors can be changed via the `xenpm set-scaling-governor` command. If there are many cores, each core may be set to operate on a separate governor. The cpufreq driver could be controlled via either dom0 or hypervisor. The number of virtual CPUs should not exceed the physical CPUs and dom0/vcpu (virtual cpu) should be pinned to a physical processor to enable dom0 based xenpm. However, the hypervisor based xenpm has no such restriction and has more control. The cpufreq governor has three components as follow:

- cpufreq core – The cpufreq core would control the clock speed of processor which affects the energy consumption. The logic is built into the core.

- cpufreq governor – There are four governors built into the system, which has different algorithms to change the processor frequency.

- cpufreq processor driver – The driver has the ability to change the frequency of the processor.

C states are controlled by the cpuidle driver. By default the cpuidle states are enabled on Xen version 4. The `get-cpuidle-states` command could be used on xenpm to get the information on C states.

17.4.3 VM Consolidation with Xenpm

We have conducted a series of experiments to identify energy consumption (efficiency) with VM consolidation in various power settings (power management governors).

17.4.3.1   Experiment Environment

Xen was selected as the experiment hypervisor due to its openness and built-in power management feature. Debian GNU/Linux squeeze with 2.6.32-5-xen-686 kernel was utilized as the operating system for both Dom0 and DomU guest operating systems; by default Xen version 4 comes with operating system distribution.

The experimental system was built around an IBM system x3250 M3 which includes an Intel Xeon X3440 with 8MB Cache, 2.53 GHz processor, 4GB DDR-3 ECC memory with SATA hard drive attached and which also includes a 350W power supply.

Raritan Dominion PX, Model: PX (DPXR8A-16), power distribution (PDU) was engaged to read the power meter readings. It comprises IP connectivity that facilitates remote controlling and monitoring. The power meter reading was read remotely via a separate server.

Simple network management protocol (SNMP) was used to gather all related information, such as processor load, memory utilization and PDU reading. A separate server was used for data gathering, which reduces the performance impact on the experiment servers [535]. The overview of our test environment is shown in Figure 17.5.

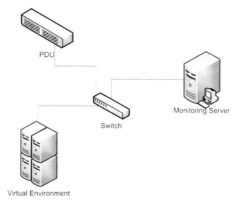

FIGURE 17.5
Test Environment.

17.4.3.2   Performance Metric: Energy Consumption

Energy can be defined as capacity to work and can exist in various forms such as kinetic, thermal, chemical, electrical, and nuclear. Each form of energy may convert to another. The unit of energy in the International System of Units (SI) is defined as a Joule [754]. Most electrical devices including computers are rated in watts. A watt is defined as a unit of power in SI which is also equal to one joule of work performed per second, or equivalent to the power dissipated in an electrical conductor carrying one ampere current between points at one volt potential difference; more formally, $W = v \cdot i$ where W is watt, v is volt and i is ampere. Then, one joule is defined as $J = 1W \cdot s$ where $W \cdot s$ is one "watt second."

17.4.3.3 Preliminary Experiments: Idle Power

Our experiment equipment indicates 65W as the idle power usage. The measurement was carried out at the state in which the system was not performing any active operations and all the CPUs are operating on the lowest frequency (P11, C3). We introduced a few idle VMs to measure the power reading fluctuation; however, there was no noticeable difference in the reading.

When the system was operating on its highest performance level, the power meter was reporting 133W. According to the above findings, 48.8% of the power was drawn to maintain the idle state.

When the system was powered down via the operating system, the system was merely consuming 5W according to the power meter reading, so 45% power saving could be achieved if the system was powered down rather than idle; however, powering up a system is a time and power consuming operation [507].

17.4.3.4 Preliminary Experiments: Power Usage (Watts) with Respect to Different Processor Frequencies

The following experiment was conducted to identify the relationship between the frequency and power usage. The frequency was increased at a predefined fixed time interval. Since the userspace governor in Xenpm has no logic to control the frequency (i.e., it allows us to alter the frequency), our experiment was carried out using that power governor. For example, the command, xenpm set-scaling-speed 2394000 sets the frequency for all the CPUs at 2394000. Our results (Figure 17.6) confirmed that power consumption is directly related to the processor P-state. Specifically, the power usage increases from 70W to 130W when the P-states change from P11 to P0.

17.4.3.5 Preliminary Experiments: Power Usage with Respect to Different Governors

It is important to thoroughly characterize the energy usage of default power governors (performance, ondemand and powersave) in Xenpm in order to help VM consolidation decisions. Specifically, the trade-off between performance and energy consumption in different power governors can be identified. For each of those three different governors, we have first measured the idle power consumption (i.e., VMs are idle) and their energy consumption with some workload. Table 17.2 shows idle power consumed by the three power governors that have some frequency control logic. The powersave and ondemand governors reported the same frequency of 1188160 kHz; that is, they both operate on the lowest P-state (P11) while the system is idling. However, the

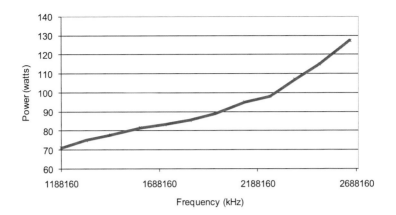

FIGURE 17.6
Average Power with Respect to Different Processor Frequencies.

ondemand governor would scale the frequency according to the load, unlike powersave that always uses the lowest frequency (or P-state). In the meantime, the performance governor has a slight change in frequency. It operates only on two frequencies, either P11 or P0. Since those figures in Table 17.2 are averages of our measurements the frequency that the performance governor reported is higher than the other two, i.e., the average of frequencies in P11 and P0.

TABLE 17.2
Idle Power with Respect to Different Default Power Governors

Governor	Ave CPU Freq (kHz)	Power reading (W)
Powersave	1188160	65.92308
Ondemand	1188160	65.72727
Performance	1619816	66.4

Now, we report the energy consumption on those three power governors with a certain workload (Figure 17.7 and Table 17.4). Although the energy consumption on the performance and powersave governors is easily predictable, it is still important to confirm such predictability. Unlike these two governors, the ondemand governor behaves a little unpredictably; and thus, the major focus in the following experiment. The experiment was conducted using a single VM with four virtual cores and 128MB of memory. A power governor is set by the Xenpm command `xenpm set-scaling-governor [cpuid]`. The VM was loaded via mysqlslap [529]. All four cores were utilized while the test was running (Table 17.3); the load is distributed evenly across all the cores.

Specifically, 1,000 queries were executed with four concurrent clients and the experiment was repeated five times. The actual command used is as follows:

```
mysqlslap -u root -padmin123 -auto-generate-sql
-number-of-queries=1000 -concurrency=4 -iterations=5
-number-char-cols=54 -number-int-cols=16
```

TABLE 17.3

Workload Distribution

CPU	%usr	%nice	%sys	%iowait	%irq	%soft	%steal	%guest	%idle
all	28.06	0.00	22.23	0.00	0.00	0.00	0.39	0.00	49.32
0	28.97	0.00	23.41	0.00	0.00	0.00	0.40	0.00	47.22
1	29.92	0.00	22.83	0.00	0.00	0.00	0.00	0.00	47.24
2	27.52	0.00	22.09	0.00	0.00	0.00	0.39	0.00	50.00
3	26.14	0.00	20.45	0.00	0.00	0.00	0.38	0.00	53.03

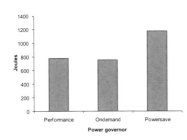 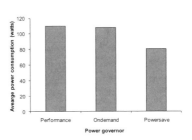

(a) total energy consumption in joules (b) average energy consumption in watts

FIGURE 17.7

Energy Consumption with Respect to Different Power Governors.

Powersave is found to be the worst performing governor in terms of both energy consumption and performance, since energy usage (Joules) is directly related to the amount of time spent to process the workload. The ondemand governor changes P-state "gradually" to the maximum state to complete the request efficiently, drops it to the lowest state, and continues in the idle state until another request arrives. Ondemand and performance governors have no noticeable difference as per the experiments since the transition between idle state and active state does not cost much time. We have observed that after completing the requests, higher power consumption continues for a while. It is quite evident that fans within the system could cause this type of power consumption due the heat generated by the components within the server. Based on these results, the performance governor may be preferable to the other power governors in that it delivers better energy efficiency directly and indirectly. In other words, idle power consumption can be much reduced by shutting down the idling server or at least putting it into a deep sleep mode.

TABLE 17.4
Energy Consumption and Performance with Respect to Different Power Governors

Governor	Performance (sec)			Energy consumption		
	average	max	min	average (W)	idle (W)	total Joules
Performance	7.07	7.19	6.99	110.20	68	778.78
Ondemand	6.98	7.06	6.89	108.10	68	754.43
Powersave	14.52	14.58	14.43	80.86	68	1174.41

17.4.3.6   Results: Energy Efficiency in VM Consolidation

To investigate the power saving factor of VM consolidation, experiments with eight different numbers of VMs from 1 to 8 were carried out. Each VM ran the same workload (operation) and the time was measured as a performance indicator. The operation was repeated twelve times and resulting completion times were averaged. Results are presented in Figure 17.8.

Each VM is configured to have a single virtual core. The resource allocation of a VM is performed using Xen's default credit scheduler, which allocates and limits resources per VM via cap and weight values. Each VM in our experiments had the same weight and cap values (creditScheduler parameters). And, 128MB of memory and 3GB of separate storage space per VM are set. A single processor was assigned to dom0 and pinned to core 0. The only active operation is VM1 with processor affinity 1 to 3; this prevents the VM from accessing the physical core assigned to dom0 and avoids any performance impact on dom0 (dom0 is just an idle VM). With our experiment system, the power meter reading when idling was 68W, whereas it was 89W when loaded with a VM. The number of VMs was increased gradually and the maximum reading was reported as 117W (Table 17.5). The maximum power meter reading was recorded with three or more VMs; that is, as per the configuration all three physical cores were utilized (the remaining core is assigned to dom0).

As the number of VMs increases, total joules increase; however, joules per VM decrease; this leads to the fact that power consumption can be reduced by accommodating more VMs (higher consolidation density). Clearly, there is a practical limitation when considering performance. The maximum power usage was recorded when the system was in operation with over six VMs. Power consumption per VM reaches the optimal reading when the system reports its maximum power usage. Overloading physical resources certainly impacts performance and energy consumption per VM. Note that scalability and resource isolation are beyond the scope of our study.

TABLE 17.5
Energy Consumption with Respect to Different VM Consolidation Densities

#VMs	W	Average Time	Joules	Joules/VM
0	65	-	-	-
1	89	31.95	2843.55	2843.55
2	102	32.72	3337.44	1668.72
3	117	34.09	3988.53	1329.51
4	117	35.51	4154.67	1038.67
5	117	36.15	4229.55	845.91
6	117	38.07	4454.07	742.35
7	117	38.23	4472.91	638.99
8	117	38.29	4479.93	559.99

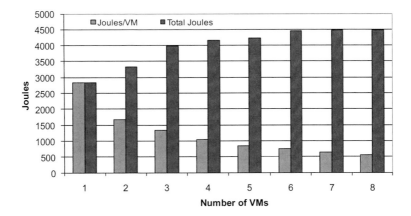

FIGURE 17.8
Energy Reduction in VM Consolidation.

17.5 Summary and Conclusion

In this chapter, we introduced energy efficiency issues in LDCSs, reviewed and discussed several alternatives to such issues with an example model in VM consolidation using the power management facility of Xen (Xenpm). Idle power consumption is approximately 50% of the total power consumption of a server; and this raises various issues particularly in LDCSs. VM consolidation is a viable energy efficiency solution in this regard. We investigated the energy efficiency of Xenpm (more specifically, its power governors) for VM consolidation. It has been identified that the best practice of improving energy efficiency using Xenpm might be the use of the highest performance state (P-state) when processing workloads. In other words, the system may be put into a near sleep state to reduce idle power. We have conducted a set of experiments using the built-in power governors in Xenpm with cpufreq and cpuidle subsystems and observed that the performance governor is most energy efficient when considering the trade-off between performance and energy consumption. Based on our experimental results, we conclude that the increase in VM consolidation density is an effective way of improving energy efficiency as long as the performance remains at an acceptable level.

18

Security, Privacy and Trust Management Issues for Cloud Computing

Sara Kadry Hamouda

University of East Anglia, UK, Department of Computer Science

John Glauert

University of East Anglia, UK, Department of Computer Science

CONTENTS

18.1   Chapter Overview

After decades of engineering development, Internet connectivity has become
a commodity product in many countries. The rapid adoption of Web 2.0
technologies such as blogging, online media sharing, social networking, and
web-based collaboration has moved enormous quantities of data onto Internet
servers. Along with this migration to web services has come a push for companies to adopt utility computing. Much like traditional infrastructure utilities,
such as gas and electricity, utility or "cloud" computing seeks to abstract the
supply of computing services from concerns of everyday users.

Today, security and privacy concerns may represent the biggest hazards to
moving services to external clouds. With cloud computing, data is stored and
delivered over the Internet. The owner of the data does not control, and typically does not even know, the location of the data. There is a real chance that
the owner's data could rest on the same resources as a competitor's application and data. In this chapter we focus on how to secure cloud computing. It
is specifically structured around three broad topics:

• Cloud Security

• Cloud Privacy

• Trust Management

18.2   Introduction

Cloud Computing has been envisioned as the next-generation information technology (IT) architecture for enterprises, as it exploits a long list of unprecedented advances in the history of IT: on-demand self-service, everywhere network access, location independent resource pooling, rapid resource flexibility, usage-based pricing and transmission of risk [721]. The field of cloud computing is still young in terms of implementation and usage, partially because it is heavily tied to technology progress and is so resource dependent that researchers in academic institutions have not had much opportunity to analyze and experiment with it [739]. However, cloud computing develops from the ambition of IT engineers to add another layer of separation in processing information. A deep understanding of cloud computing relies on the following notions: grid computing, utility computing, software as a service (SaaS), storage in the cloud and virtualization [721].

Cloud computing providers claim they are responding to customer demands for guarantees on the security of the services they provide. Some even claim that, because they know that their customers place high priority on the security of the data they own, the security of the cloud services they offer is often significantly better than that of the customer's own IT systems.

The security issues with cloud computing do not vary enormously from those facing users in any other computing environment [739]. The classic problem can be epitomised by the acronym *CIA*, representing; *Confidentiality* of data, *Integrity* of data, and *Availability* of data. Once data is out in the cloud, only users who are authenticated and authorized should be able to see the company data, ensuring confidentiality [395]. For every piece of information that is considered confidential or sensitive in some way, it is necessary know where it is stored, who has been looking at it, under what conditions they have been accessing it, and be provided with an audit trail, so that if something unplanned happens, the owner can track its cause [721].

Security is one key requirement to enable privacy. This principle specifies that personal data should be protected by reasonable security safeguards against such risks as loss, or unauthorized access, use, modification, destruction, or disclosure of data. Cloud computing changes the way we think about computing by changing the relationship of location with resources. Because cloud computing represents a new computing model, there is much uncertainty about how to achieve security at all levels (e.g., network, host, application, and data levels) [395].

18.3   What Is Cloud Computing Security?

Control of security in cloud computing is not fundamentally different from security control in any IT environment. However, because of the cloud service models employed, their operational models, and the technologies used to enable cloud services, cloud computing may introduce different risks to an organization than traditional IT solutions [738].

18.3.1   What Cloud Computing Security Is Not

- Cloud computing security is IT responsibility to secure the cloud for all customers, including enterprise security [128].

- Security as a Service (SaaS) or outsourcing management to a third party.

- It is not about securing the cloud itself. Cloud computing security is an IT procedure. A secure cloud is important to enable security cloud computing.

18.3.2   What Cloud Computing Security Is

- Utilizes the cloud for security applications such as identity management, access control.

- Leverages a service model decreasing capital disbursement and operating disbursement uncertainty.

- Enhances security systems' performance while decreasing cost related to infrastructure and technical staff item decreasing efficiency and effectiveness of security applications at the enterprise level.

18.3.3   Cloud Computing Security Fundamentals

Confidentiality

Confidentiality refers to the prevention of intentional or unintentional unauthorized disclosure of information. Confidentiality in a cloud system is related to the areas of intellectual property rights, covert channels, traffic analysis, encryption, and inference.

Integrity

The concept of cloud information integrity requires that the following three principles are met:

- Changes are not made to data by unauthorized personnel or processes.

- Unauthorized changes are not made to data by authorized personnel or processes.

- The data is internally and externally consistent — in other words, the internal information is consistent both among all sub-entities and with the real-world, external situation.

Availability

Availability ensures the reliable and timely access to cloud data [320]. Availability guarantees that the systems are functioning properly when needed. In addition, this concept guarantees that the security services of the cloud system are in working order. A denial-of-service attack is an example of a threat against availability.

The converse of confidentiality, integrity, and availability is disclosure, alteration, and removal.

Auditing

To control operational assurance, organizations use two basic methods; system audits and monitoring. These methods can be employed by the cloud customer, the cloud provider, or both, depending on asset architecture and deployment.

- A system audit is a one-time or recurrent event to evaluate security.

- Monitoring refers to progressive activity that examines either the system or the users, such as attack detection.

18.3.4   Cloud Security Advantages

- Fault tolerance and reliability

- Low cost disaster recovery and data storage solutions [738]

- Hypervisor protection against network attacks

- Data partitioning and replication

- Improved resilience

18.3.5   Cloud Security Disadvantages

- Need to trust the provider's security model

- Loss of physical control

- Inability to examine proprietary implementations

- Inflexible support for monitoring and auditing

18.3.6   Taxonomy of Security

We classified the security concerns as:

- Traditional security

- Availability

- Third-party data control

Traditional Security

Concerns involving computer and network intrusions or attacks that will be made possible or at least easier by moving to the cloud [128]. Cloud providers respond to these concerns by arguing that their security measures and processes are more mature and tested than those of the average company. If companies are worried about insider threats, it could be easier to lock down information if it is administered by a third party rather than in-house. Moreover, it may be easier to force security via contract with online service providers than via internal controls.

Availability

Concerns centering on critical applications and data being available. As with the traditional security concerns, cloud providers argue that their server uptime compares well with the availability of the cloud user's own data centers [389]. Besides just services and applications being down, this includes the concern that a third-party cloud would not scale well enough to handle certain applications.

Third-Party Data Control

Concerns the legal status of data being held by a third party: implications are complex and not well understood [320]. There is also a potential lack of control and transparency when a third party holds the data.

18.3.7   Security Benefits of the Cloud

Data Centralized

How many backup tapes? Total data size and insecure replication could be decreased by the cloud as thin client technology becomes pervasive. Small, temporary caches on mobile devices or netbook computers pose less risk than transporting core data via laptops. The advantages of thin clients can be realized today but cloud storage provides a way to centralize the data faster, more consistently, and potentially cheaper.

Incident Response

Infrastructure as a Service (IaaS) providers make it possible to build a dedicated forensic server in the same cloud used by the company and maintain it offline, ready for use when needed.

Decrease Time to Access Protected Documents

Access to temporary CPU power via cloud computing resources opens some doors [738]: If a suspect has password protected a document that is significant to an investigation, it is feasible to test a wider range of candidate passwords in less time.

Password Assurance Testing

If your organization regularly tests password strength by running password crackers you can use cloud computing resources to decrease crack time and you only pay for what you use.

Improve Log Indexing and Search

By placing your logs in the cloud, you can leverage cloud computing resources to index logs in real-time and gain the benefit of instant search results.

Ease of Testing of the Impact of Security Changes

Through IaaS, create a copy of your production environment, implement a security change and test the impact at low cost, with minimal start-up time. This removes a major barrier to developing security architecture in production environments [320].

Reduce Cost of Security Testing

A SaaS provider only passes on a portion of their security testing costs [128]. By sharing the same application as a service, you do not bear the full cost of an expensive security code review. Even with Platform as a Service (PaaS) where your developers have to write code, there are possibile cost economies of scale.

18.4   Cloud Computing Security Scenarios

Cloud computing is available in many service models (and hybrids of these models). Each delivers different levels of security management [389].

18.4.1   Cloud Security Levels

Software as a Service (SaaS) Model

- May be customized by the user [575].

- Places most of the responsibility for security management on the cloud provider.

- Provides ways to control access to the Web portal, such as the management of user identities, application level modification, and the ability to constrain access to specific IP address ranges or geographies.

Platform as a Service (PaaS) Model

- Refers to application development platforms where the development tool itself is hosted in the cloud and accessed and deployed through the Internet.

- Allows clients to assume more responsibility for managing the configuration and security [575] for middleware, database software, and application runtime environments.

The Infrastructure as a Service (IaaS) Model

- Provides fully scalable computing resources such as CPU, and storage infrastructure.

- Transfers responsibility for security is from the cloud provider to the client [575].

- Provides full access to the operating system that maintains virtual images, networking, and storage.

18.4.2   Cloud Security Scenarios

Security of cloud computing can be viewed through numerous views. One view is the perspective of stakeholders. Each of the stakeholders has their own security interest and aims. These security objectives are associated with specific services that they provide or consume. We note that a cloud is not

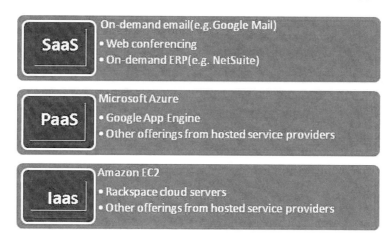

FIGURE 18.1
Cloud Computing Models.

a single entity; it can contain one or more open clouds, internal clouds or external clouds. There is no doubt that new security risks derive from the communication that goes on among cloud consumers and cloud providers; and between cloud providers and cloud infrastructure providers [610]. In order to understand this phenomenon as well as the cloud computing security concerns, we will consider the following three scenarios.

Scenario 1

An airport inspector PhotoScan, a cloud end-user, consumes computing services offered by a cloud service provider CloudA. The service is to process the traveller' digital personal photo. The confidentiality and integrity of photos are very important for PhotoScan, but are beyond its control; rather, CloudA is responsible for all these. Further, it may be that the huge numbers of the pictures are stored on several physical storage facilities owned and managed by a cloud infrastructure provider, CloudB. PhotoScan wants to know from CloudA about its specific security concerns: How

(i) confidentiality of photos is kept,

(ii) integrity is preserved, and

(iii) read-write access of others to its data is controlled.

How these are ensured at different levels of abstraction in relation to the service it consumes from CloudA.

Scenario 2

A second cloud end-user, a UK government border agency GovBorder, uses the same cloud service provider CloudA's computing services to process various photos such as personal visa photos. The confidentiality of the photos is not important for GovBorder because all photos are publicly available. However, the integrity of photos is very important for GovBorder but beyond its control; rather CloudA is responsible for protecting the photos. CloudA stores some of the photos in the devices owned and managed by the cloud infrastructure provider, CloudB. GovBorder does not care who can see and copy its photos, but it wants to know from CloudA about its specific security requirements: How

(i) integrity of its images is preserved;

(ii) authentication of CloudA is maintained, and

(iii) non-repudiation is ensured for all communications between GovBorder and CloudA.

In other words, the photos should not be allowed to be tampered with or modified by any unauthorized entities (integrity), and CloudA cannot claim later to GovBorder that it did not perform the operation.

Scenario 3

AppDev, an application developer, uses a software development platform offered by CloudA. AppDev composes software with some of its existing software components, new components and utilities provided by CloudA. The main two concerns of AppDev are:

(i) The security requirements of its application software are complied with by the security assurances of the new components and utilities available from CloudA's platform. It likes to know how the security of properties of components and software, which is supported by the provisions relating to current security provided by CloudA;

(ii) What are the guarantees demanded by the security of software and hardware facilities provided by CloudA that will always hold.

18.5   Cloud Security Challenges

One of the most important characteristics of cloud computing is that it offers "self-service" access to computing power, most likely via the Internet. In traditional data centers, administrative access to servers is controlled and restricted to direct or on-premise connections. In cloud computing, this administrative access must now be conducted via the Internet, increasing exposure and risk [610]. It is extremely important to restrict administrative access and monitor this access to maintain visibility of changes in system control.

18.5.1 Dynamic Virtual Machines

Virtual Machines (VMs) are dynamic. They can quickly and easily be reverted to previous instances, paused and restarted. They can also seamlessly move among physical servers [459]. This vital flexible organization of a network of VMs makes it difficult to achieve and maintain consistent security. Vulnerabilities or configuration errors may be unknowingly propagated [610]. Also, it is difficult to maintain an auditable record of the security state of a virtual machine at any given point in time. In cloud computing environments, it will be necessary to be able to prove the security state of a system, regardless of its location or proximity to other, potentially insecure, virtual machines.

18.5.2 Vulnerability Exploits and VM to VM Attacks

Servers of cloud computing use the same operating systems, enterprise and web applications as static virtual machines and physical servers. The ability for an attacker or malware to exploit remotely vulnerabilities in these systems and applications is a significant threat to virtualized cloud computing environments. In addition, co-location of multiple virtual machines increases the attack surface and risk of VM-to-VM compromise [459]. Invasion detection and prohibition systems need to be able to detect malicious activity at the virtual machine level, regardless of the location of the VM within the virtualized cloud environment.

18.5.3 Securing Offline Virtual Machines

In contrast to a physical machine, when a virtual machine is offline, it is still available to any application that can access the virtual machine storage across the network, and is therefore vulnerable to malware infection. However, offline VMs do not have the capability to run an anti-malware scan agent. Dormant virtual machines may exist not just on the hypervisor, but can also be backed up or archived to other servers or storage media. In cloud computing environments, responsibility for protecting and scanning offline machines remains with the cloud provider [790]. Enterprises using cloud computing should look for cloud service providers that can secure these dormant virtual machines and maintain cohesive security in the cloud.

18.5.4 Performance Impact of Traditional Security

Current content security approaches evolved before the concepts of virtualization and cloud computing, and were not designed for use in cloud environments. In a cloud environment, where virtual machines from different tenants share hardware resources, concurrent full-system scans can cause debilitating performance degradation on the underlying host machine. Cloud service providers providing a baseline of security for their hosting clients can address this problem by performing resource-intensive scans at the hypervisor level thereby eliminating this contention at the host level [505].

18.6 How to Handle Cloud Security Challenges

The challenges of cloud computing are very similar to those of any other IT architecture. Like internal IT, cloud providers are subject to internal and external threats that can be mitigated [505].

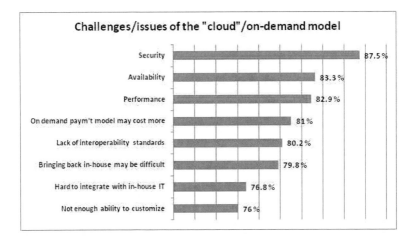

FIGURE 18.2
Results of IDC Survey Ranking Security Challenges.

Multi-Tenancy

The cloud provider builds its security to meet the higher-risk client, and all of lower-risk clients then receive better security than they would have otherwise. On the other hand, a client may be exposed to a greater level of external risk due to the business practices of the other tenants [790]. When Information

Technology is managed within an organization, risk is largely for the organization alone to bear.

Shared Risk

If a Software as a Service (SaaS) provider needs infrastructure, it may make more sense to acquire that infrastructure from an Infrastructure as a Service (IaaS) provider rather than building it. Thus cloud service provider tiers are built by layering SaaS on top of IaaS. In this type of multi-tier service provider arrangement, each party shares the risk of security issues because the risk potentially impacts all parties at all layers.

Distributed Data Centers

In theory, a cloud computing environment should be less prone to disasters because providers can provide an environment that is geographically distributed. Organizations that subscribe to cloud computing services that are not geographically distributed should require their provider to have a working and regularly tested disaster recovery plan [575].

Physical Security

Physical external threats should be analyzed carefully when choosing a cloud security provider. Do all of the cloud providers facilities have the same levels of security? Is the provider advertising its most secure facility with no guarantee that your data will actually reside there?

Coding

Every customer should make sure that the cloud provider follows secure coding practices. Also, all code should be written using a standard methodology that is documented and can be demonstrated to the customer.

Data Leakage

The cloud provider should have the ability to map its own policies to the security framework a client must comply with, and to discuss the issues. At a minimum, data under consideration should be encrypted while in flight and at rest. The cloud provider also needs to have a policy that feeds into the client security incident policy to deal with any data leakages that might happen.

18.6.1 Cloud Computing Risks for the Enterprise

Some instances of cloud computing risks for the enterprise that needs to be achieved include:

- Enterprises need to be careful in choosing a provider [459]. Prominence, history and stability, should all be factors to consider. Stability is of particular importance to ensure that services will remain available and data can be tracked.

- The cloud provider generally takes responsibility for information controlling, which is a critical part of the business. Failure to acheive agreed-upon service levels can affect not only confidentiality but also availability, severely affecting business operations.

- The dynamic nature of cloud computing may result in confusion as to where information actually resides. When information recall is required, this may create delays.

- Third-party access to sensitive information creates a risk of compromise to information confidentiality. In cloud computing, this can pose a significant threat to ensuring the protection of intellectual property (IP) and trade secrets [505].

- Public clouds enable high-availability systems to be developed at service levels often impossible to create in private networks at acceptable cost. The downside to this availability is the potential for intermingling of information assets with those of other cloud customers, including competitors. Compliance to regulations and laws in different geographic regions can be a risk for enterprises. At this time there is little legal precedent regarding responsibility in the cloud. It is critical to obtain legal advice to ensure that the contract specifies the areas where the cloud provider is responsible and liable for branches arising from potential issues [459].

- Due to the dynamic nature of the cloud, information may not immediately be located in the event of a disaster. Business continuity and disaster restoration plans must be well documented and tested. The cloud provider must know the role it plays in terms of replication, occurrence response and restoration. Restoration time objectives should be stated in the contract.

18.6.2   Assessing the Security of a Third Party Cloud Provider

Defining Security Requirements

A customer's information security requirements are derived from the organization's own policy, legal and regulatory obligations, and may carry through from other contracts the company has with its own customers [575].

Managing Cloud Supplier Risks

The outsourcing of key services to the cloud may require customer organizations to seek new and more mature approaches to risk management and accountability. While cloud computing means that services are outsourced, the risk remains with the customer and it is therefore in the customer's interest to ensure that risks are appropriately managed according to their risk exposure. Effective risk management also requires maturity both in provider relationship management processes and operational security processes [790].

18.6.3   Cloud Security Issues

- Data location: depending on contracts, some clients do not know in what country or where data is stored.

- Restoration: every provider should have a disaster restoration protocol to protect user data.

- Inquisitive support: if a client suspects faulty actions by the provider, it may have few legal ways continue an enquiry.

- Data isolation: encrypted information from multiple companies may be saved on the same hard disk, so a mechanism to isolate data should be deployed by the provider [505].

- Privileged user access: information transfer from the client over the Internet presents a certain degree of risk, because of issues of data ownership; enterprises should spend time getting to know their providers and their regulations as much as possible, ideally assigning some trivial applications first.

- Regulatory compliance: clients are responsible for the security of their solution, as they can select between providers that allow audits by third party organizations that test levels of security, and providers that do not.

18.6.4   Keeping Cloud Server Infrastructure Secure

First, we have to secure any device connected to the open Internet from unauthorized access; that is a common risk and is faced equally by any company and customers. Secondly we have the job of securing any data in transit over the network and finally we have the important task of ensuring dependable high quality networking to the cloud. The challenge of keeping the infrastructure platform secure and controlling access to any device that is connected to the open Internet is a common one and standard solutions apply.

- All devices holding "out of cloud" data (such as desktop computers, laptops etc.) must have fully encrypted hard drives.

- All operating systems must be kept patched with current security upgrades.

- Restrict access to only those needing access on a per device basis. Maintain good records of user access permissions.

- Unauthorized login attempts should be logged and monitored.

The simple discipline of having a secure password, changing that password often and providing different passwords for different devices goes a very long way to securing infrastructure that is openly accessible to the Internet. A great many security systems are created to get around the fact that people usually choose insecure passwords. It is safest to combat this effect by ensuring that everyone has very secure passwords, and work from there.

For any customer requiring a multi-server environment it is advisable to use private networking features. By adding servers to a virtual LAN, they have a second networking card added to that server. If demanded, the public networking interface can then be disabled. In this way, users can establish private networks within the cloud, not accessible via a public IP address [459]. Many customers create private clusters that sit behind gateway servers. The aim is that the cloud provides full control to create and secure organizational infrastructure in a way akin to dedicated hardware.

18.6.5 Securing Your "Data in Transit" within the Cloud

A key challenge for the cloud and Infrastructure as a Service in particular, is maintaining sufficient separation between different users on a shared infrastructure. This immediately affects the security of users of the cloud when it comes to keeping "data in transit" secure. It is important that one user is unable to view the Internet traffic of another user. This is the networking risk in a multi-tenant environment [575].

With open networking it is possible for users of the cloud to create secure encrypted VPN connections among their cluster in the cloud and corporate infrastructure. This solution is used by some large clients and allows end-to-end encryption of data.

The cloud does not actually pose meaningful challenges beyond those faced by traditional dedicated server set-ups. One might debate that the ease with which private networks and VPNs can be implemented in our cloud makes their deployment and use a lot more cost effective. Many cloud customers end up with a more robust setup from the security perspective than they had under their previous arrangements, because this ease of deployment reduces the costs of security measures [505].

18.6.6 The Challenge of Maintaining Network Quality of Service in a Public Cloud

Cloud computing at the Infrastructure as a Service level maintains quality of service of networking while experiencing a wide variety of network traffic types that are often quite unpredictable and vary over time.

In a traditional enterprise environment, the infrastructure and the network used can be well defined in terms of usage. Traffic is predictable, if not by quantity (although it is usually relatively predictable), then by type. This allows a relatively static network configuration that allows the sort of traffic expected and blocks everything else. Likewise, in such a single tenant environment, the opportunity to be affected by malicious attacks by other tenants (or against them) is of course zero.

Everything changes when we move to a multi-tenant public cloud environment; we now have diverse, unforeseeable networking traffic that can change significantly in nature from hour to hour. Consequently it is important to manage a high quality networking environment for a number of very different uses. The first thing to acknowledge is that such a general purpose network will never perform as predictably and in such an optimized way as a specialized dedicated network.

18.6.7 Securing Physical Access to the Cloud

The first task of any security framework is controlling who has physical access to the infrastructure running the cloud servers. In common with dedicated hardware, physical access to servers will negate any software level security measures that have been implemented [505]. In the case of cloud infrastructure, the control of physical access and related security measures are maintained by the cloud provider. In short, access to the physical server hosts should be maintained and monitored. A cloud provider should be open about the arrangements they make with regards to securing their infrastructure from unauthorized access and monitoring the access that is provided to authorized personnel.

18.6.8 Replicate Data

Data replication across multiple data centers is important to prevent data loss. In the event of a disaster in one area, data could still be accessed from other regions and users would be unaware of any problems [208]. For example, in the occurrence of a disaster in the Southeast, data could still be accessed from other regions. If something bad happened to the Southeast, such as a snowstorm, that switched off power, your data would be served from another data center, and no one would really know.

18.6.9 Interoperability

- Data mobility and seamless use of interoperable applications is a key consideration for all cloud users.

- Cloud providers must work together to check that interoperability and mobility are supported through open collaboration and the the use of suitable standards. Using and adopting existing standards is preferred wherever possible.

- Government agencies should allow standards for interoperability and portability to be developed in industry-led standards processes. The government should meet industry to accelerate standards development and share its user-needs with industry-led, open standard setting organizations. As the government develops and arranges cloud computing solutions it should declare its requirements to the public.

18.6.10 Steps to Ensure Your Cloud Is Secured

- Use certificates and encrypt all sensitive information.

- Deploy strong authentication for all remote users.

- Ensure isolation by using private IP address spaces and (virtual) networks.

- Provide location independence through virtual machines and networks that can be physically allocated in any data center.

- Use anti-virus software on every device.

- Use firewall technology at every point and block unused services, ports and protocols.

- Teach all users "safe Internet skills."

18.6.11 Types of Attackers

Internal Attacker

- Employed by the cloud service provider, customer or other third party provider organization supporting the operation of a cloud service.

- May have existing authorized access to cloud services, customer data or supporting infrastructure and applications, depending on their organizational role [575].

- Uses existing privileges to gain further access or support third parties in executing attacks against the confidentiality, integrity and availability of information within the cloud service.

External Attacker

- Not employed by the cloud service provider, customer or other third party provider organization supporting the operation of a cloud service.

- Has no authorized access to cloud services, customer data or supporting infrastructure and applications.

- Exploits technical, operational, process and social engineering vulnerabilities to attack a cloud service provider, customer or third party supporting organization to gain further access to propagate attacks against the confidentiality, integrity and availability of information within the cloud service.

18.6.12   Cloud Computing Threats

Threat 1: Abusive Use of Cloud Computing

Some Infrastructure as a Service (IaaS) providers do not have strong restrictions on who may sign up for their services and often offer free limited trials. As a result, spammers, malware coders, and other criminals have been able to take advantage of the services to conduct their activities [208].
Criminals continue to purchase new technologies to enhance their reach, avoid detection, and improve the effectiveness of their activities. Cloud Computing providers are actively being targeted, partly because their relatively weak registration systems allow anonymity, and their fraud detection capabilities are limited.

Threat 2: Insecure Interfaces

Cloud computing providers support many software interfaces or APIs that customers use to manage and interact with cloud services. Weak interfaces can expose an organization to many security threats.

While most providers' aim to ensure security is well integrated into their service models, it is crucial for consumers of those services to understand the security implications related to usage, management, configuration and monitoring of cloud services. Services based on a weak set of interfaces and APIs expose organizations to many security issues related to confidentiality, integrity, availability and accountability.

Threat 3: Malicious Insiders

The dangers presented by a malicious insider at any organization are well known, and the same level of risk has to be considered with cloud service providers. The effect that malicious insiders can have on an organization is considerable, given their level of access and ability to infiltrate organizations and assets. Financial impact and productivity losses are some of the ways a malicious insider can affect any operation. As organizations choose cloud services, the human element takes on even more significance [505]. It is critical therefore, that consumers of cloud services understand what providers are doing to detect and defend against the malicious insider threat.

Threat 4: Shared Technology

Cloud providers deliver their services in a scalable way by sharing infrastructure [505]. Virtualization hypervisors maintain a means of creating virtual machines or operating systems. Attacks have arisen in recent years that target the shared technology inside Cloud Computing environments. Disk partitions, CPU caches, and other shared elements were never designed for strong compartmentalization. Therefore, attackers focus on how to impact the operations of other cloud customers, and how to gain unauthorized access to data.

Threat 5: Data Loss

The threat of compromises to data increases in the cloud due to a number of underlying risks and challenges. Data loss or leakage can have a crushing impact on a business. In addition to the damage to one's brand and reputation, a loss could importantly impact employee, partner, and customer morale and trust. Loss of core intellectual property could have strategic and financial implications. Worse still, depending upon the data that is lost, there might be compliance violations and legal implications.

Threat 6: Unknown Risk Profile

Cloud computing involves outsourcing hardware and software ownership and the related maintenance. There is a risk, however, that in handing over ownership, responsibility for ensuring security procedures, policies and maintenance becomes unclear. This can result in a lack of transparency over time. When acquiring a cloud service, the features and functionality may be clearly promoted but:

- What about details on compliance to internal security procedures, modification, patching, auditing, and logging?

- How are your data and logs files saved and who has access to them?

- What information, if any, will the provider reveal in the event of a security incident?

Often such questions are not answered or are overlooked, leaving customers with an unknown risk profile that may present serious threats.

18.6.13 SSL Technology Makes Cloud Computing Secure

At a high level, SSL authenticates and encrypts data. Authentication and encryption are important to securing cloud computing environments because they keep hackers from tampering with data as it travels to the cloud over the public Internet, and they establish the identity of the site with which the data is being exchanged. Cloud service providers can acquire a SSL Certificate from a Certificate Authority that certifies that a given web domain belongs with a particular organization, and which is digitally signed by the Certificate Authority. SSL technology uses asymmetric key encryption, a system that requires two encryption keys — a public one that is freely available, and a private one that is kept confidential. All data encrypted with a public key can only be decrypted and read by the corresponding unique private key [208].

To acquire a SSL certificate for a given domain name a cloud service provider generates a public/private key pair and sends a Certificate Authority a copy of the public key, along with a certificate request signed with the private key. The Certificate Authority verifies that the public key works, and then carries out checks on the organization to ascertain that:

- It is the real owner of a domain name.

- It is a legal entity.

- The individual making the certificate request has the right to request a certificate. That means that when you connect to a server with an SSL Certificate offering a cloud service you have a means of authenticating the identity of the organization providing the service [575].

18.6.14 Assurance Considerations for Cloud Computing

Some of the key assurance issues that will need to be achieved are:

Transparency: Service providers must describe the existence of effective and robust security controls, assuring customers that their information is correctly secured against unauthorized access, change and damage. Key questions to decide are: What needs to be transparent? How much transparency is enough?

Key areas where supplier transparency is important include: What employees (of the provider) have access to customer information? Is segregation of duties between provider employees maintained? How are different customers' information segregated? What controls are in place to prevent, detect, and react to breaches?

Privacy: With privacy matters growing across the globe it will be necessary for cloud computing service providers to verify to existing and prospective customers that privacy controls are in place and describe their ability to prevent, detect, and react to breaches in a timely manner [98]. Information and communication reporting lines need to be in place and agreed on before service provisioning begins. These communication channels should be tested regularly during operations.

Compliance: Most organizations today must agree with laws, regulations and standards. There are matters with cloud computing that data may not be located in one place and may not be easily recollect. It is critical to ensure that if data are demanded by authorities, it can be provided without compromising other information. Audits completed by legal, standard and regulatory authorities themselves demonstrate that there can be plenty of overreach. When using cloud services there is no guarantee that an enterprise can get its information when needed, and some providers are even reserving the right to withhold information from authorities.

Trans-border information flow: When information can be saved any-where in the cloud, the physical location of the information can become an issue. Physical location prescribes jurisdiction and legal obligation. Country laws governing personally recognizable information vary greatly [505]. What is accepted in one country can be a violation in another.

18.7   Cloud Computing Privacy

Many of the most successful and most visible cloud applications today are consumer services such as e-mail, social networks, and virtual worlds. The companies maintaining these services collect terabytes of data, much of it sensitive information, which is then located in data centers in countries around the world. How these companies and the countries in which they operate, achieve privacy issues will be a vital factor affecting the development of cloud computing [98].

18.7.1   Privacy and Security

- Are hosted data and applications within the cloud protected by properly robust privacy rules?

- Are the cloud computing providers' technical applications, infrastructure, and processes highly secure?

- Are processes in place to support appropriate actions in the event of an occurrence that affects privacy or security?

18.7.2   What Is Privacy?

Privacy is a fundamental human right. There are various forms of privacy, including "the right to be left alone" and "control of information about our-selves." A taxonomy of privacy has been produced that focuses on the harms that arise from privacy violations, and this can provide a helpful basis on which to develop a risk/benefit analysis.

- Cloud users need a guarantee that their private information, saved, processed and communicated in the cloud, will not be used by the cloud provider in unenforceable ways [208].

- Cloud providers should create privacy rules that are appropriate for the particular cloud service they provide and business model they employ. They should make full and prominent disclosure of such policies and should give reasonable advance notification to their customers of any changes in those policies. When appropriate, they should provide customers with the opportunity to opt out of such changes.

- Governments should concede similar protections from disclosure to the government of data held by cloud providers as are currently applied to data held on a person's own computer or within a business' internal data center.

18.7.3   Privacy Risks for the Cloud

- For the cloud service user: being forced to or not to give personal information against their will, or in a way in which they feel uncomfortable.

- For the organization using the cloud service: non compliance with enterprise policies and legislation and, loss of reputation [208].

- For implementers of cloud platforms: exposure of sensitive information stored on the platforms, legal liability, loss of reputation and credibility, lack of user trust and take-up.

- For providers of applications on top of cloud platforms: legal non-compliance, loss of reputation, abuse of personal information stored on the cloud.

- For the data subject: exposure of personal information.

- Certain types of data may trigger specific obligations under national or local law.

- Vendor issues:

  - Organizations may be unaware they are even using cloud-based products.
  - Due diligence is still required as in any vendor relationship.
  - Data security is the responsibility of the customer.
  - Service level agreements need to account for access, correction and privacy rights.

- Data Transfer:

  - Cloud models may trigger international legal data transfer requirements.

18.7.4   Managing Privacy in the Cloud

- Policies and procedures must explicitly address cloud privacy risks [208].

- Information governance must be put in place that:

  - Provides tools and procedures for classifying information and assessing risk.
  - Establishes policies for cloud-based processing based upon risk and value of assets.

- Evaluate third party security and privacy capabilities before sharing confidential or sensitive information [98].

  - Thorough review and audit of providers.
  - Independent third party verification.

- Train employees and staff accordingly to mitigate security/ privacy risks in cloud computing.

  - Address from a multi-departmental perspective.

18.8   Trust Management

Cloud computing trust is more complex than in a traditional IT scenario where the information owner manages their own computers. The trust chain combines trust from the information owner's domain and trust from the service provider's platform into the virtual environment trust. The information owner has an inferred trust in the platform from a social trust relationship with the service providers [786].

There are two vital trust relationships that must be established in cloud computing from the perspective of the information owner: trust of both cloud user and service provider. Service provider trust depends on the relationship between customer and provider [306]. If the provider has a good reputation, then there is good reason for customers to trust the provider. A provider that has a questionable track record or ethics would not be as trustworthy as a provider with excellent track record and ethics. Cloud user trust derives from the trust the user places in the services achieved via the cloud. The user has to be confident that the system will protect their data, procedures, and privacy. The user's trust is a social trust of his information. The user's research must ensure that the services being provided meet the user's expectations.

In a cloud computing technology, multiple entities should trust the cloud services: the user of the cloud service or information owner, the provider of the cloud service, and third parties [98]. Owner's domain defined a new paradigm of cloud computing that separates the security responsibility among the service provider and information owner and accounted for third parties. A third party is an external entity that is providing service to or receiving services from either the user or service provider. The cloud trust model is dependent on transitive trust, which is the notion that if entity A trusts entity B and entity B trusts entity C, then entity A trusts entity C. This model enables a chain of trust to be built from a single root of trust. There two basic sources of trustworthiness in a cloud: information owner trust and hosting platform trust. By combing these two sources of trust, virtual environment trust can be established.

18.8.1   Related Work in Trusted Computing

The Trusted Computing Group (TCG) is an international computer industry standards organization that specifies and encourages trusting computing techniques. The TCG is responsible for maintaining and updating the Trusted Platform Model (TPM) specification. The TPM specification is currently at version TPM1.2. The TCG is working on the next version of the TPM specification, referred to at TPM.Next. TPM.Next will incorporate new hash algorithms and other features that will require TPM manufacturers to change the design of their TPM implementations. The TPM specification was developed with the following high-level requirements:

- T1: Securely report the environment that has been booted.

- T2: Securely stores data.

- T3: Securely identifies the use and system.

- T4: Supports standard security systems and protocols.

- T5: Supports multiple users on the same system while preserving security among them.

- T6: Can be produced inexpensively.

These requirements lead to a robust and secure design; however, design tradeoffs were made to keep costs low through reduced functionality. Each TPM has limited resources but sells for less than one dollar in production quantities [786]. Many of the TPM's limited resources cannot be shared with virtual environments; therefore, the TPM must be virtualized to support multiple virtual environments.

There are many issues that must be solved to virtualize a TPM. The limited resources of the TPM must be either shared or replicated for each virtualized TPM. Specifically, resources that cannot be shared on the TPM are the EK, Platform Configuration Registers (PCRs), and non-volatile storage. These resources must be replicated by every VTPM implementation. A common approach to virtualizing the TPM has been to emulate the TPM in software and provide an instance for each virtual environment. The VTPM can be bound to a physical TPM for additional security [306].

18.9   Recommendation

From my point of view, to secure cloud computing we should start first by securing data and then secure the infrastructure. In this section, I will mention some recommendations for secure cloud computing [786].

18.9.1   Protecting Your Data

First, it is important to categorize your data to know what rules you must follow to protect it:

- Its sensitivity: must it only be managed subject to specific trust levels? If so, which?

- Determine what level of security you want to maintain in the cloud. Different cloud models provide different layers of business service.

- Determine what types of data and processes move to the cloud.

- Determine which Cloud models are best suited to your requirements.

- Establish a data category model that is sufficiently straightforward for all originators of data to use.

- Engage with a trusted partner who has knowledge of both enterprise cloud computing and data centers.

- Consider applications derived outside of the firewall as first candidates for the cloud.

- Build Trust Management mechanisms to improve data confidence.

Analyze the Data Flow

This calls for charting the lifecycle of the relevant data assets, from development to their destruction. IT managers must know where data is at all times so they can confirm that it is being stored and shared in compliance with local laws and industry regulations at appropriate levels of IT security [306].

18.9.2 Protecting the Users

Cloud computing or any type of online application structure must account for protecting its users. Developers should make sure that data corresponding to the user cannot be misused and could be extracted just by one. There are two main ways to enhance cloud computing security: limiting user access and employing certification. Limited access could come from simple username/password challenge to strengthen login procedures, but applications in cloud computing should not depend on these challenges. User recess and IP specific applications are only some of the security measures that should be implemented [489].

The challenge in limiting user access is to restrict the access privilege of the user appropriately. Each user will have to be appointed manually with security clearance to ensure restriction of access to different files.

Certification is also important for user certification. Developers have to open their application to security professionals or companies that provide certifications for security. This is one method of promising users that the application has been fully tested against different types of attacks [596]. This is often an issue for cloud computing, as external security checks might open the company secrets held in the cloud. But this risk has to be accepted to assure the security of users.

Security against Hackers

The first security matter in any computing environment is the threat from hackers. Cloud computing in a shared environment generates new opportunities for hackers to discover vulnerabilities, which may ultimately allow them to deny service or gain unauthorized access. Moreover, security experts have detailed approaches for attacking cloud infrastructure from the inside by running hacker tools in the cloud itself [489].

Once a hacker has access in a public cloud, the hacker gets greater visibility inside the cloud. The hacker then uses this inside information to more effectively explore the system and plan attacks. To fight this vulnerability, a private cloud limits access to its resources to authorized users and administrators only. By avoiding potential hackers from gaining an inside view, private clouds allow businesses to control access to the entire environment [596].

Assurances over Security in the Cloud

- Define the organization's security requirements and the security requirements applicable to information assets in the cloud.

- Carry out appropriate due diligence investigations of cloud providers. Manage risks associated with using cloud services.

Assess Cloud Providers Controls for Segregating Data between Cloud Users

- Encryption can protect data stored in the cloud if the cloud provider can demonstrate robust key management and security.

- Assess the cloud provider's mechanisms for segregating data when unencrypted (for example, when being processed).

Assess Effectiveness of Protective Monitoring of the Cloud

- Tracing actions back to individual users may be impossible without tight integration of the cloud users' and cloud providers' protective monitoring.

18.10   Summary

Cloud computing is the most popular notion in IT today; Reservations about cloud computing largely stem from the perceived loss of control of sensitive data. Cloud computing allows consumers choice of software, hardware, and computing environment. The ultimate cost for a service will be based on the choice of service as well as the choice of security method. The multi-tenant environment inherent in public cloud networks is clearly the highest challenge specific to IaaS cloud providers.

In practice, the security of cloud networks comes down to a combination of robust company policies, well thought out network software/architecture and having an ability to evolve and react quickly where necessary to threats as they present themselves. To take full advantage of the cloud, users must be given reliable assurances regarding the privacy and security of their online data. In addition, several regulatory, juridical, and public policy issues remain to be solved in order for online computing to thrive.

In order to secure cloud computing, we should answer the following questions:

1. Who is accountable for the security of our data and to whom do they report?

2. Might the provider lose our data through misuse, or theft or fraud, for example? If so, what recovery plans do we have?

3. What are our obligations regarding data protection versus those of a cloud services provider?

4. What is our policy for which staff are authorized to deposit and store data with a cloud provider?

5. How do cloud providers assist customers with their compliance requirements?

6. What kinds of physical segregation of virtual machines are available for customers?

18.11   Glossary

Access control: The ability to selectively control who can get at or manipulate information in, for example, a web server.

APIs: Application Programming Interfaces.

Asymmetric key: Employs two keys. One of these is publicly known and the other is held privately. To derive a public key from a private key, any would be hacker would need to factor a very large number, and this is computationally infeasible for such derivation.

Audit: Is an evaluation of a person, organization, system, process, enterprise, project or product. Audits are performed to ascertain the validity and reliability of information.

Authentication: The process of verifying to a reasonable degree of certainty that an entity (for example, a person, a corporation, or a computer system) is the entity it represents itself to be.

Authorization: A process for controlling user activities within a system to only actions defined as appropriate based on the user's role.

Availability: Ensures that data continues to be available at a required level of performance in situations ranging from "normal" through "disastrous."

Certificate Authority: Is an entity that issues digital certificates for use by other parties. It is an example of a trusted third party.

Cloud Storage: Is the persistent storage allotment shared on a cloud provider's network.

Compliance: Describes the goal that corporations or public agencies aspire to in their efforts to ensure that personnel are aware of and take steps to comply with relevant laws and regulations.

Confidentiality: Ensuring that information is accessible only to those authorized to have access. One of the cornerstones of information security.

CPU: The part of a computer (a microprocessor chip) that does most of the data processing; "the CPU and the memory form the central part of a computer to which the peripherals are attached."

Cracker: Software cracking is the modification of software to remove protection methods: copy protection, trial/demo version, serial number, hardware key, date checks, CD check or software annoyances like nag screens.

Data Aggregation: Is the ability to get a more complete picture of the information by collecting and analyzing several different types of records from various channels at once.

Data Fragmentation: Is a phenomenon in which storage space is used inefficiently, reducing storage capacity and in most cases performance. The term is also used to denote the wasted space itself.

Denial-of-service attack: Is an attempt to make a computer resource unavailable to its intended users.

Encryption: Is the process of transforming information (referred to as plaintext) using an algorithm (called cipher) to make it unreadable to anyone except those possessing special knowledge, usually referred to as a key.

External cloud: The cloud infrastructure that is made available to the general public or a large industry group and is owned by an organization selling cloud services.

Fault tolerance: The built-in capability of a system to provide continued correct execution in the presence of one or more failures.

Firewall: Is a part of a computer system or network that is designed to block unauthorized access while permitting authorized communications. It is a device or set of devices which is configured to permit or deny computer application based upon a set of rules and other criteria.

Grid computing: Is the combination of computer resources from multiple administrative domains for a common goal.

Host: Is a utility for performing Domain Name System lookups. It was developed by the Internet Systems Consortium (ISC), and is released under the ISC license, a permissive free software license.

Hypervisor: A layer between software environments and physical hardware that virtualizes the system's hardware.

Identity management: Is a broad administrative area that deals with identifying individuals in a system (such as a country, a network or an organisation) and controlling the access to the resources in that system by placing restrictions on the established identities.

Infrastructure as a Service (IaaS): Provides informatics resources, such as servers, connections, storage and other necessary tools to construct an application design prepared to meet different needs of multiple organizations, making it quick and, easy.

Integrity: Is data that have a complete or whole structure. All characteristics of the data including business rules, rules for how pieces of data relate dates, definitions and lineage must be correct for data to be complete.

Internal cloud: The cloud infrastructure is operated solely for an organization. It may be managed by the organization or a third party and may exist on premise or off premise.

IP address: Is a numerical label that is assigned to devices participating in a computer network that uses the Internet Protocol for communication between its nodes.

IT: Information Technology: the branch of engineering that deals with the use of computers and telecommunications to retrieve and store and transmit information.

Monitor: Check, track, or observe by means of a receiver.

Net book: A rapidly evolving category of small, light and inexpensive laptop computers suited for general computing and accessing web-based applications.

Platform as a Service (PaaS): Generates all facilities required to support the complete cycle of construction and delivery of web-based applications wholly available in Internet without the need of downloading software or special installations by developers.

Resilient: The capacity of a system, community or society potentially exposed to hazards to adapt, by resisting or changing in order to reach and maintain an acceptable level of functioning and structure.

Resource pooling: Management technique where a collection of previously used resources (that is, a resource pool), such as connections to a data source, are typically stored and reused by an application as they are needed.

Server: A computer that provides client stations with access to files and printers as shared resources to a computer network.

Software as a Service(SaaS): Is a software delivery method that provides access to the software and its functions as a web-based service. SaaS allows organizations to access business functionality at a lower cost, without the heavy technology implementation requirements.

SSL: Secure Sockets Layer: cryptographic protocols which provide secure communications on the Internet; Start-Stop Logic.

Thin client: Is a computer or a computer program which depends heavily on some other computer (its server) to fulfil its traditional computational roles.

Threat: Possibility that vulnerability may be exploited to cause harm to a system, environment, or personnel.

Utility computing: Is the packaging of computing resources, such as computation and storage, as a metered service similar to a traditional public utility.

Virtualization: The ability to separate the physical layout of a network and its devices from how uses are organized into workgroups (knows a logical configuration).

Web 2.0: Refers to the second generation of the web, which enables people with no specialized technical knowledge to create their own websites, to self-publish, create and upload audio and video files, share photos and information and complete a variety of other tasks.

Web portal: Website considered as an entry point to other websites.

Web Services: Is the umbrella term for a group of loosely related web-based resources and components that may be used by other web applications. Those resources could include anything from phone directory data to weather data to sports results.

Part III

Case Studies, Applications, and Future Directions

19

Fundamentals of Cloud Application Architecture

Justin Y. Shi

Temple University

CONTENTS

19.1 Introduction

The scale of economy of cloud computing has ignited wide-spread imaginations. With the promises of mighty computing power at unbelievable prices, all applications are poised to gain mission critical status. The fundamentals on cloud application architecture, however, are surprisingly short. For example, how can mission critical applications leverage the cloud resources? Would non-mission critical applications gain mission critical status by moving to the cloud? This chapter attempts to fill in the fundamentals for mission critical networked computing applications.

The definition of "mission criticalness" has also evolved. Decades ago, few applications were considered "mission critical." They were mostly related to military, nuclear power plants, financial institutions, air craft controls, etc. Today, it is harder to find applications that are not mission critical. Evidently the application requirements grew faster than technologies could deliver. The availability of cloud computing simply adds fuel on the fire. In other words, cloud computing has not only offered a cost-effective way of computing, it also has forced the performance, information security and assurance issues to surface.

This chapter discusses the fundamentals of mission critical cloud applications. In particular, we focus on three objectives: maximal application survivability, unlimited application performance scalability, and zero data losses. These properties are desirable for all mission critical applications. The availability of cloud computing resources has inspired all applications to attain the mission critical status.

Application survivability is a long standing practical (and research) issue. Although processor virtualization has shortened service interruptions, by quicker restart of check-pointed virtual machines, the fundamental challenges of non-stop service and maximal survivability remain the same.

Application performance scalability is another nagging issue. Each application needs to serve increasingly larger numbers of clients without degrading performance. The low costs of cloud services have reduced the budget barriers. However, except for a few "stateless" applications, such as web crawling, file sharing and mass mailing, most other applications face scalability challenges, regardless the computing platform.

The most puzzling is perhaps the data losses in networked computing applications. The traditional synchronous and asynchronous replication methods have been proven incapable of delivering performance and survivability at the same time. The puzzling fact is that we are losing data while using the lossless TCP protocol. Why?

This chapter discusses the generic "DNA" sequence of the maximally survivable systems. For cloud ready mission critical applications, we advocate a focus on the networked computing application architecture — a semantic network architecture formed via an application programming interface (API) and the logical communication channels within. Unlike traditional computer architecture and data communication network studies, our focus is a holistic, application-centric networked computing application development methodology. In particular, we focus on embedding the maximally survivable "DNA" sequence in the networked application architectures.

This chapter is organized as follows: Section 19.2 covers the historical background of the "DNA" sequence of maximally survivable systems, the necessary and sufficient conditions for the maximally survivable systems, and the introduction of Networked Computing Applications Architectures (NCA^2). Section 19.3 defines the Unit of Transmission (UT) for four basic NCAs: messaging, storage, computing, and transaction processing. As the first example, Section 19.4 describes the detailed steps for embedding the maximally survivable sequence for the messaging system: Enterprize Service Bus (ESB). Section 19.5 describes the steps required for embedding the maximally survivable sequence for a transaction processing system. Section 19.6 describes the design of a maximally scalable high performance computing system. Section 19.7 is the summary.

19.2   Necessary and Sufficient Conditions

Approximately fifty years ago, Mr. Paul Baran was given the task of designing a survivable network for the United States Air Force. It was in the middle of the Cold War. The objective was simple: build a dependable network that can survive a nuclear attack.

The packet switching network concept was born. However, deeply invested in analog technologies (circuit-switching networks), major research institutions refused to embrace the concept. Telecom carriers refused to adopt the technology.

The key arguments were "poor performance" and "unknown stability." Comparing packet switching to circuit switching networks, on the surface, the "store-and-forward" nature of packet switching protocol did seem counterproductive for transmitting voice signals in real time.

Fast forward to today, a twenty-year old would have a tough time identifying the existence of circuit switching networks. The Internet is a seamless "wrap" of packet switching circuits on top of circuit switching networks. Although there is still much to improve on the IP addressing scheme and service qualities, the two-tier Internet architecture seems destined to stay. And the TCP protocol remains a lossless protocol. In other words, the packet switching concept has stood the test of time.

The performance scalability benefit can only be realized after the packet switching networks reach enough scale. Indeed, it took the world quite a few years to change the "world-wide wait" to the World Wide Web we enjoy today. Although circuit switching networks still play a critical role in today's Internet, the lesson was that performance centric architectures may not sustain in larger scales. The packet switching network was one of those counterintuitive ideas. Much research has been conducted on the packet switching protocols. The results are captured in continually improving Internet services for higher speed, better reliability and lower costs. These improvements have brought the power of networked computing to the masses.

Now, we are confronted with the nagging survivability issues for networked computing applications (NCAs). Server virtualization, the key technology that has made cloud computing possible, although cost-effective, merely shifts the service locations. The technological challenges remain.

It seems that the history is repeating itself. The same problems we solved four decades ago for the telecommunication industry have come back to challenge us at a higher level for NCAs.

We ask: what is the generic "DNA" sequence in the packet switching network that has made it possible to create the maximally survivable systems? How can this sequence be "inherited" by the networked computer applications to achieve the similar results? What are the necessary and sufficient conditions for the maximally survivable systems?

19.2.1   Necessary Conditions

Any survivable system must rely on redundancy [114]. Theoretically there are two types of redundancies: temporal (or re-transmission), and spatial (or replication). There has been no clear direction as when to apply which type of redundancy and for what purposes.

"Store-and-forward," or, "statistical multiplexing," characterizes the essence of packet switching architecture. Unlike circuit switching networks, voice signals are transmitted via discrete packets by interconnected routers and switches (Figure 19.1).

The "DNA" sequence of this "store-and-forward" architecture has four components:

1. A well-defined unit of transmission (UT),

2. Transient UT storage,

3. Forward with once-only UT re-transmission logic, and

4. Passive (stateless) spatial UT redundancy.

To date, the "store-and-forward" network has been proven the most survivable architecture. The Internet has inherited this "DNA" sequence by wrapping packet switching circuits around circuit switching networks.

Looking deeper, the "store-and-forward" packet switching protocol is a clever utilization of multiple redundancy types for the communication system: with

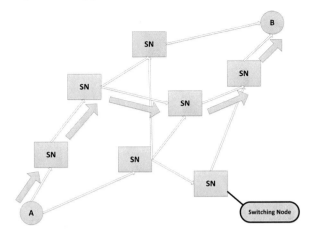

FIGURE 19.1
Packet Switching Network.

a given unit of transmission (a data packet), a packet switching network supports transient spatial redundancy (store), temporal redundancy (forward with re-transmission) and passive spatial redundancy (routers and switches). These essential statistical multiplexing methods afford: a) the ability to mitigate massive component failures, b) the potential to leverage parallel processing for performance scalability; and c) provably least cost fault tolerance.

The primary winning argument for the packet switching network was the provably least cost and maximal survivability design. The unlimited performance scalability came as a welcome after-effect — a benefit that can only be delivered after the system has reached enough scale.

The maximal survivability feature allowed decades of continued innovations. After quite a few years, the "world-wide-wait" network has eventually become the world wide web that has poised to change the landscape of our lives.

19.2.2 NCA and Sufficient Conditions

Networked Computing Application (NCA) represents the vast majority computer applications today. Invariably, all NCAs face performance scalability, service availability and data loss challenges. For example, upscaling a database server is a non-trivial challenge. Providing non-stop service is another nagging issue. The most puzzling is that none of NCAs would promise lossless service, even though almost all are riding atop the lossless TCP (Transport Control Protocol).

All cloud-computing applications are NCAs.

The common feature of all NCAs is the need for networking. Indeed, each basic NCA has its own higher level communication protocol on top of TCP/IP, such as messaging, transaction processing, online storage and parallel computing.

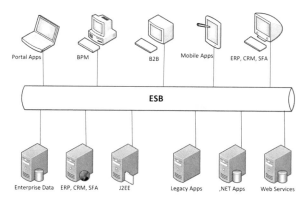

FIGURE 19.2
Enterprize Service Bus.

It should not have been a surprise that the maximally survivable "DNA" is probably needed for all NCAs.

A key observation is the maximal entropy between communication protocol layers. This means that the benefits of a packet level network are not automatic for higher layers. Therefore, the necessary conditions for gaining the same benefits at a higher level should be somewhat similar.

The sufficient conditions include all those required for the correct respective NCA processing. These are application dependant conditions.

Unlike the low-level data communications, the NCA semantic network typically carries complex application dependencies and must satisfy non-trivial semantics. Low-level data packets can be transmitted and re-transmitted without violating the semantics of data communication; it is not clear how to define the application dependent UTs that can be supported at architecture level without breaking the application processing correctness. Poor "packetization" could violate the crucial application dependencies (too little state information) or incur excessive communication overheads (too much state information). It becomes a non-trivial exercise to embed the generic "DNA" sequence in any NCA.

For example, Enterprize Service Bus (ESB) is the core of Service Oriented Architectures (SOA). It consists of a set of message servers dispatching requests and replies between clients and an assortment of heterogeneous servers (Figure 19.2).

A sustainable cloud-ready ESB should meet the following (sufficient) requirements:

1. Adding processing and communication components should increase overall performance.

2. Adding processing and communication components should increase both data and service availabilities.

3. Adding processing and communication components should decrease the probability of permanent failures.

4. There should be zero message loss at all times.

Simply moving an existing ESB to the cloud can not meet all requirements. Since each message is typically partitioned into multiple packets, and each packet is routed individually, on the surface it seems that meeting all the requirements can be cost prohibitive.

Ironically, the packet switching network had met all the above requirements for data transmission for many years. What did we miss?

19.3   Unit of Transmission (UT)

The first step in embedding the "store-and-forward" sequence in a target NCA is defining the unit of transmission (UT).

For example, we are interested in the following basic NCAs:

1. The UT of an Enterprize Service Bus is a message.

2. The UT of a transaction processing system is a transaction. For simplicity, we consider a simple read-only query also a transaction.

3. The UT of a mission critical high performance computing system is a computation task.

4. The UT of a mission critical storage system is the disk transaction. Similarly, we consider a storage retrieval query a disk transaction.

These UT definitions represent the four basic service layers of modern NCAs: messaging, transaction, storage and computing. Once we understand how to build these basic NCA architectures, we should be able to construct robust higher level applications with full mission critical support down to the physical medium.

19.4   Mission Critical Application Architecture: A First Example

As mentioned earlier, the basic NCA architectures include messaging, transaction, storage and computing. As the first example, this section describes the construction steps for a mission critical ESB.

We have three operational objectives:

1. Maximal survivability.

2. Unlimited performance scalability.

3. Zero data losses.

Traditional replication technologies can only address one objective. It seems cost prohibitive, if not impossible, to meet all the objectives using traditional methods.

For a mission critical messaging architecture, UT is a message. The UT of TCP is a packet. A message is typically decomposed into multiple packets. Although the lossless feature of TCP guarantees the delivery of every packet, the actual delivery of each message is only known to the application. This is because each packet is routed individually. Since packets can be delivered at different times, only the application could know if a message is delivered consistently on the semantic network.

It is interesting to observe the absence of re-transmission API in existing messaging systems, as if every message transmission must either succeed or fail.

There is a third state: *timeout*. It happens more often than we would desire. The problem is the lack of direction as what to do with timeouts. Printing error messages is the most common practice. Textbooks simply avoid this very subject. This is, however, the cause of information losses.

A timeout message is in an "unknown" state since we are not certain if the message has ever been delivered. Dealing with a timeout is actually a non-trivial challenge since not all messages can be safely re-transmitted: consider a timeout message that carried a missile firing command.

The irony is that because of the maximal entropy between protocol abstraction layers, only the application knows the exact semantics of each message. Therefore, unlike the popular designs, the only possible layer for implementing this non-trivial re-transmission logic is the application layer. The lower layers simply do not have enough information. Optimization is possible, leveraging multiple lower level protocol layers. But methods like "group communication" cannot address the fundamental survivability and information loss issues.

Having the correct re-transmission logic at the application layer is not enough. Without transient storage and passive spatial redundancy supported in the architecture, repeated transmissions will not deliver appreciable survivability improvements. For the messaging systems, it is necessary to install passive redundant ESB hardware and allow it be discovered at runtime (Figure 19.4). This is because each application (a messaging client) holds the transient storage of its own message. The transient storage is automatically cleared after the confirmation of the message delivery. Therefore, with the help of any passive messaging hardware, this new infrastructure (Figure 19.4) ensures maximal survivability (defined by the degree of hardware redundancy) and there can be no message losses.

For comparison, without showing the service servers, Figure 19.3 shows a conceptual diagram of a mission critical ESB system with synchronous and

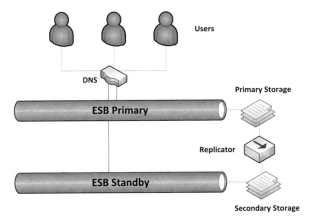

FIGURE 19.3
ESB with Spatial Redundancy.

asynchronous storage replication via fiberoptic channels. Figure 19.4 shows a conceptual lossless ESB system with a re-transmission API at all clients without the disk replication harness.

Although the differences are subtle, the benefits are non-trivial.

In Figure 19.3, the disk replication harness adds significant overheads to each message transmission, either synchronously or asynchronously. In asynchronous mode, message losses cannot be avoided and the ESB overall speed must be throttled to avoid the replication queue overflow. In synchronous mode, the two-phase-commit replication protocol imposes severe performance penalty to each message transfer. Adding message queue managers increases the replication overheads linearly. Among the three operating objectives, only one can be attempted by any production system. This problem persists even if Figure 19.3 is moved to a computing cloud.

In Figure 19.4, the disk replication harness is removed. There will be zero replication overhead. This enables unlimited performance scalability: one can add as many messaging servers as the application requires. There will be zero message losses since all clients hold the transient storage of their own messages until confirmed delivery. Further, the re-transmission logic will automatically discover alternative message routes and recover lost messages. All we need to supply is the passive ESB redundancy in different location(s). Passive ESB systems need much less maintenance (similar to a network router) than active systems. Properly configured, this system can automatically survive multiple simultaneous component failures.

The critical observation is that messaging storages are all transient. Replication of transient messages on permanent storage is absolutely unnecessary.

This scheme can be further improved if we enhance the re-transmission protocol to leverage the passive ESB hardware. For example, we can build an interconnecting network of ESB servers where each server runs an enhanced

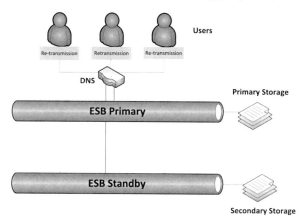

FIGURE 19.4
ESB with Re-Transmission API and Passive Redundancy.

message store-and-forward protocol at each stage. The result is an extreme scale ESB infrastructure that can actually meet all three operating objectives simultaneously. Implementing the improved ESB in a computing cloud can result in a provably optimal messaging architecture.

It is also worth mentioning that not all messages need confirmed delivery. Non-mission critical messages can be sent without delivery confirmation. In other words, there should be two kinds of messaging services in its API: lossless with automatic re-transmission (like TCP), and others (like UDP).

Since none of the existing messaging systems was built as shown in Figure 19.4, these applications are really using UDP-like messaging without delivery guarantees.

This first example illustrates the non-obviousness for introducing statistic multiplexing into the common messaging task. In this case, both the client and the messaging servers provide the transient UT storages. Therefore, having the correct embedding of "DNA" sequence satisfies both the necessary and sufficient conditions for the maximally survivable messaging system.

This example also alludes to the potential difficulties and significant benefits of introducing statistical multiplexing to other basic NCA architectures. The next sections describe the challenges and solutions for embedding the "store-and-forward" sequence for transaction processing and high performance computing systems.

19.5 Maximally Survivable Transaction Processing

If a message contains a transaction, would the above lossless messaging architecture deliver lossless transaction processing?

The answer is NO. This is because of the maximal entropy between the transaction processing and messaging protocol layers. The units of transmission are different.

A transaction processing system must meet the ACID (Atomic, Consistency, Isolation and Durability) requirements. The messaging service layer simply does not have enough information regarding transactions. As will be shown in the following section, embedding transient storage and re-transmission in a transaction processing system is much harder. It must ensure ACID properties, zero single-point of failures, transaction fault tolerance, transaction integrity, non-stop service and unlimited performance scalability. The results are equally non-trivial.

Another interesting case is the high performance computing architecture, where performance scalability and fault tolerance issues have long troubled researchers and practitioners alike. Next, Section 19.6 describes a possible DNA sequence embedding for multiprocessor architectures that can deliver maximally survivable applications with unlimited performance scalability.

A high performance cloud-ready lossless storage network is a simplified transaction processing system. The details are omitted for brevity.

19.5.1 Maximal Survivability and Performance Scalability

Transaction processing systems (databases) provide the fundamental layer for all electronic information processing applications. Databases are well known for their scalability limitations in performance and service/data availability. Embedding the maximally survivable DNA sequence could potentially eliminate all these scalability limitations.

This section describes the DB^x (database extension gateway) project. It is the only commercial transaction processing architecture with statistical multiplexing known to the author at the time of this writing.

A DB^x cluster uses statistical multiplexing to link multiple redundant database servers. Like the Internet, it does not have performance scalability limitation. Therefore it is a candidate architecture for VLDB (Very Largescale DataBase) systems.

DB^x is a joint work of the author and his former students at Temple University (www.temple.edu) and later the Parallel Computers Technology Inc. (www.pcticorp.com). Section 19.5.2 describes the theoretical foundations of clustering database servers for unlimited scalability in performance and availability.

19.5.2 Transaction Processing Failure Model

We define a transaction as:

Definition 1 *A collection of conditional data change requests that uniquely defines resulting dataset(s) when the conditions are satisfied at runtime.*

A committed transaction is a collection of data change requests with confirmed positive result (transaction state). A committed transaction should result in a unique persistent dataset until modified by other transactions. All data changes in a transaction are assumed be either all committed or all canceled. There should be no partial updates. All modern database engines today meet these requirements.

Theoretically, each transaction can be in any one of the following four states:

1. A *commit success* is a transaction that is committed to the processing harness and the user has received successful acknowledgement.

2. A *commit failure* is a transaction that has failed to commit in the processing harness and the user has received error notification.

3. An *unknown transaction* is a timeout transaction whose status is not yet verified.

4. A *lost transaction* is a committed successful transaction whose presence cannot be confirmed afterwards.

A correct transaction processing system must meet the following:

- Maximally possible committed transactions,

- Zero unknown transactions, and

- Zero lost transactions.

The number of *commit failures* is a function of the quality of database/application design and implementation. Commit failures do not contribute to unknown and lost transactions.

Like the ESB project, using a single updatable database in the processing harness, it makes little sense to re-transmit the timeout transaction. For statistical multiplexing, spatial data redundancy is necessary. Without statistical multiplexing, database applications can only generate error messages for unknown transactions. For high value transactions, such as for banks and stock exchanges, the unknown transactions are manually repaired, if found. For others, error messages are the only artifacts left behind.

The lack of statistical multiplexing also inhibits performance scalability since the processing harness simply does not have the option of exploiting alternative resources during peak moments (especially for read-only queries).

Existing transaction replication methods are designed only for protecting service uptimes [517] [554]. As in ESB, the synchronous mode uses the 2-Phase-Commit (2PC) protocol that serializes all transactions for the purpose of replication. Failures in any phase of the protocol require rolling back the entire transaction, which has serious performance and downtime consequences. But it ensures the replication of all transactions in the exact same order on multiple targets (therefore there is no transaction loss but plenty of service time losses).

Asynchronous transaction replication uses either an explicit or implicit (log scan) replication queue to ease the immediate performance degradation. The replication queue affords a time buffer for the queue to be emptied. However, replication is strictly serial since it must ensure the same commit order on the replication target(s). According to the Little's formula [474] for queued services, the primary database server must be throttled below the serial transaction replication speed to avoid overwhelming the replication queue and potential service shutdown. In practice, a multicore processor with multi-spindle storage can easily outperform any high speed serial replication. The system must shut down to enter a recovery mode if the replication queue is persistently overwhelmed.

Transfer service from the primary to the secondary risks transaction losses. Failures in the replication target can also force extended service shutdown if the replication queue (and its manager) is not manually reconfigured quickly. These deficiencies exist in all current transaction processing systems [85] [68]. In this sense, databases are the weakest links in all existing IT infrastructures. For banks, lost transactions can potentially make money when small account owners do not balance their accounts. For traders, a lost trade may never be recovered. For service providers, millions of revenue can be lost annually due to unrecorded usages. For mission critical applications, the consequences can be serious if a critical transaction is forever lost. Identifying what was actually lost is theoretically impossible. Many losses can cause irreparable damages.

Due to the lack of transaction failure model support, large scale transaction clouds can only exacerbate this problem.

As the importance and scale of transaction processing applications grow, these limitations impose serious threats to the higher level social, economic and political infrastructures. Large scale computing clouds only accelerate the exposure of scalability and reliability issues.

It is worth mentioning that the same problems exist in storage systems [365], [366], and [684]. Storage data updates are also transactions, since the changes must be persistent. They also require a single updatable persistent image.

19.5.3 Parallel Synchronous Transaction Replication

Currently, most people believe transaction replication is dangerous [393]. In theory, synchronous transaction replication can eliminate transaction losses at a cost. The strictly sequential, "all-or-nothing" 2PC protocol causes severe performance degradation at runtime. Since very few applications today can afford synchronous transaction replication, the vast majority of transac-

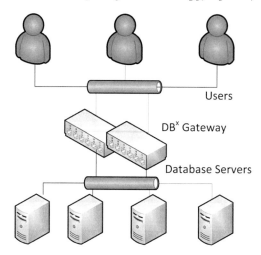

FIGURE 19.5
Conceptual Diagram of DB^x.

tion processing systems today use "leaky solutions" powered by asynchronous replication and human intervention.

Careful examination of practical applications reveals that most concurrent transactions do not have update conflicts; serializing only the concurrent update conflicts can alleviate much of the performance losses.

Further, it is also highly desirable in practice to continue a transaction when only a subset of replication targets fail. In other words, dynamic serialization and best effort fault tolerance can drastically enhance the quality of transaction processing. Further, for read-only queries, there is no need to replicate. Dynamically load balancing read-only queries can deliver more scalable performance.

Synchronous replication with dynamic transaction serialization and dynamic load balancing requires a database communication protocol gateway to capture all concurrent transactions in transit. We call this the database extension (DB^x) gateway [663] [645]. The new statistically multiplexed database cluster is capable of parallel synchronous transaction replication where serialization is optimized only for concurrent update conflicts. Figure 19.5 shows a conceptual diagram of a DB^x cluster.

In Figure 19.5, the UT is a transaction. DB^x is the transient storage and the database clients are assumed to contain automatic re-transmission logic.

Unlike ESB, where all messages are transient in the messaging servers, passive ESB redundancy enables unlimited survivability; database transactions leave persistent data changes that can only be protected by active spatial redundancy. Combined with the automatic re-transmission logic, the statistically multiplexed spatial redundancy not only provides unlimited survivability but also the potential for performance scalability.

Similar to ESB, there are two kinds of transaction services: lossless with confirmed delivery (TCP-like) and others (UDP-like).

However, to make a practically useful cloud capable VLDB system, we must also address the following fundamental questions (sufficient conditions):

1. What is the transaction trust model?

2. How to deliver non-stop service when servers with very large datasets become out-of-sync?

3. How to maintain database internals without shutting down service?

4. What is the cluster failure model? How to eliminate single-point-failures?

5. How to ensure transaction ACID properties?

6. How to eliminate transaction losses?

7. How to deliver unlimited performance scalability?

19.5.4 Transaction Trust Model

From a database application's perspective, each transaction must result in a database state that is consistent with all data changes issued from the application. If multiple databases reply differently for the same transaction, each reply is valid, since each individual transaction processing context is semantically identical to a single database environment. However, the data inconsistency cannot be tolerated (caused by different processing orders on different servers).

This means that in practice, we can simply designate any server as the "primary." Servers that reply with different results than the primary can be removed from service and resynchronized at a later time.

This trust model ensures a single consistent data view with the designated "primary" server for all clients. Like a re-transmitted packet, Byzantine failure makes little sense in this context. There is no need for a quorum.

19.5.5 Non-Stop Resynchronization — 2PCr Protocol

When one or more databases goes out of sync for any reason with respect to the primary server, deactivation is trivial. Bringing them back in service with reconciled contents and without service downtime is a non-trivial challenge. Instead of trying to reconcile the datasets by finding their differences, the critical network position of DB^x affords the execution of a "non-stop" resynchronization algorithm [645]. The resynchronization algorithm is capable of reconciling multiple out-of-sync datasets (of arbitrary sizes) for at most 60 second service downtime:

1. Assume that multiple database servers are synchronously replicated. In the same cluster, one or more database servers are deactivated for hardware/software errors or for scheduled maintenance. They become instantly out-of-sync.

2. Start a full backup of any synchronously replicated server, say S. S continues to be updated while backing up.

3. Restore the backup set to the group of out-of-sync servers G in parallel.

4. Obtain transaction differentials from the transaction log on S.

5. If the transaction differential is empty, enable all members of G. They are all in sync.

6. Apply the transaction differentials. Wait T. Repeat (4).

7. If not terminate, pause DB^x gateway and disconnect all clients. Repeat (4).

The worst-case downtime comes from step (7) where 60 seconds can cover multiple shorter scan intervals (typically 10 seconds each). The better cases can have zero downtime if the DB^x gateway could find a quiet network time. Step (7) is necessary to guard against the inherent speed differences between parallel transaction processing and the sequential nature of the backup and restore processes.

The resynchronization and dynamic serialization algorithms can be viewed as an optimistic two-phase-commit protocol (2PCr). 2PCr allows any server to be taken out of the cluster and put back at a later time, while providing virtually non-stop service.

19.5.6 ACID Properties and RML

Ensuring ACID properties for all transactions is probably the most difficult task with multiple replicated databases.

For a single database, the Atomic property is ensured by the database engine using the transactional two-phase-commit (2PC) protocol — if any update in a transaction is not complete, the entire transaction rolls back.

For multiple databases in the DB^x cluster, the Atomic property is also ensured by the same protocol, since the gateway simply replicates the semantically identical sequence of commands transparently. The transaction trust model guards against the potential inconsistencies. Dynamic serialization does not affect the execution of the transactional 2PC protocol.

Transaction Durability is much enhanced in a DB^x cluster since there are multiple synchronously replicated datasets.

The Isolation and Consistency properties are intertwined. In practice, application programmers often relax the isolation level in exchange for higher throughput (less locking and higher concurrency). ANSI/ISO defines the following isolation levels:

1. Serializable

2. Repeatable Read

3. Read Committed

4. Read Uncommitted (dirty read)

In [123], a snapshot isolation level is also defined. Serializable is the highest isolation level that ensures that all concurrent applications will see a consistent database at runtime. Others are progressively more relaxed to increase the degree of concurrency.

In a DB^x cluster, dynamic load balancing will alter the isolation behaviors subtly. For example, unless explicitly directed using RML (Replication Markup Language), dynamic load balancing will break the assumed isolation behaviors, since the reading target may not be the same server with the proceeding updates. For many applications, this can be tolerated; since the delay to value stabilization is small and the isolation levels are often relaxed anyway. Dynamic load balancing can be disabled if the application requires strict adherence to isolation levels.

With a single database target, concurrency issues cause deadlocks. With transaction replication onto multiple servers, the chance of deadlocks is increased, multiple servers can lock different transactions thus cause deadlocks and data inconsistencies.

These race conditions can be eliminated if conflicting concurrent updates are *globally synchronized*. The discipline is identical to composing "thread safe" programs. For "cloud safe" transaction applications, the programmer should have the knowledge of concurrent update zones and should be able to mark up the concurrent updates with correct "cluster locks."

The DB^x gateway will use the named "cluster locks" for dynamic serialization of subsequent operations.

Dynamic serialization with cluster locks ensures the delivery of the Consistency property to all "cloud safe" transaction applications.

Like database locks, the "cluster locks" can have shared and exclusive levels; and with different granularity — database, table and row. Unlike database locks where the locks are typically acquired one at a time, a single cluster lock can synchronize multiple objects. For example, for a transaction updating three rows, say r1, r2 and r3, it can create a single cluster lock named r123. All other concurrent updates on r1, r2 or r3 can synchronize via the same lock. In this way we eliminate all potential deadlocks.

A "cluster lock" is really a named "mutex" that can be associated with a table, a row, a field value or an arbitrary symbol. Smaller lock sets can give better performance but increase the risk of deadlocks. The discipline of cluster lock optimization is exactly the same as database locks.

For practical purposes, we will also need two more tags to complete RML. Here is the complete list:

1. Lock (name, level)/unlock (name). The Lock tag sets up a named

mutex with lock level designation. The Unlock tag frees the named mutex. Proper use of this pair can prevent all potential deadlocks and race conditions.

2. LoadBalance on/off. There are three purposes. It can instruct DB^x gateway to load balance read-only stored procedures. It can also force the gateway to load balance a complex query that the parser fails to analyze. This tag can also be used to enforce transaction isolation levels by turning off dynamic load balancing.

3. NoRep on/off. This tag suppresses the replication functionality. This is necessary for cluster administration.

These tags are inserted in application programs as SQL comments. Correct use of RML ensures the Consistency property for all "cloud-safe" transaction processing applications, including applications using stored procedures, triggers, and runtime generated SQL dialects. Transaction isolation levels can be preserved at the expense of processing concurrency. RML also allows performance optimization and effective cloud administration, since not all administrative tasks need cloud-wide replication. The formal ACID property proof is inductive; that is omitted here for brevity.

In summary, this section has described the treatments for meeting ACID properties using statistical database multiplexing. The basic challenge is that the DB^x gateway is not a true transient storage for complete transactions. It is only capable of "seeing" one query at a time. Correct use of the locking tags (Section 19.5.9) and automatic transaction re-transmission (Section 19.5.8) can satisfy the ACID requirements and transient storage requirement simultaneously. Like the ESB project, the database clients are responsible for the transient transaction storages until their statuses are confirmed.

19.5.7 Cluster Failure Model

Since the DB^x gateway does not maintain transaction states, like the ESB messaging servers, DB^x failures can be masked by simple physical replacement of passive DB^x gateways (IP-takeover or DNS takeover). For the same reason, transaction re-transmits can automatically discover alternative resources and can ensure unique execution of mission critical transactions. This very feature also allows DB^x gateways to upscale processing performance by horizontally distributing transaction loads (active redundancy).

Failback from a repaired gateway is trickier. Non-stop service can be accomplished using a virtual IP and the following algorithm.

Let X = virtual IP address that is known to all database clients, G = Group of 2 DB^x gateways supporting the services provided via X. Each DB^x gateway has a real IP address G_i.

The zero downtime failover and failback sequence is as follows:

1. Select a primary gateway by binding G_i with X Set G_j to monitor and failover to X.

2. When G_j detects G_i's failure, G_j takes over by binding G_j with X.

3. When failback, remove X from G_i binding, set G_i to monitor and failover to X.

This simple algorithm ensures the elimination of all single-point-failures in a DB^x cluster and absolute zero downtime for multiple consecutive DB^x gateway failures.

19.5.8 Lossless Transaction Processing

As in ESB, we use the term "lossless transaction processing" analogous to lossless packet switching protocol. Like TCP/IP, applications with automatic re-transmits represent the "best effort" lossless transaction processing protocol that exploits infrastructure redundancy for overall transaction processing reliability and performance gains. Permanent error occurs only after all component redundancies are exhausted.

The key to every transaction is the database updates. There are two types of database updates:

1. Updates that cannot tolerate redundant commits. Additive updates, such as $X = X + 100$, are examples. These updates (or transactions) must have unique IDs for tracking to maintain the "commit once" property.

2. Updates that can tolerate redundant commits or are *idempotent*. For *idempotent* updates and simple inserts, tracking IDs may not be necessary.

For the lossless transaction processing protocol, the only difference between (1) and (2) is the status verification of the last timed out transaction. Figure 19.6 shows the pseudo code for automatic transaction re-transmission.

It is worth mentioning that the probability of transaction permanent error decreases exponentially as the number of redundant database servers increases. For example, if the probability of permanent failure is 10^{-6} for a single database server, then the probability of permanent failure using P servers is 10^{-6p}.

If all transactions are protected by this re-transmission logic, as in ESB (Figure 19.4), we can claim lossless transaction processing allowing arbitrary failures of hardware, software and network components.

19.5.9 Cloud-Ready VLDB Application Development

One of the last sufficient condition for the maximally survivable transaction processing system is that the VLDB applications must satisfy the ACID prop-

```
Count=0;
Tid=GetTransID();

Repeat:
    Begin Tran:
        Update1;   // Insert Tid to sys$trans$table
        Update2;
        ---        // Other operations
    End Tran;

catch (timeout | TransactionException):
{
    if (!Exist(Tid))
    {
        // Only necessary for once-only re-transmit.
        if ((Count++) > Max)
            FatalError("Fatal Error:", Tid, QueryBuffer);
    }
    goto Repeat;
}

int function Exist(Tid int)
{
    sprintf(QueryBuffer,
        "select * from sys$trans$table where Tid=\%d", Tid);
    SQL_exec(DBH, QueryBuffer);
    bind(ReturnBuffer, DBH);
    while (!Empty(ReturnBuffer))
    {
        if (Tid == ReturnBuffer) return (1);
        next;
    }
    return (0);
}
```

FIGURE 19.6
Automatic Re-Transmission.

erties. RML and automatic re-transmission are two important additions for the making of cloud-ready lossless transaction processing applications (OLTP). There are four steps:

1. Move all server-specific backend data inserts (such as $GUID$ and $Timestamp$) to client-end. This eliminates data inconsistencies by the known sources.

2. Markup all concurrent update conflicts with cluster locks.

3. Markup all concurrent reads for acceptable isolation behavior.

4. Add automatic re-transmit logic: There should be a new transaction processing API: RCommit, where RCommit(0) bypasses the last transaction status verification otherwise it automatically adds a unique transaction ID, transmits the transaction until all alternatives are exhausted. In other words:

   (a) RCommit(1) should be used for all updates requires tracking IDs (TCP-like).

   (b) RCommit(0) should be used for all other updates without tracking IDs (UDP-like).

Note that transactions with *idempotent* updates can gain performance advantages by not using tracking IDs.

As in ESB, the cloud-ready VLDB applications will generate a permanent error for each re-transmission that exceeds a pre-defined threshold (Max). The permanent error log will contain the failed transaction details so it can be manually executed at a later time. The probability of permanent error decreases exponentially as the degree of spatial redundancy increases.

The cloud-ready VLDB applications can also enjoy accelerated asynchronous replication on remote targets by employing dynamic serialization on the far-flung replication targets. Although the queuing effects remain the same (throttle is still needed) the same buffer size will tolerate much heavier update loads compared to existing asynchronous replication schemes.

19.5.10 Unlimited Performance Scalability

There are three ways we can leverage multiple real time synchronized databases for performance scalability:

1. Dynamic load balancing by distributing read-only queries (and stored procedures) among all database servers in the cluster. This benefit is delivered by compromising the default isolation behavior.

2. Session-based load balancing by distributing "sticky" connections to the database servers in the cluster. Without any code changes, all OLAP applications are "cloud safe" applications in this mode.

3. Data partitioning with DB^x replication.

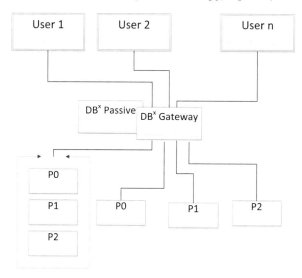

FIGURE 19.7
Replicated Partitioned Database ($P=3$, $R=2$).

Item (1) delivers localized performance benefits for OLTP applications with relaxed isolation requirements. Item (2) enables plug-and-play performance boost for all OLAP applications. Item (3) enables unlimited performance scalability for OLTP applications.

These benefits were predicted for share-nothing database clusters many years ago [677].

To see item (3), consider database partition without using DB^x. To date, database partition has been used to meet performance demands at the expense of reduced overall system availability. The overall system reliability is adversely affected as the number of partitions increases; higher number of partitions represents more potential failure points. With DB^x, we can meet arbitrary high performance and reliability demands by creating large number of partitions (P) while maintaining a smaller degree of redundancy (R) (Figure 19.7).

Any heavily accessed data portion can be partitioned and distributed to multiple servers. The growth of data tables can be regulated automatically via DB^x gateway [771]. The small degree of redundancy eliminates availability concerns with added dynamic load balancing benefits. The entire system can deliver virtually non-stop service, allowing any number of subsets to be repaired/patched/resynchronized. DB^x gateway also conveniently affords database structural changes without shutting down service for extended periods.

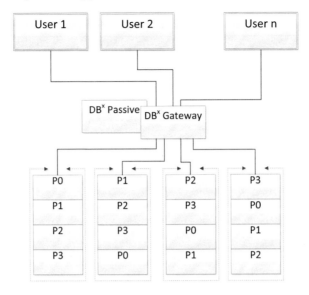

FIGURE 19.8
K-Order Shift Mirroring, $K=P=4$.

For further acceleration of heavy SELECT queries, a k-order shift mirroring (KSM) scheme can be deployed, where $1 < k \leq P$. The idea is to create a RAID-like database cluster as shown in Figure 19.8.

KSM is suitable for extremely demanding OLAP applications with light updates. A higher degree of redundancy can unleash the full parallel processing potentials of all processors. KSM uses the same ideas as in the RAID storage system to implement a parallel transaction processing architecture with full parallel I/O support [645].

In Figure 19.8, "join" two partitioned tables across four servers can be done in two steps:

1. Parallel partial joins. For example, for tables A and B, there are 16 partial join pairs $[A_0 * B_0]$, $[A_0 * B_1]$, $[A_0 * B_2]$, $[A_0 * B_3]$, $[A_1 * B_0]$, $[A_1 * B_1]$,...,$[A_{15} * B_{15}]$. All 16 pairs can be executed in parallel on four servers.

2. Merge and post processing.

Linear speedups can be expected.

For extreme scale OLTP applications, since R-degree of synchronous transaction replication can be as small as 2 and P is unrestricted, therefore DB^X clusters can deliver unlimited performance scalability.

19.6   Maximally Survivable High Performance Computing

Scientific computing is another basic NCA that can benefit from this statistical multiplexing magic. Unlike the transaction processing system where a lost transaction may not be recovered, given identical inputs, every lost computation task can always be re-computed to deliver semantically identical results, even for non-deterministic applications. Therefore, if every HPC (High Performance Computing) application can be decomposed into stages of "bag of tasks" (Figure 19.9), cheap fault tolerance and non-stop HPC is potentially possible. In theory, unlimited performance scalability should also be within reach.

This section describes the high level concepts of the Synergy Project [646] at Temple University. The Synergy Project was designed to address:

1. Maximally survivable HPC applications (Sections 19.6.1–19.6.3).

2. Unlimited performance scalability with heterogeneous processor types including GPGPU, DSP, single and multicore processors (Sections 19.6.5–19.6.6).

Section 19.6.4 introduces a practical parallel performance modeling method. A maximally survivable HPC application does not have the information loss issue. It has, however, a higher dimension scalability issue that cannot be addressed by performance scalability and application survivability alone: parallel programmability (Section 19.6.6).

Each HPC program requires combined skills in processing and communication hardware, domain knowledge and computer programming. While job security is granted to the few talented individuals, the productivity is notoriously low. Section 19.6.6 reports an attempt to address this high order scalability issue using a Parallel Markup Language (PML).

19.6.1   Protection for the "Bag of Tasks"

Each HPC application contains two types of programs: a) sequential (not parallelizable); and b) parallelizable. There are three types of parallelism [275]: SIMD (Single Instruction Multiple Data), MIMD (Multiple Instruction Multiple Data) and pipeline.

In literature, MIMD is further divided into SPMD (Single Program Multiple Data) and MPMD (Multiple Program Multiple Data). Due to the topological similarity, we consider SIMD and SPMD synonymous.

By far, SPMD/SIMD is the simplest and the most dominant parallelism type that delivers high performance using massively many identical processing elements. In legacy parallel programming environments, such as message passing

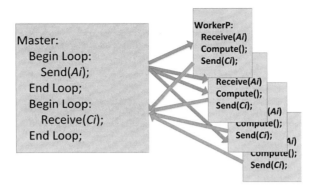

FIGURE 19.9
Message-Based "Bag of Tasks" Parallel Processing.

and shared memory, exploiting MPMD and pipeline parallelisms requires inventive programming. For data parallel processing, these parallelism types are automatically exploited [231] [550].

Each SPMD component can be implemented as a "bag of tasks." Each "bag of tasks" group consists of two kinds of programs: a master that distributes working assignments and collects the results; and replicated workers that continuously work for available assignments. The natural dataflow between the master and workers automatically exploits MPMD and pipeline parallelisms (Figure 19.9).

For fault tolerance, it is apparent that master failure can only be protected by check-pointing. The failure of workers, however, may be recovered by re-issuing the lost working assignments.

Specifically, master failures may only be recovered via either global or local check-pointing. Global check-pointing requires freezing and saving the state of the entire computation (often with tens of thousands of processing elements), thus it can be very costly. Local check-point has a much smaller "foot print" but suffers the difficulty of consistent global state reconstruction [174]. It is possible, however, to execute synchronized local check-points and recover a consistent global state with the help of a distributed synchronized termination algorithm [688].

In order to understand the potential performance benefits of protecting the SPMD/SIMD components (workers), for simplicity, the following discussion assumes the same overhead for global and local master check-pointing.

We define the expected computing time with failure using the following model [317]:

t_0 : Interval of application-wide check-point

α : Density function of any processing element failure

K_0 : Time needed to create a check-point

K_1 : Time needed to read and recover a check-point

T : Time needed to run the application without check-points.

Further, we define

α_1 : Density function of critical (non-worker) element failure

α_2 : Density function of non-critical (worker) element failure

Thus,

$\alpha = \alpha_1 + \alpha_2$.

Assuming failure occurs only once per check-point interval and all failures are independent, then the expected running time, E, per check-point interval with any processing element failure is

$$E = (1 - \alpha t_0)(K_0 + t_0) + \alpha t_0(K_0 + t_0 + K_1 + t_0/2) \qquad (19.1)$$

The expected running time per check-point interval with worker failure tolerance will be:

$$E' = (1 - \alpha t_0)(K_0 + t_0) + \alpha_1 t_0(K_0 + t_0 + K_1 + t_0/2) + \alpha_2 t_0(K_0 + t_0 + X), \qquad (19.2)$$

where $X =$ recovery time for worker time losses. We can then compute the differences $E' - E$, as follows:
Since:

$\alpha_2 = \alpha - \alpha_1$

$E = (1 - \alpha t_0)(K_0 + t_0) + \alpha t_0(K_0 + t_0 + K_1 + t_0/2)$

$E' = (1 - \alpha t_0)(K_0 + t_0) + \alpha_1 t_0(K_0 + t_0 + K_1 + t_0/2) + \alpha_2 t_0(K_0 + t_0 + X)$

$E - E'$

$= (\alpha - \alpha_1)t_0(K_0 + t_0 + K_1 + t_0/2) - \alpha_2 t_0(K_0 + t_0 + X)$

$= \alpha_2 t_0(K_0 + t_0 + K_1 + t_0/2 - K_0 - t_0 - X)$

$= \alpha_2 t_0(K_1 + t_0/2 - X)$

In other words, the savings equal to the product of the probability of partial (worker) failure and the sum of check-point reading time and master lost time with an offset of lost worker time. Since the number of workers is typically very large, the savings are substantial.

The total expected application running time E_T without worker fault tolerance is:

$$E_T = \frac{T}{t_0}(K_0 + t_0 + \alpha(t_0 K_1 + \frac{t_0^2}{2})) \tag{19.3}$$

We can compute the optimal check-point interval as:

$$\frac{dE_T}{t_0} = T(-\frac{K_0}{t_0^2} + \frac{\alpha}{2})$$

$$0 = -\frac{K_0}{t_0} + \frac{\alpha}{2}$$

$$\frac{K_0}{t_0^2} = \frac{\alpha}{2}$$

$$t_0 = \sqrt{\frac{2K_0}{\alpha}} \tag{19.4}$$

The total application running time E_T with worker fault tolerance is:

$$
\begin{aligned}
E_T' &= \frac{T}{t_0}\left(K_0 + t_0 + \alpha t_0 K_1 + \frac{\alpha t_0^2}{2} - \alpha_2 t_0 K_1 - \frac{\alpha_2 t_0^2}{2} + \alpha_2 t_0 X\right) \\
&= T\left(1 + \frac{K_0}{t_0} + \alpha K_1 + \frac{\alpha t_0}{2} - \alpha_2 K_1 - \frac{\alpha_2 t_0}{2} + \alpha_2 X\right) \tag{19.5}
\end{aligned}
$$

The optimal check-point interval with worker fault tolerance is:

$$\frac{dE_T}{t_0} = T(-\frac{K_0}{t_0} + \frac{\alpha - \alpha_2}{2})$$

$$0 = -\frac{K_0}{t_0} + \frac{\alpha - \alpha_2}{2}$$

$$\frac{K_0}{t_0} = \frac{\alpha - \alpha_2}{2}$$

$$t_0 = \sqrt{\frac{2K_0}{\alpha - \alpha_2}} \tag{19.6}$$

For example, if we set the check-point interval $t_0 = 60$ minutes, the check-point reading and writing time $K_0 = K_1 = 10$ minutes, and the average worker failure delay $X = 30$ sec $= 0.5$ minute, the expected savings per check-point under any single worker failure is about 39.5 minutes.

$E - E'$

$= (\alpha - \alpha_1)t_0(K_0 + t_0 + K_1 + t_0/2) - \alpha_2 t_0(K_0 + t_0 + X)$

$= \alpha_2 t_0(K_0 + t_0 + K_1 + t_0/2 - K_0 - t_0 - X)$

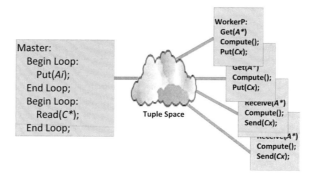

FIGURE 19.10
Parallel Processing Using Tuple Space.

$$= \alpha_2 t_0 (K_1 + t_0/2 - X) \mid \text{Since } \alpha_2 T_0 = 1 \text{ (any worker failure)}$$

$$= (10 + 30 - 0.5)$$

$$= 39.5 \text{ minutes.}$$

Furthermore, if the mean time between failure (MTBF) is 3 hours in a system of 1024 processors, this gives $\alpha t_0 = \alpha 180 = 1$ or $\alpha = 1/180$. Thus, $\alpha_1 = 1/(180P) = 1/184,320$. The optimal check-point interval (in minutes) is:

$$t_0 = \sqrt{\frac{2K_0}{\alpha - \alpha_2}} = \sqrt{2 \cdot 10 \cdot 184,320} = 1,920 \qquad (19.7)$$

This means that it is not necessary to check-point the masters unless the application running time T is greater than 30 hours.
Implementing the "worker fault tolerance" scheme requires:

1. All workers must be free from global side-effects. Otherwise the correctness of the parallel programs may not be maintained.

2. Dynamic sender-receiver binding at runtime. Otherwise worker failure recovery cannot be done.

Running side-effect free parallel programs also has the benefits of reduced check-point sizes.

19.6.2   Data Parallel Programming Using Tuple Space

Among all existing HPC programming paradigms, such as message passing (MPI [276]), share memory (OpenMP [173]), Tuple Space (Linda [165] and Synergy [646]), only the Tuple Space paradigm seems a natural fit (Figure 19.10).
A Tuple Space supports three operators:

1. *Put*(TupleName, value);

2. *Read*(NamePattern, &buffer);

3. *Get*(NamePattern, &buffer);

The *Read* operator fetches the value from a matching tuple. The *Get* operator extracts the value from a matching tuple and destroys the tuple. Both operators are "blocking" that they suspend the calling program indefinitely until a matching tuple is found. Tuple name matching allows the same tuple to be fetched by different programs. This allows "shadow tuple" implementation (transient tuple storage).

Specifically, under the Tuple Space parallel processing paradigm, the master simply *puts* working assignments into a tuple space. The massively many workers simply repeat the same loop: retrieve the working assignments, compute and return the results. The Tuple Space runtime system can easily support automatic worker failure detection and recovery by implementing "shadow tuples" — tuples masked invisible after being retrieved via the "get" operator, but recovered if the corresponding worker crashes, and destroyed only after the corresponding result is delivered [646].

The problem was that the tuple Space parallel programming paradigm requires indirect communication and implicit parallel programming. These ideas seemed counterproductive against the performance centric designs of HPC systems where programmers are used to manipulating program structures to match hardware features. To pursue higher performance, some earlier Tuple Space efforts even tried to manipulate the compiler's code generator [165] [55]. For mitigating massive component failures and for attaining unlimited scalability, the transient tuple storage is not only necessary but also desirable due to statistic multiplexing needs. This argument can only be persuasive if a physical Tuple Space architecture could demonstrate the feasibility.

As discussed in Section 19.3, for this type of NCA service, the UT (Unit of Transmission) is a working assignment. The worker's work-seeking code is naturally amenable to the re-try logic with "shadow tuple" support from the potential Tuple Space architecture. The theoretical framework is identical to data parallel computing ([231] [99] and [55]). The Tuple Space runtime (centralized or distributed) implementations could support the tuple transient storage with spatial hardware redundancy.

Theoretically, these treatments transform the seemingly insurmountable HPC performance, scalability and reliability issues into solvable communication problems.

19.6.3   Stateless Parallel Processing Machine

The failure of any path in the interconnection network of a multiprocessor will cause the HPC application to halt. Therefore the interconnection network is the single point-of-failure in all HPC systems. The problem was caused by the lack of statistical multiplexing on the semantic network. Without a properly

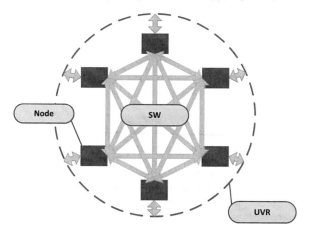

FIGURE 19.11
Stateless Parallel Processor.

defined UT (Unit of Transmission), interconnection network virtualization would only enhance data transmission with little help to the HPC applications scalability.

Further, explicit parallel programming paradigms, such as message passing (MPI) and share memory (OpenMP), rely on direct low-level resource binding to deliver performance. Semantic network statistical multiplexing is not possible.

These observations inspired the design for a Stateless Parallel Processor (SPP) [647].

1. Architecture: The maximally survivable HPC architecture will feature a UT defined statistic multiplexed data parallel interconnection network. It is matched with a UT defined high level data parallel API and a distributed transient UT storage implementation with passive spatial redundancy. The conceptual diagram SPP is depicted in Figure 19.11 [647].

   In Figure 19.11, SW represents the passive spatial redundancy using multiple redundant interconnection networks. The number of interconnection networks must match the number of interfaces on each compute node.

   UVR stands for Unidirectional Virtual Ring — a virtualized interconnection network responsible for transporting application data and system administration requests. The dotted ring represents the virtual paths that data requests travel. The solid lines represent the physical data travel paths. Further, not shown in Figure 19.11, the HPC application programming interface (API) supports the three

Tuple Space operators. Together, these are a higher level interconnection network responsible for the robust UT operations.

In this design, since application API and the supporting architecture are parts of statistical multiplexing for the defined UT, there is no single point-of-failure for any application's workers. The masters are only application single point-of-failures. Masters can be protected by check-pointing and recovery (CPR).

In other words, failure of computing or communication components can be masked by transient storage (for workers) and check-points (for masters). The earlier discussion had made it clear that the master check-point overhead can be minimized drastically if worker fault tolerance is supported.

The only scalability concern is the round-trip overhead of UVR. Assuming multiple redundant physical switched networks, the worst-case UVR round trip latency will be on the order of $\lg_k(P)$ [85], where P is the maximal number of computing nodes per application and k is the fan-out factor (degree of parallel communication). For example, for an application running on a million nodes with k=2, at most 20 hops are required for all participating nodes to communicate their data needs. Actual data transfers will be in parallel using the multiple redundant interconnection networks [648] [647].

UVR is only required for each HPC application in order to maintain a single consistent system image. Therefore, there can be multiple UVRs for multiple applications. There is no limit on the number of computing nodes. Like the Internet, theoretically there is also no limit on the number of networks to be included, even though each node does have a physical limit on the number of installed network interfaces.

2. Implementation: A Data Routing Agent (DRA) implements the UVR using the passive spatially redundant interconnection networks. DRA will also support the UT-based communications and synchronization (Tuple Space semantics) [648] [647].

Unlike other tuple space implementations [165] [120], DRA is a distributed tuple space mechanism with automatic failure detection and recovery capabilities. The DRA has a T-shaped communication pattern (see "Compute Node" in Figure 19.11). It is responsible for three tasks:

(a) Local Tuple Space operations: tuple storage, matching and program activation, tuple recovery upon notification of retriever's failure, program activation, control and monitor, program check-point and re-start.

(b) Global Tuple Space services: data request service, provision-

ing and fast parallel forwarding. It must account for potential
downstream node and link failures.

(c) Administrative services: User account management, application
activation, control and monitor.

3. Automated Master Protection: Even though the masters are not
natively protected by the SPP architecture, the worker fault tolerance feature has reduced master check-point sizes. Unlike application level check-pointing, where the programmer is responsible to
find the most economic location of the global state saving points,
it is possible to implement a non-blocking system level check-point-
restart protocol automatically with little overhead. The technical
challenge of migrating communication stacks between heterogeneous compute nodes can be handled by a "precision surgery" of
the TCP/IP kernel stack [776]. The theoretical difficulty [653] [452]
of reconstructing a consistent global state from multiple independently check-pointed masters can be resolved using a synchronized
distributed termination algorithm [688].

Under the SPP paradigm, the "put," "read" and "get" operations will
block/unblock requesting programs to automatically form application dependent SPMD, MPMD and pipeline clusters at runtime.
The three Tuple Space operations allow application programs to express arbitrary communication patterns including broadcasts, neighbor-to-neighbors
and one-to-one conversations.
Each HPC application builds its own semantic network at runtime. Runtime
statistical multiplexing allows automatic protection of a massive number of
workers. SPP applications can deliver virtually non-stop service regardless of
multiple computing and communication component failures.

19.6.4   Extreme Parallel Processing Efficiency

Unlike explicit parallel programming paradigms where overlapping computation and communication requires inventive programming, finding the optimal
processing grain size in a data parallel application can automatically maximize
concurrency. Processors form dynamic SPMD, MPMD, and pipeline clusters
automatically for a data parallel application. Optimal granularity ensures zero
synchronization overhead, thus maximal overlapping of computing and communication.
Optimal granularity can be found using a coarse-to-fine linear search method.
There are two steps: a) find the most promising loop depth; and b) find the
optimal grouping factor. The first step is labor intensive since each change
in partition loop depth requires reprogramming. This is the very reason that
has prevented practical HPC applications from being optimized. Once the
loop depth is found, optimal grouping factors can be calculated statically or
heuristically at runtime [583] [356].

Since deeper loop decomposition (finer processing grain) risks higher communication overheads, maximizing the degree of parallelism can adversely impact performance. The optimal loop depth gives the best possibility to deliver the optimal parallel performance.

Finding the optimal loop depth can be done experimentally via some back-of-envelop calculation or aided by an analytical tool we call timing models [649] [651].

The timing model for a compute intense loop is an equation containing estimation models for major time consuming elements, such as computing, communication, synchronization, and disk I/O. Setting the synchronization time to zero makes it possible to find the performance upper bound that all parallel tasks complete in time. In reality, the synchronization time may be negative when computing overlaps with communications.

For example, let

$$T_{seq}(n) = \frac{cf(n)}{\omega'(n)} \tag{19.8}$$

be the running time of a program of input size n, where $f(n)$ is the time complexity, $c > 0$ captures the efficiency losses (instruction to algorithmic step ratio) of the programmer, compiler, and operating system scheduler, and $\omega'(n)$ is the processor speed measured in instructions processed per second. Since both c and $\omega'(n)$ are hard to obtain in practice, we introduce

$$\omega(n) = \frac{\omega'(n)}{c} \tag{19.9}$$

measured in algorithmic steps per second. Thus $\omega(n)$ can be obtained from program instrumentation:

$$\omega(n) = \frac{f(n)}{T_{seq}(n)} \tag{19.10}$$

$\omega(n)$ is typically smaller than the manufacturer's peak performance claim. $\omega(n)$ will also vary according to problem size (n) due to memory handing overheads. A typical memory intense application will exhibit $\omega(n)$ characteristics (in MOPS = Millions Operations Per Second) as shown in Figure 19.12. Depending upon the partitioned problem sizes (n), Figure 19.12 gives the entire spectrum of $\omega(n)$. This is useful in speedup calculations.

Ignoring the disk IO and synchronization times, a parallel application's timing model contains the following:

$$T_{par}(n, p) = T_{comp}(n, p) + T_{comm}(n, p). \tag{19.11}$$

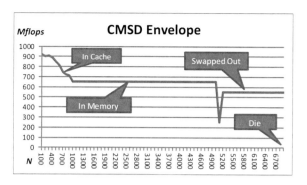

FIGURE 19.12
Application Dependent CMSD Envelope.

Let μ be the network capability in bytes per second; a parallel matrix multiplication timing model is:

$$T_{par}(n,p) = \frac{n^3}{p\omega(n)} + \frac{\delta n^2(p+1)}{\mu}; \tag{19.12}$$

where n = problem size, p = number of processors, μ = the application dependent network capability in bytes per second, and δ = matrix cell size in bytes. This model represents a row or column partitioning strategy but not both (tiling).
The sequential timing model is:

$$T_{seq}(n) = \frac{n^3}{\omega(n)}. \tag{19.13}$$

The speed up is:

$$Sp = \frac{T_{seq}(n)}{T_{par}(n,p)} = \frac{p}{1 + \dfrac{\delta\omega(n)(p^2+p)}{n\mu}} \tag{19.14}$$

Given $\delta = 8$ bytes (double precision), $\omega(n) = 300$ MOPS and $\mu = 120$ MBPS, Figure 19.13 shows the performance map of parallel matrix multiplication for a small number of processors and problem sizes.
A timing model helps to determine the optimal partition depth. For example, if $n = 10,000$, for the matrix multiplication program with the processing environment characterized by $\omega(n) = 300$ MOPS and $\mu = 120$ MBPS; it is not a good idea to pursue anything deeper than the top-level parallelization since the speed up will not be greater than 12. As shown in Figure 19.13,

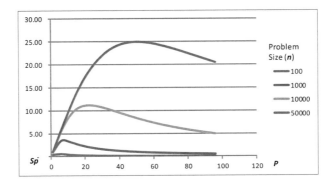

FIGURE 19.13
Parallel Performance Map of Matrix Multiplication.

spreading calculations onto too many processors can have severe adverse performance effects. However, deeper loop partitioning should be re-evaluated if the processing environment changes, such as slower processors, larger problem sizes, faster network, or a combination of these. For simulation applications, the optimal processing grain size should be identified using the same process within each simulated time period. This is because the time marching loop does not impact computing versus communication ratio.

Theoretically, the Timing Model method overcomes the impossibility of program performance prediction (due to Turing's Halting Theorem [716]) by introducing application dependent $\omega(n)$. $\omega(n)$ is obtained via program instrumentation. The rest is straightforward.

19.6.5   Performance Scalability

The statistically multiplexed interconnection network not only eliminates the single point-of-failure but also allows the applications to leverage multiple redundant networks simultaneously. Higher network speed allows for better performance scalability. This is the enabling factor.

The ultimate performance scalability can be achieved using a hybrid approach analogous to the Internet. The idea is to wrap the data parallel processing layer on top of multiple "coarse grain workers." Each coarse grain worker is a piece of HPC code running in the conventional message passing or share memory environment (analogous to the circuit-switching networks). The overall application is a multi-scaling algorithm [30] employing multiple coarse grain workers integrated via the Stateless Parallel Processing virtual machine. The SPP properties will deliver the extreme dependability and efficiency for these very large scale parallel applications.

Multi-scale modeling and simulation [30] is an emerging field involving mathematics, chemistry, physics, engineering, environmental science and computer science. Traditional mono-scale simulation has been proven inadequate due to prohibitively high costs of interconnection communication. The focus is on the fundamental modeling and computational principles of underlying multi-scale methods. These studies have been supported by observations of multi-scale phenomena.

Since the maximal survivability framework is based on the Unit of Transmission (tuples), as long as the supporting architecture can find a way to transmit tuples, it can mitigate multiple simultaneous runtime (computing and communication) component failures for all applications. Non-stop HPC service has thus become possible by allowing online component crashes, repair and reboots (since the application architecture supports UT fault tolerance). Like the Internet, a HPC application's performance and reliability is only bound by the maximal degree of passive spatial redundancy.

19.6.6 Automatic Parallel Program Generation – The Higher Dimension

Traditional parallel programming requires a combination of three hard to acquire skills: domain knowledge, programming and processing/communication hardware architecture. Although very high performances have been delivered through carefully optimized codes, the productivity of these systems is notoriously low. A typical HPC production system takes many years to mature. Once it has reached a usable stage, the underlying sciences could have evolved to a point that the value of the entire system may be in question.

Recently, using markup language to aid automated parallel program generation has achieved varying degrees of successes ([670], [796] and [637]). This section reports a similar attempt but focused in Tuple Space parallel programming.

Parallel programming using Tuple Space relies on tuples to communicate and synchronize parallel programs. This is a five step process:

1. Identify the computing intensive parts and separate them into independent sub-programs.

2. Identify data dependencies among all parts. This defines the tuples to connect the sub-programs.

3. For each computing intensive sub-program, decide a partition loop depth.

4. Design a "bag of tasks" scheme for each loop partition.

5. Develop a complete sequential program for each sub-program. This will produce the necessary masters and workers.

The resulting parallel programs will run in a Tuple Space supported environment, such as Synergy [646] [650] where the worker will be automatically replicated to run on multiple compute nodes.

```
<reference></reference>
<parallel>
        <reference></reference>
        <master>
            <send> or <read>
            <worker>
            <send> or <read>
            <target>
                    the loop to be parallelized
            </target>
            <send> or <read>
            </worker>
            <send> or <read>
        </master>
</parallel>
```

FIGURE 19.14
PML Tag Structure.

For performance optimization, each partition depth change requires re-coding steps 3–5. This is labor intensive. PML (Parallel Markup Language) [678] was developed to ease the finding of optimal processing granularity and for the ease of parallel programming in general.

Like all other parallel program markup efforts ([670], [637] and [796]), the core concept of PML is to eliminate automated dependency analysis — a common component in traditional parallel compilers. Instead, the user is required to perform dependency analysis and mark the results in the sequential source code using PML tags.

PML is a XML-like language designed for generating data parallel programs from sequential programs. It contains seven tags (Figure 19.14).

The "reference" tag marks program segments for direct source-to-source copy in their relative positions. The "master" tag marks the range of the parallel master. The "send" or "read" tags define the master-worker interface based on their data exchange formats. The "worker" tag marks the compute intense segment of the program that is to be parallelized. The "target" tag defines the actual partitioning strategy based on loop subscripts, such as tiling (2D), striping (1D) or wave-front (2D). The coarse-to-fine grain size search is to place "target" tags in an outer loop first and then gradually drive into deeper loop(s) if the timing model indicates there are speedup advantages in the deeper loops.

Figure 19.15 shows the PML marked sequential matrix multiplication program.

In this example, the variable declarations of i, j and k will be copied exactly.

```
/* <reference id="123"> */
int i, j, k;
/* </reference> */
```

The first three lines of "*master*" tag define the output data interface to workers. The two "*put*" tags insert two tuples named "B" and "A" with double precision $[N, N]$ cells to the space.

```
/* <master id="123"> */
/* <put var="B" type="double[N][N]" opt="ONCE" /> */
/* <put var="A" type="double[N][N]"/> */
```

The "*worker*" tags define how the tuples are to be accessed. Tuple "B" will be read "$ONCE$." Tuple "A" must be retrieved along the i subscript for N times.

```
/* <worker> */
/* <read var="B" type="double[N][N]" opt="ONCE"/> */
/* <get  var="A" type="double[N(i)][N]"/> */
```

The "*target*" tag defines the partition depth and grouping factor. In this case, the partition will happen at subscript i, within the range of $[0, N]$ step 1 and group G. This is the 1st order partition.

```
/* <target index="i" limits="(0,N,1)" chunk="G" order="1"> */
  for (i = 0; i < N; i++)
/* </target> */
```

The "worker" body concludes with a single output tag describing the overall dimensions and the partitioned dimension $[N(i)]$.

```
/* <put var="C" type="double[N(i)][N]"/> */
/* </worker> */
```

Finally, the "*master*" is responsible for collecting the results with a single *read* tag.

```
/* <put var="C" type="double[N][N]"/> */
/* </master> */
```

In the "*target*" tag, the "*type*" attribute design allows complex loop (subscript) manipulations. In this example, the expression $[N(i), N]$ indicates an one dimension (leftmost) parallelization. The grouping factor is described by the

```
/* <parallel appname="matrix"> */
main(int argc, char **argv[]) {
    /* <reference id="123"> */
    int i, j, k;
    /* </reference> */

    /* <master id="123"> */
    /* <put var="B" type="double[N][N]" opt="ONCE" /> */
    /* <put var="A" type="double[N][N]" /> */

    /* <worker> */
    /* <read var="B" type="double[N][N]" opt="ONCE"/> */
    /* <get  var="A" type="double[N(i)][N]"/> */
    /* <target index="i" limits="(0,N,1)" chunk="G" order="1"> */
    for (i = 0; i < N; i++)
    /* </target> */
    {
        for (k = 0; k < N; k++)
            for (j = 0; j < N; j++)
                C[i][j] += A[i][k]*B[k][j];
    }
    /* <put var="C" type="double[N(i)][N]"/> */
    /* </worker> */
    /* <put var="C" type="double[N][N]"/> */

    /* </master> */
    exit(0);
}
/* </parallel> */
```

FIGURE 19.15
PML Marked Matrix Program.

"chunk" attribute. Expression $[N(i), N(j)]$ indicates a two-dimension paral-lelization. As mentioned earlier, deeper parallelization typically require higher communication volume. It should be used with care. Similarly, sliding windows, equal partitions, and wave-forms can all be implemented via similar mechanisms.

In the "*master*" tag, it is also possible to insert check-point instructions.

To show the feasibility of the PML approach, a PML compiler was constructed [678]. The PML compiler generates two parallel programs: a master and a worker. We tested the manually crafted and the generated programs in the Synergy parallel processing environment [646] [649].

We also installed MPICH2-0.971 [276]. It was compiled with -enable-fast switch. A MPI parallel matrix application program was also obtained.

All test programs were compiled with gcc (version 2.95.3) -O3 switch. All tests were conducted using a Solaris cluster consisted of 25 Sun Blade500 processors connected via 100 Mbps switches. All nodes are of exactly identical configuration.

The Synergy and PML experiments were timed with worker fault tolerance turned on. The Synergy master and MPICH2 programs have no fault tolerance protection.

Figure 19.16 shows the recorded performance results comparing Synergy hand crafted programs, PML generated programs and MPI parallel programs.

This study [678] also included other common computational algorithms, such as Laplacian solver using Gauss-Seidel iteration, block LU factorization, convolution and others. This study revealed that a) the tuple space can be augmented to reduce code generation complexities; and b) the PML tags can be very flexible to accommodate arbitrarily complex partition patterns. Like other markup languages, [670], [796] and [637], practice and memorization help with coding efficiency. In comparison, the scale of efforts is still much less than coding data parallel applications directly.

For extreme scale applications, we could develop new PML tags to include the support for multi-scale algorithms [30] with automatic check-point generation. The new features would allow programmers to compose sequential multi-scale algorithms and use existing MPI or OpenMP codes as coarse-grain workers directly. PML tags can parallelize the multi-scale program with direct references (via the "reference" tag) to the legacy MPI or OpenMP codes.

19.7   Summary

Cloud computing has brought the "mission critical features" within the reach of all applications. Reaping the full cloud benefits, however, requires non-trivial efforts. However, none of the challenges is new. This chapter has described

Nodes(P)	Size(n)	PML	Manual	MPICH2	Sequential
2	600	6.7	5(G=25)	5.12	8.9
2	800	15.3	12.2(G=200)	11.87	21.6
2	1000	28.3	23.4(G=63)	22.86	42.4
2	1600	118.3	95(G=100)	95	181.7
2	2000	231.3	187.6(G=75)	186	358
4	600	4.5	3.6(G=16)	3.3	8.9
4	800	10	7.8(G=12)	7.2	21.6
4	1000	17.3	14.1(G=13)	13.4	42.4
4	1600	66.7	53(G=23)	53	181.7
4	2000	128.3	101.2(G=21)	100	358.7

FIGURE 19.16
PML Performance Comparisons.

the fundamentals of mission critical applications for scalability, availability and information assurance goals.

Starting from the packet switching networks, we identified the generic "DNA" sequence that is essential to the robustness of extreme data communication systems. Using information theory, we argued that extreme scale networked computing applications (NCA) are also possible if statistic multiplexing is introduced to the application semantic networks.

By the end-to-end principle [26], we also argued the necessary condition of the maximally survivable NCA to contain four elements: Unit of Transmission (UT), transient spatial redundancy, temporal redundancy, and passive spatial redundancy. The sufficient conditions include maximal NCA service/data availability, unlimited NCA scalability and loss free NCA processing. NCA correctness and integrity are also assumed.

This chapter also describes specific steps toward meeting the necessary and sufficient conditions of mission critical applications. We have included detailed "DNA" embedding sequence for three basic NCAs: mission critical Enterprise Service Bus (ESB), lossless mission critical transaction processing and nonstop high performance computing.

Although the results are non-trivial, somewhat experimental still; it can be hard to argue with the technological direction.

We have only scratched the surface. The basic NCA services: messaging, storage, transaction processing and computing, suggest the direction for the development of next generation networking equipment and distributed application APIs. Differentiating from the traditional silo development processes, these are integrating frameworks based on the NCA semantic networks. The next generation systems are likely to include more advanced support for the basic NCA services.

The holistic NCA development methodology has brought the essential but non-functional factors to the surface. We hope this chapter could help the application API designers, system developers, application integrators and network developers to better understand the critical factors for solving some of the most difficult non-functional networked computing problems.

19.8   Acknowledgments

The author wishes to thank many students and colleagues who contributed in different ways to the body of work described here. In particular, contributions of Kostas Blathras, John Dougherty, David Muchler, Suntian Song, Feijian Sun, Yijian Yang, and Fanfan Xiong are essential to the philosophical developments in the "decoupling" school of thought.

The projects described here are partially supported by the National Science Foundation, the Office of Naval Research, Temple University Vice Provost Office for Research and Parallel Computers Technology Inc. Special thanks to Professors David Clark (MIT) and Chip Elliott (GENI) for the informal discussions after their Distinguished Speaker talks on TCP/IP protocols, transaction losses and information entropies.

This manuscript was particularly inspired by recent invigorating discussions with author's new colleagues Drs. Abdallah Khreishah, Shan Ken Lin and Ph.D. student Moussa Taifi, on fault tolerance modeling and information theory.

20

An Ontology for the Cloud in mOSAIC

Francesco Moscato
Department of European and Mediterranean Studies
Second University of Naples, Italy

Rocco Aversa, Beniamino Di Martino, Massimiliano Rak and Salvatore Venticinque
Department of Information Engineering
Second University of Naples, Italy

Dana Petcu
Computer Science Department
Western University of Timisoara

CONTENTS

Cloud Computing has emerged as a model to provide access to large amount of data and computational resources by using simple interfaces. Computing is being transformed to a model consisting of services and they are delivered without regard to where and how services are hosted. The ease of using and configuring resources has made this model widespread and several enterprises now offer Cloud-based services to end users and businesses. Anyway, Cloud Computing is still an emerging architecture, and new providers with different services rise every month. In addition, it is very hard to find a single provider which offers all services needed by end users. For this reason, a way to provide a common access to Cloud services and to discover and use required services in Cloud federations is appealing. This is a problem, since different Cloud systems and vendors have different ways to describe and invoke their services, to specify requirements, and to communicate. The mOSAIC project addresses these problems by defining a common ontology and it aims to develop an open-source platform that enables applications to negotiate Cloud services as requested by users. The main problem in defining the mOSAIC ontology is in the heterogeneity of terms used by Clouds vendors, and in the number of standards which refer to Cloud Systems with different terminology. In this work, the mOSAIC Cloud Ontology is described. It has been built by analyzing Cloud standards and proposals. The Ontology has been then refined by introducing individuals from real Cloud systems.

Keywords: Ontology, Semantics, Cloud, API

20.1 Introduction

Cloud Computing is an emerging model for distributed systems. It refers to both applications delivered as services and to hardware, middleware and other software systems needed to provide these services. Nowadays the Cloud is drawing attention from the Information and Communication Technology (ICT) thanks to the appearance of a set of services with common characteristics which are provided by industrial vendors. Even if the Cloud is a new concept, it is based upon several technologies and models which are not new and are built upon decades of research in virtualization, service oriented architecture, grid computing, utility computing or distributed computing ([359, 518, 756]). The variety of technologies and architectures makes the Cloud overall picture confusing [359]. Cloud service providers make resources accessible from the Internet to users, presenting them *as a service*. The computing resources (like processing units or data storages) are provided through virtualization. *Ad-hoc* systems can be built based on users' requests and presented as services (*Infrastructure as a Service*, IaaS). An additional abstraction level is offered for supplying software platforms on virtualized infrastructure (*Platform as a Service*, PaaS). Finally, software services can be executed on distributed platforms of the previous level (*Software as a Service*, SaaS). Except for these concepts, several definitions of Cloud Computing exist ([154, 292, 391, 510, 571, 619]), but each definition focuses only on particular aspects of the technology. Cloud computing can play a significant role in a variety of areas including innovations, virtual worlds, e-business, social networks, or search engines but it is actually still in its early stages, with consistent experimentation to come and standardization actions to effort. Cloud computing solutions are currently used in settings where they have been developed, without addressing a common programming model, open standard interfaces, adequate service level agreements or portability of applications. In this scenario, vendors provide different Cloud services at different levels, usually providing their own interfaces to users and Application Program Interfaces (APIs) to developers. This results in several problems for end-users that perform different operations for requesting Cloud services provided by different vendors, using different interfaces, languages and APIs. Since it is usually difficult to find providers that fully address all users' needs, interoperability among services of different vendors is appealing.

Cloud computing solutions are currently used in settings where they have been developed without addressing a common programming model, open standard interfaces, adequate service level agreements, or portability of applications. Neglecting these issues, current Cloud computing offers force people to be stranded into locked, proprietary systems. Developers making an effort in Cloudifying their applications cannot port them elsewhere.

In this scenario the mOSAIC project (EU FP7-ICT program, project under grant #256910) aims at improving state of the art in Cloud computing by creating, promoting and exploiting an open-source Cloud application programming interface and a platform targeted for developing multi-Cloud oriented applications. One of the main goals is that of obtaining transparent and simple access to heterogeneous Cloud computing resources and to avoid lock-in proprietary solutions.

In order to attain this objective, a common interface for users has to be designed and implemented, which should be able to wrap existing services, and also to enable intelligent service discovery. The keystone to fulfill this goal in mOSAIC is the definition of an ontology able to describe services and their (wrapped) interfaces.

20.2   The mOSAIC Project

The Open Cloud Manifesto [553] identifies five main challenges for a Cloud:

- data and application interoperability;

- data and application portability;

- governance and management;

- metering and monitoring;

- security.

Actually, the main problem in Cloud computing is the lack of unified standards. Market needs drive commercial vendors to offer Cloud services with their own interfaces, since no standards are available at the moment. Vendors' solutions have arisen as commonly used interfaces for Cloud services but interoperability remains a hard challenge, like the portability of developed services on different platforms. In addition, vendors and open Cloud initiatives spent few efforts in offering services with a negotiated quality level.

The mOSAIC project tries to fully address the first two challenges and partially addresses the next two by providing a platform which:

- enables interoperability among different Cloud services,

- eases the portability of developed services on different platforms,

- enables intelligent discovery of services,

- enables services' composition,

- allows for management of Service Levels Agreements (SLA).

The architecture of mOSAIC platform is depicted in Fig.20.1:

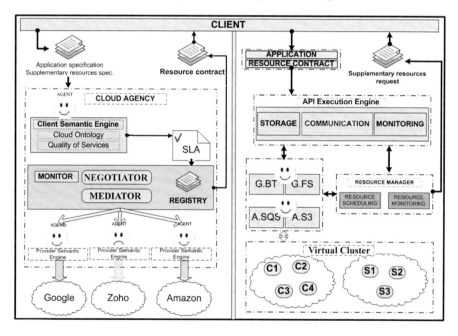

FIGURE 20.1
mOSAIC Architecture.

it provides facilities both for end-users (at the left of Fig.20.1) and for services developers and managers (depicted on the right side of Fig.20.1)
From the end-users' point of view, the main component is the *Cloud Agency*. This consist in a core set of software agents which implement the basic services of this component. Basic services include:

- negotiation of SLAs;

- deployment of Cloud services;

- discovery and brokering of Cloud services.

In particular, the *Client Agent* is responsible for collecting users' application requirements, for creating and updating the SLAs in order to grant always the best QoS. The *Negotiator* manages SLAs and mediates between the user and the broker; it selects protocols for agreements, negotiates SLA creation, and it handles fulfillment and violation. The *Mediator* selects vendor agents able to deploy services with the specified user requirements; it also interfaces with services deployed by different vendors' providers. The *Provider Agent* interacts with virtual or physical resources at provider side. In mOSAIC the Cloud Agency was built upon the MAGDA [108] toolset, which provides all the facilities to design, develop and deploy agent-based services. The *semantic engine* uses information in the Cloud Ontology to implement a semantic-based Cloud services discovery, exploiting semantic, syntactic and structural schema matching for searches.

In the Cloud developers' and managers' perspective, the main components of mOSAIC Architecture are the *API execution engine* and the *Resource Manager*. The first component offers a unique API to use Cloud Services from different vendors in using and developing other services. The API execution engine is able to wrap storage, communication and monitoring features of Cloud platforms. In particular, Virtual Clusters (VC) [252] are used as resource management facilities. Virtual Clusters are configured by software agents in order to let a user setup his configuration of required services. A *Resource contract* will grant users' requirements and the Resource Manager will assign physical resources to VC on the basis of a contract.

In this architecture, the bonding element which allows for interoperability and resources description is the *Cloud Ontology*. It is the base for Cloud services and resources description and it contains all information needed to characterize API also from a semantic point of view.

The Cloud Ontology is based on several Cloud taxonomies proposed in literature [93, 184, 447, 464, 694, 695]. It is developed in OWL [504] and OWL-S languages [499]. The benefit of using an ontology language is that it acts as a general method for the conceptual description or modeling of information that is implemented by actual resources. mOSAIC aims at developing ontologies that would offer the main building block to describe services of the three delivery models of Cloud Computing (i.e., IaaS, PaaS, SaaS).

20.3   Languages for Ontologies Definition

As mentioned before, the mOSAIC project aims at providing interoperability among Cloud systems and standards. The keystone for interoperability is the definition of an ontology able to provide a common definition of concepts related to Cloud domains. The main problem in the definition of this ontology is in the heterogeneity of Cloud levels and components. The Cloud Ontology must be able to define concepts from Cloud domains, and also to describe Cloud components like infrastructures, platforms, and services.

Many languages have been defined in literature for ontologies description, each language with its own features and expressiveness.

We can distinguish ontology languages into two main classes:

- languages based on First Order Logic (FOL) and Description Logic (DL);

- rule-based languages.

The two classes approach the problem of knowledge representation in a somewhat different way. Languages in the first class are used to define all possible representations (or models) of the domain, and axioms are used to eliminate the models that do not satisfy them. Rule-based languages, instead, define a set of rules that are exploited in order to *compute* the model, based on a data set. Integrity constraints are used to check if the model satisfies certain conditions.

Several languages in the first class are based on the Resource Description Framework (RDF) [549], which is a framework defined by W3C for representing information in the web. It is recommended by W3C as base of languages for the Semantic Web and allows for the definition of graph based data models which manage information in a minimal and flexible way. RDF and RDF Schema (RDFS) allow the representation of information by means of modeling primitives which concern the organization of vocabularies in typed hierarchies. Some important features are missing in the RDF/RDFS, like the ability of defining ranges for properties and relationships, or of defining disjoint classes of concepts, cardinality, etc.

OWL [504] overcomes RDF problems by defining proper extensions for RDF and RDF Schema. Three sublanguages (OWL Lite, DL and FULL) are provided in order to cope with expressiveness and the complexity of reasoning on OWL ontologies. Indeed only OWL Full is an extension of RDF and RDF Schema, but decidability problems on description logics have led to define new languages based on restrictions on RDF and RDF Schema. OWL lite was intended to support the definition of simple hierarchies and constraints in order to manage thesauri and taxonomies. For example, it restricts the cardinality values to 0 or 1. Indeed, the development of tools supporting OWL Lite needs the same effort used to develop OWL DL tools, hence OWL Lite sublanguage is not widely used. OWL DL, instead, was create to provide the best trade off between expressiveness, computational completeness, and decidability in order to allow for definition of practical reasoning algorithms. The result was the development of a number of reasoners, including Fact++ [714], Pellet [659], RACER [326] and HermiT [644], and of a number of tools which support OWL ontology creation, like Protégé [542] or Swoop [407].

Anyway, experience with OWL has shown that its most expressive but still decidable sublanguage lacks of several constructs that are necessary for modeling complex domains [314].

These problems have been solved with OWL 2 [314] which is still based on RDF semantics and still can be translated into RDF graphs, but allows also the definition of qualified number restrictions and increases Datatype expressiveness. OWL 2 has three sublanguages too, called profiles: OWL2 EL, OWL2 QL and OWL2 RL. The first one has polynomial time reasoning complexity; the second one is designed to enable easy access to data stored in databases, by means of a proper query language, while the last one has been designed to implement reasoning as a set of rules in a forward-chaining rule system.

From the Rule-based languages' perspective, RuleML [132] and Datalog [159] have been widely used for knowledge representation. RuleML allows for the definition transformation and reaction rules to manage information. Reaction rules are used to define actions to execute when some information is retrieved or derived from data sets, while transformation rules define derivation rules managing facts, queries, and integrity constraints. Datalog allows the addition of a number of deductive rules to a database. These rules complement the model with additional information that can be inferred by applying rules to a given database instance, leading to a domain representation with additional information.

Several attempts have been made to merge the two kinds of language. An example is the Semantic Web Rule Language (SWRL) [364], which is a combination of OWL DL and OWL Lite with Datalog RuleML. In SWRL, OWL has been extended with a set of axioms which are able to specify Horn-like rules that can be combined with OWL knowledge bases.

The necessity of ontologies for sharing information among different agents is particularly felt in the web. Semantic Web aims at providing semantic-based searches and services by using ontologies to describe services' features and interfaces and to represent information to be used. In particular, languages for semantic description of services have been developed for Service Oriented Architecture (SOA) standards. Web Services Description Language (WSDL) [245] is a well established building block for web services description. Hence several languages for semantic description of Web Services uses WSDL to specify the syntax of the input and output messages of a basic service, as well as other details needed for the invocation of the service. For this reason, several languages for semantic description of web services are based on WSDL. OWL for Services (OWL-S) [499] is an OWL ontology with three related ontologies, known as: profile, process model, and grounding. The profile is used to give a semantic definition of service functionalities in order to give information needed to build the service requests and to perform matchmaking. The process model is used to describe how the service works, enabling service invocation, activation and monitoring, providing a set of concepts to specify capabilities of services. The grounding deals with message formats and protocols (which are usually expressed in WSDL). OWL-S characterizes services in terms of its inputs, outputs, preconditions and effects (IOPE). Input and output are described semantically by OWL concepts and properties are then *grounded* to WSDL descriptions. Preconditions are logical formulae that need to be satisfied before the execution of a service while effects state what will be true upon the successful execution of the service. Preconditions and effects are expressed by RuleML-based languages.

Another WSDL-based language is Web Service Semantics (WSDL-S) [593]. It is basically a language for annotating existing WSDL with semantics. The domain ontology representations are external to the annotation, which is embedded in the WSDL description. In this way it is possible to choose the language to use for ontology description. Like OWL-S, WSDL-S also describes semantics of services' preconditions and effects.

Similarly to OWL-S, Semantic Web Services Language (SWSL) [625] uses a logic-based language (SWSL-FOL) to express ontologies and a rule-based language(SWSL-Rules) for supporting reasoning. The first language is used to define domain concepts and relationships, the second one is used for profile specification, services discovery and policy specification. As in other languages, grounding is defined for WSDL.

A slightly different approach is the one that is adopted by Web Service Modeling Ontology (WSMO). It describes ontologies for providing a vocabulary to use with other WSMO descriptions. Postconditions and effects of web services are defined by WSMO Goals. They consist of the objectives that a client would achieve when he is consulting a web service. In addition to preconditions and effects (called postconditions in WSMO), which are modeled in OWL-S too, WSMO also allows separate modeling of the state space of the environment where services are executed. In addition, composition of services is supported by WSMO by *interface* definition, that gives details about composed services by choreography or orchestration. Finally heterogeneous components in WSMO are linked together by *mediators* that manage mappings, transformations or reductions between linked elements.

20.4 Cloud Standards and Proposals

Nowadays several Cloud computing systems are available, both from commercial and open source communities. Some examples are Amazon EC2 [65], Google's App Engine [312], Microsoft Azure [109], GoGrid [307], 3Tera [40], Open Nebula [667], Eucalyptus [544] and Nimbus [590]. Cloud systems and services offered by various vendors differ and overlap in complicated ways. Each solution provides different services. For example Amazon EC2, GoGrid, 3Tera, Open Nebula and Eucalyptus are basically IaaS Clouds offering entire instances of virtual machines to customers; Google's Apps and Microsoft Azure offer SaaS applications also providing API for development and monitoring, offering a PaaS Cloud. Nimbus was developed as an IaaS for scientific applications. Main vendors' platforms and services have become standards *de facto* for Cloud computing, but several different solutions exist at different

Cloud layers and interoperability is still a distant goal. In this scenario, some attempts have been made to make order in the chaos of Cloud systems by trying to propose standards for them.

The Cloud Computing Interoperability Forum (CCIF) [168] aims at defining an open, vendor neutral and standardized Cloud interface for the unification of various Cloud APIs. This should be done creating an API, wrapping other, existent APIs. CCIF proposes to define an OWL/RDF ontology to describe a semantic Cloud data model in order to address Cloud resources uniquely. The ontology is still under development and no draft version are available at the moment.

Similarly, the Open Grid Forum (OGF) [551] is another open initiative which aims at the creation of a practical solution to interface existing Cloud IaaS. OGF is defining interfaces (the Open Cloud Computing Interface: OCCI [551]) to provide unified access to existing IaaS resources. The main goal of OCCI is the creation of hybrid Clouds operating environments. OCCI has released white papers where an attempt to define Cloud resources and APIS has been made. The main formalism used to define the Cloud model is UML, and the work is still in a preliminary stage.

The National Institute of Standards and Technology (NIST) is also working at Cloud standards definitions. At the moment, it basically gives a definition of Cloud Service Models (i.e., IaaS, PaaS, SaaS) and introduces Cloud Deployment models (Private, Community, Public and Hybrid Clouds).

The need for a good, complete definition of Cloud components is really felt by scientific community. In particular, in [783] the need for an Ontology defining Cloud-related concepts and relationships is outlined. In the paper an ontology for the Cloud is proposed *in natural language*. Cloud layers have been defined and organized in an architectural view. The ontology starts with firmware and hardware as its foundation, eventually delivering to Cloud applications. The paper also defines elements which belong to different layers, like resources, virtual machines, etc. No formal representation of the Ontology is reported in the paper.

Another similar taxonomy for Cloud Systems has been presented in [609], where only Cloud layers and some requirements like fault tolerance and security have been discussed.

A more detailed taxonomy has been described in [94]. It is a simple taxonomy, with only the main concepts related to Cloud Computing defined in a graphical schema. This work in progress is more complete than the previous ones.

One of the few attempts to provide a formal ontology for Cloud System comes from Unified Cloud Interface (UCI) Initiative that has released a very simple OWL ontology for Cloud Systems but it consists only of few concepts.

20.5   mOSAIC Ontology

An analysis of the state of the art in Cloud computing reveals the need for an ontology for Cloud Computing. This is the missing piece for enabling interoperability among Cloud Systems and to allow for semantics-driven searches of Cloud services and resources. In order to be able to address interoperability issues, the following requirements have to be satisfied:

- the ontology has to be defined in a formal language;

- not only Cloud resources have to be considered, but also other features like quality of services, security etc.;

- a way for semantically describing Cloud services interfaces and APIs has to be defined;

- existing standards and proposals have to be addressed defining a proper methodology for retrieving concepts and relationships from specification documents;

- existing platforms and services have to be included in the ontology as individuals and they have to be properly semantically annotated;

- the specific domain of offered services has to be specified in the ontology in each different case;

- in order to trace the source documents for concepts and relationships, proper annotations have to be introduced in the ontology.

The language chosen for mOSAIC ontology development is OWL2. The reason for this choice has to be sought in the fact that OWL is one of the most used languages for defining formal ontologies. It is supported by several tools for ontology development and many libraries have been developed to manage OWL files. In addition, some reasoners exist for OWL and OWL2 ontologies which can be used by mOSAIC Semantic Engine. Several annotation models exist for OWL ontologies, and this is another reason for choosing OWL. In particular, an external annotation model is used in mOSAIC to annotate services and API. The choice of OWL is also dictated by the need for semantically describing Cloud services interfaces and API. This allows OWL-S to be used for services and APIs description, which in turn are modeled in mOSAIC *as a service*. Finally, OWL allows for annotation of ontology's classes, properties and individuals, meeting the tracing requirements.

A proper methodology is defined to build the ontology. It retrieves information from other documents and proposals, maintaining a good degree of compatibility with them. In addition, the ontology has to contain information about real commercial vendors' solutions and open source systems. This is managed by introducing proper individuals in the OWL ontology.

Quality of Services, security, and other requirements (with Service Level Agreement -SLA) information will be codified in a proper ontology. For domain description concerns, proper ontology will be defined for each case.

In the following the methodology used for ontology definition is sketched. Then the main structure of the mOSAIC ontology is described.

20.5.1 The mOSAIC Methodology for Building Cloud Ontology

The methodology used to build the Cloud Ontology can be resumed in the following steps:

1. Collection: Documents from standards proposals, scientific literature, manuals, white papers etc. are collected. Documents are stored in proper repositories in order to be easily indexed and retrieved.

2. Analysis: Documents are analyzed (by humans or by tools for automatic retrieval of ontologies) in order to:

   - Identify concepts (i.e., OWL classes);

   - Identify relationships (i.e., OWL properties);

   - Identify concept instances (i.e., OWL individuals).

3. Creation: For each identified element an ontology element is created;

4. Ontology Annotation: the newly created element is associated to an annotation reporting the document and the position of the sentence or the figure from which the element has been created.

5. Instance Annotation: a semantic annotation is produced with the model presented in [525] and associated to the retrieved instance.

In Fig.20.2 an example an annotated document from OCCI is reported. This is the core document where the definition of a Cloud *Resource* is reported. In this document, a resource is defined as: *Any resource exposed through OCCI ... A resource can be e.g a virtual machine, a job ..., a user etc. ....* In the figure, the concepts identified in the sentence have been surrounded by a circle. They are: *Resource, Virtual Machine, Job.* A *superclass/subclass* relationship can be retrieved from the text (a resource *can be* a virtual machine).

In Fig.20.3 the creation in Protegé of the class Resource and Virtual Machine with the subclass property is shown. The annotation of the class reporting the name of the document and the sentences to which the class is related is shown in the box on the right.

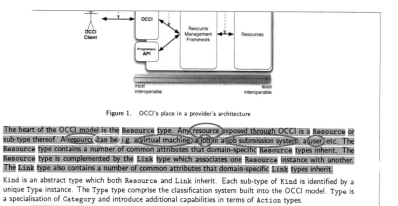

Figure 1. OCCI's place in a provider's architecture

The heart of the OCCI model is the Resource type. Any resource exposed through OCCI is a Resource or sub-type thereof. A resource can be e.g. a virtual machine, a job in a job submission system, a user, etc. The Resource type contains a number of common attributes that domain-specific Resource types inherit. The Resource type is complemented by the Link type which associates one Resource instance with another. The Link type also contains a number of common attributes that domain-specific Link types inherit.

Kind is an abstract type which both Resource and Link inherit. Each sub-type of Kind is identified by a unique Type instance. The Type type comprise the classification system built into the OCCI model. Type is a specialisation of Category and introduce additional capabilities in terms of Action types.

FIGURE 20.2
Analysis of a Document.

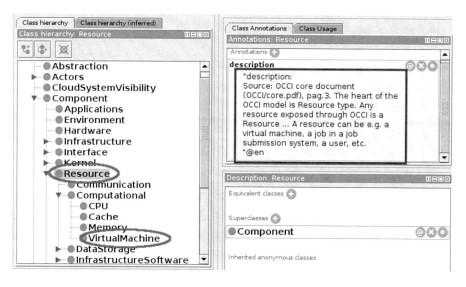

FIGURE 20.3
Class Creation.

20.5.2 Cloud Ontology Main Structure and Components

The mOSAIC Cloud Ontology is structured as in Fig.20.4.

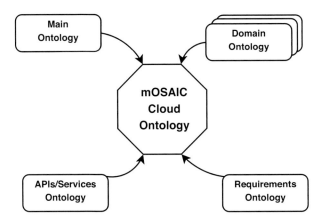

FIGURE 20.4
Main Structure of mOSAIC Cloud Ontology.

It is the result of the merge of four kinds of ontologies:

- Main Ontology,

- Domain Ontologies,

- APIs and Services Ontology,

- Requirements Ontology.

The **Main Ontology** is expressed in OWL. It contains the definition of general concepts and relationships related to Cloud systems. For example, the concept of Cloud Resource, of virtual machine, of service layer etc. are specified in this ontology. This ontology will be described later in depth.

Several **Domain Ontologies** are created. Each domain has its own ontology. This is needed because usually different domains have different vocabularies: different concepts may not exist in all domains, and it is also possible that the same term (used to identify a concept) has different meanings (or connections with other concepts) in different domains. These ontologies are expressed in OWL too.

APIs and Services Ontology describe the model, the profile and the grounding for Cloud API and services. Here they are expressed in OWL-S because they can be managed as web services from this point of view. For example, let us assume that we are developing a service by using the *blobstore* [131] JAVA API for Google's App Engine. The Blobstores are used to manage large amounts of data. The JAVA API for blobstores uses a Factory pattern (*BlobstoreServiceFactory*) to create a Blobstore service (*BlobstoreSer-vice*) that will be used to index, upload and download data. The only method in BlobstoreServiceFactory is *getBlobstoreService()* with no input and a Blob-storeService as output. The OWL-S description of this method will have the *hasInput* part empty; the *hasOutput* part filled with the description of the datatype BlobStoreService and the grounding will bind the service to the *op-eration: getBlobStoreService()* where proper message parts defining the seri-alization of BlobstoreService class are used as input. Proper namespaces sub-stitute for the usual *wsdl* namespace when defining grounding for non-wsdl services.

Requirements Ontology has the main goal of modeling Service Level Agree-ments (SLA) between Cloud consumers and providers. Due to the dynamic na-ture of Clouds, continuous monitoring on Quality of Services (QoS) attributes is necessary to enforce SLAs. This ontology models concepts and relationships needed for the definition of requested and provided SLAs.

The top level of the Main Ontology is shown in Fig.20.5.

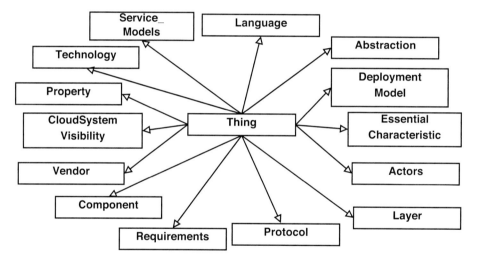

FIGURE 20.5
Main Ontology Top Level.

Notice that not all the classes will be shown during the explanation, for brevity's sake.

In brief, the *Language* class contains instances of languages used for APIs im-plementation. Examples are JAVA and Python languages for Google's Apps Engine. The **Abstraction** class models the abstraction level to which ser-vices are provided by Cloud providers [783]. Cloud services belong to the same layer if they have an equivalent level of abstraction. For example if we consider Cloud's users' point of view, all Cloud software environments target programmers, while Cloud applications target end users. Therefore, Cloud software environments would be classified in a different abstraction layer than Cloud applications. **Deployment Model** and **Essential Characteristic** are concepts inherited from the NIST Cloud definition [540].

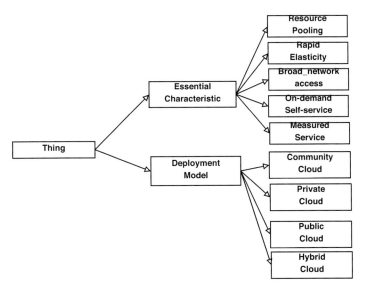

FIGURE 20.6
Main Ontology Based on NIST Classes.

In Fig.20.6 subclasses of these last concepts are reported. Essential Characteristic defines minimal characteristics and QoS that any Cloud System has to provide. They comprise Rapid Elasticity for scalability; Measured Service to control and optimize resources; Resource pooling to serve multiple consumers using a multi-tenant model with different physical and virtual resources; Broad network access to promote heterogeneous client platforms; On-demand self-service to allow transparent provisioning of computing capabilities. On the other hand, Deployment Model defines the types of deployment for the Cloud System. Deployment can address Private or Public Clouds, Hybrid Clouds from composition of two or more Clouds, and Community Clouds sharing several organizations and supporting a specific goal.

The **Actors** class identifies Cloud actors like end-users and programmers; the **Layer** superclass is shown in Fig.20.7.

Layers distinctions are defined in [783] and can be divided into: Firmware and Hardware Infrastructure; Software Kernel for Cloud platforms; Software Infrastructure; Software Environment with monitoring tools and APIs, and Application layer with all Cloud services for end-users.

The **Protocols** class models all protocols used to manage Cloud applications, services and components. Some individuals from this class are *REST* and *SOAP* protocols. The **Requirement** and *Property* classes are links to Requirements Ontology and will be described later.

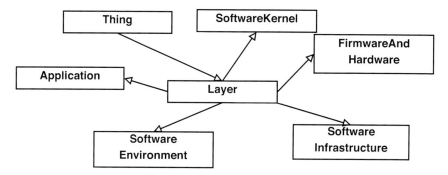

FIGURE 20.7
Layer Subclasses.

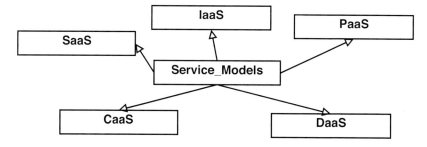

FIGURE 20.8
Service Models Subclasses.

Service Models include all kinds of services provided by Cloud Systems. In Fig.20.8 its subclasses are reported. They include the elements introduced in [783] and in [540]: Hardware as a Service (HaaS), Communication as a Service (CaaS) Data as a Service (DaaS), Infrastructure as a Service (IaaS), Platform as a Service (PaaS), Service as a Service (SaaS). The **Technology** class includes all technologies used in Cloud Systems, like virtualization etc. The **Vendor** superclass organizes Cloud providers. They are divided into vendors for: Applications, Infrastructures, Platforms and Services and the taxonomy is further expanded depending on type of service or platform provided etc. **Cloud System Visibility** defines visibility depending on Deployment Models.

The **Component** subtree is more articulated. It is reported in Fig.20.9. Cloud components are: Cloud Environment; Cloud Services, Resources, Infrastructure, Applications, Interfaces, Hardware, and Cloud Kernel. In particular, the Kernel class includes Cloud and other middlewares (like Grid Middleware in scientific Clouds); Monitor systems for Cloud resources, and Operating systems for Cloud platforms.

Infrastructures include Cloud systems, Computing Grid, Data Grid and Vir-

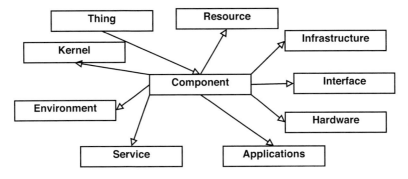

FIGURE 20.9
Component Subclasses

tual appliances, while Interfaces include APIs, services interfaces and all interfaces to access and manage Cloud components. Instances in this class include for example, Amazon EC2 infrastructure services.

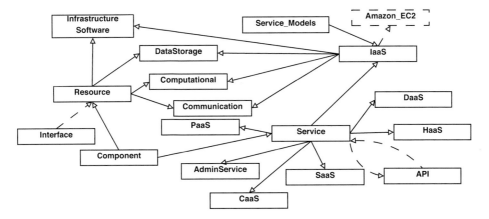

FIGURE 20.10
Resource Subclasses.

Fig.20.10 shows Resource and Services classes in mOSAIC Cloud Ontology. The **Resource** class includes Computational and Data Storage resources as well as Infrastructure software, Communication Resources and, in general, any other Cloud Components (as specified in OCCI documents). **Services** are classified into IaaS, CaaS, DaaS, PaaS, Saas and HaaS as shown for Service Models. In addition, services to administer and manage Cloud resources are considered here.

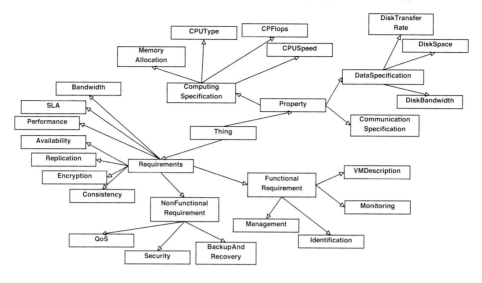

FIGURE 20.11
Requirements and Properties.

Requirements and Properties concepts are listed in Fig.20.11. Requirements can be divided into functional and non-functional requirements. Examples for requirements a Cloud system may meet are QoS, Security, Backup and Recovery features, Performances, Availability, Replication, Monitoring, etc.

For what Cloud components concern, Properties can be divided into computing, data, and communication related properties. Some examples of computing properties are the type of the CPU, the measure of its FLOPS, the memory allocated at run-time, the total amount of memory, etc. Data properties include disk space, disk transfer rate, and disk bandwidth.

20.6   Conclusions

In this work we propose a detailed ontology for Cloud systems that can be used to improve interoperability among existing Cloud Solutions, platforms and services, both from the end-user and the developer side. The ontology has been developed in OWL and can be used for semantic retrieval and composition of Cloud services in the mOSAIC project. It will be released with an open license. Several attempts have been made in the past to introduce a Cloud ontology. The ontology presented in this chapter also maintains compatibility with previous works, because it is built upon existing standards and proposals analyses and it results in a more comprehensive description of all Cloud-related aspects. The Ontology has been populated with individuals from real Cloud

Systems services and APIs, and new individuals and elements are going to be included in the ontology with an incremental design approach.

Acknowledgements: This work was supported by the mOSAIC project; EU FP7-ICT programme, project under grant #256910.

21

On the Spectrum of Web Scale Data Management

Liang Zhao
University of New South Wales, Australia
National ICT Australia

Sherif Sakr
University of New South Wales, Australia
National ICT Australia

Anna Liu
University of New South Wales, Australia
National ICT Australia

CONTENTS

Over the past decade, rapidly growing Internet-based services have substantially redefined the way of data persistence and retrieval. Relational database management systems (RDBMS) have been considered as the *one-size-fits-all* solution for data persistence and retrieval for decades. However, ever-increasing needs for scalability and new application requirements have created high challenges for these systems. Therefore, recently, a new generation of low-cost, high-performance database software has emerged to challenge dominance of RDBMS named as *NoSQL* (Not Only SQL). The main features of these systems include: ability to horizontally scale, supporting weaker consistency models, using flexible schemas and data models, and supporting simple low-level query interfaces. In this chapter, we explore the recent advancements and the new approaches of the web scale data management. We discuss the advantages and the disadvantages of several recently introduced approaches and their suitability to support a certain class of applications and end-users. Finally, we present and discuss some of the current challenges and open research problems to be tackled in order to improve the current state-of-the-art.

21.1   Introduction

Over the past decade, rapidly growing Internet-based services have substantially redefined the way of data persistence and retrieval. By taking the recent advances in the web technology, content has been made easy for any user to provide and consume in any form. For example, building a personal web page (e.g., Google Sites[1]), starting a blog (e.g., WordPress[2], Blogger[3], and LiveJournal[4]) and making both publicly searchable for users all over the world have become a commodity. Arguably, the main goal of the next wave is to facilitate the job of implementing every application as a distributed, scalable, and widely-accessible service on the web. Services such as Facebook[5] Flickr[6], YouTube[7], Zoho[8], and Linkedin[9] are currently leading this approach. Such applications are both *data-intensive* and very *interactive*. For example, the Facebook social network contains 500 million users[10]. Each user has an average 130 friendship relations. Moreover, there are about 900 million objects that registered users interact with such as: pages, groups, events, and community pages. Other smaller scale social networks such as Linkedin, which is mainly for the professional has more than 80 million registered users. Therefore, it becomes an ultimate goal to make it easy for everybody to achieve high scalability and availability goals with minimum effort.

In general, relational database management systems (e.g., MySQL, PostgreSQL, SQL Server, Oracle) have been considered as the *one-size-fits-all*

---

[1]http://sites.google.com/
[2]http://wordpress.org/
[3]http://www.blogger.com/
[4]http://www.livejournal.com/
[5]http://www.facebook.com/
[6]http://www.flickr.com/
[7]http://www.youtube.com/
[8]http://www.zoho.com/
[9]http://www.linkedin.com/
[10]http://www.facebook.com/press/info.php?statistics

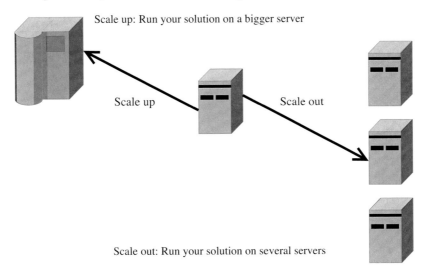

FIGURE 21.1
Database Scalability Options.

solution for data persistence and retrieval for decades. They have matured after extensive research and development efforts and very successfully created a large market and solutions in different business domains. However, ever-increasing needs for scalability and new application requirements have created new challenges for traditional RDBMS. Therefore, recently, there has been some dissatisfaction with this *one-size-fits-all* approach in some web scale applications. Typically, a three-tier approach, including the web server layer, the application server layer, and the data layer, is the most common architecture for building enterprise web applications. In practice, data partitioning [559] and data replication [423] are two well-known strategies to achieve the availability, scalability, and performance improvement goals in the distributed data management world. In particular, when the application load increases, there are two main options for achieving scalability at the database tier and making the application able to cope with more client requests (Figure 21.1):

1. *Scaling up*: aims at allocating a bigger machine to act as a database server.

2. *Scaling out*: aims at *replicating* and *partitioning* data across more machines.

In fact, the scaling up option has the main drawback that large machines are often very expensive and eventually a physical limit is reached, where a more powerful machine cannot be purchased at any cost. Alternatively, it is both extensible and economical — especially in a dynamic workload environment

— to scale out by adding storage space or buying another commodity server, which fits well with the new *pay-per-use* philosophy of cloud computing.

Recently, a new generation of database software with low-cost and high-performance has emerged to challenge dominance of relational database management systems. An important reason for this movement, named *NoSQL* (Not Only SQL), is that database requirements of web, enterprise, and cloud computing applications may vary because of different implementations (e.g., strong data consistency is not necessary for all applications). For example, scalability and high availability are essential requirements that can not be compromised for high-volume web sites (e.g., eBay, Amazon, Twitter, Facebook). For these applications, even the slightest breakdown can cause significant financial consequences and affect customer trust. The *CAP* theorem [143, 304] shows that a distributed database system can at most satisfy two out of three properties: Consistency, Availability and tolerance to Partitions. Therefore, most of these systems decide to compromise the strict consistency requirement. In particular, they apply a relaxed consistency policy called *eventual consistency* [731], which guarantees that if there is no new update for a period of time, eventually all retrievals will return the last updated value. If no failures happen, based on factors such as communication delays, the load on the system and the number of replicas involved in the replication scheme, the maximum size of the inconsistency window can be determined [731]. In particular, these new NoSQL systems share a number of common design features such as horizontal scalability, simple interface, weak consistency model, distributed indexes and RAM and semi-structured data schema to achieve the following system goals:

- *Availability*: They must always be accessible even during a network failure or when a whole datacenter has gone offline.

- *Scalability*: They must be able to serve very large databases under very high throughput rates at very low latency.

- *Elasticity*: They must be able to satisfy changing application requirements in both directions (scaling up or scaling down). Moreover, the system must be able to gracefully respond to these changing requirements and quickly recover its steady state.

- *Load Balancing*: They must be able to automatically move load from overloaded servers to underloaded ones so that most of the hardware resources are effectively utilized and avoid any resource overloading situations.

- *Fault Tolerance*: They must be able to deal with the situation when the rarest hardware problems go from being freak events to eventualities. While hardware failure is still a serious concern, this concern needs to be addressed at the architectural level of the database, rather than requiring developers, administrators and operations staff to build their own redundant solutions.

- *Ability to run in a heterogeneous environment*: The number of nodes can be increased to hundreds or thousands in a scaling out environment; however, the homogeneous performance across nodes is hardly guaranteed due to hardware performance degradation caused by parts failures. Hence, the system should be designed to run in a heterogeneous environment and must take appropriate measures to prevent performance degrading due to parallel processing on distributed nodes [46].

This chapter explores the recent advancements and the new approaches of the web scale data management. We discuss the advantages and the disadvantages of each approach, and its suitability to support a certain class of applications and end-users. Section 21.2 describes the NoSQL systems, which are introduced and used internally in the big players: Google, Yahoo and Amazon respectively. Section 21.3 provides an overview of a set of open source projects, which have been designed following the main principles of the NoSQL systems. Section 21.4 discusses the notion of providing database management as a service and gives an overview of the main representative systems and their challenges. The web scale data management trade-offs and open research challenges are discussed in Section 21.5, before we conclude the chapter in Section 21.6.

21.2 NoSQL Key Systems

This section provides an overview of the main NoSQL systems, which has been introduced and internally used by three of the big players in the web scale data management domain: Google, Yahoo and Amazon.

21.2.1 Google: Bigtable

Bigtable is used as a scalable, distributed storage system [176] in Google for a great number of Google products and projects such as: Google Docs[11], Google Earth[12], Google Finance[13], Google search engine[14], and Orkut[15]. These products can configure Bigtable for a variety of usages, supporting workloads from throughput-oriented job processing to latency-sensitive data serving, spanning servers from a handful number to thousands of commodity servers, and scaling data from a small amount to a size of petabytes.

The data model designed in Bigtable is not a relational data model, but a

---

[11]http://docs.google.com/
[12]http://earth.google.com/
[13]http://www.google.com/finance
[14]http://www.google.com/
[15]http://www.orkut.com/

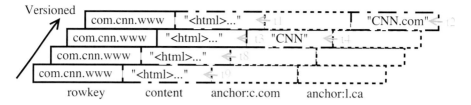

FIGURE 21.2
Sample BigTable Structure [176].

simple data model with dynamic control. Thus, users can change data layout and data format without being restricted by data schemas. In particular, Bigtable uses a sparse, multidimensional, sorted map to store data. Each cell in the map can be located by a row key, a column name, and a timestamp. A concrete example that reflects some of the main design decisions of Bigtable is the scenario of storing a collection of web pages. Figure 21.2 illustrates an example of this scenario where $URLs$ are used as row keys and various web elements as column names. Values of web elements such as contents and anchors of the web page were in versioned cells under the timestamps when they were fetched.

The row keys are sorted in lexicographic order in Bigtable. Every single row key is an atomic unit of read or write. Usually, ranges of row keys, named *tablets*, can dynamically span multiple partitions for distribution and load balancing. Therefore, a table with multiple ranges can be processed in parallel on a number of servers. Each row can have an unlimited number of columns. Sets of them are grouped into *column families* for access control rights. Each cell is versioned and indexed by timestamps. The number of n versions of a cell can be declared, so that only recent n versions are kept in decreasing timestamp order.

The Bigtable provides low-level APIs for following functions: creating, deleting, and changing tables and column families; updating configurations of cluster and column family metadata; adding, removing, and searching values from individual rows or a range of rows in a table. However, Bigtable does not support general transactions across row keys. Only atomic read-modify-write sequences on a single row, known as *single-row* transactions, are allowed.

On the physical level, the distributed Google File System (GFS) [303] is used to store Bigtable log and data files. The data is in Google *SSTable* file format, which offers an ordered, immutable, keys to values map for persistence. Bigtable relies on a distributed lock service called *Chubby* [148], which only runs if a majority of five composing replicas are accessible to each other. Among the five replicas, one of them is voted as *master*, initiatively serving all requests and balancing workloads across tablet servers. Each Bigtable has to be allocated to one master server and a number of tablet servers to be available. Hence, Bigtable can not work properly without Chubby, as it is

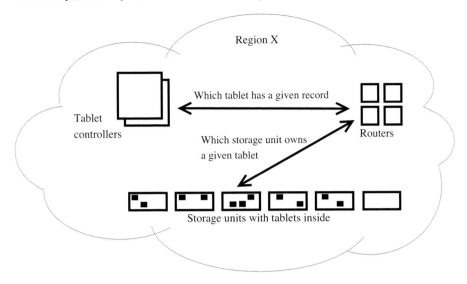

FIGURE 21.3
PNUTS System Components [205].

necessary for keeping the master server running, as well as storing information for Bigtable, such as bootstrap locations, schemas, and access control lists.

21.2.2   Yahoo: PNUTS

The *PNUTS* system (renamed later to Sherpa) is a scalable database system, storing tables of records with attributes to support web applications internally in Yahoo! [205]. The main goal of the system is serving data. Therefore, a list of functions is enhanced. First, a simple relational model is supported, avoiding complex queries. Second, *blob* is validated as a main data type, storing arbitrary structures in a record, in addition to large binary objects like image or audio. Third, the data schema of tables is enforced in a flexible way, allowing adding attributes at any time and keeping values of attributes empty in a record.

Figure 21.3 illustrates the system components of PNUTS. A region is a basic unit, which contains complementary system components such as storage units, tablets, tablet controllers, and routers, as well as a full copy of tables. In practice, the PNUTS system consists of multiple geographically distributed regions. On the physical level, *tablets* that are horizontal partitions of data tables are scattered across storage units in many servers. In each server, the number of tablets is variable, due to workload balancing, which shifts tablets from overloaded servers to underloaded ones. Hence, hundreds to thousands of tablets can be achieved in a server. The *router* (Figure 21.3) can determine the location of a given record in two steps. It resolves which tablet has a given record in first place, by querying cached interval mapping, which defines tablet boundaries, and maintains mapping correlations of tablets and storage units. Then, it determines which storage unit owns a given tablet, by applying mapping correlations to the given tablet. The *tablet controller* is the owner of interval mapping. It is also in charge of tablet management, such as moving a tablet across storage units for workload balancing or data recovery, or splitting a large tablet.

As mentioned above, the system is designed for data serving that consists mainly of queries of single record or small groups of records. The query model is designed to keep simple in mind. Thus, it provides selection and projection of a single table, but join operation is too expensive to provide. It also allows updating and deleting operations only on a primary key basis. Moreover, for reading multiple records, it supports a *multiget* operation for retrieving data in parallel.

PNUTS provides a consistency model that supports a variety of levels from general serializability to eventual consistency [731]. The model is driven by the fact that web applications normally operate one record at a time, whereas different records may be manipulated in different geographic areas. Thus, the model defines *per-record timeline* consistency; for a given record, all updates to the record are applied in the same order across replicas. Specifically, for each record, if one replica receives the most write for a specific record, the one among all replicas is elected as the master that maintains the update timeline of the record. The per-record timeline consistency model can be divided into various levels of consistency guarantees.

- *Read-any*: Read any version of the record. It is possible to return a stale version.

- *Read-critical (required version)*: Read a version of the record that is newer than, or the same as the *required version*.

- *Read-latest*: Read the latest version of the record that all writes have succeeded.

- *Write*: Write a record without reading value in advance. It may cause blind writes.

- *Test-and-set-write (required version)*: Write a record if and only if the current version is equivalent to the requirement version. It can be used as an incremental counter.

21.2.3 Amazon: Dynamo

As a high-volume web site, reliability is essential to Amazon, because even the slightest breakdown can cause significant financial consequences and affect customer trust. Amazon Dynamo originates from Amazon, aiming to serve

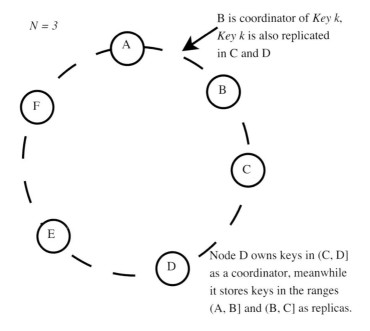

FIGURE 21.4
Partitioning and Replication of Keys in a Dynamo Ring [228].

tens of millions customers with tens of thousands of servers geographically distributed all over the world.

The Dynamo system [228] is a highly available and scalable distributed key-value based datastore implemented for *internal* Amazon applications. The design of the Dynamo system is based on two concerns of using a relational database. On one hand, although the relational database can provide complex data schemas, in practice, many applications in Amazon only require simple primary key access. Thus, the query model of the Dynamo system is key-based single read and write operations. There is no operation that spans multiple data items. On the other hand, a relational database tends to be limited in scalability and availability according to common patterns, therefore, the Dynamo system implements a Dynamo ring to enhance replications.

In order to distribute workload across multiple hosts, Dynamo uses a variant of a consistent hashing mechanism [413] to do the partitioning. This mechanism defines a fixed circular space or ring first as the output range of a hash function. Then, a random value in the range of the space is assigned to each node, known as the "position" of the node on the ring. Hence, each data item is stored in a node whose position is clockwise closest to the data item's position, which is determined by hashing the item's key. Thus, each node is only in charge of the range of the ring from it to its previous node, while adding or removing a node on the ring has no impact on other nodes except its neighbors.

In the Dynamo system, each data item identified by a key, k, is assigned to a coordinator and $(N-1)$ clockwise successor nodes for replication, where N is a configurable parameter. The coordinator owns the data items falling in its range, and takes responsibility of the replication of them. As a result, each node stores data items in the range of the ring from it to its Nth predecessor. As illustrated in Figure 21.4, node B owns a copy of the key k locally, as well as replicates it at nodes C and D. Node D stores the keys within the ranges $(A, B]$, $(B, C]$, and $(C, D]$, and takes care of the keys that fall within the range of $(C, D]$.

21.3   NoSQL Open Source Projects

In practice, most the NoSQL data management systems which are introduced by the key players (e.g., BigTable, Dynamo, PNUTS) are for their internal use and not available for public users. Therefore, many open source projects have been built to implement the concepts of these systems and make it available for public users. These systems started to have a lot of interest from the research community. Not many details have been published about the implementation of most of these systems. In general, the NoSQL open source project can be broadly classified into the following categories:

- *Key-value stores*: These systems use the simplest data model which is a collection of objects where each object has a unique key and a set of attribute/value pairs.

- *Extensible record stores*: They provide variable-width tables (Column Families) that can be partitioned vertically and horizontally across multiple nodes.

- *Document stores*: Where the data model consists of objects with a variable number of attributes with a possibility of having nested objects.

Here, we give a brief introduction about some of these projects. For the full list, we refer the reader to the NoSQL database web site[16].

Cassandra[17] is known as highly scalable, eventually consistent, distributed, structured key-value store [448, 449]. It was initially designed as an inbox storage service in Facebook, and then open sourced in 2008. One of its authors is also an author of Amazon's Dynamo. Hence, Cassandra combines the distribution technology from Dynamo with the data model from Google Bigtable. This results in the system that comes with Dynamo's eventual consistentcy and Bigtable's ColumnFamily-based data model. The data model comes with

---

[16]http://NoSQL-database.org/
[17]http://cassandra.apache.org/

four basic concepts. The basic unit of the data model is *column* including a name, a value and a timestamp. A *column family* groups multiple columns together, comparable with the table of a relational database. Column families can be composed into a *keyspace*, which can be considered a schema to a relational database, typically, one keyspace per application. *Super columns* represent columns that themselves have subcolumns (e.g., Maps). Cassandra offers various levels of consistency models that are suitable for specific applications. In particular, for every *read* and *write* operation, one can choose the following consistency level: a) *ONE*: It ensures that at least one replica has been retrieved or committed to logs and memory before responding to the client. b) *QUORUM*: It ensures that a majority of replicas ($N/2 + 1$) have been reported where N is the total number of replicas. e) *ALL*: It ensures that all N replicas have to be contacted. Moreover, for the write operation, two more levels are supported, namely a) *ZERO*: Nothing ensures in this level. The write operation is asynchronously executed in the background. b) *ANY*: It ensures that the data has been committed to at least one node. Starting from 0.7, two new quorum options are available, *LOCAL_QUORUM* and *EACH_QUORUM*.

HBase[18] is another project based on the ideas of the BigTable system. It builds on top of the Hadoop Distributed File System (HDFS)[19] as its data storage engine. The advantage of this approach is that HBase does not need to worry about data replication, data consistency and resiliency because HDFS is already considering it. However, the downside is that it becomes constrained by the characteristics of HDFS, which is not optimized for random read access. In the HBase architecture, data is stored in a farm of Region Servers. A *key-to-server* mapping is used to locate the corresponding server. The in-memory data storage is implemented using a distributed memory object caching system called *Memcache*[20] while the on-disk data storage is implemented as a HDFS file residing in a Hadoop data node server.

The HyperTable[21] project is designed to achieve a high performance, scalable, distributed storage and processing system for structured and unstructured data. The same as HBase, Hypertable also runs on top of HDFS that offers automatic data replication, data consistency and resiliency. In HyperTable, the data model is represented as multi-dimensional tables. The system supports creating, modifing, and querying data via low-level APIs or Hypertable Query Language (HQL). Data processing can be executed in parallel to increase the performance.

CouchDB[22] is a document-oriented database. A document object identified by a unique identity is the primary data unit, consisting of named fields and typed field values such as strings, numbers, dates, or even ordered lists and

---

[18]http://hbase.apache.org/
[19]http://hadoop.apache.org/hdfs/
[20]http://memcached.org/
[21]http://hypertable.org/
[22]http://couchdb.apache.org/

associative maps. Data query is accessed via a RESTful HTTP API that offers read, update, add, and delete operations. The system is lockless and optimistic, and there are no partially edited documents saved in system. If two clients try to save the same document, an edit conflict error happens to one client on updating. The system resolves the conflict by reopening the latest document version and reappling all updates. The document update can either be all, (succeeding entirely) or none, (failing completely).

Many other variant projects recently started, to follow the NoSQL movement and support different types of data stores such as: key-value stores (e.g., Voldemort[23], Dynomite[24]), document stores (e.g., MongoDB[25], Riak[26]) and graph stores (e.g., Neo4j[27], DEX[28]).

21.4   Database-as-a-Service

Database-as-a-service (DaaS) is an emerging paradigm for data management in which a third party service provider hosts a database as a service [53, 329]. The service providers charge customers on pay-per-use basis, and in return for offering hardware and software, managing system and software upgrades, and maintaining administrative and maintenance tasks, since the cost of an external database service is comparatively low, considering the promising reliable, scalable, and elastic data storage. It is an attractive solution for various purposes such as archive, development and test, and startup companies. In this section, we give an overview of the-state-of-the-art of different options of DaaS from the key players Google, Amazon and Microsoft.

21.4.1   Google Datastore

Google App Engine Datastore[29] is not externally accessible, as it is the scalable schemaless object data storage sitting behind Google App Engine. The data object is called *entities*, composed of a unique identity and a number of *properties* where one property can hold a typed value or refer to another entity. A *kind* is a container of entities, analogous to the table in a relational database. However, entities are schemaless just as two entities can have different properties or even different types of the same properties.

---

[23]http://project-voldemort.com/
[24]http://wiki.github.com/cliffmoon/dynomite/dynomite-framework
[25]http://www.mongodb.org/
[26]http://wiki.basho.com/display/RIAK/Riak
[27]http://neo4j.org/
[28]http://www.dama.upc.edu/technology-transfer/dex
[29]http://code.google.com/appengine/docs/python/datastore/

```
[frame=single]
    SELECT [* | __key__] FROM <kind>
    [WHERE <condition> [AND <condition> ...]]
    [ORDER BY <property> [ASC | DESC] [,<property> [ASC | DESC]...]]
    [LIMIT [<offset>,]<count>]
    [OFFSET <offset>]

    <condition> := <property> {< | <= | > | >= | = | != } <value>
    <condition> := <property> IN <list>
    <condition> := ANCESTOR IS <entity or key>
```

FIGURE 21.5
Basic GQL Syntax.

The Google App Engine datastore provides APIs in Python[30] and Java[31] versions. For the Python interface, it includes a rich data modeling API and a SQL-like query language called Google Query Language[32] (GQL). Figure 21.5 depicts the basic syntax of GQL. For the Java interface, it supports two API standards for modeling and querying, namely Java Data Objects[33] (JDO) and Java Persistence API[34] (JPA). An entity can be retrieved with its identity or by querying its properties. A query can return 0 to maximum 1000 sorted-by-property-values results for the sake of memory and runtime limitations. In principle, join is not supported in the query.

The Google App Engine datastore supports transactions. A transaction ensures that operations in a transaction succeed entirely or fail completely. A single operation of creating, updating or deleting an entity happens in a transaction implicitly. Meanwhile, a group of operations can be explicitly defined as a transaction. The datastore manages transactions in an optimistic manner. The datastore replicates data to multiple locations. Among all replicas, one is selected as the primary to keep the view of the data consistent by replicating delta data to other locations. In the case of failures, the datastore can wait for the primary to become available, or can continue accessing data from an alternative replica, depending on the selection of read policies: 1) *strong consistency* means reading from the primary. 2) *eventual consistency* [731] means reading from an alternate replica when the primary location is unavailable.

21.4.2 Amazon: S3/SimpleDB/Amazon RDS

Amazon Simple Storage Service (S3) is an online public storage web service offered by Amazon Web Services. Conceptually, S3 is an infinite store for

---

[30]http://www.python.org/
[31]http://www.java.com/
[32]http://code.google.com/appengine/docs/python/datastore/gqlreference.html
[33]http://code.google.com/appengine/docs/java/datastore/jdo/
[34]http://code.google.com/appengine/docs/java/datastore/jpa/

objects of variable sizes. Each object is a container of bytes. It is identified by a URI. With the specified URI, clients are able to access via SOAP or REST-based interface remotely, for example, *get(uri)* returns an object and *put(uri, bytestream)* writes a new version of the object. Ideally, S3 can be considered as an online backup solution or for archiving large objects which are not frequently updated.

Amazon has not revealed details on the implementation of Amazon S3 yet. However, Brantner et al. [139] have presented initial efforts of building web-based database applications on top of S3. They described various protocols in order to operate S3 in a manner of a relational database. In their system, the *record manager* component is designed to create, read, update and scan records where each record contains a key and payload data. The size of a record must be no larger than a page size, as a page is a container of records, and each page is physically stored in S3 as a single object. In addition to record manager, a buffer pool is also implemented in a *page manager* component. It interacts with S3 like normal buffer pool in any standard database system: reading pages from S3, pinning the pages in the buffer pool, updating the pages in the buffer pool, and marking the pages as updated. The page manager is mainly in charge of commit and abort transactions. Moreover, they implemented standard B-tree indexes on top of the page manager and basic redo log records. However, there are still many database-specific issues that have not been addressed yet by this work, for example, DB-style strict consistency and transaction mechanisms. Furthermore, as addressed in the paper, more functionalities can be devised: query processing techniques (e.g., join algorithms and query optimization techniques) and traditional database functionalities (e.g., bulkloading a database, creating indexes, and dropping a whole collection).

Similar to S3, Amazon has not published the details of its other two products: SimpleDB and RDS. Generally, SimpleDB is designed for running queries on structured data. In SimpleDB, data in is organized into *domains* (i.e., tables) within which we can put data, get data or run queries. Each domain consists of *items* (i.e., records) which are described by pairs of *attribute* names and values. It is not necessary to pre-define all of the schema information, as new attributes can be added to the stored dataset when needed. Thus, the approach is similar to that of a spreadsheet and does not follow the traditional relational model. SimpleDB provides a small group of API calls that enable the core functionality to build client applications such as: *CreateDomain*, *DeleteDomain*, *PutAttributes*, *DeleteAttributes*, *GetAttributes* and *Select*. The main focus of SimpleDB is fast reading. Therefore, query operations are designed to run on a single domain. SimpleDB keeps multiple copies of each domain where a successful write operation guarantees that all copies of the domain will durably persist. In particular, SimpleDB supports two read consistency options: eventually consistent read [731] and consistent read.

Amazon Relational Database Service (RDS) is a new service, which gives access to the full capabilities of a familiar MySQL database. Hence, the code, applications, and tools, which are already designed on existing MySQL databases can work seamlessly with Amazon RDS. Once the database instance is running, Amazon RDS can automate common administrative tasks such as performing backups or patching the database software. Amazon RDS can also manage synchronizing data replication and automatic failover management.

21.4.3 Microsoft SQL Azure

Microsoft has recently released the Microsoft SQL Azure Database system[35]. Not much detail has been published on the implementation this project. However, it is announced as a cloud-based relational database service, which has been built on Microsoft SQL Server technologies. So, applications can almost move whatever available operations in SQL Server to SQL Azure such as creating, accessing, and manipulating tables, views, indexes, roles, stored procedures, triggers, and functions. It can execute complex queries and joins across multiple tables. It also supports Transact-SQL (T-SQL), native ODBC and ADO.NET data access[36]. In particular, SQL Azure service can be seen as running an instance of SQL server in a cloud-hosted server, which is automatically managed by Microsoft instead of running an on-premise managed server. In SQL Azure, the size of each hosted database can not exceed the limit of 50 GB.

21.4.4 Challenges

In general, the service level agreements (SLA) of the commercial DaaS products are focused on providing their customers with high availability (99.99%) to the hosted databases. However, they are not providing any promises or guarantee on the performance and scalability aspects. In particular, for each hosted database, the cloud provider stores 3 replicas in the same data center with the main guarantee that they have no single point of failure (e.g., physical server, network). However, these providers are currently not providing the management of geo-replica(s) that can be used for recovery in the case of a physical disaster to the hosting data centre. One of the 3 replicas hosted by the service providers is selected to be a *primary* copy, which is used to serve all read and write requests, while the other 2 replicas are used as a *standby replicas* (hot backups) which are only used to save the situation under any failure circumstances that may happen to the primary copy. These hot backups are not used for serving any read/write operations or load balancing purposes. Moreover, currently, these services are supporting very simple application-aware (non-transparent) data partitioning strategies. Such limited data partitioning and data replication strategies force the application (customer) to take care of additional responsibilities and challenges in order

---

[35]http://www.microsoft.com/windowsazure/sqlazure/
[36]http://msdn.microsoft.com/en-us/library/h43ks021(VS.71).aspx

to achieve performance improvement and scalability goals for many condition and situation such as:

- *Achieving Performance Aspects of Defined SLAs*: In the Data centers, each physical server hosts a number of databases where each database is allocated a specific portion of the resources of that server, assuming that these allocated resources have the limit to serve a number of application requests (L) according to a specific application-defined *SLA* performance requirement. It is the responsibility of the application to achieve the same *SLA* requirements with any increasing workload that might exceed the limit L (e.g., volume spike). This problem can be solved by adding an additional replica for the hosted database and distributing the workload between the available replicas in a balanced way (*scale Out*). However, this action should be done in a transparent way for the application side. The same behavior should be achieved in the situation where *scaling down* is required, in order to reduce the cost where it could be possible to achieve the specified *SLA* requirements with fewer replicas.

- *Data Spike*: In the previous situation, replicating the whole database can deal with the *volume spike* situation in order to achieve the performance requirements of a specific *SLA*. However, in the *data spike* situation (increasing volume to specific object or table), may require just replicating a specific shard (partition) of the database in order to tackle the problem in a more efficient, effective, and economical way. Such replication of specific partitions should be also done transparently to the application code.

- *Distributed Transactions*: Due to the size limit on a single database or performance requirements, it would be common to run transactions over multiple partitions (databases). Currently, these cloud database services *support transactions execution in a single partition*. There is no support for execution distributed transactions over different partitions, even in the same data center. It is the application, responsibility to deal with such situations.

Another main challenge for the DaaS products is that the service provider needs to guarantee that the data is *secure*, not only being secure in results of queries, but also being secure to the data provider. Some research efforts have considered the problem of how to index and query encrypted data [52, 328, 329, 410]. However, querying encrypted data is known to be computationally expensive. Therefore, as an alternative, providing an efficient trust mechanism has emerged to be the solution for turning data outsourcing into a viable paradigm. Agrawal et al. [53] described privacy preserving algorithms for data outsourcing. Instead of encryption, they distribute data to multiple data provider sites and information theoretically proven secret sharing algorithms as the basis for privacy preserving outsourcing. However, more research and development efforts are still required to find effective solutions for data security and data privacy issues in order to encourage more customers to rely on these new data management services.

TABLE 21.1

Design Decisions of Various Web Scale Data Management Systems

System	Data Model	Query	Consis.	CAP	License
BigTable	Col. Families	API	Strict	CP	Inter@Google
Datastore	Col. Families	API/GQL	Strict	CP	Commercial
PNUTS	Key-Value	API	Timeline	AP	Inter@Yahoo
Dynamo	Key-Value	API	Eventual	AP	Inter@Amazon
S3	Large Obj.	API	Eventual	AP	Commercial
SimpleDB	Key-Value	API	Multiple	AP	Commercial
RDS	Relational	SQL	Strict	CA	Commercial
SQL Azure	Relational	SQL	Strict	CA	Commercial
Cassandra	Col. Families	API	Tunable	AP	Apache
Hypertable	Mul-dim. Tab	API/HQL	Eventual	AP	GNU
CouchDB	Document	API	Eventual	AP	Apache

21.5   Web Scale Data Management: Trade-Offs

An important issue in designing large scale data management applications is to avoid the mistake of trying to be "*everything for everyone.*" Because different systems make various tradeoffs to optimize for different purposes, while there is no single system that can best suit all kinds of workloads. Therefore, the most challenging aspects in these applications is to identify the most important features of the target application domain and to decide about the various design tradeoffs, which immediately lead to performance trade-offs. To tackle this problem, Jim Gray came up with the heuristic rule of "*20 queries*" [348]. The main idea of this heuristic is that on each project, we need to identify the 20 most important questions the user wants the data system to answer. He said that five questions are not enough to see a broader pattern and a hundred questions would result in a shortage of focus.

Table 21.1 summarizes the design decisions of our surveyed systems. In general, it is difficult to guarantee ACID properties for replicated data over large geographic distances. The CAP theorem [143, 304] shows that a shared-data system can at most satisfy two out of three properties: *Consistency* (all records are the same in all replicas), *availability* (a replica failure does not prevent the system from continuing to operate), and *tolerance to partitions* (the system still functions when distributed replicas cannot talk to each other). When data is replicated over networks, this essentially just leaves a system only one selection between consistency and availability. Thus, the C (consistency) part is typically compromised to yield reasonable system availability [44]. Therefore, most of the cloud data management applications relax data consistency to overcome the difficulties of distributed replication. In particular, they implement various forms of weaker consistency models (e.g., eventual consistency, timeline consistency, session consistency [693]) so that not all replicas have to agree on the same value of a data item at every moment of time. Therefore, transactional data management applications (e.g., banking, stock trading, supply chain management), which rely on the strict consistency of databases' offerings, tend to be fairly write-intensive or require microsecond precision, so are less obvious candidates for the cloud environment until the cost and latency of wide-area data transfer decrease. Cooper et al. [206] discussed the tradeoffs facing cloud data management applications as follows:

- *Read performance versus write performance*: An update to a record can either attach the delta to the existing record, or completely overwrite the existing one. The former write is efficient, as it costs the write only modified bytes. However, the former read is inefficient, which is contrary to the write; as for the former read, there is a cost of reconstruction of deltas.

- *Latency versus durability*: Synchronizing writes immediately to disk before responding success takes a longer time than storing writes in memory and synchronizing them later to disk. The latter approach avoids costly disk I/O operations to reduce write latency. However, the unsynchronized data could lose if system failures happen before the next synchronizing.

- *Synchronous versus asynchronous replication*: Synchronous replication keeps all replicas up to date during the time, but potentially incurs high latency on updates. Furthermore, availability of the system may be affected if synchronization is suspended due to offline of some replicas. Asynchronous replication avoids high write latency over networks but allows stale data. Moreover, data loss may occur if an updated replica goes offline before propagating data.

- *Data partitioning*: Data can be partitioned strictly on row basis or on column basis. Row-based partitioning allows efficient access to an entire record. Hence it is ideal for accessing a few records in their entirety. Column-based storage is more efficient for accessing a subset of the columns, particularly when multiple records are accessed.

Kraska et al. [443] have argued that finding the right balance between cost, consistency, and availability, is not a trivial task. High consistency implies high cost per transaction and, in some situations, reduced availability, but avoids penalty costs. Low consistency leads to lower costs per operation but might result in higher penalty costs. Hence, they presented a mechanism that not only allows designers to define the consistency guarantees based on the data at the transaction level but also allows to automatically switch consistency guarantees at runtime. They described a dynamic consistency strategy, called *Consistency Rationing*, to reduce the consistency requirements when possible (i.e., the penalty cost is low) and raise them when it matters (i.e., the penalty costs would be too high). The adaptation is driven by a cost model and different strategies that dictate how the system should behave. In particular, they divide the data items into three categories (A, B, C) and treat each category differently depending on the consistency level provided. The A category represents data items for which we need to ensure strong consistency guarantees as any consistency violation would result in large penalty costs; the C category represents data items that can be treated using session consistency as temporary inconsistency is acceptable, while the B category comprises all the data items where the consistency requirements vary over time depending on the actual availability of an item. Therefore, the data of this category is handled with either strong or session consistency depending on a statistical-based policy for decision making.

Florescu and Kossmann [274] argued that in cloud the environment, the main metric that needs to be optimized is the cost as measured in dollars. Therefore, the big challenge of data management applications is to determine the right number of machines to meet the performance requirements of a particular workload under an acceptable cost. Hence, performance requirements such as how fast a database workload can be executed, or whether a particular throughput can be achieved, is no longer the main metric any more. This argument fits well with a rule of thumb calculation which has been proposed by Jim Gray regarding the opportunity costs of distributed computing in the Internet as opposed to local computations [315]. Gray reasons that for outsourcing computing tasks, network traffic fees may outnumber savings in processing power. In principle, it is useful to involve economies into trade-off calculation between basic computing services. This method can easily be applied to the pricing schemes of cloud computing providers (e.g Amazon, Google). Florescu and Kossmann [274] have also argued that in the new large scale web applications, the requirement of providing full read and write availability for all users has surpassed the importance of the ACID paradigm in data consistency. In this circumstance, blocking a user is not ever allowed. Therefore, in order to minimize the cost of resolving inconsistencies, it is better to design a system that deals with resolving inconsistencies rather than having a system that prevents inconsistencies under all circumstances.

Kossmann et al. [441] conducted an end-to-end performance and cost evaluation on alternative cloud services (e.g., RDS, SimpleDB, S3, Google AppEngine, Azure) with OLTP workloads. The results of the experiments showed that the alternative services differed from each other greatly both in cost and performance. Most services had significant scalability issues. They confirmed the observation that public clouds lack support to upload large data volumes. It was difficult for them to upload 1 TB or more of raw data through the APIs provided by the providers. With regard to cost, they concluded that Google seems to be more interested in small applications with light workloads whereas Azure is currently the most affordable service for medium to large services. With the goal of facilitating performance comparisons of the trade-offs in cloud data management systems, Cooper et al. [206] have presented the Yahoo! Cloud Serving Benchmark (YCSB) framework and a core set of benchmarks. The benchmark tool has been made available via open source[37] in order to encourage expansion of cloud benchmark suites that represent different classes of applications, as well as include different cloud data management systems.

21.6 Discussion and Conclusions

For more than thirty years, the relational database management systems (RDBMS) have been recognized as the dominant solution for data persistence requirements. In particular, they provide a simple but extremely powerful interface for storing and accessing data. In addition, they have been shown to be wildly successful in many business domains such as: financial, business and Internet applications. However, with the new trends of web scale data management, they have started to suffer from some serious limitations [213]:

- *Database systems are difficult to scale.* In practice, each database system has a maximum limit which they can not easily scale beyond. When the application workloads hit this scalability limit, a set of time consuming and manually expensive data partitioning, data migration, and load balancing tasks are required to tackle this challenge.

- *Database systems are difficult to configure and maintain.* In general, getting a good performance out of most commercial relational database systems requires highly experienced professionals. Therefore, administrative cost represents a significant fraction of the total cost of ownership of a database system.

- *Diversification in available systems complicates selection.* Recently, specialized database systems for specific types of applications have been entering the market (e.g., main memory systems for OLTP or column-stores for OLAP). Such a situation makes the system selection process quite complex, especially for customers with different application workloads that are not neatly classified.

- *Peak provisioning leads to unneeded costs.* The workloads of large scale web applications are often bursty and dynamic in nature. Thus, a peak provisioning process is usually applied to deal with this challenge. Therefore, inefficient resource utilization during the off-peak times usually happens, which consequently causes unneeded costs.

Recently, a new wave of NoSQL systems has started to gain some mindshares as an alternative model for database management. In principle, some of the main advantages of NoSQL systems can be summarized as follows:

---

[37]http://wiki.github.com/brianfrankcooper/YCSB/

- *Elastic Scaling*: For years, the *scale up* approach has been considered as the favorite approach to rely on rather than the *scale out* approach for achieving the scalability goal. However, with the continuous increase in the transaction rates and high availability requirements, the economic advantages of the scaling out approach on commodity hardware becomes very attractive. In practice, it is not easy to scale out RDBMS on commodity clusters. Therefore, the NoSQL systems have considered the ability to expand transparently as one of their main requirements in order to take advantage of the addition of any new nodes.

- *Less Administration*: Over the years, RDBMS vendors have introduced many manageability improvements. However, it is still very expensive to maintain high-end RDBMS without the assistance of expensive and highly trained database administrators. DBAs are intimately involved in the design, installation and ongoing tuning of high-end RDBMS systems. On the contrary, NoSQL database systems are designed from the beginning with the ability to be maintained with less expertise and effort.

- *Better Economics*: RDBMS tends to use expensive proprietary servers and storage systems. On the contrary, NoSQL systems tend to rely on clusters of cheap commodity servers in dealing with the increasing data and transaction rates. Therefore, the cost per gigabyte or transaction/second for NoSQL can be many times less than the cost for RDBMS, which allows one to store and process more data with a much lower price. Moreover, when an application uses data distributed across hundreds or even thousands of servers, simple economics points to using no-cost server software as opposed to paying per-processor license fees. Once freed from license fees, an application can safely scale horizontally with complete avoidance of the capital expenses.

- *Flexible Data Models*: Even small changes to the schema of a large production relational database have to be carefully considered and may require downtime or degraded service levels. NoSQL databases have more relaxed (if any) data model restrictions. Therefore, any application change or database schema change can be more softly managed.

These advantages have given the NoSQL systems a lot of attractions. However, enterprises are still very cautious about relying on these systems, because there are many limitations that still need to be addressed such as[38]:

- *Programming Model*: NoSQL databases offer limited support for ad-hoc querying and analysis operations. Therefore, significant programming expertise are usually required even for a simple query. In addition, missing the support of declaratively expressing the important join operation has been always considered one of the main limitations of these systems.

---

[38]http://blogs.techrepublic.com/10things/?p=1772

- *Transaction Support*: Transaction management is one of the powerful features of RDBMS. The current limited support (if any) of the transaction notion from NoSQL database systems is considered to be a big obstacle toward their acceptance in implementing mission critical systems.

- *Maturity*: RDBMS are well-known for their high stability and rich functionalities. In contrast, most NoSQL systems are open source projects or in preproduction stages where many key features are either not stable enough or still under development. Therefore, enterprises are still very cautious about dealing with this new wave of database systems.

- *Support*: Enterprises look for the assurance that if the system fails, they will be able to get timely and competent support. All RDBMS vendors have great experience in providing a high level of enterprise support. In contrast, most NoSQL systems are open source projects. Although there are a few firms offering support for NoSQL database systems, these companies are still small start-ups without the global reach, support resources, or the credibility of an Oracle, Microsoft or IBM.

- *Expertise*: There are millions of developers around the world who are familiar with RDBMS concepts and programming models in every business domain. On the contrary, almost every NoSQL developer is still in a learning mode. It is natural that this limitation will be addressed over time. However, currently, it is far easier to find experienced RDBMS programmers or administrators than a NoSQL expert.

Currently, there is a big debate between the NoSQL and RDBMS campuses, which is centered around the right choice for implementing online transaction processing systems. RDBMS proponents think that the NoSQL camp has not spent enough time to understand the theoretical foundation of the transaction processing model. For example, the eventual consistency model is still not well-defined and there are different implementations which may significantly differ with each other. Therefore, it is the responsibility of the application developer to figure out all the inconsistent behavior that may arise, which makes their task very much harder. On the other side, the NoSQL camp argues that this is actually a benefit as it provides the application developers with domain-specific optimization opportunities where they are no longer constrained by the one-size-fits-all model. However, they admit a lot of experience is required for making optimization decisions; otherwise they can be very error-prone and make dangerous decisions.

In principle, we believe that it is not expected that the new wave of NoSQL data management systems will provide a complete replacement of the relational data management systems. Moreover, there will be no a single winner (one-size-fits-all) solution. However, it is more expected that different data management solutions will coexist in the same time for a single application (Figure 21.6). For example, we can imagine an application which uses different datastores for different purposes as follows:

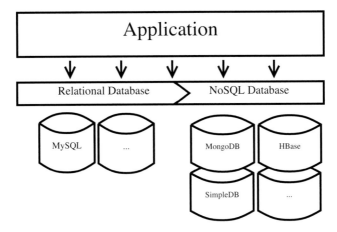

FIGURE 21.6
Coexistence of Multiple Data Management Solutions in One Application.

- MySQL for high-value and low-volume data such as billing information or user profiles.

- A key value store (e.g., Hbase) for low-value and high-volume data such as log files or hit counts.

- Amazon S3 for user-uploaded data objects such as photos, sound files and big binary files.

- MongoDB for storing the application documents (e.g., bills).

Finally, we believe that there are still huge required research and development efforts needed for improving the current state-of-the-art in order to tackle the current limitations in both campuses: NoSQL database systems, data management service providers and traditional relational database management systems.

22

Leasing Videoconference Resources on Hybrid Clouds

Javier Cerviño

ETS de Ingenieros de Telecomunicación, Universidad Politécnica de Madrid, Spain

Fernando Escribano

ETS de Ingenieros de Telecomunicación, Universidad Politécnica de Madrid, Spain

Pedro Rodríguez

ETS de Ingenieros de Telecomunicación, Universidad Politécnica de Madrid, Spain

Irena Trajkovska

ETS de Ingenieros de Telecomunicación, Universidad Politécnica de Madrid, Spain

Joaquín Salvachúa

ETS de Ingenieros de Telecomunicación, Universidad Politécnica de Madrid, Spain

CONTENTS

22.1 Introduction

Since a few years ago developers have been increasing the number of applications and services that offer to their users the possibility of establishing videoconference sessions of multiple users through a web portal. Some of these systems allow the recording of their sessions. We can find examples of these portals such as WebEx [1], FlashMeeting [2] or Marte [172]. On the other hand, during this period some Internet portals have arisen to allow the users to upload their own recorded videos of lessons, conferences, etc. In these other portals the videos can also be accompanied with slides or presentations that have been used in the session. Examples of these services are Youtube [3], RTV-Doc [4] or iTunesU [5].

The great majority of these systems use Cloud Computing infrastructures in order to offer their services, either in the recording or in the storage of the generated videos. However, none of them offer a wide solution for creating, storing and distributing videos using Cloud technologies. At least at the moment of writing this paper we did not find an example of this kind of system.

On the other hand, there are some works about systems that offer the possibility of creating hybrid clouds, such as OpenNebula [667], Zimory [6] or VMWare vCloud [7]. The most common use case is a scenario where the developer uses any of the previous systems to provide the user with services which can be accessed through both public and private networks of an enterprise. More examples can be found in [718]. This chapter gives another reason for using hybrid clouds in the creation of services, that is to enhance the usage of resources efficiently. We think that there are a lot of systems that can be improved by deploying them on hybrid clouds instead of single provider clouds.

We want to validate this concept by presenting a system that follows this principle. This work is part of the Global Project [8] and it is based on a service

---

[1] http://www.webex.com
[2] http://flashmeeting.open.ac.uk/home.html
[3] http://www.youtube.com
[4] http://www.ucm.es/info/tvdoc/
[5] http://www.apple.com/education/itunes-u/
[6] http://www.zimory.com/
[7] http://www.vmware.com/products/vcloud/
[8] http://www.global-project.eu/

which offers to the users the possibility of scheduling videoconference sessions through a web portal (named GlobalPlaza [9]). The users can attend each of these sessions through the web and even take part in the meetings that can be created with this application. With this tool several users should be able to join to a single videoconference session and control it. This control is based on the interaction modes that are set up at different moments (presentations, conferences, questions, etc.). The system presented takes into account the different resources needed in order to establish a videoconference and reacts accordingly. In this project, a new service that offers videoconferencing which can be accessed through different technologies has been created. These technologies can be summarized as SIP [334], web browser and Isabel application access, the latest being an application developed in our research group. We think that this work can be used by anyone who is interested in working on a different kind of hybrid cloud, because of the experience provided in this system. We propose this work as a starting point of Cloud videoconference systems which are aware of the use of the available resources and the costs that are generated by them. Section 22.3 introduces the main motivation of this research, explaining how hybrid architectures can work better in some scenarios. Section 22.4 gives us an example of the concepts presented in the previous section; this example is an implementation of a videoconference system that we have developed in our research group. Section 22.5 explains the validation of the system in terms of cost and resource usage. Finally, some conclusions are presented in Section 22.6.

22.2 Related Work

We have designed, developed and made some tests of a new architecture for a session based videoconference system. This system presents a web application in which users can schedule, delete and modify videoconference sessions in which they are going to participate with many other users. It is intended to be used in many different scenarios, such as classrooms, government, tutorials, meetings, etc.

Furthermore, the system is focused on the optimal usage of the available resources. To do so we studied the existent hybrid infrastructures that were used for any other purposes rather than videoconference. And we also based our work on similar videoconference systems that do not use any type of Cloud infrastructures.

In this section, we are going to present the related work which we have based our research on.

---

[9]http://globalplaza.org

22.2.1   Hybrid Infrastructures

There is much research done on this topic, and when we started our work we could check some Cloud technologies that allowed hybrid architectures. From our point of view the most important were OpenNebula, and Eucalyptus.

OpenNebula is an IaaS (Infrastructure as a Service) application. The main objective of OpenNebula is to allow the creation of a private Cloud in a data center. But it also has an API which can directly manage virtual machines on Amazon EC2, and another standard API named OCCI, so it is very interesting for hybrid architectures that use both third party resources and their own. In our case the most important feature of OpenNebula is that it can schedule the creation/deletion of instances of virtual machines, but this feature does not work in the case of Amazon EC2 instances.

Eucalyptus is also an IaaS application, with features that are similar to the ones from OpenNebula. This service lacks a scheduler that can start and stop instances of the virtual machines, so we think that it is less interesting for our purposes.

There are other important systems that offer similar features to the user. There is a very interesting table comparing different tools in [667].

On one hand they both are very interesting for future purposes even though they achieve the objective of hybrid clouds. On the other hand they do not meet all our needs of scheduling, so we finally decided to create a service that worked with Amazon EC2 directly and also with our current virtual machine solution, that was VMWare.

Furthermore, there are some research studies like the one in [465] that proposes a measurement tool for comparing different Cloud providers in order to select the best-performing provider for a given application of a Cloud customer. This tool can be used to perform a comprehensive measurement study over four major cloud providers, namely, Amazon AWS, Microsoft Azure, Google AppEngine, and Rackspace CloudServers. And it says that the results of such study is ephemeral due to the periodic changes that every Cloud provider introduces in their software and hardware, and because the user demands vary over time.

22.2.2   Videoconference Systems

Nowadays there are some videoconference systems that allow users to schedule web videoconference sessions or to participate through their web browsers. We can see FlashMeeting, Adobe Connect, WebEx, GoToMeeting, Skype, Marte 3.0 [172] and so on. Table 22.1 shows some of the features that are present on a single videoconference session of the Conference Manager service and checks if they are present in any of the other systems.

Examples like WebEx that are running on Cisco private Clouds are very similar to the Conference Manager. But here in this work we propose the use of Hybrid Clouds. In other words, we want to use different infrastructures from

TABLE 22.1

Comparison of Videoconferencing Systems

Feature	FlashMeet.	GoToMeet.	WebEX	AdobeCon.	Skype
Web	yes	yes	yes	yes	no
Scheduling	no	yes	yes	yes	no
Recording	yes	yes	yes	yes	no
Streaming	yes	no	no	yes	no

different Cloud providers in order to enhance the global performance of our videoconference application.

22.3 Motivation

A videoconference system that allows a great number of users per conference, multiple simultaneous conferences, different client software (requiring transcoding of audio and video flows) and provides a service like automatic recording, like the one we are trying to build, requires a lot of computing resources. Figure 22.1 shows the videoconference scenario we are trying to build. This scenario includes several videoconference clients. Some are connected through a MCU and others participate via Flash or SIP. In both cases transcoding of the data flows is necessary. The scenario also includes a RTMP server for the flash clients and a SIP server for the SIP clients. In order to allow cost effective scaling of our videoconference system, the use of cloud computing resources appears as a natural approach, since they provide the

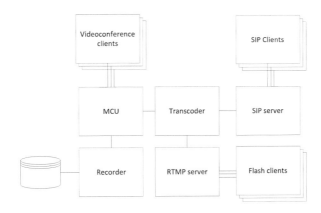

FIGURE 22.1

Videoconference Scenario.

illusion of infinite computing resources available on demand and the ability to pay for use of computing resources on a short-term basis as needed [95]. However, the use of cloud computing resources from a single provider comes with several disadvantages as shown in [95], [455]. Critical problems that can benefit from an hybrid cloud architecture are listed below in no particular order.

- *Geographical location and legal issues.* It may be useful to start some services in a specific location for performance or legal reasons. The use of different providers will give us access to more locations or will allow us to start some services in our private cloud that may be more suitable for sensible data.

- *Cost and lock-in.* Different providers may offer different services at a different price. Furthermore this price may change over time. By using several providers we can use this to our advantage. In addition the use of a single provider may result in a lock-in problem.

- *Availability.* Cloud Computing service by a single company is a single point of failure. By using different providers we can achieve better availability.

- *Wasting of existing resources.* In some environments a lot of resources are already available to use. By moving all services to the cloud we are wasting these resources. The use of hybrid private/public clouds can avoid this problem.

In light of the problems listed above, the use of resources from different providers as well as private resources can help us to provide a service with better performance, lower cost and avoiding or mitigating most of the problems of cloud computing. This will be proved for our videoconference service in Section 22.5 of this chapter.

To be able to effectively make use of hybrid clouds we need two things. First, we need to make use of a virtual infrastructure manager [665] to provide a uniform and homogeneous view of virtualized resources, regardless of the underlying virtualization platform. Second, we need to split our service into three parts:

- CPU intensive modules. Parts of the application that consume most of the CPU cycles needed to provide a service. In our case, we have identified the transcoding and recording modules of our videoconference system as the CPU intensive modules.

- Bandwidth intensive modules. Modules that consume most of the bandwidth. In our videoconference system, the MCUs and RTMP servers are bandwidth intensive components.

- Storage intensive modules. Disk servers and databases fall into this category. In our case, the recorded conferences are stored in a NFS server.

This division gives us the opportunity of placing the modules that need a specific kind of resource where they better serve our needs and objectives. We have named this partition *Cloud computing Resource Oriented Partition* or CROP. An example of this partition widely is widely used in video streaming systems, where we have the video encoder (CPU intensive) sending the encoded video to a streaming server (bandwidth intensive) and storing the video on a disk server (storage intensive). The videoconference system described in the next section fulfills these requirements and results in a scalable and cost-effective system.

22.4 Implementation

The work we are going to explain throughout this section is based on a system that allows to schedule several videoconference sessions. These sessions are going to be created by the users and they are going to use the available resources in each moment. First, there is an enumeration of the objectives of this work and then there is the description of the system.

22.4.1 Objectives

This implementation has two main objectives:

- To develop a videoconference system which can schedule, record and stream videoconference events. This system has an API that can be used by external applications in order to offer the services to their users. The access to the service could be done through different ways: SIP [617], Web browser (with Flash Player) and Isabel [592] and [591].

- A new resource reservation system which takes into account the different scenarios, technologies and resources that are going to be used. This is the objective that gives us the hybrid cloud perspective.

22.4.2 Conference Manager

Figure 22.2 shows a general architecture of the Conference Manager, which is the name of the videoconference system, where only those components that are important to better understand the function of the hybrid cloud scheduler are present. This section describes all the components of the Conference Manager, that is divided into three parts:

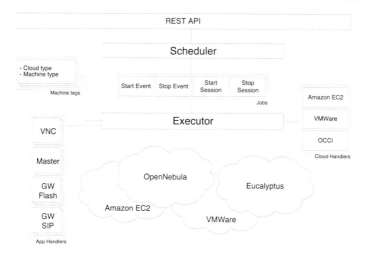

FIGURE 22.2
Conference Manager Architecture.

22.4.2.1   API

This is the interface between the scheduler and third party services. This API is based on the REST methodology [269], so it represents conference events and sessions as resources that are accessible through HTTP requests. All these requests have to be authenticated via a mechanism that has already been implemented in [614].

There are many requests that are forwarded to the scheduler in order to create, modify or delete some videoconference events or sessions. Each event consists of one or more sessions that are going to be recorded or not, depending on what the service is requesting. So the way that an event with several sessions has to be created is by first requesting the creation of such an event and then the creation of all of its sessions, one by one.

22.4.2.2   Scheduler

The next one is responsible for scheduling all the videoconference events so they are executed by the next component. Each event consists of several jobs that have to be performed in a given order to correctly start or stop the event and its sessions.

The scheduler (which is based on the Quartz scheduling framework [167]) has to check the correctness of the system so that it can not use more than a maximum limit of resources, even if they are on the cloud. When there is a request for creating a new event or session it checks if there are enough available resources, throughout its duration.

It is also responsible for deciding at which cloud provider the machines are going to be executed. This decision is based on three parameters that depict the starting of each of these resources: The current price of the resources that are required in each cloud provider in order to execute the machine, the geographical region where the session is going to take place, and the type of machine that is going to be started by the executor. All the information about the cloud provider that is going to host the machine is stored at the machine tags.

There are many differences between start and stop jobs and also between event and session jobs. For each event there are two associated jobs; the first is the job that prepares all the necessary resources for such event. The latter does the opposite, it stops all the machines and releases all the resources that are used. It is the same with the session jobs, but the difference between events and sessions is the type of resources they are going to hold and release. In the context of this chapter event jobs are more interesting, since most of the resources they are going to manage are machines that can be hosted on different clouds.

22.4.2.3   Executor

This component executes the jobs scheduled by the previous component. This is done at the time scheduled with the dispatch of the required event and the information stored within it. For example, a new start event job could be scheduled in order to run the required resources for a given event. Once an event start is dispatched, this component starts the machines, and then the videoconference application with a specific configuration. This application is usually Isabel, since this is the core videoconference service of this system.

As we have already mentioned above, there are two types of event jobs, the start and the stop of the event. When the executor initializes a start event job it follows the next steps:

1. Retrieval of the event configuration from the database. It is useful in order to choose and get the machine tags which are going to be started.

2. Initialization of the machines on the given clouds. In order to do this, the executor uses the corresponding cloud handlers, which are tools that can communicate with the API of its cloud provider. In this step all machines are going to be started, and it does not end until all machines are running.

3. Execution of Isabel (or any other application such as VNC server). Once all the machines are running the executor starts Isabel on each machine, with different parameters. Depending on this configuration Isabel can operate as Master, Flash gateway or SIP gateway depending on the functions that are required. The next section explains in detail each of these functions.

22.5   Validation of the Hybrid Cloud

In this section, we are going to validate the system explained in the previous one in the terms introduced in Section 22.3. Subsection 22.5.1 gives details about Isabel resources and their costs, while in Subsection 22.5.2 we analyze the costs of these resources on different cloud typologies.

22.5.1   Isabel Resource Usage

22.5.1.1   CPU Cost

The Isabel videoconference system uses a windowing system in order to represent accurately the process of the conference, showing all videos, applications, and whiteboard on the screen. In a traditional Isabel conference there can be different clients connected through the network. This network can be organized as a tree or mesh overlay, depending on the requirements of each scenario. Each Isabel node of this network can perform different functions: the main function would be the interactive mode in which users can take part of the session. Another function is the MCU that only forwards multimedia streams between Isabel terminals; the SIP transcoder transforms the traffic between Isabel and SIP clients using the SIP server. Finally, the Web Isabel transcoder does the same thing offering the users access from a web browser; in this particular case there is a RTMP server that forwards multimedia data using RTMP protocol to Adobe Flash clients.

Depending on what type of Isabel is running on a machine, it could use more or less CPU. The use of CPU increases with the number of users for each type of Isabel node. The reason for this CPU usage on the SIP and web transcoders is that the main task of the application is to generate a single video showing the conference session, so it has to render all the videos and encode them with the corresponding codec.

22.5.1.2   Bandwidth Cost

Something similar occurs with the bandwidth, because depending on the way in which the users are connected to the conference, it needs more or less bandwidth. The bandwidth increases with the number of web users. It always reaches a limit in a way that we will explain later in this section. Again, in this case the MCU uses more bandwidth than in an Isabel session, most of all when its topology follows a star network. In these networks the MCU is going to forward all the traffic between nodes. This topology is not the most recommended but it is preferred by the users because it is easier to configure and it requires fewer machines. The SIP transcoder also consumes a lot of bandwidth, as it is the entry gate to the videoconference for any SIP phone, so usually all SIP phones are connected to it directly. Finally, the RTMP server is the third of all by bandwidth consumption as a consequence of the number of users that are directly connected to it. It is also the responsible for sending the video to all web users.

For the calculus of the web bandwidth that a user consumes in a videoconference session we have taken into account that there is always a top limit in the number of video and audio streams. This is the maximum number of user videos that are shown in the screen at each moment. In our case, and based on a great number of recorded videoconference sessions in the past, we propose Equation 22.1 that references the real number of users who are sending video and audio streams each moment. It can be summarized by saying that there is always a maximum number of videos that are shown on the screen and that the number of videos sent to the system is going to increase up to this maximum.

$$
\alpha_{web} \cdot N_{web-users} = \begin{cases} N_{web-users} & if N_{web-users} < N_{max} \\ N_{max} & if N_{web-users} \geq N_{max} \end{cases} \tag{22.1}
$$

Multiplying the number of videos that will be sent at a given time to the system by the bandwidth of each of these videos (it will be the same for all of them) we get the overall bandwidth of the system.

$$
BW_{web-in} = \alpha_{web} \cdot N_{web-users} \cdot BW_{web-user} \tag{22.2}
$$

$$
BW_{web-out} = BW_{web-isabel} \cdot N_{web-users} \tag{22.3}
$$

Both the CPU usage and the bandwidth are critical when we have to choose the right topology in order to make an optimal use of the resources which are available in the network and in the system. In Subsection 22.5.2 we can see the definition of some of the variables that are part of the previous equations. Figure 22.1 shows the architecture of a typical session in which users can be connected through Isabel as well as through the web transcoder. In order to simplify the equations SIP traffic will be left out of this scope. This architecture also has a system that records the video generated from the session.

22.5.1.3   Costs of an Isabel Session in the University

The requirement for the above described resources as well as others that are not very critical, entails a usage in the university that is proportional to the number of machines used in each session. Most systems usually deal with this kind of problem and nowadays there are Cloud Computing based solutions that solve them.

This solutions are focused on the variation of the demand for some services (whether it is in large periods of time such as months or in short periods such as hours). The workload of these services and the resources needed usually vary along the time. The consequence is that during the workload peaks the system has less available resources than needed and that during periods of lower workload the number of available resources will increase.

In Spain, all universities are directly connected with a high bandwidth network named RedIRIS, so in these scenarios the bandwidth consumption is not usually a problem. In the case of CPU usage, higher investments in machines are required depending on the number of the offered simultaneous sessions during the periods of low workload and during peaks, so any investment saving on this resources will always be welcome. This is the reason why the case of CPU usage is a different problem and we have to use different approach to deal with it.

22.5.2   Calculus of the Cost of Cloud Computing Architectures

In the cost calculus we intended to abstract details of the provider of the IaaS (Infrastructure as a Service). To do this we have defined a set of constants that refer to the cost of each resource. Nowadays the best known example is Amazon EC2[10] in which there are CPU and bandwidth usage costs. In the case of CPU the costs can vary according to the type of machine that is going to be used and its amount of memory. In the case of bandwidth usage it refers to the outgoing traffic as well as the incoming traffic to the Cloud Infrastructure. Specifically we have defined the next cost constants:

- C_{mcu}, C_{tr}: They represent the cost of a machine running an Isabel application that acts as MCU and transcoder. These constants are measured in $/hour.

- $C_{T_{in}}, C_{T_{out}}, C_{T_{internal}}$: These are the costs of the incoming, outgoing, and internal traffic data respectivelly, measured in $/bit consumed per hour.

- $BW_{isabel-user}$: This is the bandwidth used by an Isabel user in each direction.

- $BW_{web-isabel}$: This is the bandwidth used between the RTMP transcoder and the RTMP server.

- $BW_{web-user}$: This is the bandwidth used by a RTMP client in each direction.

- $N_{isabel-users}$: This is the number of Isabel clients that are connected to the session.

- $N_{web-users}$: The number of RTMP clients (or web users) that are connected to the session.

Next we present two different architectures for the same system. In the first, all the components that are part of the videoconference are executed in the Cloud Computing system where we are measuring the cost. In the latter, we have divided the components separating those which make an intensive use of CPU and those which are associated with the bandwidth consumption.

---

[10]http://aws.amazon.com/ec2/

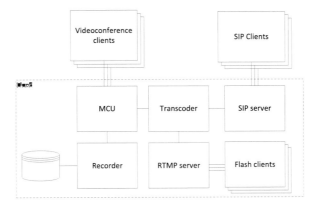

FIGURE 22.3
Single Cloud Architecture.

22.5.2.1   Costs of the Single Cloud Architecture

The first architecture contains all of the components in an external Cloud provider, such as Amazon EC2. All the costs can be calculated knowing the provider rates, but we will first show the formulas that relate the resource usage (CPU and bandwidth) with the cost that is associated with each of them. Having the formulas facilitates the reader to calculate the cost with different Cloud vendors. The CPU cost is the result of summing the machines that are needed for an Isabel session. Although according to the Figure 22.3 a session could be composed of more machines than those included in the formula, some of them are not used only by a single session. They are shared among different sessions, so the value for any scenario in which there are several simultaneous sessions is negligible. In that way, we have three machines which have different requirements according to CPU and memory.

$$C_{cpu} = (C_{mcu} + C_{tr}) \cdot t$$

The result represents a single second duration cost of the videoconference. The second formula is the one which works out the cost of the traffic generated by the internal, outgoing and incoming Isabel transmissions that are sent through the Cloud infrastructure. In the case of Isabel we could assume that the consumed bandwidth will always be the same for each connection between Isabel terminals, so the formula will be the following:

$$C_{BW_{isabel}} = (N_{isabel-users} \cdot (C_{T_{in}} + C_{T_{out}}) + 2 \cdot C_{T_{internal}})$$
$$\cdot BW_{isabel-user} \cdot t \cdot 3600$$

The case of the bandwidth consumed by the web users depends on the number of users who are connected to the session. Here we can see the variables defined in (22.2) and (22.3).

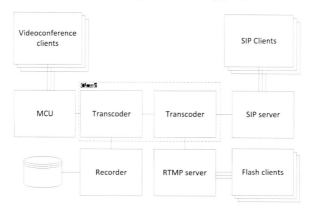

FIGURE 22.4
Hybrid Architecture.

$$C_{BW_{rtmp}} = (C_{T_{out}} \cdot BW_{web-out} + C_{T_{in}} \cdot BW_{web-in}) \cdot t \cdot 3600$$
$$C_{BW_{recording}} = C_{T_{out}} \cdot t \cdot BW_{web-isabel} \cdot 3600$$

22.5.2.2 Costs of the Hybrid Architecture

The second architecture is oriented to reduce the cost of separating those components that use more CPU from those that consume more bandwidth. Mainly, we have to take into account the increasing number of users. The CPU cost is the result of summing all the machines that are required in a session. Although there are more machines in the global system, the rest of them are going to be started in the internal datacenter of the University, so they are part of the cost of Cloud resources. Consequently we have three machines that have different requirements of CPU and memory. We can see this on Figure 22.4.

$$C_{cpu} = 2 \cdot C_{tr} \cdot t$$

The result is the cost of all the CPUs per second in a videoconference session. The cost of Isabel traffic in this case is the main consequence of all the internal traffic in the Cloud infrastructure, plus an outgoing connection that lets one transcoder and the MCU be connected and forward the traffic towards other Isabel clients.

$$C_{BW_{isabel}} = (C_{T_{in}} + C_{T_{out}} + 2 \cdot C_{T_{internal}}) \cdot BW_{isabel-user} \cdot t \cdot 3600$$

The cost of the traffic originated by other web clients falls because in this case there is only one web connection.

$$C_{BW_{rtmp}} = (C_{T_{out}} \cdot BW_{web-isabel} + C_{T_{in}} \cdot BW_{web-in}) \cdot t \cdot 3600$$

The cost of the traffic that is generated in order to record the session only has one direction as shown in the following formula:

$$C_{BW_{recording}} = C_{T_{out}} \cdot BW_{web-out} \cdot t \cdot 3600$$

Figure 22.5 depicts the differences between both architectures and how the cost remains constant in spite of the increasing number of web users for the hybrid case, which comes as a result of applying the methodology presented in section 22.3. This behavior is due to the deployment of part of the resources in the data center of our university, in which we have some old machines that can do light work but that are connected via a high bandwidth connection to the Internet. It shows how the resources used in a session can be reallocated in different clouds or data centers in order to develop a cost effective architecture. For the calculus we have taken the price of the Amazon EC2 resources applicable at the time of writing this chapter.

22.6   Conclusion

In this chapter we have shown how a videoconference system can be greatly improved with usage of hybrid cloud resources. We have gone through this by establishing the advantages of partitioning the system into CPU, bandwidth and store intensive parts. Finally we have validated our claims with a videoconference scenario where the use of an hybrid cloud architecture results in much lower cost.

We think that other systems can benefit from this idea by using a virtual infrastructure manager and a scheduler that starts the resources required for a service in the most appropriate provider based on location, price, and resource type.

We intend to further develop this architecture and make it as general as possible and abstract away the details of the system being managed by the scheduler and the executor.

At the moment of writing this chapter we have used this architecture in several real case scenarios, such as videoconferencing events with users participating from different countries. There were two main types of participation: the first one was joining the session in order to talk and make presentations; the second one was just watching the conference through web video streaming.

Finally, Figures 22.6(a) and 22.6(b) show several connections made around the world in order to assist the Conferencia Rails 2010 and an Isabel Workshop. These were real life scenarios that used the described architecture.

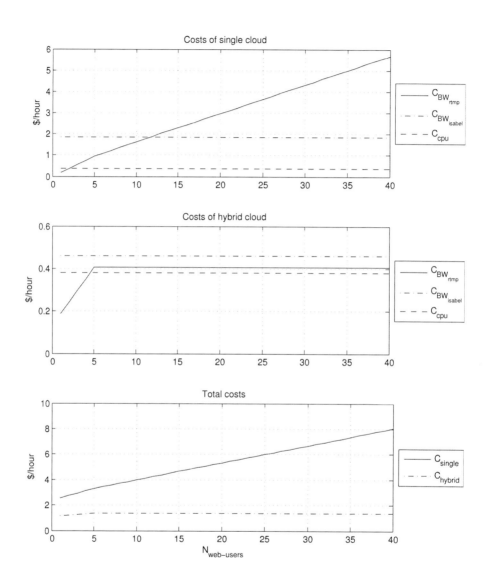

FIGURE 22.5
Cost Comparison.

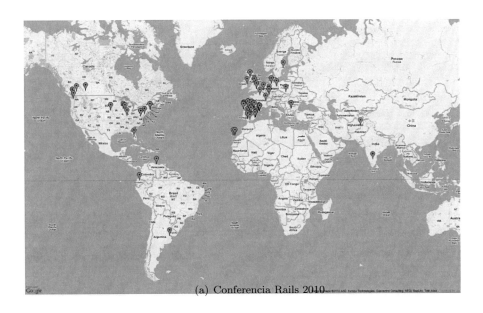

(a) Conferencia Rails 2010

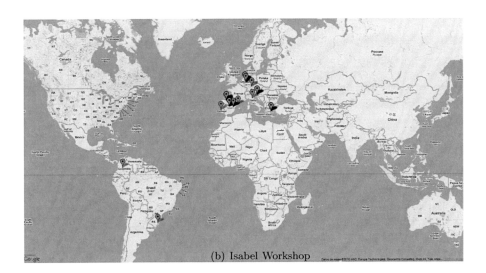

(b) Isabel Workshop

FIGURE 22.6
Videoconference Real Scenarios.

23

Advanced Computing Services for Radiotherapy Treatment Planning on Cloud

Luis M. Carril, Zahara Martín-Rodríguez, Carlos Mouriño, Andrés Gómez, Rubén Díaz, and Carlos Fernández

CESGA, Galicia, Spain

CONTENTS

Abstract

The eIMRT platform is a remote distributed computing tool that provides users Internet access to several services such as CRT and IMRT virtual verification of a planned treatment using highly accurate Monte Carlo simulations. Verification of the IMRT treatments is a compulsory procedure, which involves experimental measurements using a phantom. It is a time-consuming process, in terms of an accelerator's operational time and human resources. The virtual verification stage plays a significant role in the radiotherapy treatment chain as it allows comparison between the dose distributions calculated for a given treatment plan using an in-hospital Treatment Planning System (TPS) and those calculated employing Monte Carlo simulations. Goals for this comparison are to detect possible regions of dose disagreement, their causes, and to assess consequences to the patient.

These services, designed as a Software-as-a-Service (SaaS), are accessible through a user-friendly and platform-independent web page as well as through web services. Its flexible and scalable design focuses on providing the final users with services rather than a collection of software pieces. The service can contribute to the improvement and quality assurance of treatments in medical institutions, with no need for investments in new equipment and software components.

Monte Carlo simulations demand a lot of computer time. To overcome this high demand, the service provider can obtain a highly flexible and scalable resource pool using Cloud computing. The number of resources can be increased by provider aggregation and varies dynamically, avoiding unwanted overloads. Using Cloud computing, the platform returns the full Monte Carlo simulation in few hours, which is an acceptable time-to-solution level to be used on a daily basis. It also reduces the cost, as the accelerator's operational downtime to check the treatment is not required. Therefore, it is an open possibility for development in the clinical environment.

Keywords: *Cloud computing, HTC, SaaS, treatment verification, CRT & IMRT planning, Monte Carlo simulation*

23.1 Introduction

Cancer represents the second largest cause of death in Europe [73]. Over 40,000 people per 10 million inhabitants are diagnosed with cancer annually in developed countries [74]. Radiotherapy is frequently used to treat it, either on its own or combined with other methods. During a radiotherapy treatment, cancerous growth is exposed to electron beams, x–rays or gamma rays, which eradicate cancerous cells. The effectiveness of the radiotherapy treatment relies on the enhanced sensitivity of tumor cells to radiation in comparison to healthy cells. In addition, the harmful effects of radiation can be minimized by focusing only on the area to be treated and sparing the surrounding healthy tissues. There are two main types of radiotherapy treatments: external and internal radiation therapy. In external–beam radiation therapy, radiation comes from an external source, an electron accelerator (Linac) and is most often delivered in the form of x–rays or gamma rays. One of the most common types of external–beam radiation therapy is called three–dimensional conformal radiotherapy (3D–CRT) [80], in which radiation is precisely delivered to shaped targets areas. Many other methods of external–beam radiation therapy have been also developed and are currently used in cancer treatment. These methods include: Intensity–modulated radiation therapy (IMRT) [76] [75],

Image–guided radiation therapy (IGRIT) [82], Tomotherapy [234], Stereotactic radiosurgery (SRS) [79], Stereotactic body radiation therapy (SBRT) [79], and proton therapy [78] [77]. IMRT uses either stationary or dynamic collimators, which allow the intensity of radiation to vary during a treatment session. In this manner, different areas of the tumor or nearby tissues receive different doses of radiation. The goal of IMRT is to increase the radiation dose to the tumor and reduce radiation exposure to specific sensitive areas of surrounding normal tissue. In IGRT, repeated imaging scans are performed during treatment. They help to identify variations in tumor's size and location due to treatment and allow patient–positioning or planned radiation to be adjusted during treatment as needed. Repeated imaging can increase the accuracy of radiation treatment and may permit reduction in the planned volume of tissue to be treated, thereby decreasing the total radiation dose to healthy tissue. Tomotherapy is a type of image-guided IMRT. A tomotherapy machine is a hybrid between a computed tomography (CT) imaging scanner and a Linac. The tomotherapy machine delivers radiation for both imaging and treatment, rotating completely around the patient. It acquires CT images of the patient's tumor immediately before the treatment session to allow for very accurate tumor targeting and sparing of normal tissue. SRS and SBRT are types of image-guided radiotherapy. They are commonly used for the treatment of small tumors with well–defined edges located in the brain or spinal cord.

Internal radiation therapy (also known as brachytherapy) is radiation delivered from radiation sources placed inside or on the body [71]. The radioactive isotopes are sealed in pellets placed in patients by using delivery devices. As the radioactive isotopes decay, radiation is emitted, damaging nearby cancer cells. Several brachytherapy techniques are used in cancer treatment. Interstitial brachytherapy uses a radiation source located within the tumor tissue. Intracavitary brachytherapy uses a source positioned within a surgical or body cavity. Brachytherapy may be able to deliver higher doses of radiation than external-beam radiation therapy while causing less damage to normal tissue.

For each of the methods described previously, radiotherapy treatment planning is individually defined for each patient. Radiotherapy treatment planning requires an exhaustive dose calculation before radiation is delivered to tumor cells. Doses must be high enough to exterminate cancerous cells but below a threshold in order to protect healthy tissues and evade side effects. The oncologist initially prescribes the dose to be received by the tumor. Before delivery, medical physicists follow a strict planning protocol to establish the final doses. This process is carried out using an in-house treatment planning system (TPS) and ensures quality and effectiveness of the dose delivered during treatment. It is desirable for the treatment planning to be obtained the soonest for the actual treatment to begin. Frequently, a second calculation is required for quality control or, in complex cases, experimental verification. Therefore, a complete treatment planning is very time-consuming and highly expensive. Decreasing the time required to verify or improve dose delivery calculations will benefit quality, efficiency, and accuracy of this procedure.

In response to these requirements the eIMRT platform has been developed. The eIMRT [69] platform consists of a suite of remote tools used to help medical physicists in the definition of treatment plans and their verification. The eIMRT platform provides access to Monte Carlo verification, optimization of treatments, and a prototype of a database of interesting treatment/clinical cases through a user friendly platform and browser-independent web page. Also, it is provided with interfaces based on web services to allow other applications (i.e., TPS) to use these services. Design, implementation and execution of the verification and optimization algorithms are hidden to the user. This permits a robust and unified handling of the software and hardware necessary for these computational intensive services.

This approach is advantageous for the final user, as the implementation and maintenance aspects are transparent. Thus, there is no need to purchase specific hardware or software. eIMRT is an offline tool although the elapsed time between requesting and obtaining the results is very short. After calculations are completed, the user is informed by email and can re-login and review results. The eIMRT web client provides them with a set of tools to visualize the results, compare dose distributions in the case of the verification, and download them in DICOM format for archiving and reuse inside the hospital. Monte Carlo simulations present a high demand in terms of computing resources, which is overcome using Cloud computing. The platform returns the full Monte Carlo simulation in a few hours, which is a suitable time-to-solution level to be used on a daily basis. It also reduces the cost, as the accelerator's operational downtime for treatment inspection is not required.

This chapter focuses on the feasibility of using a Cloud for such demanding services, using treatment verification as an example.

This chapter is organized as follows: A first section describing IMRT verification. A second section provides a description of the architecture of the platform, followed by the analysis of Cloud posibilities in the eIMRT services. Experiments carried out on two different testbeds, locally, and on Amazon are then described. Finally, discussions and conclusions are drawn at the end of the chapter.

23.2   IMRT Verification

The verification tool provides a virtual checking of the treatment plan. Hospital protocols usually include a cross-checking of the doses calculated during planning treatment in order to detect possible errors, which can result in detriment to the patient. This is usually carried out experimentally using a phantom, which emulates the patient. The phantom is provided with dosimeters to measure doses at specific control points. A treatment plan is fully delivered and the recorded doses are analyzed and compared with those calculated by the TPS. Only when both calculations are in agreement is the treatment plan considered to be valid. Experimental verification is laborious and expensive and requires the Linac to be down for treatment delivery to patients during data acquisition. Therefore, medical physicists demand new software tools for accurate verification of treatment plans. Monte Carlo simulation methods are considered the best solution. Although these methods are very complex, not only from a technical viewpoint, but also because they need a large amount of CPU cycles, which make them impractical and almost impossible to be performed on site with the current computing infrastructure of hospitals. A suitable solution is software decoupling, i.e., the user's interface executes in local machines at the hospital and the computing core is taken to institutions that can effectively manage it.

Monte Carlo treatment verification consists of the comparison of the doses calculated by in-house TPS and the dose distribution calculated by Monte Carlo simulations for the same radiation treatment plan. The goals of such a comparison are to detect regions of dose disagreement and to analyze possible causes and consequences to the patient. This method is particularly suitable for detecting underdose or overdose regions with steep density gradients. The eIMRT has implemented the Monte Carlo verification process using the BEAMnrc [67] and DOSXYZnrc packages [72]. BEAMnrc permits simulation of the accelerator's head from the bremsstrahlung target to the bottom of a multileaf collimator (Figure 23.1). Dose deposition inside a patient is calculated using the DOSXYZnrc code. Input data files such as the patient Computerized Tomography (CT) images and the treatment plan in standard formats are required (DICOM CT for the patient's images and DICOM RT-Plan for the treatment plan). Optionally, a DICOM RTdose file containing the TPS–calculated dose distribution can also be provided, allowing the calculation of figures of merit such as gamma maps [480] and histograms with both dose distributions. Before the verification is submitted, the tomograph and the linear accelerator are commissioned. Commissioning is only carried out once, in order to characterize the electron source. In the first case, the user provides the data to transform the CT units into material (bone, air, water, etc.). In the latter, the platform provides the protocol to calculate the initial energy and spatial distribution of electrons generated by the linear accelerator, which collide with the bremsstrahlung target [70]. This characterization process finally compares experimental data acquired using a water phantom with the corresponding Monte Carlo simulation performed. It involves comparison of lateral profiles (Figure 23.2) and percentage-depth-dose (PDD) curves obtained (Figure 23.3). Lateral profiles display the radiation pattern on a plane perpendicular to the radiation beam. PDD curves show this radiation pattern on a parallel plane to the radiation beam and its variation with depth inside the water phantom.

This procedure is repeated for every available energy and configuration. Com-

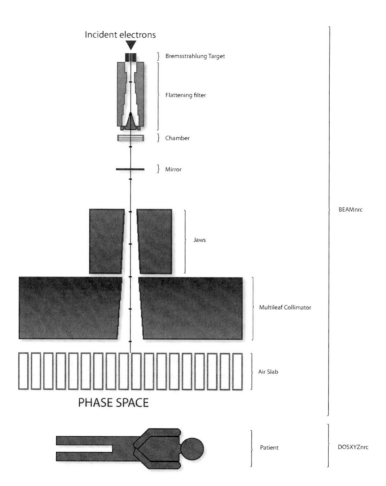

FIGURE 23.1
Schematic View of the Treatment Setup. BEAMnrc is Used to Simulate the
Linac from the Target Down to the Patient. DOSXYZnrc is Used to Calculate
Dose Deposition Inside the Patient.

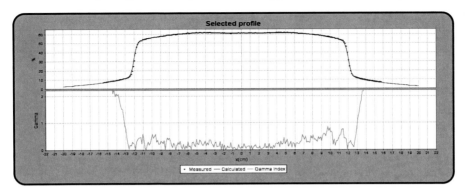

FIGURE 23.2
Comparison of Lateral Profiles for Experimental and Simulated Measurements.

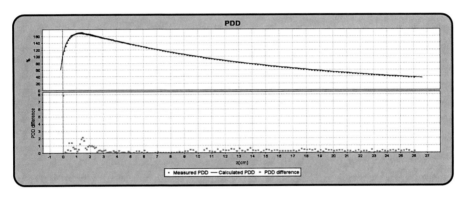

FIGURE 23.3
Comparison of PDD for Experimental and Simulated Measurements.

plete characterization of the Linac involves comparison of lateral profiles and PDDs for experimental and simulated data. Agreement between both takes place when maximum differences are not greater than 2%. Then, the commissioned Linac is available for verification. This process is individually carried out for an acquired Linac at a given hospital.

The implemented virtual verification process comprise five phases:

Phase 1 corresponds to accelerator simulation setup. During this phase, the linear accelerator's geometry, radiation source and treatment to verify (Linac and coach angles, positions of the multileaf collimator, times of radiation for each position, etc.) are acquired. It basically collects this information from the treatment's DICOM RT plan, and input files are generated in the format required by the BEAMnrc Monte Carlo code. These are:

- For CRT treatments: One single file for each radiation field, i.e., the shape of radiation field is adjusted to fit the profile of the tumor viewed as the beam's eye. This means a single file per incident angle.

- For IMRT treatments: A single file is generated for each segment in IMRT step-and-shoot treatments. The leaves of the collimator move from one position to the next, while the irradiation is stopped. The Linac is then back on to irradiate when they form a determined shape. For each control point in IMRT dynamic-MLC treatments, the time collimator leaves move continuously; meanwhile the accelerator is active.

As a result, few files are produced for CRT treatments (one per incident angle) and up to several hundreds for IMRT treatments. Each input file includes the number of particles (or histories) to simulate based on the planned monitor units (MU is a measurement of the radiation delivered) for each segment. Phase 2 executes the Monte Carlo simulation for each of the generated input files. Optimization of variance reduction techniques is carried out in order to maximize particle production and scoring for a given radiation field shape. A simulation is executed using the BEAMnrc and creates one output file for each input file. Phase 3 comprises patient simulation setup. Consistency tests are carried out at this stage in order to ensure that no particle loss occurred during previous steps. Patient's computed tomography (CT) data is converted into densities and materials for further calculation of the dose delivered considering characterization data of CT. Output of this phase is input for Phase 4, which entails calculating the dose-inside-the-patient using DOSXYZnrc Monte Carlo code. Since this task is highly parallelizable, can be divided into many different jobs. Finally, dose collection is performed at Phase 5, results are merged into a final single dose distribution. This dose distribution is normalized per unit of primary fluency and converted into absolute or relative dose distribution. Once the Monte Carlo process is completed, medical physicists can independently compare the calculated dose with doses obtained using the Monte Carlo provided the DICOM RTdose file generated by the TPS. For this purpose, a special service has been developed. Gamma maps and histograms are generated taking dose distributions as input. This task only requires a few seconds of CPU and the results are graphically displayed on the client's end (Figure 23.4 and Figure 23.5). Not a single value is produced to assess the quality of the treatment plan. This is a decision to be made by the medical physicist.

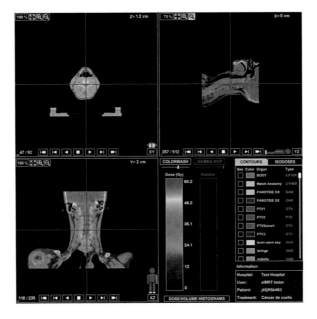

FIGURE 23.4
Dose Distribution Scheme for a Given Treatment.

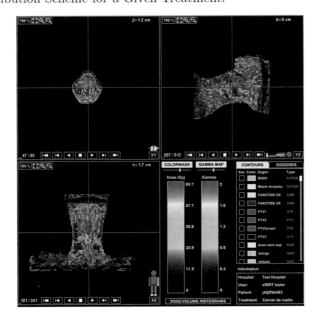

FIGURE 23.5
Gamma Map Dose Distribution for a Given Treatment.

23.3 Architecture

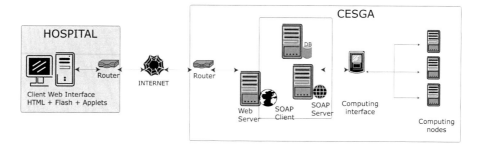

FIGURE 23.6
Overview of the eIMRT Architecture.

The eIMRT platform system has been designed following the Software Oriented Architecture paradigm in a three-layer model (Figure 23.6). The first layer is a client based on a web interface only requiring a web browser with Java applets and Flash enabled. This user's interface is used by medical physicists to analyze output data and define the parameters of the service request. These parameters are reference values for comparison between both dose distributions at the verification stage. Java applets are used to upload and download treatment files and patient's information as DICOM files. Patient's personal details are removed from these files in order to increase security and comply with the European regulations on health data protection. Finally the files are compressed to decrease transfer time.

The web client communicates with a web portal, which translates the request into a suitable message to call the second layer consisting of a set of web services. These services expose the main functionality of the platform. There are specific services for the different supported operations, such as user data management, file management, process request or results visualization [41]. These are the unique ways to access the stored information and to control the status of the requests. Recently, the backend stores information about each user and treatment plan and manages the execution of the different processes. Alternatively to the web client, the hospital is able to call directly to the web services from its own application. Then, the hospital is responsible of removing patient's personal details from the files. The last layer, known as the backend, is composed by the storage service and the computing interface. The platform executes calculations on different computing infrastructures: CESGA computing facilities, GRID, etc. The computing infrastructure behind the system is completely hidden from the final user.

There are three types of users in the platform: the main administrator who manages the platform, including algorithm and user/hospital management;

the hospital leader or administrator, who manages the information about the hospital facilities, as the model and parameters of the Linacs or the tomographs, their commissioning, and also manages the hospital's users. No other user type is provided by the platform. Patients and doctors are not expected to use it.

23.4   eIMRT as SaaS in a Cloud Infrastructure

The eIMRT platform has been analyzed as a Software as a Service, which should be capable of scaling to thousands of users and service requests per day, as will be discussed later. Therefore, eIMRT benefits from the Cloud infrastructure services (IaaS) which allow a service provision with a high quality. Three main possibilities have been identified (Figure 23.7). The platform can be divided in a web front end, a data and storage infrastructure and the backend computing facility, as previously described. Each can be managed independently and virtualized, and transferred to a Cloud environment, due to their independence.

The front end, as a web server, can be replicated several times and managed by a load balancer, which invokes new instances, as the access demand increases. It also avoids having a single point of failure and deploys in different geographic sites. Therefore, it could offer a faster response to the users. Storage and data management can also be transferred to a Cloud environment, which allows replication of data with almost unlimited capacity, and independent storage per customer. Although patients' details are removed from images, a high level of security is necessary to prevent unauthorized access, damage, or loss of the uploaded images. Lack of security would decrease significantly the trustworthiness of the platform. However, the requirement of a large computing infrastructure is a current major limitation to providing a wider service and being able to execute several verifications simultaneously. For this reason, the first priority has been to examine if Cloud can be used to execute the simulations, leaving the other two parts for future research.

The implemented solution for transfering the computing backend to a Cloud is the generation of a single virtual cluster for each request. This guarantees that each request is processed independently solving previous issues detected on local clusters, such as the interference between two simultaneous requests. This leads to an increase in time-to-solution and consequently a decrease of the service quality. The same effect is observed on Grid, even magnified, due to sharing with other users, and transfer of large files. This virtual cluster consists of a suite of virtual machines, one of them acting as a head node. The cluster nodes share a common file system and execute the Sun Grid Engine (SGE) as job scheduler.

Testing of the platform operating over a Cloud environment was carried out

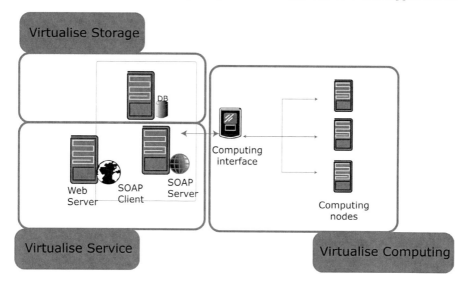

FIGURE 23.7
Virtualization of the Different Sections of the Architecture.

using two testbeds: a local testbed composed of a hardware server, and the
Amazon Elastic Compute Cloud service. These testbeds are described later.

23.4.1   Verification Life Cycle

The previously described Monte Carlo verification process requires a high
computational capacity. The execution of hundreds of short jobs (one per segment) is carried out for one IMRT treatment verification in Phase 2. These
jobs take approximately three minutes in a 2.27GHz Nehalem core as illustrated in Figure 23.9. In Phase 4, tens of larger jobs are run, consuming 42
minutes each (Figure 23.10). Adding all the phases, tens of hours of CPU are
consumed.

The Cloud is able to achieve the desired amount of computing power through
the invocation of multiple virtual machines working as a typical computing
cluster and sharing the tasks. Therefore, a virtual cluster architecture is created and deployed to process the verification.

Each one of the described phases of the treatment verification process has
different computational requirements. Consequently, they will be executed in
different contexts. They will be described in detail, from the computational
point of view.

Previous to the verification process, the virtual cluster infrastructure is deployed. Once the master node is started, the verification process is initialized.
The computing nodes are also deployed in the meantime. The minimum time
to start operating the cluster is around five minutes for both testing environ-

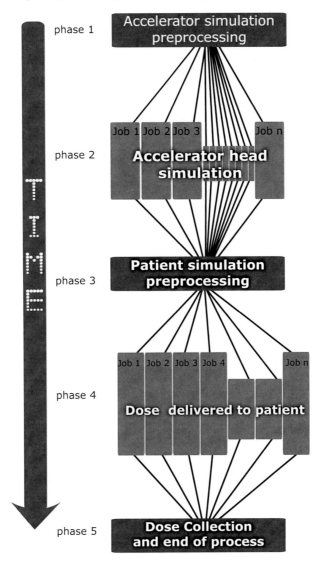

FIGURE 23.8
Verification Workflow.

ments. SSH commands are used for the entire workflow invocation, and data transfer is carried out through SCP.

The software required to run the simulation is uploaded after the master node's deployment. The eIMRT backend software consists of a set of Bash shell scripts to control the workflow and C programs to generate and process input and output files, and BEAMnrc and DOSXYZnrc packages (approximately 30MB

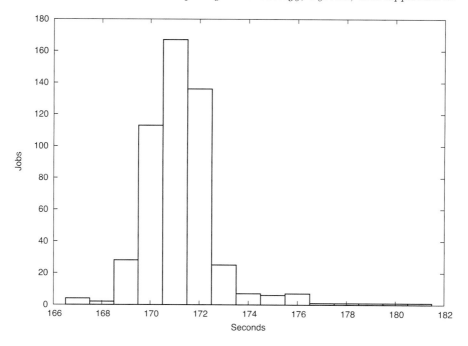

FIGURE 23.9
Elapsed Times for Phase 2 Jobs.

in total). These binaries are compiled in 32 bit version to be executed over 32 and 64 bits.

At this stage, the input files are sent to the cluster. These files carry information about the configuration, accelerator description, and patient's DICOM images. Once the files reach the cluster, the worflow process is started in the master node. Also, a monitoring script is initialized in the eIMRT backend computer to periodically examine the process status.

Phase 1 is involved with the generation of the input files to be executed in the master node. Jobs for Phase 2 are submitted as job arrays to SGE to prevent scheduling problems. Each job acquires the input data from the shared volume to its local temporary disk at startup to develop the execution locally. It returns the output data to the shared volume after execution.

Once the jobs in Phase 2 are completed satisfactorily, Phase 3 is executed in the master node. It collects the output data from the accelerator's simulation and creates new input files for patient simulation. Then, the jobs are submitted to Phase 4. Again, each job copies the input files to local disk and stores the output files back in the shared file system.

Finally, all the output files are postprocessed in the master node in Phase 5. They are merged in a single dose file (of a few tens of MB), finishing the

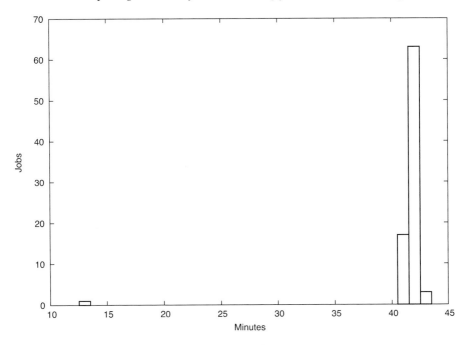

FIGURE 23.10
Elapsed Times for Phase 4 Jobs. Notice that One Job is Especially Short, as
It Contains the Remaining Monte Carlo Stories Simulated.

verification workflow script. The monitoring system detects the end of the
verification, downloads the final output file, and retires the cluster.

23.5   Testbeds

Testing of the platform operating over a Cloud environment was carried out
using two testbeds: a local testbed composed of commodity hardware, and the
Amazon Elastic Compute Cloud service.

23.5.1   Local Testbeds and Backend Virtualization Cluster

The local testbed comprises a set of forty double quadcore Nehalem servers
with a clock of 2.27GHz, each one provided with 16GB of memory and 0.5TB
of disk. They are interconnected by a gigabit ethernet network. Each node
is automatically installed and configured by an installation server to run a

CentOS 5.5 [667] with the Xen 3.0.3 hypervisor [770], in which the virtual machines will be executed in paravirtual mode.

In order to deploy and manage the virtual machines over the infrastructure, the OpenNebula 2.0 cloud toolkit [81] is used. OpenNebula is a fully open-source toolkit to build any type of IaaS Cloud: private, public and hybrid. It orchestrates storage, network, virtualization, monitoring, and security technologies to enable the dynamic placement of multitier services (groups of interconnected virtual machines) on distributed infrastructures, combining both datacenter resources and remote cloud resources, according to allocation policies.

OpenNebula is installed in one node and the rest of the infrastructure remains as resources. It stores virtual machines images; and under command (following a user-created template) copies the desired image, creates dynamically additional disks, makes the Xen configuration file and launches the virtual machine remotely, through SSH commands.

As OpenNebula works at a virtual machine level, it does not manage the possibility of service definitions or a virtual cluster definition.

Therefore, a virtual cluster architecture working over OpenNebula has been built for eIMRT. It is provided with a master node and up to 254 computing nodes. Limitation of the number of nodes is chosen by design but can be expanded. Using its feature for creating virtual subnets, each virtual cluster includes an exclusive subnet to which all nodes are connected. The master node has also a public connection to be used as cluster frontend; meanwhile, the compute nodes are isolated.

The master node exports a NFS volume (user's home) to all the private subnet, providing an easy mechanism to share specific application binaries to all nodes with only one installation and a central storage volume for each node's final results. The shared volume is created dynamically simultaneously to deployment in the physical node. Thus, no transferring is needed.

At startup, the compute nodes import the user's home from the shared volume and connect to the queue system as SGE nodes. They have a temporal volume for scratch data, which is also created dynamically following the same procedure used to create the shared disk.

One benefit of having all the cluster nodes in the same and exclusive subnet and using SGE is the elasticity of the cluster as a feature, although limited. Once the initial cluster is deployed and running, if new nodes are required, no additional configuration is needed. Launching the new ones with the standard compute node configuration for this cluster deployment is enough. It automatically imports shared directories, and SGE manages the addition to the resource pool.

A set of scripts that deploy and undeploy the cluster, add and remove computing nodes, has been developed to manage all the life cycle of the cluster over OpenNebula. The deployment of the cluster allows a total configuration of virtual hardware resources: number of CPUs, memory, shared volume size, temporal volume size, import of SSH-keys, user creation, etc.

Although the characteristics of de virtual cluster can be modified easily, in this testbed usually the virtual machines were composed by 1 CPU (one physical core), 1GB of memory, 6GB of shared disk and 512MB of scratch disk. The OS of the virtual machines is a CentOS 5.5 and the version of the SGE is 6.1u6.

23.5.2 Amazon Testbed

In order to analyze the platform over a public cloud (in opposition to the private local cloud) the Amazon EC2 has been used.

The Amazon EC2 [64] is a service of Amazon Web Services that provides the clients with on–demand virtual machines. The clients can define their own virtual images, and instantiate them over a set of predefined hardware configurations. Hence, paying only for the amount of hours used per instance. The Amazon WS includes additional tools to use with EC2, as alternative storage systems (EBS and S3), monitoring, elasticity, etc.

As OpenNebula the Amazon EC2 operates on a virtual machine basis, allowing the user the task of building a more complex architecture. As before, the eIMRT virtual cluster has to be constructed over a set of virtual machines. StarCluster has been used for its development.

The StarCluster toolkit [751] is a utility for creating and managing general purpose computing clusters hosted on Amazon EC2. It minimizes the administrative overhead associated with obtaining, configuring, and managing a traditional computing cluster used in research labs or for general distributed computing applications. StarCluster utilizes Amazon's EC2 web service to create and destroy clusters of Linux virtual machines on demand.

This tool generates clusters with the required attributes, although it does not support cluster elasticity, but reduction can be achieved manually. A set of Amazon images (AMIs) is predefined, and can be tuned to generate new AMIs with more preloaded software. These AMIs are Ubuntu 9.04 based and include SGE 6.1u5. As the different Amazon instance types are 32 bit or 64 bit, two versions of the AMIs exists, one for each architecture.

Amazon provides different sets of hardware combinations, defined as instances. Three instances have been used in these tests:

- m1.small: one 32 bit virtual CPU of 1 ECU, 1.7GB of memory, a medium I/O performance, and 160GB of storage. Price per hour $0.085. It is the default instance type.

- c1.medium: two 32 bit virtual CPUs of 2.5 ECUs each one, 1.7GB of memory, a medium I/O performance, and 350GB of storage. Price per hour $0.17.

- c1.xlarge: eight 64 bit virtual CPUs of 2.5 ECUs each one, and 7GB of memory, a high I/O performance, and 1690GB of storage. Price per hour $0.68.

The Amazon ECU (EC2 Compute Unit) is a measure of computing power that equates in performance to a 1.0-1.2GHz 2007 Opteron or 2007 Xeon processor. The prices correspond to the US North Virginia region. The experiment was carried out during October and November of 2010.

By default the instances make use of an instance storage, disk space assigned to the virtual machine during the instantiation process. These disks are volatile, only existing as long as the virtual machine exists. Another sort of storage available in Amazon is the Elastic Block Storage (EBS). EBS is a persistent block storage with a high I/O performance; it is not bound to any instance but can be attached to any virtual machine. The EBS implies an additional cost of $0.10 per GB-month of persistent storage and $0.10 per million I/O requests.

Additionally, Amazon charges for the I/O data transfer to and from the Amazon's infrastructure, but not the internal transfer, if in the same region. The input price is $0.10 per GB and the output price is $0 for the first GB per month. From this first GB onwards the cost is $0.15 per GB and decreases as the number of GB transferred increases. It is necessary to notice that the additional costs that imply EBS and the I/O operations has little impact on the total price of a verification execution.

23.6 Experimental Results

The experiments carried out on the testbeds, previously described, were performed using a phantom. A monochromatic energy beam of 6MeV with incidence along the z axis was used for a circular electron source.

Figure 23.11 displays the scalability for the local testbed. It is observed, as the number of nodes increases, the execution time is reduced. The phases 2 and 4 are highly paralellizable and the entire numbers of tasks are completely independent; 84 nodes appears as limit in terms of scalability. This is because the number of jobs in Phase 4 for the examined treatment is 84. Also the reason of the poor scalability for more than 16 nodes.

During parallel phases 2 and 4, the process waits for the entire number of jobs to finish. The remaining number of jobs are not divisible and therefore it is not possible to distribute them to the available nodes. Bottleneck is the result of a series of idle nodes at the end of each phase. This effect is especially important in Phase 4 as the execution time is approximately tens of minutes per job, but is not as relevant in Phase 2.

Table 23.1 illustrates different configurations for computing clusters with different instance types in Amazon EC2. Attention is paid to the relation between total ECUs in the deployment, the number of CPUs, and the ECUs per CPU. Tests show that one CPU with 2.5 ECUs is similar to one node of the local testbed. Consequently, scalability results and times are quite similar to local

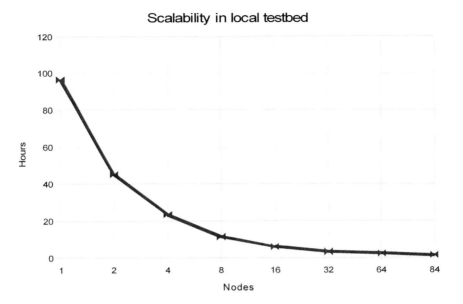

FIGURE 23.11
Scalability in Local Testbed.

testbed results. The results display that as the number of ECUs increases per CPU the execution time decreases, as expected. It is also relevant to observe using an EBS volume has better performance in I/O operations leading to a reduction of the execution time. The effect is acute when a few processors are used, as previously reported [381].

The cost also appears to decrease as the ECUs per CPU increase. This is due to a better performance than the price incrementation. Such as incrementation is because entire hours are actually charged. For example, if the duration of a process is 2.1h, actually 3h will be charged, wasting 0.9h for large and expensive nodes.

TABLE 23.1
Results in Different Instance Types in Amazon EC2.

Inst. type	m1.small		m1.medium		c1.xlarge		
# Instances	2	7	7		2		6
ECUS	2	7	35		40		120
CPUS	2	7	14		16		48
Shared vol.	Inst.	Inst.	Inst.	EBS	Inst.	EBS	Inst.	EBS
Time (hours)	95.2	27.5	8.3	6.9	8.8	5.9	2.1	2.1
Cost ($)	13.3	16.7	10.7	8.3	12.2	8.2	12.1	12.1

23.7 Discussion

The results of the experiments show the feasibility of using a Cloud to provide radiotherapy services which demand high computing capacity. In fact, the main limitation to launch such a service to the community is the amount of computing resources, which the service provider needs. For example, the Directory of Radiotherapy Centers (DIRAC) of the International Atomic Energy Agency (IAEA) [50] has registered 7,278 radiotherapy centers in the world with 9,474 clinical linear accelerators. Assuming that only a small fraction of the treatments will be virtually verified, the Table 23.2 lists the number of computing hours of m1.small instances in Amazon. If only one treatment per day is executed in 10% of these institutions, more than 69,000 computing hours are used. This number is equivalent to having a local infrastructure of 2,881 cores, which could be affordable for a small company or institution. However, if the percentage of interested institutions or the number of treatments per day increases, the backend infrastructure must be increased significantly; up to more than 150,000 cores, which is not now feasible for an small company if QoS needs to be guaranteed. However, the service can be provided with such requirements of scalability, executing the simulations in the cloud environment, by renting on demand the resources necessary from IaaS vendors. Two models are valid in this case: a hybrid model, in which the provider mixes its own local computing resources plus external cloud virtual machines (following a hybrid cloud model), or really passing all the computing needs onto external providers.

Other authors [428] have used the same virtual cluster model on Amazon as external resources of a single hospital. Although the simulation workflow is different, they have demonstrated that the model is economically feasible for the hospital, as AWS costs for their demands is only 20% of the annual costs of an in-house cluster. In the case of eIMRT, which is designed to provide services to hospitals, issues are more related to the scalability of the solution rather than the number of customers. Recently, CESGA has invested 150,000 Euro (plus VAT) in a 1,200 AMD cores cluster (125 Euro/core). Each one consumes, including air conditioning, around 20W priced 0.11 Euro/kWh. Therefore, adding amortization, electric consumption and maintenance personnel (two full time technicians) and assuming that the equipment is fully used, the cost per hour is around 1 cent of Euro (or around $0.013 per hour) which is really competitive with the price of Amazon EC2 ($0.085/hour of a m1.small). In the virtual cluster model implemented, the communication costs for sending the data to and receiving from Amazon is negligible, leading to a small increase of a few cents in the initial price. There are other hidden costs, not included (assurance, building space, etc.), which will increase it. Even including these items, the final cost could be competitive with Amazon EC2, but it exposes the service provider to other problems, as it can not guarantee a Quality of

TABLE 23.2
Number of Instances and Computing Hours Needed for Service Provision.
The left column represents the number of treatments to validate per day and
the first row indicates the oercentage of radiotherapy centers which use the
service.

	10%	30%	50%	80%	100%
# hours					
1	69,141	207,423	345,705	553,128	691,410
2	138,282	414,846	691,410	1,106,256	1,382,820
3	207,423	622,269	1,037,115	1,659,384	2,074,230
4	276,564	829,692	1,382,820	2,212,512	2,765,640
5	345,705	1,037,115	1,728,525	2,765,640	3,457,050
6	414,846	1,244,538	2,074,230	3,318,768	4,148,460
# 24h Instances					
1	2,881	8,643	14,404	23,047	28,809
2	5,762	17,285	28,809	46,094	57,618
3	8,643	25,928	43,213	69,141	86,426
4	11,524	34,571	57,618	92,188	115,235
5	14,404	43,213	72,022	115,235	144,044
6	17,285	51,856	86,426	138,282	172,853

Service to all its customers. If the demand of the service in peak hours exceeds
the computing capacity, the time-to-solution will degrade. As a result, the
company should make a large investment, which will be only competitive if
sold 24 hours a day. Otherwise, the cost per hour will increase dramatically.
Although the price per CPU hour could be competitive, to provide this sort
of service an external cloud provider will be desirable to reduce the associated
risks and large initial investment in infrastructure.

23.8 Future Work

At the site of the results, some aspects have been identified for improvement.
Scalability issues could be solved, improving granularity and duration of the
jobs in Phase 4. During this phase, tens of jobs of approximately 45 minutes
are executed. However, these jobs could be split, resulting in hundreds of
independent jobs of a few minutes each, leading to a finer granularity.
Further improvements can be achieved by eliminating sequential Phase 3. This
could be possible, if dose deposition is individually calculated for each phase
space calculated in Phase 2. It would imply saving time spent on files' transfer
and due to the bottleneck barrier.
Another possible improvement is cluster reutilization. A virtual cluster is de-

ployed for the calculation of a single treatment. The Amazon EC2 Cloud provider charges for entire hours; a lot of the charged computing time is then wasted. Therefore, reusing the cluster for other treatments' calculation could be beneficial in order to prevent a waste of CPU time and resources. This might require a metascheduler in order to manage all deployed clusters.

Currently, verification of radiotherapy treatments has been implemented. There are other processes, as the treatment optimization to be implemented in the near future. Optimization also requires great number of computational resources. Therefore, this process will be submitted to the Cloud as the verification process.

Also, virtualization of the web server and storage service can be carried out. In this manner, not only the calculations developed in the backend would be executed in the Cloud. Moreover, if there is a significant increase in the number of users, a load balancer and a varying numbers of web servers would be used in an Elastic Computing paradigm.

23.9   Conclusions

In this chapter, a Cloud solution for radiotherapy calculations based in SaaS has been presented. The future of radiotherapy planning and verification shall be based on solutions, which would be supplied through Internet and require a high computing capacity. Both services can be provided on-demand with service quality. The proposed model has shown that these services demand a high number of CPUs to afford results in a reasonable time.

The eIMRT platform backend calculations benefit from the cloud paradigm:

- Decreases the market entry cost, the eIMRT no longer needs access to a local computing cluster. It can borrow at any instant the resources needed from the cloud and pay only for the amount of time used in the simulations.

- The capability of selecting the number of resources used at one time and change them dynamically.

- The flexibility in the deployment of virtual clusters allows the creation of one infrastructure for each backend calculation, completely isolating the executions, producing a better control over each process and avoiding interferences, especially by software failures.

In summary, the eIMRT platform is a good case for the provision of external services following a utility model in the health sector.

Acknowledgments

This could not have been developed without the collaboration of many people and institutions. The authors want to thank Xunta de Galicia for the funding (Project R&D Grant 09SIN007CT) and Spanish Government (Project R&D Grant TSI–020301–2009–30). Finally, the experiment would not have been possible without the usage of the Cloud computing infrastructure of Centro de Supercomputación de Galicia (CESGA) and their economic support for Amazon costs.

24

Cloud Security Requirements Analysis and Security Policy Development Using HOOMT

Kenneth Kofi Fletcher

Missouri University of Science and Technology

Xiaoqing (Frank) Liu

Missouri University of Science and Technology

CONTENTS

Security continues to be a major challenge for cloud computing, and it is one that must be addressed if cloud computing is to be fully accepted. Most technological means of securing non-cloud computing systems can be either applied directly or modified to secure a cloud; however, no integrated model-based methodology is yet available to analyze cloud security requirements and develop policies to deal with both internal and external security challenges. This work proposes just such a methodology and demonstrates its application with cases of use. Cloud assets are represented by high-order object models, and misuse cases together with mal-activity swimlane diagrams are developed to assess security threats hierarchically. Cloud security requirements are then specified, and policies are developed to meet them. Examples show how the methodology can be used to elicit, identify, analyze, and develop cloud security requirements and policies using a structured approach, and a case study evaluates its application. Finally, the work shows how the prevention and mitigation security policies presented here can be conveniently incorporated into the normal functionality of a cloud computing system.

24.1   Introduction

Over the past few years, cloud computing has gained a lot of attention as widespread use of its applications and services has increased. As promising as it is, cloud computing also brings forth many new challenges. Security is one of such major concerns. In order to minimize the security risks and to maintain a secure computing environment, cloud computing systems must be bound by good security policies. Regardless of its extensive utilization and growth potential, cloud computing makes it more difficult to protect confidential information [524]. This is due to the fact that most organizations lack the right policies and procedures that ensure that this confidential and sensitive information put in the cloud remains secure. Cloud security policies provide many benefits and are worth the time and effort needed to develop them. Security policies define allowed and disallowed behaviors and also set the framework for security implementation. Despite these advantages, good and robust security policy development becomes an issue due to the complex nature of the cloud environment.

To effectively perform its role of safeguarding, it is important to develop comprehensive cloud security policies based on security requirements [331]. The analysis of security requirements on which policies are developed must consider the complex nature of the cloud architecture. The analysis method must identify security requirements at multiple levels of the cloud. It is crucial, therefore, to use a top-down approach based on a clear policy to analyze security requirements and develop effective security policies.

24.1.1   Research Rationale

Although data stored in the cloud and other compute capabilities are not actually in the "cloud" but reside in data centers housing hundreds of servers, networking cables, and other physical devices, physical threats are among the greatest dangers to the cloud. Also, because its applications and services are delivered through the internet, cloud computing is prone to various kinds of external security risks such as denial-of-service (DoS) and man-in-the-middle

(MiTM) attacks. The relative anonymity of registration and usage models in cloud computing encourages spammers, malicious code authors, and other misusers to conduct their activities with relative impunity [194].

Communication with the hypervisor contains vital information, including account names and passwords, which must be secure. Hypervisors are not sufficiently robust because researchers have focused on the notion that security should be built around applications and not virtual machines. Finally, in a virtualized environment, it is relatively easy to steal an entire virtual server, along with its data, without anyone noticing.

These and many more security concerns posed to cloud computing systems in addition to its complex environment demands that security requirements be analyzed and policies to address them be developed early in the development process, using a comprehensive approach that considers the entire cloud. Although security policies themselves do not solve problems, and in fact can actually complicate things if they are not clearly written and consistently observed, comprehensive cloud security policies can effectively address security risks and provide a foundation for preventing or mitigating security concerns [687]. They also provide an essential role of bridging the gap between security requirements and implementation [734].

The unified modeling language (UML) [622] that is most often employed to elicit requirements was not initially designed to capture nonfunctional requirements such as security. Existing methods to analyze security requirements do not consider both internal and external threats in a structured manner. They focus entirely on external misusers, and they rely only on security technologies such as network monitoring systems, intrusion detection and prevention systems, firewalls, antivirus systems, and data leakage protection.

Internal threats have steadily increased over the past few years, and cloud computing is not necessarily any more secure internally than non-cloud computing environments. Internal misusers generally have more knowledge of and access to data and applications than do external misusers. Although internal threats cannot be entirely eliminated, some effective barriers can be developed to mitigate them.

Therefore, a systematic methodology and process are necessary to analyze security requirements and develop security policies for cloud computing systems. This methodology must identify security requirements at multiple levels to address threats posed by both internal and external misusers in order to develop clear cloud security policies and thus build a secure cloud environment. The process presented here employs the high-order object-oriented modeling technique [475] together with use cases [622], misuse cases [655] and mal-activity swimlane diagrams [654].

24.1.2   Background and Hierarchical Architecture of Cloud Computing

This section presents background information and the various architectural elements that form the basis for cloud computing. Cloud computing has emerged in recent years as a new and important computing paradigm; it is gaining increased attention in the service computing community. According to the National Institute of Standards and Technology, the cloud computing model grants convenient, on-demand network access to a shared pool of configurable computing resources (e.g., networks, servers, storage, applications, and services) that can be rapidly provisioned and released with minimal management effort or service provider interaction [534]. Some essential characteristics include on-demand self-service, scalability, network access, measured services, and resource pooling. CSPs offer three basic services: infrastructure as a service (IaaS), platform as a service (PaaS), and software as a service (SaaS). All of these can be deployed through private, public, community, or hybrid cloud deployment modules.

Figure 24.1 shows a hierarchical design of cloud computing architecture. The figure is best explained from the bottom up. At the bottom is the system level, which serves as a foundation and the backbone of the cloud. It consists of a collection of data centers that supply the computing power in the cloud environment. At this level, there exist enormous physical resources such as storage disks, CPUs, and memories.

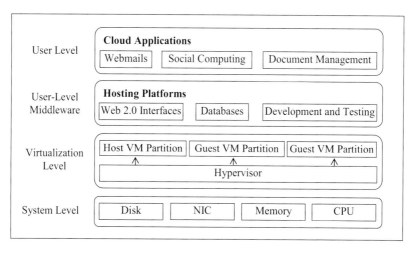

FIGURE 24.1
Hierarchical Cloud Computing Architecture.

Just above the system level is the virtualization level. Virtualization, the factor that facilitates cloud computing, is an abstraction of applications and services from the underlying physical services. It is achieved with the help of a hypervisor, a software or hardware that serves as a bridge between physical devices and virtual applications. This abstraction ensures that no application or service is tied directly on the hardware resources. This level manages the physical resources and allows sharing of their capacity among virtual instances of servers, which can be enabled or destroyed on demand. The physical cloud resources and their virtualization capabilities form the basis for delivering IaaS.

The user-level middleware includes software-hosting platforms such as Web 2.0 Interfaces that permit developers to create rich, cost-effective user interfaces for web-based applications. It also provides the programming environments and tools that ease the creation, deployment, and execution of applications in clouds. This level aims at providing PaaS capabilities.

The top user level focuses, as its name suggests, on providing application services by making use of services provided by the lower levels. It provides SaaS capabilities. SaaS or PaaS services are often developed and provided by a third party distinct from the IaaS provider.

24.1.3   The High-Order Object-Oriented Modeling Technique (HOOMT)

The HOOMT addresses a challenge faced by requirement analysts and software engineers to develop well-structured object-oriented software systems [475]. It incorporates the object-oriented paradigm seamlessly into a structured analysis [475]. It also permits the development of object, functional, and dynamic models hierarchically according to their abstraction levels. The process eliminates incompatibility between a flat object model, in which all modeling elements are analyzed at a single level of abstraction, and hierarchical functional and dynamic models, in which modeling elements are analyzed at multiple levels of abstraction [624]. It uses hierarchical decomposition in the analysis and design of object functionality and dynamic behavior. HOOMT also has a unique starting point and incorporates nonfunctional requirements. It has three models: the high-order object model (HOOM), the hierarchical object information flow model (HOIFM), and the hierarchical state transition model (HSTM). This work uses HOOM extensively to model the assets of the target system (i.e., the cloud) hierarchically. Liu, Lin and Dong [475] described HOOMT notation in detail.

24.2   Related Work

Although there has been much discussion of cloud computing security concerns, very few have focused on security requirements and security policies specific to the cloud.

Wahsheh and Alves-Foss [734] proposed a security policy engineering methodology that provides system security managers with a procedural process to develop security policies in high assurance computer systems. They established the fact that an engineering approach to policy forms the foundation of security [734]. Their security policy development life cycle starts with a clearly stated policy requirements analysis and goes through policy design, policy implementation, policy enforcement, and finally policy enhancement. Policies are usually designed not only to guide information access, but also to control conflicts and cooperation of security policies of different security enclaves [734]. They considered security as a vital issue in computer systems and therefore focused on integrating security and software engineering.

Kee [422] proposed a process for creating security policies. This process is not cloud specific and focuses on protecting only information in an organization. But effective security policies should not only protect information but also devices and employees as well. He uses ISO 17799 as a security checklist to help develop the policies. He suggested the following steps for developing security policies but there was no defined methodology for accomplishing these tasks.

1. Identify all the assets that we are trying to protect.

2. Identify all the vulnerabilities and threats and the likeliness of the threats happening.

3. Decide which measures will protect the assets in a cost-effective manner.

4. Communicate findings and results to the appropriate parties.

5. Monitor and review the process continuously for improvement.

Kadam [404] proposed a security policy development process similar to Kee [422] but his focus was on capturing the essentials of information security as applicable to an organization. In his process, before identifying all the threats and vulnerabilities and deciding on which action plans to take, a top-level information security policy (a statement of intention) is developed to answer questions with respect to confidentiality, integrity, and availability.

1. What value does the information have?

2. Why should that information be secured?

3. How can the information be secured?

4. Who is responsible for this security?

5. Where to deploy resources for such security? and

6. When are the security measures successful?

Hanna [335] proposed a streamlined security analysis process to capture and analyze security requirements in cloud computing. His method identifies the assets to be protected and the attacks that could be mounted against these assets. It then identifies countermeasures. The process prevents or mitigates threats posed to the cloud by external misusers; however, it gives little consideration to threats posed by internal misusers, especially those who have authorized access.

A number of proposals, not specific to cloud computing, address security concerns early in the development life cycle. Ware, Bowles and Eastman [750] offer a methodology to elicit security requirements using common criteria and use cases. Their work extends existing UML case notation used to model requirements so that it can capture actor threats. Their approach identifies potential threats by developing actor profiles and identifying threats based on relationships among actors in a use case [750]. Sindre and Opdahl [655] also extend use cases, which describe what a system should do, to misuse cases, concentrating on what should not happen in a system. Their approach combines both use case diagrams and misuse case diagrams in a single diagram and introduces new relationships like prevents and detects. Sindre [654] has also developed mal-activity swimlane diagrams, using them to capture attacks that could complement misuse cases, thus permitting early elicitation of security requirements. His technique permits the inclusion of both hostile and legitimate activities.

24.3   The Approach

24.3.1   Framework of the Structured Development of Cloud Security Policies Based on Security Requirements Using HOOMT

This section describes the approach used here to analyze security requirements and develop security policies in a cloud computing environment. It involves two phases: First, cloud security requirements are analyzed. Second, cloud security policies are developed, and measures are put in place to communicate and enforce them. Figure 24.2 shows a high-level view of the approach.

As noted above, the HOOMT, which is a major aspect of this approach to the analysis of cloud security policies, provides a structured object-oriented design methodology based on hierarchical model development. HOOMT allows every object in the cloud to be modeled comprehensively and verified systematically for completeness. The analysis process introduced here integrates use cases, misuse cases, and malactivity swimlane diagrams with the HOOM. The malactivity swimlane diagrams decompose misuse cases, revealing in detail the activities of misusers. Also, detailed investigation of each incidence of malactivity permits development of more ways to prevent or mitigate such malactivity. This technique serves as a countermeasure for identified threats. Moreover, more threats can be identified this way; making possible the development of comprehensive cloud security policies. The result is a more efficient

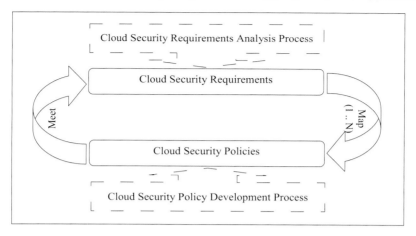

FIGURE 24.2
High-Level View of the Approach.

way to discover threats posed to cloud computing systems, both internally and externally. The structured development of the cloud security policies together with the relationships among the various diagrams at each level is shown in figure 24.3.

24.3.2   Cloud Security Requirements Analysis

Figure 24.4 outlines the process of analyzing cloud security requirements. The process begins with the development of a context object diagram (COD) for the cloud computing system; this is considered as a high-order object. This COD represents the entire cloud computing system and shows its interactions with external objects such as users, either internal or external. The COD also serves as the starting point for the analysis process.

The next step is to identify use cases that describe how the cloud computing system responds to requests from users. These cases capture the behavioral requirements of the cloud computing system with detailed scenarios derived from the cloud's functionalities. Next, each use case is analyzed thoroughly to determine how it could be subverted. Based on this analysis, misuse cases and misusers, either internal or external, that can harm the cloud computing system, are identified. The misuse cases also reveal the various threats posed to the cloud at each level of the hierarchical model.

To identify security requirements that can serve as countermeasures to these misuse cases, the actions taken by misusers must be understood in detail. Malactivity swimlane diagrams can be used to further decompose misuse cases. Decomposition reveals the details of such misuse events and thus permits identification of more threats. It also permits the inclusion of both hostile and

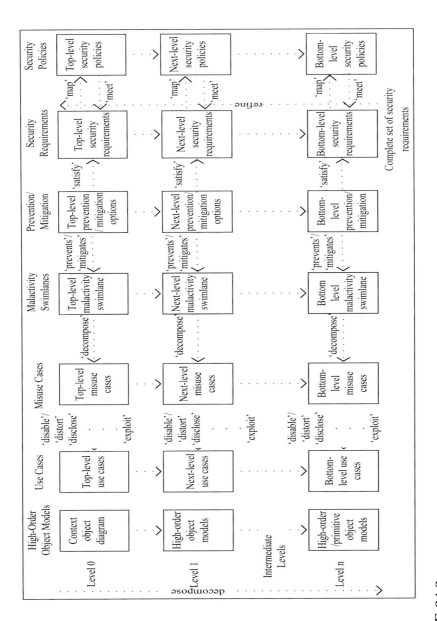

FIGURE 24.3

Framework of the Structured Development of Cloud Security Policies.

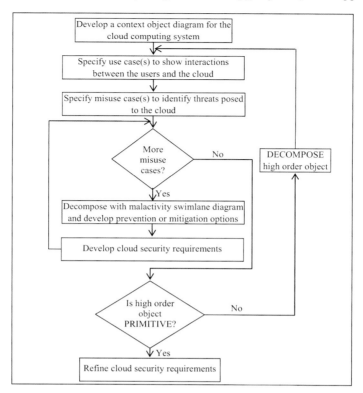

FIGURE 24.4
Cloud Security Requirements Process.

legitimate activities and determines the point at which prevention and mitigation options can be added to these activities to serve as countermeasures. Based on the countermeasures, security requirements are specified.

The COD is further decomposed and the cycle repeated, generating cloud security requirements at the end of every cycle. The term decompose refers to a process that reveals the subcomponents of the cloud object at a lower level [493]. The decomposition and security requirements analysis process continues until a stage is reached at which the cloud objects are primitive and corresponding use and misuse cases are fully explored [493]. At that point, the cloud security requirements are refined by checking for inconsistencies and ambiguities. They serve as a deliverable at the end of the first phase of the approach.

24.3.3 Cloud Security Policy Development, Communication and Enforcement Based on Security Requirements

In this work, the security policies for cloud computing systems are based on the cloud security requirements through the security requirements analysis process. Policies and requirements are not necessarily mapped one-to-one. Usually, one requirement can be satisfied by a set of security policies. These requirements are high-level statements of countermeasures that will adequately prevent or mitigate identified misuse cases and are dependent on rigorous analysis of threats to the cloud at each level, as described above. Consequently, security policies are developed and integrated into the development of the cloud computing system. This approach provides a framework of best practices for CSPs and makes security policies tenable. The policies ensure that risk is minimized and that any security incidents are met with an effective response. The process of developing these policies permits authorized security personnel to monitor and probe security breaches and other issues pertaining to cloud security. Figure 24.5 illustrates the process for developing security policies for cloud computing systems. The process begins with a statement articulating the motivation for developing such a policy, describing the malactivities to be governed by it, and listing the cloud assets to be protected. The problem the policy is designed to resolve is articulated. In general, the overall benefit of the policy is described. Next, those individuals or groups who must understand and observe this policy in order to perform their job are identified. Any exceptions to this policy are also noted.

At this point, the policy itself is articulated, including a description of what is actually covered by the policy, the responsibilities of the various individuals or groups involved, and the technical requirements that each individual or device must meet to comply with the policy.

Finally, once cloud security policies have been developed, they must be disseminated to users, staff, management, vendors, third party processors, and support personnel. The complexity of the cloud environment demands that some, if not all, policies be communicated to consumers. Enforcing these policies is also an essential part of the process. This is accomplished by establishing a record that those involved have read, understood, and agreed to abide by the policies, and by discussing how violations will be handled.

24.4 Illustrative Examples

The example described here illustrates how the proposed approach analyzes security requirements and develops security policies for cloud computing systems. This example involves a company that wants to create a cloud computing system to provide data hosting and processing services for the healthcare

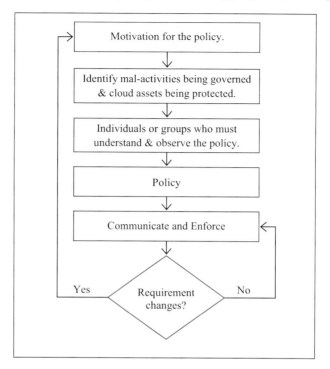

FIGURE 24.5
Cloud Policy Development Process.

industry throughout the United States. As a CSP, this company understands the importance of secure and timely access to data for such an industry. The company also wants to maintain its own secure, state-of-the-art data center to house the servers, networking equipment, backup power systems, and other tools necessary to deliver fast, secure, and effective data services. The approach described here was used to develop a security policy document for this potential CSP.

First, the cloud was considered an object, and a COD was developed for it. The COD shows the relationship between the cloud object (i.e., the target system) and external objects including the CSP, the contingency, and the cloud end user (CEU). Natural contingencies like tornadoes, floods, and earthquakes can affect the availability of the cloud, as can human (intentional) actions like terrorist attacks. At this point, the cloud object is considered a high-order object; therefore, it can be decomposed into two or more high-order and/or primitive objects. Figure 24.6 shows the COD of the cloud.

Next, the cloud object is decomposed to reveal its constituent objects. This represents the first level of the process, the point at which analysis of security requirements begins and the associated security policies are developed. The

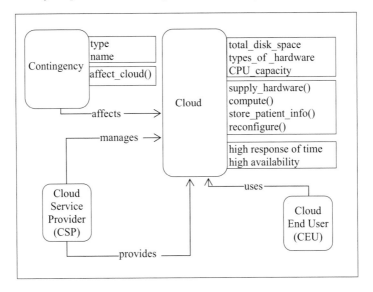

FIGURE 24.6
The COD of the Cloud Object.

cloud object is decomposed into three high-order objects and one primitive object. The high-order objects are an application and related services, a hardware system, and virtualization. The only primitive object is the service management. Decomposition of the cloud object reveals not only its constituent objects but also shows the relationships among them. See Figure 24.7.

At this point, both use cases and misuse cases are specified. Figure 24.8 represents the use case-misuse case diagram of the cloud object. At this level, the misusers, whether contingency or intentional, trigger the following four misuse cases: destroy hardware system, change hardware settings, DDoS attack, and unauthorized data access. These misuse cases disable or distort the provisioning or consumption of the cloud and involve both internal and external misusers.

A malactivity swimlane diagram is developed for each misuse case to further decompose them. Sindre [654] offers a detailed description of the malactivity swimlane diagram notation. For demonstration purposes, the misuse case of unauthorized data access (an internal threat) will be decomposed at this level (see Figure 24.9). In this scenario, an employee of the CSP with administrative privileges to the data storage infrastructure may attempt to access patient information of one of their clients using the client's own login credentials which the CSP administrator has access to. Similar to email accounts, storage buckets are accessed by using a username and password pair. With such a setup, the CSP can gain access to user accounts and in turn access their confidential information. Hence adopting a simple bank locker security mechanism where

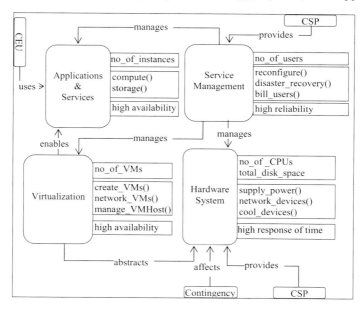

FIGURE 24.7
Decomposition of the Cloud Object.

a locker needs to be opened with multiple keys kept by different bank employees, a "seal" is created. The seal, composed of two separate keys, one managed by the CSP and the other by the client, acts as a locker to each storage bucket and both will be needed to access a storage bucket.

As shown in Figure 24.9, decomposing misuse cases with malactivity swimlanes reveals the details of activities performed by the misuser. Thus, it is possible to determine the point in the process at which mitigation or prevention can be added. For instance, in order to prevent the CSP administrator from gaining access to client's confidential information, the CSP can implement a seal on every storage bucket such that it plays a part in the login process.

Once all misuse cases are decomposed and their respective mitigation and prevention options specified, security requirements are also developed. Figure 24.10 shows the top-level security requirements for the cloud object.

With these security requirements, it should now be possible to determine what kind of security policies must be developed. This is done such that every security requirement is met by at least one associated security policy. Figure 24.11 shows an example of a security policy that meets CSR 1.5.

The cloud security requirement analysis and policy development process then continues at the second level. The virtualization object is of particular interest in this research since virtualization is the main driver of cloud applications and services. Figure 24.12 shows the decomposition of this object into four

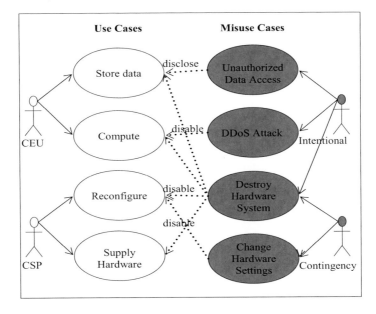

FIGURE 24.8
Use-Case/Misuse-Case Diagram at the Cloud Level.

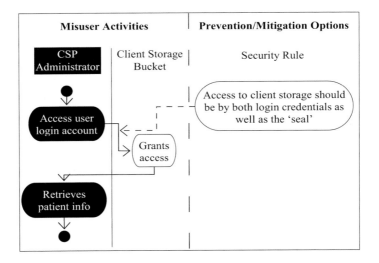

FIGURE 24.9
Mal-Activity Swimlane Diagram for the Unauthorized Data Access Misuse Case together with the Prevention or Mitigation Options.

Top-Level Cloud Security Requirements (CSR):

CSR 1.1: The system must provide physical protection to all physical hardware.

CSR 1.2: The system must employ multifactor authentication with a one-time password for CEUs to prevent intrusion.

CSR 1.3: The system must monitor network requests so that any kind of distributed denial of service (DDoS) attack can be detected.

CSR 1.4: The system must audit and log CEUs, recording who logs in, when, and from where in order to recover from a breach.

CSR 1.5: The system must encrypt data in transit in order to prevent vital data from reaching unauthorized users.

FIGURE 24.10
Security Requirements at the Cloud Object Level.

Data-in-Transit Encryption Policy

1.0 Purpose
This document describes the encryption of data in transit to ensure the information security of the cloud. Encryption is designed to prevent unauthorized disclosure of vital information.

2.0 Scope
This policy applies to any data in transit.

3.0 Policy
All data in transit must be encrypted, and such data must be protected to prevent their unauthorized disclosure and subsequent fraudulent use.

4.0 Enforcement
Any employee found to have violated this policy may be subject to disciplinary action, up to and including termination of employment.

5.0 Definitions
Data in transit refers to any data transferred in the cloud.

6.0 Revision History
09/24/2010 – 1.0 initial policy version, Kenneth Fletcher

FIGURE 24.11
Security Policy to Meet CSR 1.5.

primitive objects, and the relationships among them. Also at this level, use cases and misuse cases are specified. Two use cases were identified: *createVM*, and *manageVMs*. The misusers, either contingency or intentional, initiate *VMescape*, *DoSattack* and *changeVMsettings* misuse cases.

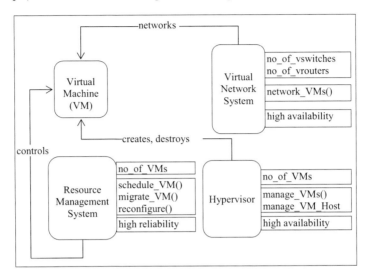

FIGURE 24.12
Decomposition of the Virtualization Object.

Figure 24.13 shows the use case-misuse case diagram. The VM escape misuse case is further decomposed with the malactivity swimlane diagram. Security prevention and mitigation options are specified in the decomposition as shown in Figure 24.14. The remaining misuse cases are also decomposed and their respective security prevention and or mitigation options specified. Security requirements and policies to meet them are also specified; these are shown in Figure 24.15 and Figure 24.16 respectively.

24.5 Case Study–Application Example

The previous section explained how cloud security requirements can be analyzed and security policies developed. Here, the proposed approach is applied to a real case study involving a cloud service provider whose name has been omitted due to confidential reasons. The objective is to analyze the company's current security state and provide advice on strengthening the security of its cloud.

The U.S. company offers highly available business solutions, including colocation, cloud computing, managed services, and insourcing in a carrier-class

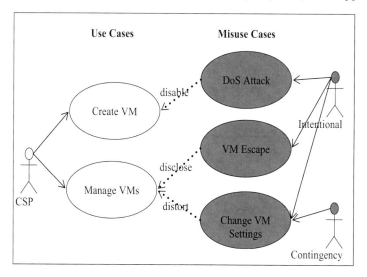

FIGURE 24.13
Use-Case/Misuse-Case Diagram for the Virtualization Object.

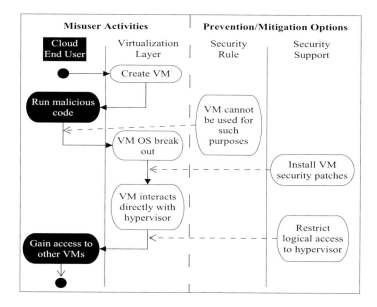

FIGURE 24.14
Mal-Activity Swimlane Diagram for the VM Escape Misuse Case together
with the Prevention or Mitigation Options.

Second-Level Cloud Security Requirements (CSR):

CSR 2.1: The system must restrict physical and logical access to the hypervisor to prevent VM from having direct interaction with hypervisor.

CSR 2.2: The system must employ efficient load balancing techniques to prevent VMs from causing denial of service (DoS) attacks.

CSR 2.3: The system must implement authentication of network flow such that a guest VM cannot monitor other VMs.

CSR 2.4: The system must monitor guest-host VM interaction for improper configuration changes, and in the event of any such incident it should report to the network manager.

FIGURE 24.15
Security Requirements at the Virtualization Object Level.

Hypervisor Access Policy

1.0 Purpose
This document describes cloud information security's required encryption of data in transit. This is designed to prevent unauthorized disclosure of vital information.

2.0 Scope
This policy applies to all nonhost virtual machines in the cloud.

3.0 Policy
All data in transit must be encrypted, and data covered by this policy must be protected to prevent their unauthorized disclosure and subsequent fraudulent use.

4.0 Enforcement
Any employee found to have violated this policy may be subject to disciplinary action, up to and including termination of employment.

5.0 Definitions
Data in transit – Data transferred in the cloud.

6.0 Revision History
09/24/2010 – 1.0 initial policy version, Kenneth Fletcher

FIGURE 24.16
Security Policy to Meet CSR 2.1.

data center facility. The private cloud computing environment provides access to resources from storage, virtual servers, and desktops to email and mobile devices, all on an as-needed basis. These systems are powered from their own platform supported by a 30,000-square-foot state-of-the-art data center. In order to provide a geographically diverse redundancy system as a backup for the primary data center, the service provider operates another data center elsewhere in the country. The private cloud offerings of the company fall primarily in the IaaS space, although it offers a number of applications that are delivered and consumed by clients on a variable per-use basis.

The security requirements analysis process begins by developing the COD of the private cloud. Because the primary data center location is an earthquake zone, the cloud is vulnerable to natural contingencies. Figure 24.17 shows the COD and the relationships among the objects.

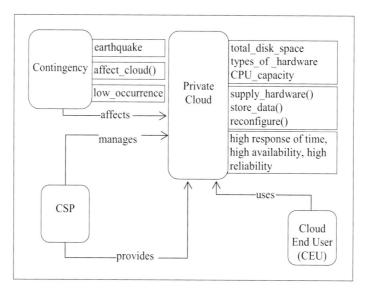

FIGURE 24.17
The COD of the Private Cloud Object.

The private cloud object is decomposed into four subobjects: three high-order objects and a primitive object. The high-order objects are services, hardware resources, and the VMware vSphere. The only primitive object is service management. Decomposition of the cloud object reveals not only its constituent objects but also the relationship among them (see Figure 24.18).

Figure 24.19 represents the use-case/ misuse-case diagram of the private cloud at the cloud level. The following three misuse cases were identified: destroy hardware, change hardware settings, and unauthorized data access. These misuse cases subvert the supply hardware, reconfigure system, and store data use cases, respectively.

The cloud service provider for this case study understands its data center as belonging to its clients and therefore permits clients access to it. With this

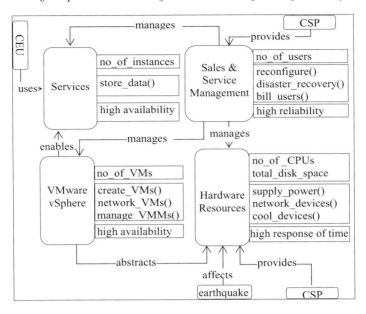

FIGURE 24.18
Decomposition of the Private Cloud Object.

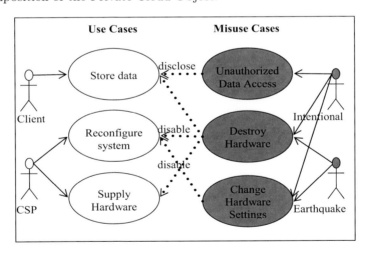

FIGURE 24.19
Use-Case/Misuse-Case Diagram for the Private Cloud Object.

setup, internal threats are likely to be the main security issue at this level. Therefore, the unauthorized data access misuse case is decomposed here to determine how such a setup could be compromised and develop prevention or mitigation options to serve as countermeasures.

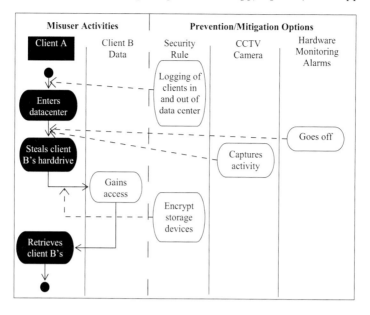

FIGURE 24.20
Mal-Activity Swimlane Diagram for the Unauthorized Data Access Misuse
Case together with the Prevention or Mitigation Options.

Figure 24.20 is a malactivity swimlane diagram describing a scenario in which
one client (client A) goes into the data center to steal another client's (client
B) hard drive and access confidential data on it. The decomposition clarifies
the activities of client A and makes it easier to prevent them. The prevention
or mitigation options specified in Figure 24.20 are translated into security
requirements, which are shown in Figure 24.21. These are the security re-

---

Private Cloud Object-Level Cloud Security Requirements (CSR):

CSR 1.1: The system must provide hardware monitoring alarms for all physical
hardware.

CSR 1.2: The system must audit and log client and visitor access to the data center,
recording who logs in and when in order to recover from a breach.

CSR 1.3: The system must encrypt data at rest in order to prevent vital data from
reaching unauthorized users.

FIGURE 24.21
Security Requirements at the Private Cloud Level.

quirements specified at the first level of the private cloud. Figure 24.22 shows a security policy to meet CSR 1.2.

Customer and Visitor Data Center Access Policy

1.0 Purpose
The purpose of this document is to provide guidance for customers and visitors to the data center, as well as for employees sponsoring visitors.

2.0 Scope
This policy applies to all customers and visitors to the data center and to employees who sponsor visitors.

3.0 Policy
3.1 Check-In
All visitors must arrive at a designated check-in entrance (i.e., the main reception desk) and present government-issue photo identification at time of check-in.
All visitors must be met by their employee sponsor at the time of check-in. Visitors must sign a "Visitor Agreement." All visitor electronics will be checked in as well.

3.2 Badges
Customer and visitor badges must be worn at all times. Employees are instructed to immediately report anyone not wearing a customer, visitor, or employee badge.
Visitors requiring access to areas controlled by swipe card access locks should be assisted by their sponsoring employee.

3.3 Photographs and Cameras
Customers and visitors are not permitted to take photographs inside the data center, without specific prior arrangement with sponsoring employees.

3.4 Check-Out
Visitors will check out at the same station where they arrived. All visitor electronics will be checked out.

3.5 Exit Inspection
Visitors may be subject to a brief search of their laptop bags or other luggage as they exit the data center.

4.0 Enforcement
Violation of any part of this policy by any employee will result in suitable disciplinary action, up to and including prosecution and or termination.
Violation of any part of this policy by any visitor can result in similar disciplinary action against the sponsoring employee, and can also result in termination of services or prosecution in the case of criminal activity.

6.0 Revision History
09/24/2010 – 1.0 initial policy version, Kenneth Fletcher

FIGURE 24.22
Security Policy to Meet CSR 1.2.

At the second level of the security requirements analysis process, the VMware vSphere object (the virtualization layer) is analyzed. Figure 24.23 shows its decomposition and the relationship existing among its four constituent primitive objects. The objects are VMware ESXi hypervisor, vCenter server, virtual machine, and application services.

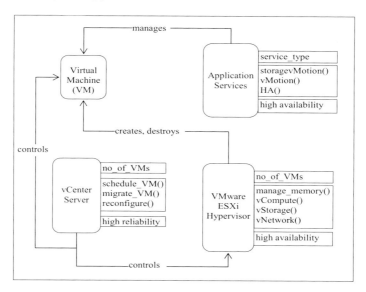

FIGURE 24.23
Decomposition of the VMware vSphere Object.

Also at this level, use cases and misuse cases are specified. Three use cases were identified: create VM, vstorage, and manage VM host. The intentional misuser, whether a cloud user or the CSP itself, initiates DoS attack, VM escape, MITM attack, and redirect packets misuse cases. Figure 24.24 shows the use-case/misuse-case diagram. At this level, the MITM attack misuse case was further decomposed with a malactivity swimlane diagram.

Figure 24.25 shows this decomposition together with the prevention or mitigation options. This attack occurs when a victim thinks the attacker is the default gateway and the actual default gateway thinks otherwise.

During the course of this attack the victim sends packets to the attacker (default gateway) who then copies the information, stops it, or at worst changes the contents of the frame itself. The modified or copied frame is sent to the unsuspecting default gateway (actual) for further processing. When the receiving packet returns, the data can be similarly intercepted. Other misuse cases identified here were also decomposed and their respective security prevention or mitigation options specified. Security requirements for this level are shown in Figure 24.26. Finally at this level, security policy to meet CSR 2.1 is developed as shown in Figure 24.27.

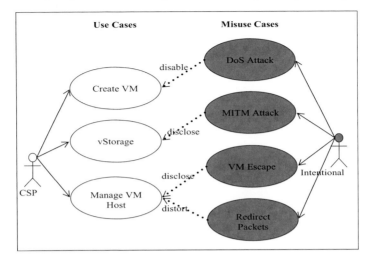

FIGURE 24.24
Use-Case/Misuse-Case Diagram for the VMware vSphere Object.

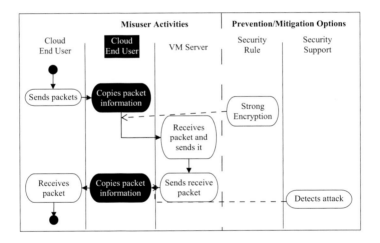

FIGURE 24.25
Mal-Activity Swimlane Diagram for the MITM Attack Misuse Case together with the Prevention or Mitigation Options.

Finally, the hardware resources object was analyzed and further decomposed into the following four primitive objects: hardware devices, network management system, cooling system, and power system. The high-order object model for the hardware resources object and the relationship between its primitive objects are represented in Figure 24.28.

VMware vSphere Object-Level Cloud Security Requirements (CSR):

CSR 2.1: The system must isolate all traffic to and from storage repositories (data-in-motion) from other nonstorage traffic.

CSR 2.2: The system must not allow VM repositories or datastores to be accessible to other VMs except for the VM host servers.

CSR 2.3: The system must restrict physical and logical access to the hypervisor to prevent VM from having direct interaction with hypervisor.

FIGURE 24.26
Cloud Security Requirements at the VMware vSphere Object Level.

Data-in-Motion Isolation Policy

1.0 Purpose
The purpose of this policy is to meet CSR 2.1, which defines the isolation of data in motion in order to prevent MITM attacks when using and managing virtualization with VMware vSphere technologies.

2.0 Scope
This policy applies to any data-in-motion

3.0 Policy
All data in transit must be isolated by employing storage area network (SAN) connectivity, that is, a network of servers and storage devices independent of the ethernet network.

4.0 Enforcement
Any employee found to have violated this policy may be subject to disciplinary action, up to and including termination of employment.

5.0 Definitions
Data in transit refers to data transferred in the VMware vSphere vStorage medium.

6.0 Revision History
09/24/2010 – 1.0 initial policy version, Kenneth Fletcher

FIGURE 24.27
Security Policy to Meet CSR 2.1.

Supply power, network hardware devices, and supply hardware are three use cases specified at the hardware resources level. Threats at this level of the private cloud are mostly physical. The misusers, contingency and intentional,

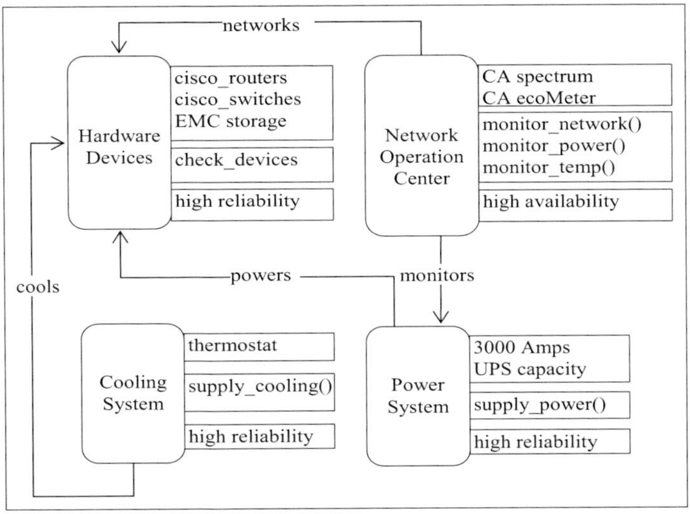

FIGURE 24.28
Decomposition of the Hardware Resources Object.

initiate five misuse cases, including destroy network devices, destroy power devices, change power configuration, destroy cooling systems, and change temperature configuration. Figure 24.29 shows the use-case/ misuse-case diagram for this level.

Here, the destroy power devices misuse case was decomposed further with a malactivity swimlane diagram. Figure 24.30 shows this decomposition together with the security prevention or mitigation options for this threat. Figure 24.31 shows the security requirements that were developed at level 3. Also, Figure 24.32 shows the security policy developed to meet CSR 3.1

24.6   Conclusion

Cloud computing is becoming popular and represents the future of computing. Before it can be embraced by individuals and enterprises, however, the issue of security must be addressed. Early consideration of security in cloud computing systems places it on a par with other functional requirements of the system and significantly improves the security of the system. Due to the essential role that security policies play in bridging the gap between security requirements and implementation, it should also be taken into account early in the development process [734]. This work has successfully developed a methodology and process to determine security requirements and develop

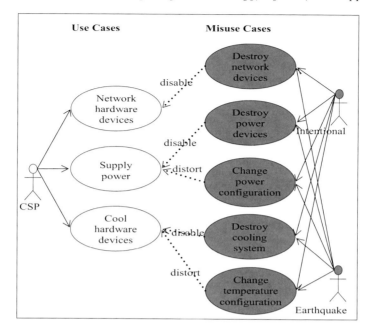

FIGURE 24.29
Use-Case/Misuse-Case Diagram for the Hardware Resources Object.

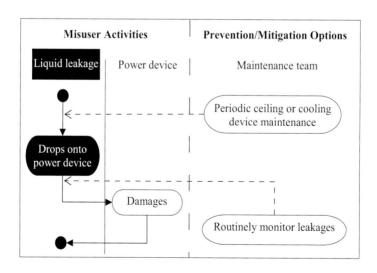

FIGURE 24.30
Mal-Activity Swimlane Diagram for the Destroy Power Devices Misuse Case
together with the Prevention or Mitigation Options.

Hardware Resources Object-Level Cloud Security Requirements (CSR):

CSR 3.1: The system must routinely monitor power quality and load in order to detect any change in power configuration.

CSR 3.2: The system must routinely monitor temperature in order to detect any change in temperature configuration and maintain constant cooling of hardware devices.

CSR 3.3: The system must routinely monitor and detect coolant or water leaks in order to prevent destruction of power devices, cooling systems, cables, and other hardware.

FIGURE 24.31
Security Requirements at the Hardware Resources Object Level.

policies for a cloud computing system level-by-level in a structured manner. The methodology analyzes security requirements by identifying threats posed by misusers both external and internal to a system. The process was applied to typical cloud architecture to demonstrate its function and it was further applied to an actual case study of a cloud service provider. In each case, misuse cases at different levels were identified. Malactivity swimlane diagrams for these misuse cases were generated, permiting development of countermeasures for prevention or mitigation. Security requirements were then derived based on the prevention or mitigation options. Finally, security policies were developed to meet each requirement.

Developing comprehensive cloud-specific security policies is a very difficult task that requires collaboration and insight from many individuals in various areas of expertise. Security policies can actually complicate things if they are not clearly written and consistently observed [687]. Enforcing cloud security policies are difficult and require management, employee, and user support. Also, It is harder to evaluate security compliance in cloud computing.

Periodic Maintenance Policy

1.0 Purpose
The purpose of this document is to define the standards for effective maintenance of the private cloud's assets so that equipment remains safe at all times.

2.0 Scope
This policy applies to all equipment serving the CSP's data center.

3.0 Policy

3.1 Maintenance Standards
Each piece of equipment will be allocated an importance rating of 1 to 5. Maintenance standards will vary depending on the importance of the facility, per the guide below:
1. Not important: Carry out only essential maintenance.
2. Low importance: Defer non-essential maintenance where possible.
3. Fair importance: Carry out maintenance based on risk assessment.
4. Important: Maintain to the best standard that resources allow.
5. Very important: Maintain to a very high standard.

3.2 Maintenance Categories
Each piece of equipment must be categorized as one of the following: preventive maintenance, statutory maintenance, corrective maintenance, or backlog maintenance.

4.0 Enforcement
Violation of any part of this policy by any employee will result in suitable disciplinary action, up to and including prosecution and or termination.

5.0 Revision History
09/24/2010 – 1.0 initial policy version, Kenneth Fletcher

FIGURE 24.32
Security Policy to Meet CSR 3.2.

25

Exploring the Use of Hybrid HPC-Grids/Clouds Infrastructure for Science and Engineering

Hyunjoo Kim

Center for Autonomic Computing, Department of Electrical & Computer Engineering, Rutgers, The State University of New Jersey

Yaakoub El-Khamra

Texas Advanced Computing Center, The University of Texas, Austin Texas

Shantenu Jha

Center for Computation & Technology and Department of Computer Science, Louisiana State University

Manish Parashar

Center for Autonomic Computing, Department of Electrical & Computer Engineering, Rutgers, The State University of New Jersey

CONTENTS

25.1 Introduction

Significant investments and technological advances have established high-performance computational (HPC) Grid [1] infrastructures as dominant platforms for large-scale parallel/distributed computing in science and engineering. Infrastructures such as the TeraGrid, EGEE and DEISA integrate high-end computing and storage systems via high-speed interconnects, and support traditional, batch-queue-based computationally and data intensive high-performance applications.

More recently, Cloud services have been playing an increasingly important role in computational research, and are poised to become an integral part of computational research infrastructures. Clouds support a different although complementary provisioning modes as compared to HPC Grids — one that is based on on-demand access to computing utilities, an abstraction of unlimited computing resources, and a usage-based payment mode where users essentially "rent" virtual resources and pay for what they use. Underneath these Cloud services are consolidated and virtualized data centers that provide virtual machine (VM) containers hosting applications from large numbers of distributed users.

It is now clear that production computational Grid infrastructures will dynamically integrate these two paradigms, providing a hybrid computing envi-

---

[1] Note that this work addresses High-Performance Grids rather than High-Throughput Grids.

ronment that integrates traditional HPC Grid services with on-demand Cloud services. While the provisioning modes of these resources are clear: scheduled request (HPC Grids) and on-demand (Clouds), the usage modes of such a hybrid infrastructure, as well as frameworks for supporting these usage modes, are still not as clear. There are application profiles that are better suited to HPC Grids (e.g., large QM calculations), and others that are more appropriate for Clouds. However, there are also large numbers of applications that have *interesting* workload characteristics and resource requirements, and can potentially benefit from a hybrid infrastructure to better meet application/user objectives. For example, it is possible to reduce cost, time-to-solution, and susceptibility to unexpected resource downtime with hybrid infrastructure.

Furthermore, developing and running applications in such a hybrid and dynamic computational infrastructure presents new and significant challenges. These include the need for programming systems that can express the hybrid usage modes and associated runtime trade-offs and necessary adaptivity, as well as coordination and management infrastructures that can implement them in an efficient and scalable manner. The set of required features includes decomposing applications, components and workflows, determining and provisioning the appropriate mix of Grid/Cloud resources, and dynamically scheduling them across hybrid execution environments while satisfying/balancing multiple objectives for performance, resilience, budgets and so on.

In this chapter, we experimentally investigate, from an applications perspective, interesting usage modes and scenarios for integrating HPC Grids and Clouds, and how they can be effectively enabled using an autonomic scheduler. Our investigation and analysis are driven by a real-world application that is the basis for a large number of physical and engineering science problems. Specifically, we use a reservoir characterization workflow, which uses Ensemble Kalman Filters (EnKF) for history matching, as the driving application and investigate how Amazon EC2 commercial Cloud resources can be used to complement a TeraGrid resource. The EnKF workflow presents an interesting use-case due to the heterogeneous computational requirements of the individual ensemble members as well as the dynamic nature of the overall workflow. We investigate how clouds can be effectively used to address changing computational requirements as well as changing Quality of Service (QoS) constraints (e.g., deadline) for a dynamic application workflow.

We then focus on the acceleration usage mode, and investigate how an autonomic scheduler can be used to improve overall workflow performance. We decided to investigate adaptivity and its impact on workflow performance when we analyzed the static, non-adaptive platform utilization and saw room for improvement. The scheduler supports infrastructure-level adaptivity as well as application-level adaptivity. When both adaptivity levels are enabled, the overall performance of the application workflow is improved, along with the hybrid platform utilization.

The research presented in this chapter is based on the CometCloud engine [430, 431] and its autonomic scheduler and autonomic application management framework. CometCloud is an autonomic Cloud engine and enables applications on federated Cloud infrastructures with re-sizable computing capacity through policy-driven autonomic Cloud bridging (on-the-fly integration of Grids, commercial and community Clouds and local computational environments) and Cloudbursts (dynamic scale-out to address dynamic workloads, spikes in demands, and other extreme requirements). This chapter also demonstrates how CometCloud can support heterogeneous application requirements as well as hybrid Grid-Cloud usage modes.

The rest of this chapter is organized as follows. We describe the hybrid HPC-Grids/Clouds infrastructure in Section 25.2 and introduce CometCloud in Section 25.3. In Section 25.4 we present the scientific application workflow that is used to drive the research. We experimentally investigate usage modes for the hybrid Grids/Clouds infrastructure in Section 25.5 and explore infrastructure and application adaptivity in Section 25.6. We present concluding remarks in Section 25.7.

25.2   The Hybrid HPC-Grids/Clouds Infrastructure

Computing infrastructures are moving towards hybridization, integrating different types of resource classes such as public/private clouds and grids from distributed locations [497, 557, 723]. Our simple federated hybrid computing infrastructure consists of an HPC TeraGrid resource and Amazon EC2 clouds. As the infrastructure is dynamic and can contain a wide array of resource classes with different characteristics and capabilities, it is important to be able to dynamically provision the appropriate mix of resources based on changing application objectives and requirements. Furthermore, resource states may change, for example, due to workload surges, system failures or emergency system maintenance, and as a result, it is necessary to adapt the provisioning to match these changes.

25.2.1   TeraGrid

The TeraGrid [699] is an open computational science infrastructure distributed across eleven sites in the continental United States. These sites provide high-performance computing resources as well as data storage resources and are connected through a high performance network. The TeraGrid resources include more than 2.75 petaflops of computing capacity and 50 petabytes of archival data storage. Most of the compute power in the TeraGrid is provided through distributed memory HPC clusters. Due to the high demand for compute power, most of these machines have a scheduler and a queue policy. With the exception of special reservations and requests, most TeraGrid sites operate on a "fair–share" policy. User submitted jobs have an escalating factor of priority, depending on the size/duration of the job, weighted historical usage and the queue where the job was submitted to. Very frequent submissions can induce a longer queue wait time as the scheduler attempts to give other users a chance to run their jobs. The utilized system units (core-hours) are then billed to the user's allocation. Allocations are granted to users based on a competitive allocation request proposal that is reviewed by the TeraGrid Resource Allocations Committee for merit and impact.

25.2.2 Amazon Elastic Compute Cloud (EC2)

Amazon EC2 allows users to rent virtual computers on which to run their own computer applications. EC2 allows scalable deployment of applications by providing a web service through which a user can boot an Amazon Machine Image to create a virtual machine, which is referred to as an *instance*. A user can create, launch, and terminate server instances as needed, paying by the hour for them, hence the term *elastic*.

EC2 provides various types of instances [246]; 1) standard instances which are suited for most applications, 2) micro instances which provide a small amount of consistent CPU resources and allow users to burst CPU capacity when additional cycles are available, 3) high memory instances which offer large memory sizes for high throughput applications including databases and memory caching applications, 4) high CPU instances which have a high core count and are suited for compute-intensive applications, 5) cluster compute instances which provide high CPU with increased network performance and are suited for HPC applications and other network-bound applications, and 6) cluster GPU instances which provide GPUs.

25.3 Autonomic Application Management Using Comet-Cloud

25.3.1 An Overview of CometCloud

CometCloud is an autonomic computing engine for Cloud and Grid environments. It is based on the Comet [471] decentralized coordination substrate, and supports highly heterogeneous and dynamic Cloud/Grid infrastructures, integration of public/private Clouds and autonomic Cloudbursts. Conceptually, CometCloud is composed of a programming layer, service layer, and infrastructure layer. The infrastructure layer uses the Chord self-organizing overlay [676], and the Squid [635] information discovery and content-based routing substrate built on top of Chord. The routing engine supports flexible content-based routing and complex querying using partial keywords, wildcards, or ranges. It also guarantees that all peer nodes with data elements that match a query/message will be located.

The service layer provides a range of services to support autonomics at the programming and application level. This layer supports a Linda-like [166] tuple space coordination model, and provides a virtual shared-space abstraction as well as associative access primitives. Dynamically constructed transient spaces are also supported to allow applications to explicitly exploit context locality to improve system performance. Asynchronous (publish/subscribe) messaging and event services are also provided by this layer.

The programming layer provides the basic framework for application development and management. It supports a range of paradigms including the master/worker/Bag-Of-Tasks. Masters generate tasks and workers consume them. Masters and workers can communicate via virtual shared space or using a direct connection. Scheduling and monitoring of tasks are supported by the application framework. The task consistency service handles lost/failed tasks. Other supported paradigms include workflow-based applications and MapReduce.

25.3.2 Autonomic Management Framework

A schematic overview of the CometCloud-based autonomic application management framework for enabling hybrid HPC Grids-Cloud usage modes is presented in Figure 25.1. The framework is composed of the autonomic manager that coordinates using Comet coordination spaces [471] that span the integrated execution environment and the adaptivity manager that is responsible for the resource adaptation not to violate user objective. The key components of the management framework are described below.

Workflow Manager: The workflow manager is responsible for coordinating the execution of the overall application workflow, based on user-defined polices, using Comet spaces. It includes a workflow planner as well as task monitors/managers. The workflow planner determines the computational tasks that can be scheduled at each stage of the workflow. Once computational tasks are identified, appropriate metadata describing the tasks (hints about complexities, data dependencies, affinities, etc.) is inserted into Comet space and the autonomic schedule is notified. The task monitors and task manager then monitor and manage the execution of each of these tasks and determine when a stage has completed so the next stage(s) can be initiated.

Estimator: The cost estimator is responsible for translating hints about computational complexity, possibly provided by the application, into runtime and/or cost estimates on a specific resource. For example, a Cloud resource such as time/cost estimate may be based on historical data related to specific VM configurations or a simple model. A similar approach can also be used for resources on the HPC Grids using specifications of the HPC Grids' nodes. However, in the HPC Grids case, waiting time in the batch queue must also be estimated. To that end we use BQP [142], the Batch Queue Prediction tool

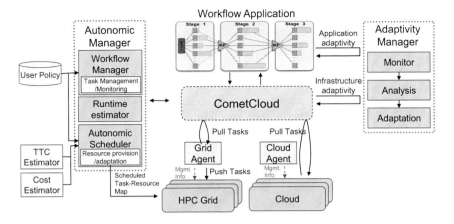

FIGURE 25.1
Architectural Overview of the Autonomic Application Management Framework.

that estimates the queue wait time for a job of a given size and duration on a given resource with a confidence factor.

Autonomic Scheduler: The autonomic scheduler performs key autonomic management tasks. First, it uses estimates of relative complexities of tasks to be scheduled, and clusters tasks to identify potential scheduling blocks. It then uses an estimator module to compute anticipated run times for these blocks on available resource classes. The estimates are used to determine the initial hybrid mix HPC Grids/Cloud resources based on user/system-defined objectives, policies, and constraints. The autonomic scheduler also communicates resource-class-specific scheduling policies to the agents. For example, tasks that are computationally intensive may be more suitable for a HPC Grids resource, while tasks that require quick turn-around may be more suitable for a Cloud resource. Note that the allocation as well as the scheduling policy can change at runtime.

Grid/Cloud Agents: The Grid/Cloud agents are responsible for provisioning the resources on their specific platforms, configuring *workers* as execution agents on these resources, and appropriately assigning tasks to these workers. The primary task assignment approach supported by the framework is *pull-based*, where workers pull tasks from the Comet space based on directives from respective agents. This works perfectly well for Clouds where a worker on a Cloud VM may only pull task with computational/memory requirements below some threshold. This model can be implemented in different ways. In contrast, on a typical HPC Grid resource with a batch queuing system a combined push-pull model is used. Specifically, we insert smart *pilot-jobs* [702] containing workers into the batch queues of the HPC Grids systems, which then pull tasks from the Comet space when they are scheduled to run by the queuing system. As is the case with VM workers, pilot-jobs can use local policy and be driven by overall objectives to determine the best tasks to take on.

Monitor: The monitor observes the tasks' runtimes which workers report to the master in the results. If the time difference between the estimated runtime and the task runtime on a resource is large, it could be that the runtime estimator underestimated or overestimated the task requirements, or the application/resource performance has changed. We can distinguish between those cases (estimation error versus performance change) if all of the tasks running on the resource in question show the same tendency. However, if the time difference between the estimated runtime and the task runtime on the resource increases after a certain point for most tasks running on the resource, then we can evaluate that the performance of the resource is becoming degraded. The monitor makes a function call to the analyzer to check if the changing resource status still meets the user objective when the time difference becomes larger than a certain threshold.

Analyzer: The analyzer is responsible for re-estimating the runtime of remaining tasks based on the historical data gathered from the results and evaluating the expected TTC of the application and total cost. If the analyzer observes the possibility for violating the user objective, then it calls for task rescheduling by the Adapter.

Adapter: When the adapter receives a request for rescheduling, it calls the autonomic manager, in particular the autonomic scheduler, and retrieves a new schedule for the remaining tasks. The adapter is responsible for launching more workers or terminating existing workers based on the new schedule.

25.4   Scientific Application Workflow

25.4.1   Reservoir Characterization: EnKF-Based History Matching

Direct information about any given reservoir is usually gathered through logging and measurement tools including core samples, thus restricted to a small portion of the actual reservoir size, namely the well-bore. For this reason, *history matching* techniques have been developed to *match* actual reservoir production with simulated reservoir production, therefore obtaining a more *satisfactory* set of reservoir models. Ensemble Kalman filters (EnKF) represent a promising approach to history matching [318, 319, 406, 470].

In history matching, we attempt to match actual production history (i.e., from the field) with the simulated production we obtain using computer models. In this process, we continually modify the reservoir models to obtain an acceptable match. In EnKF based history matching, we use the reservoir simulator

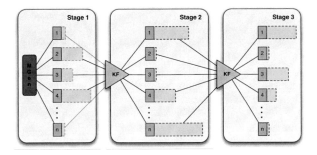

FIGURE 25.2
Schematic Illustrating the Variability between Stages of a Typical Ensemble Kalman Filter-Based Simulation. The End-to-End Application Consists of Several Stages; In General, At Each Stage the Number of Models Generated Varies in Size and Duration.

to advance the state of the reservoir models in time and the EnKF application to assimilate the historical production data.

Since the reservoir model varies from one ensemble to another, the run-time characteristics of the ensemble simulation are irregular and hard to predict. Furthermore, during simulations when real historical data is available, all the data from the different ensembles at that simulation time must be compared to the actual production data, before the simulations are allowed to proceed. This translates into a global synchronization point for all ensemble-members in any given stage as seen in Figure 25.2.

Due to this fundamental limit on task-parallelism, wildly varying computational requirements between stages and for different tasks in a stage, performing large scale studies for complex reservoirs in a reasonable amount of time would benefit greatly from the use of a wide range of distributed, high-performance and throughput, as well as on-demand computing resources.

The simulation components are:

- The Reservoir Simulator: The BlackOil reservoir simulator solves the equations for multiphase fluid flow through porous media, allowing us to simulate the movement of oil and gas in subsurface formations. It is based on the Cactus Code [157] high performance scientific computing framework and the Portable Extensible Toolkit for Scientific Computing: PETSc [111].

- The Ensemble Kalman filter: Also based on Cactus and PETSc, computes the Kalman gain matrix and updates the model parameters of the ensembles. The Kalman filter requires live production data from the reservoir for it to update the reservoir models in real-time, and launch the subsequent long-term forecast, enhanced oil recovery, and CO_2 sequestration studies.

25.5   An Experimental Investigation of HPC Grids–Cloud Hybrid Usage Modes

25.5.1   Autonomic Execution of EnKF

The relative complexity of each ensemble member is estimated and this information bundled with the ensemble member as a task that is inserted into the Comet space. This is repeated at every stage of the EnKF workflow. We will restrict our experiments to a two-stage push-pull dynamic model. Specifically, in the first stage, there is a decision to be made about how many workers should be employed and how to distribute these workers, to a possibly varying/different number of workers.

Once the tasks to be scheduled within a stage have been identified, the autonomic scheduler analyzes the tasks and their complexities to determine the appropriate mix of TeraGrid and EC2 resources that should be provisioned. This is achieved by (1) clustering tasks based on their complexities to generate blocks of tasks for scheduling, (2) estimating the runtime of each block on the available resources using the cost estimator service and (3) determining the allocations as well as scheduling policies for the TeraGrid and EC2 based on runtime estimates as well as overall objectives and resource specific policies and constraints (e.g., budgets). In case of the EnKF workflow, the cost estimation consists of a simple function obtained using *a priori* experimentation, which maps the computational complexity of the task as provided by the application and estimated runtime. Additionally, in case of the TeraGrid, the BQP [142] service is used to obtain estimates of queue waiting times and to select appropriate size and runtime duration of the TeraGrid request (which are the major determinants of overall queue wait-time).

After the resource allocations on the TeraGrid and EC2 have been determined, the desired resources are provisioned and *ensemble-workers* are launched. On the EC2, this consists of launching appropriate VMs with loading custom images. On the TeraGrid, ensemble-workers are essentially *pilot jobs* [702] that are inserted into the batch queue. Once these ensemble-workers start executing, they can directly access the Comet space and retrieve tasks from the space based on the enforced scheduling policy. The policy we employ is simple: TeraGrid workers are allowed to pull the largest tasks first, while EC2 workers pull the smallest tasks. As the number of tasks remaining in the space decreases, if there are TeraGrid resources still available, the autonomic scheduler may decide to throttle (i.e., lower the priority) EC2 workers to prevent them from becoming the bottleneck, since EC2 nodes are much slower than TeraGrid compute nodes. While this policy is not optimal, it was sufficient for our study.

During the execution, the workflow manager monitors the executions of the tasks (using the task monitor) to determine progress and to orchestrate the execution of the overall workflow. The autonomic scheduler also monitors the state of the Comet space as well as the status of the resources (using the agents), and determines progress to ensure that the scheduling objectives and policies/constraints are being satisfied, and can dynamically change the resources allocation and scheduling policy as required. Specific implementations of policy and variations of objectives form the basis of our experiments. For example, if the allocation on the TeraGrid is not sufficient, the scheduler may increase the number of EC2 nodes allocated on-the-fly to compensate, and modify the scheduling policy accordingly. Such a scenario is illustrated in the experiments below.

25.5.2 User Objective

We investigate the following scenarios (or autonomic objectives) for the integration of HPC Grids and Clouds and how an autonomic framework can support them:

- Acceleration: This use case explores how Clouds can be used as accelerators to improve the application time-to-completion by, for example, using Cloud resources to alleviate the impact of queue wait times or exploit an additionally level of parallelism by offloading appropriate tasks to Cloud resources, given appropriate budget constraints.

- Conservation: This use case investigates how Clouds can be used to conserve HPC Grid allocations, given appropriate runtime and budget constraints.

- Resilience: This use case will investigate how Clouds can be used to handle unexpected situations such as an unanticipated HPC Grid downtime, inadequate allocations or unanticipated queue delays.

25.5.3 Experiment Background and Setup

The goal of the experiments presented in this section is to investigate how possible usage modes for hybrid HPC Grids-Cloud infrastructure can be supported by a simple policy-based autonomic scheduler. Specifically, we investigate experimentally, implementations of three usage modes — acceleration, conservation and resilience, which are the different objectives of the autonomic scheduler.

Our experiments use a single stage EnKF workflow with 128 ensemble members (tasks) with heterogeneous computational requirements. The heterogeneity is illustrated in Figure 25.3, which is a histogram of the runtimes of the 128 ensemble members within a stage on 1 node of a TeraGrid compute system (Ranger), and 1 EC2 core (a small VM instance, 1.7 GB memory, 1 virtual core,160 GB instance storage, 32-bit platform) respectively. The distribution of tasks is almost Gaussian, with a few significant exceptions. These plots also demonstrate the relative computational capabilities of the two platforms.

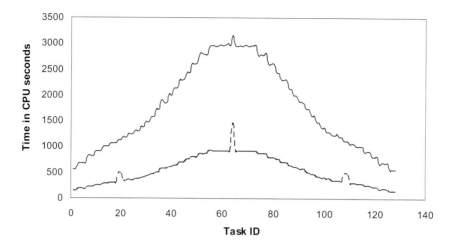

FIGURE 25.3
The Distribution of Runtimes of Ensemble Members (tasks) on 1 Node (16 processors) of a TeraGrid Compute System (Ranger) and One VM on EC2.

Note that when a task is assigned to a TeraGrid compute node, it runs as a parallel application across the node's 16 cores with linear scaling. However, on an EC2 node, it runs as a sequential simulation, which (obviously) will run for longer.

We use two key metrics in our experiments: *Total Time to Completion (TTC)*, which is the wall-clock time for the entire (1-stage) EnKF workflow (i.e., all the 128 ensemble members) to complete, and the results are consumed by the KF stage, and may include both TeraGrid and EC2 execution. The *Total Cost of Completion (TCC)* is the total EC2 cost for the entire EnKF workflow.

Our experiments are based on the assumption that for tasks that can use 16-way parallelism, the TeraGrid is the platform of choice for the application, and gives the best performance, but is also the relatively more restricted resource. Furthermore, users have fixed allocation on this expensive resource, which they might want to conserve for tasks that require greater node counts. On the other hand, the EC2 is relatively more freely available, but is not as capable.

Note that the motivation of our experiments is to understand each of the usage scenarios and their feasibility, behaviors and benefits, and not to optimize the performance of any one scenario (or experiment). In other words, we are trying to establish a proof-of-concept, rather than a systematic performance analysis.

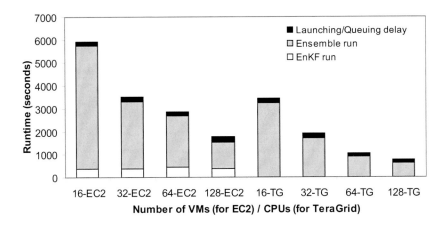

FIGURE 25.4
Baseline TTC for EC2 and TeraGrid for a 1-Stage, 128 Ensemble-Member
EnKF Run. The First 4 Bars Represent the TTC as the Number of EC2 VMs
Increase; the Next 4 Bars Represent the TTC as the Number of CPUs (nodes)
Used Increases.

25.5.4 Establishing Baseline Performance

The goal of the first set of experiments is to establish a performance baseline for the two platforms considered. The TTC for a 1-stage, 128 ensemble-
member EnKF run and for different numbers of VMs on the EC2 and different
number of CPUs on the TeraGrid are plotted in Figure 25.4. We can see from
the plots that the TTC decreases essentially linearly as the number of EC2
VMs and the number of TeraGrid CPUs increases. The TTC has 3 components — the time that it takes to generate the ensemble-members, the VM
start-up time in case of the EC2 or the TeraGrid queuing time, and the dominant run-time of the ensemble-members. In case of EC2, the VM start-up is
about 160 seconds and remains constant in this experiment, which is because,
the VMs are launched in parallel. Note that, in general, there can be large
variability, both in the launching times as well as performance of EC2 VM instances. In case of the TeraGrid, the queuing time was only about 170 seconds
(essentially the best case scenario) since we used the development queue for
these experiments. The time required for ensemble member generation also
remained constant, and is a small fraction of the TTC. Another important
observation is that the TTC on the TeraGrid is consistently lower than on
EC2 (as expected).

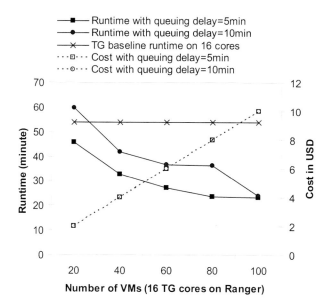

FIGURE 25.5
The TTC and TCC for Objective 1 with 16 TeraGrid CPUs and Queuing Times Set to 5 and 10 Minutes. As Expected, the More VMs that Are Made Available, the Greater the Acceleration, i.e., Lower the TTC. The Reduction in TTC is Roughly Linear, But is Not Perfectly so, Because of a Complex Interplay Between the Tasks in the Work Load and Resource Availability.

25.5.5   Objective 1: Using Clouds as *Accelerators* for HPC Grids

In this usage scenario, we explore how Clouds can be used as accelerators for HPC Grid work-loads. Specifically, we explore how EC2 can be used to accelerate an application running on the TeraGrid. The workflow manager inserts ensemble tasks into the Comet space and ensemble workers, both on the TeraGrid and EC2, pull tasks based on the defined policy described above and execute them.

In these experiments, we used 16 TeraGrid CPUs (1 node on Ranger) and varied the number of EC2 nodes from 20 to 100 in steps of 20. The average queuing time was set to 5 and 10 minutes. These values can be conservatively considered as the typical wait time for a job of this size, on the normal production queue of the TeraGrid [142]. The VM start up time on EC2 was once again about 160 seconds. The resulting TTC for hybrid usage mode are plotted in Figure 25.5. The plots also show the *best case* TeraGrid TTC for 16 CPUs (1 TeraGrid compute node) using the development queue and the 170 second wait time. The plots clearly show acceleration for both wait times; that is, the hybrid-mode TTC is lower than the TTC when only the TeraGrid is used. The exception is the experiment with 20 VMs and 10 minute wait time, where we see a small slow down that is due to the fact the some long-running task was scheduled onto the EC2 nodes causing the TeraGrid nodes to be starved. Another interesting observation is that acceleration is greater for the 5 minute wait time as compared to the 10 minute wait time. This is once again because when the queuing delay is 10 minutes, fewer tasks are executed by the more powerful TeraGrid CPUs. Also note that for the 100 VM case, the TTC is the same for both the 5 minute and 10 minute wait times, because in these cases most of the tasks are consumed by EC2 nodes.

These observations clearly indicate that the acceleration achieved is sensitive to the relative performance of HPC Grids and cloud resources and the number of HPC Grid resources used and the queuing time. For example, when we increased the number of CPUs to 64 and used a 5 minute queuing time, no acceleration was observed, as TeraGrid CPUs are significantly faster than the EC2 nodes and 64 TeraGrid CPUs are capable of finishing all the ensemble tasks before the EC2 nodes can finish *any* task. Note that the figure also presents the TCC for each case. This represents the actual cost as function of the CPU time used; the billing time may be different. For example, on the EC2, the granularity for billing is CPU-hours, so the billed cost can be higher. It is important to understand that the acceleration arising from the ability to utilize Clouds and Grids is not just a case of "throwing more resources" at the problem, but rather demonstrates how two different resource types and underlying resource management paradigms can complement one another and results in a lowering of the overall TTC.

25.5.6   Objective 2: Using Clouds for *Conserving* CPU-Time on the TeraGrid

When using specialized and expensive HPC Grid resources such as the TeraGrid, one often encounters the situation that a research group has a fixed allocation for scientific exploration/computational experiment, and typically this is in the form of a fixed allocation of the number of CPU-hours on a machine. The question then is, can one use Cloud resources to offload tasks that perhaps do not need the specialized capabilities of the HPC Grid resource to conserve such an allocation, and what is the impact of such an offloading on the TTC and TCC? Note that one could also look at conservation from the other side and conserve expenditure on the Cloud resources based on a budget, but here we investigate just the former.

Specifically, in the experiments in this scenario, we constrain the number of CPU-minutes assigned to a run and use EC2 nodes to compensate and enable the application to run to completion. The aim is to determine the number of tasks as a consequence of the constraint, that will run on either the TeraGrid, and what number will be taken up by EC2 *when attempting the quickest*

TABLE 25.1
Distribution of Tasks across EC2 and TeraGrid, TTC and TCC, as the CPU-Minute Allocation on the TeraGrid is Increased.

CPU-Time Limit (Mins)	25	50	100	200	300
TeraGrid (Tasks)	1	3	6	14	19
EC2 (Tasks)	127	125	122	115	109
EC2 (Nodes (VMs))	90	88	85	78	74
EC2 (TTC (Mins.))	28.94	28.57	27.83	26.48	26.10
EC2 (TCC (USD))	8.9	8.8	8.5	7.7	7.3

solution possible, and what the impact is on the TTC and TCC. We repeat the experiment increasing the number of CPU-minutes assigned. The results of the experiments are presented in Table 25.1.

Interestingly, given several tasks that take less than 1 minute on 1 compute node on the TeraGrid (which requires 16 CPU-mins, the unit of allocation), when the number of CPU-minute allocated is 25 mins, the TeraGrid gets exactly 1 task, and as we increase the allocation, the number of tasks pulled by TeraGrid CPUs increases in increments of 1 task per 16 CPU-minute (or 1 node-minute). These performance figures provide the basis for determining the sensitivity (of a given workload) to maximum CPU-mins (on TeraGrid). For example, we show that a factor of 12 decrease in max CPU-minutes can (300 reduced to 25), thanks to hybrid mode, lead to only a 10% increase in TTC (26.10 to 28.94).

25.5.7   Objective 3: Response to Changing Operating Conditions (*Resilience*)

Usage mode III, or *resilience*, addresses the situation where resources that were initially planned for, become unavailable at runtime, either in part or in entirety. For example, on several TeraGrid machines, there are instances when jobs with a higher-priority have to be given priority (i.e., right-of-way [673]), or perhaps it could just be that there is some unscheduled maintenance needed. Typically, when such situations arise, applications either just wait (significantly) longer in the queue or may be unceremoniously terminated and have to be rerun. The objective of this usage mode is to investigate how Cloud services can be used to address such situations and allow the system/application to respond to a dynamic change in availability of resources.

This scenario also addresses another situation that is becoming more relevant as applications have more complexity and dynamic. For such applications, it is often not possible to accurately estimate the CPU-time required to run to completion, and users either have to rely on trial and error and allow several jobs to terminate due to inadequate CPU-time, or overestimate the

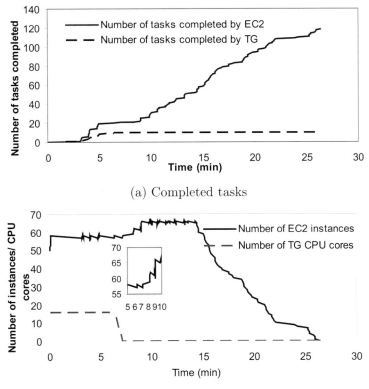

(a) Completed tasks

(b) EC2 instances and TeraGrid CPU cores used

FIGURE 25.6
Usage Mode III, Resilience. (a) Allocation of Tasks to TeraGrid CPUs and EC2 Nodes. As the 16 Allocated TeraGrid CPUs Become Unavailable After Only 70 Minutes Rather Than the Planned 800 Minutes, the Bulk of the Tasks are Completed by EC2 Nodes. (b) Number of TeraGrid Cores and EC2 Nodes as a Function of Time. Note that the TeraGrid CPU Allocation Goes to Zero After about 70 Minutes, Causing the Autonomic Scheduler to Increase the EC2 Nodes by 8.

required time and request more time than is actually needed. Both options lead to poor utilization. The *resilience* usage mode can also be applied to this scenario, where a shortfall in requested CPU-time can be compensated using Cloud resources.

In the set of experiments conducted for this usage mode, we start by requesting an adequate amount of TeraGrid CPU-time. However, at runtime, we trigger an event indicating that the available CPU-time on the TeraGrid has changed, causing the autonomic scheduler to re-plan and reschedule. Specifically, in our experiments, the autonomic scheduler begins by requesting 16 TeraGrid CPUs

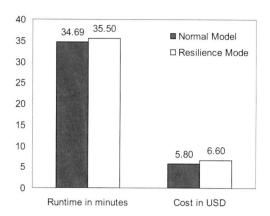

FIGURE 25.7
Overheads of Resilience on TTC and TCC.

for 800 minutes. However, after about 50 minutes of execution (i.e., 3 Tasks were completed on the TeraGrid), the scheduler is notified that only 20 CPU-minutes remain, causing it to re-plan and as a result increase the allocation of EC2 nodes to maintain acceptable TTC.

The results of the experiments are plotted in Figure 25.6 and Figure 25.7. Figure 25.6 (a) shows the cumulative number of tasks completed by TeraGrid and EC2 over time. As per the original plan, the expected distribution of tasks was 63:65 for TeraGrid:EC2. However, this distribution changes as the TeraGrid resource becomes unavailable, as shown in Figure 25.6 (b), causing EC2 nodes to take up a much large proportion of the tasks. Based on the scheduling policy used, the autonomic scheduler decided that the best TTC will be achieved by increasing the number of EC2 nodes by 8, from 58 originally allocated to 64. The resulting TTC and TCC are plotted in Figure 25.7.

As is probably obvious, the ability to support resilience is a first step in the direction towards graceful fault-tolerance, but we will discuss fault-tolerance in a separate work.

25.6   Acceleration Usage Mode: Application and Infrastructure Adaptivity

25.6.1 Dynamic Execution and Adaptivity of EnKF

We consider two types of adaptivity in this section: infrastructure and application adaptivity. Infrastructure adaptivity allows the autonomic selection of the appropriate number and type of resources based on requirements and constraints. Application adaptivity adjusts the application behavior based on application/system characteristics (e.g., the size of ensemble members, problem size and application configuration) and runtime state.

Infrastructure adaptivity is achieved by estimating each ensemble member's runtime on available resources and selecting the most appropriate resources for them. To estimate runtime on each different resource class, the CometCloud autonomic scheduler asks a worker per resource class to run the runtime estimation module. A worker running on each resource class pulls the task, executes it, and returns the estimated runtime back to the scheduler. If there are no active workers on a resource class, the scheduler launches one. The computational cost of running an estimation task is roughly 5% of that of an actual task. However, if the scheduler needs to start a new worker for estimation, it can incur additional some time delays. A good example would be the overhead of launching a new EC2 instance, or the waiting time in a queue after submitting a pilot job for TeraGrid. This runtime estimation is performed at the beginning of every stage because stages are non-uniform and the runtime of the a stage can not be used as an estimate for the subsequent stage. Once the autonomic scheduler gathers estimated runtimes from all resource classes, it maps the ensemble members (encapsulated as asks) to the most appropriate available resource class based on the defined policy. Policies determine whether runs are made with deadline-based or cost-based (i.e., with budget limits) objectives. The scheduler then decides the number of nodes/workers for each resource class and the appropriate mix of resources. Naturally, workers can consume more than one task and the number of workers is typically smaller than the number of tasks.

Application adaptivity, on the other hand, relies heavily on the application infrastructure. Since the reservoir simulator is based on PETSc [111], we have access to a wide variety of direct and iterative solvers and preconditioners. The selection of an optimal solver and preconditioner combination depends on the problem (stiffness, linearity, properties of the system of equations, etc...) as well as the underlying infrastructure. Since the simulator needs to perform several iterations, the first few iterations are performed with several solver/preconditioner combinations. This *optimization* study is performed with ensemble member rank 0 only, typically a reference or *base–case* from which all other ensemble members are generated. The combination with the best performance (shortest wall-clock time) is then selected and passed on to the next stage to reduce simulation runtime.

The overall system architecture used in the experiments is shown in Figure 25.1. Every stage of the application workflow is heterogeneous, and as a result, the selection of infrastructure and application configurations for each stage can be different. At every stage, the autonomic manager collects information about both the infrastructure and the application, and analyzes this information to decide on appropriate resources and application configuration. These decisions affect both current stages (infrastructure adaptivity) as well as subsequent stages (application adaptivity). After reaching a decision on the most efficient infrastructure/application configurations and mix of resources, resources are provisioned and *ensemble-member-workers* are executed. On the EC2, this translates to launching appropriate VMs running custom images. On the TeraGrid, ensemble-member-workers are essentially *pilot jobs* [702] that are inserted into the queue. The workflow manager inserts tasks into CometCloud and ensemble-workers directly access CometCloud and retrieve tasks based on the enforced scheduling policy. TeraGrid workers are allowed to pull the largest tasks first, while EC2 workers pull the smallest tasks. While this policy is not optimal, it was sufficient for our study. During the execution, the workflow manager monitors the executions of the tasks to determine progress and to orchestrate the execution of the overall workflow. The autonomic scheduler also monitors the status of the resources (using the agents), and determines progress to ensure that the scheduling objectives and policies/constraints are being satisfied, and can dynamically change resource allocation if they cannot be satisfied.

It is possible that a defined user objective (e.g., meet a tight deadline) cannot be met. This could be due to insufficient resources for the computational load as a consequence of the deadline imposed; or it could be due to autonomic scheduling efficiency. In addition to the efficiency of the autonomic scheduler (which we do not analyze here), the relative capacity of the TeraGrid and EC2 resources will determine the maximum value of acceleration possible for a given workload. In other words, with the addition of sufficiently large number of cloud resources — possibly of different types of clouds, any imposed deadline will be met for a given workload.

25.6.2 Experiment Background and Setup

Our experiments are organized with the aim of understanding how application and infrastructure adaptivity facilitate desired objectives to be met. Specifically, we investigate how adaptivity — application, infrastructure, or both, enable lowering of the TTC, i.e., acceleration. We explore, (1) how the autonomic scheduler reduces TTC when a deadline is specified, and 2) how adaptation helps achieve the desired objective by facilitating an optimized application and/or infrastructure configuration, or via autonomic scheduling decisions when multiple types of cloud resources are available.

We use a small number of stages of the EnKF workflow with a finite difference reservoir simulation of problem size $20\times20\times20$ grid points and 128 ensemble members with heterogeneous computational requirements. Our experiments are performed on the TeraGrid (specifically Ranger) and several instance types of EC2. Table 25.2 shows the EC2 instance types used for experiments. We

TABLE 25.2

EC2 Instance Types Used in Experiments

Instance type	Cores	Memory(GB)	Platform(bit)	cost($/hour)
m1.small	1	1.7	32	0.1
m1.large	2	7.5	64	0.34
m1.xlarge	4	15	64	0.68
c1.medium	2	1.7	32	0.17
c1.xlarge	8	7	64	0.68

assume that a task assigned to a TeraGrid node runs on all 16 cores for that node (for Ranger).

TeraGrid provides better performance than EC2, but is also the relatively more restricted resource — in that there are often queuing delays. Hence, we use EC2 which is available immediately at a reasonable cost to accelerate the solution of the problem. A task pulled by an EC2 node runs sequentially (in case of m1.small which has a single core), or in parallel (other instance types with multiple cores) inside a single VM. To enable MPI runs on multi-core EC2 instance, we created a MPI-based image on EC2. This image included the latest versions of compilers, MPICH2, PETSc and HDF5 and was configured for performance above all else. Although we experimented with using MPI across multiple VMs as described in Section 25.6.5, we exclude using data from experiments involving running MPI across VMs in the analysis of understanding the effect of adaptivity; as is expected, due to communication overheads, it displays poor performance as compared to MPI within a single VM.

25.6.3 Baseline: Autonomic Scheduling in Response to Deadlines

At first, the scheduler estimates TTC assuming immediate availability of infinite TeraGrid resources. The scheduler then decides how much of the workload should be migrated to EC2 in order to meet the given deadline. If the deadline can be met given the estimated TTC (assuming usage of only TeraGrid resources), the number of tasks which should be off-loaded onto EC2 is decided and the autonomic scheduler selects the appropriate EC2 instance types to be launched. The autonomic scheduler has no knowledge of runtime characteristics of tasks on different resources. The scheduler makes decisions based on estimated runtimes obtained from the runtime estimator module. The runtime estimator is a simple utility that launches a full-size ensemble member simulation with a reduced number of iterations to minimize cost. The results of various ensemble members are tabulated and used to predict the full runtime cost of a complete ensemble member simulation.

In this experiment set, we limit EC2 instance types to m1.small and c1.medium, and run the EnKF with 2 stages, 128 ensemble members each.

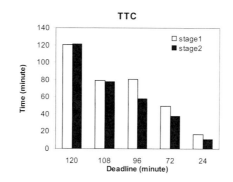

(a) Time To Completion

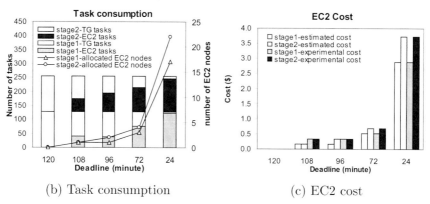

(b) Task consumption   (c) EC2 cost

FIGURE 25.8
Results from Baseline Experiments (without adaptivity) but with a Specified Deadline. We Run the EnKF with 2 Stages, 128 Ensemble Members and Limit EC2 Instance Types to m1.small and c1.medium. Tasks are Completed within a Given Deadline. The Shorter the Deadline, the More EC2 Nodes are Allocated.

For baseline experiments/performance numbers, we disabled adaptivity based optimization for both application and infrastructure.
Figure 25.8 depicts the TTC for different values of imposed deadlines and thus gives quantitative information on acceleration provided by the autonomic scheduler to meet deadlines. The imposed deadline takes the values of 120, 108, 96, 72 and 24 minutes respectively; this corresponds to 0%, 10%, 20%, 40% and 80% acceleration, based upon the assumption that a deadline of 120 minutes (0% acceleration) corresponds to a TTC when only TeraGrid nodes are used. Figure 25.8 (a) shows TTC for Stage 1 and Stage 2 where each stage is heterogeneous and all tasks are completed by the deadline (for all cases).
Figure 25.8 (b) shows the number of tasks completed by the TeraGrid and number off-loaded onto EC2, as well as the number of allocated EC2 nodes

for each stage. As the deadline becomes shorter, more EC2 nodes are allocated, hence, the number of tasks consumed by EC2 increases. Since m1.small instances have relatively poorer performance compared to c1.medium, the autonomic scheduler only allocates c1.medium instances for both stages. Figure 25.8 (c) shows costs incurred on EC2. As more EC2 nodes are used for shorter deadlines, the EC2 cost incurred increases. The results also show that most tasks are off-loaded onto EC2 in order to meet the "tight" deadline of finishing within 24 minutes.

25.6.4   Track 1: Infrastructure Adaptations

The goal of this experiment, involving infrastructure-level adaptivity is to investigate advantages — performance or otherwise, that may arise from the ability to dynamically select appropriate infrastructure, and possibly vary them between the heterogeneous stages (of the application workflow). Specifically, we want to see if any additional acceleration (compared to the baseline experiments) can be obtained. In order to provide greater variety in resource selection, we include all EC2 instance types from Table 25.2; these can be selected in Stage 2 of the EnKF workflow.

Figure 25.9 shows TTC and EC2 cost, with and without infrastructure adaptivity. Overall, TTC decreases when more resource types are available and infrastructure adaptivity is applied; this can be understood by the fact that the autonomic scheduler can now select more appropriate resource types to utilize. There is no decrease in the TTC when using infrastructure adaptivity for a deadline of 120 because all tasks are still completed by TeraGrid node (since it represents 0% acceleration).

The difference between the TTC with and without infrastructure-level adaptivity, decreases as the deadline becomes tighter; results show almost no savings with the 24 minutes deadline. This is mostly due to the fact that with a 24 minute deadline, a large number of nodes are allocated, thus no further gain can be obtained through infrastructure adaptivity.

Figure 25.9 (b) shows the number of nodes allocated for each run and the cost of their usage. The autonomic scheduler selects only c1.xlarge to use for infrastructure-adaptive runs because the runtime estimator predicts that all tasks will run fastest on c1.xlarge. On the other hand, the scheduler selects only c1.medium for non-adaptive runs. Infrastructure-adaptive runs cost more than non-adaptive runs, roughly 2.5 times more at the 24 minute deadline even though the number of nodes for non-adaptive runs is larger than the number of nodes for infrastructure-adaptive runs. This is because the hourly cost of c1.xlarge is much higher than c1.medium (see Table 25.2) and both TTCs are rather short (less than half hour). Since we used deadline-based policy, the autonomic scheduler selects the best performing resource regardless of cost; however, when we switch policies to include economic factors in the autonomic scheduler decision making (i.e., considering TTC as well as cost for the 24 minute deadline), the scheduler selects c1.medium instead of c1.xlarge.

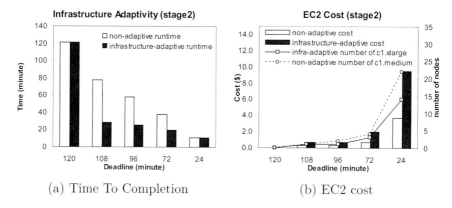

(a) Time To Completion   (b) EC2 cost

FIGURE 25.9
Experiments with Infrastructure Adaptivity. We Limit EC2 Instance Types to m1.small and c1.medium for the Non-Adaptive Run and Use All Types Described in Table 25.2 for the Infrastructure-Adaptive Run. The TTC is Reduced with Infrastructure Adaptivity at Additional Cost.

25.6.5   Track 2: Adaptations in Application Execution

In this experiment, we run ensemble member rank 0 with variations in solvers/preconditioners and infrastructure. In each case, ensemble rank 0 was run with a solver (generalized minimal residual method GMRES, conjugate gradient CG or biconjugate gradient BiCG), a preconditioner (block Jacobi or no preconditioner) for a given problem size with a varying number of cores (1 through 8). Two infrastructure solutions were available: a single c1.xlarge or 4 c1.medium instances.

Figure 25.10 shows the time to completion of an individual ensemble member, in this case the ensemble rank 0 with various solver/preconditioner combinations. We varied over two different types of infrastructure, each with 4 different core counts $(1, 2, 4$ and $8)$ and three problem sizes. We investigated six combinations of solvers and preconditioners over the range of infrastructure. Naturally, there is no one, single, solver/preconditioner combination that works for all problem sizes, infrastructures and core counts. Note the experiments were carried out on different infrastructure, but the infrastructure was not *adaptively* varied. For example, Figure 25.10 (c) shows that a problem of size $40{\times}20{\times}20$ is best solved on 4 cores in a c1.xlarge instance with a BiCG solver and a block Jacobi preconditioner. This is different from the 2 core, BiCG solver and no preconditioner combination for a $20{\times}10{\times}10$ problem on a c1.medium VM as seen in Figure 25.10 (a).

Basic profiling of the application suggests that most of the time is spent in the solver routines, which are communication intensive. As there is no dedicated, high bandwidth, low latency interconnect across instances, MPI performance will suffer, and subsequently MPI intensive solvers. The collective operations

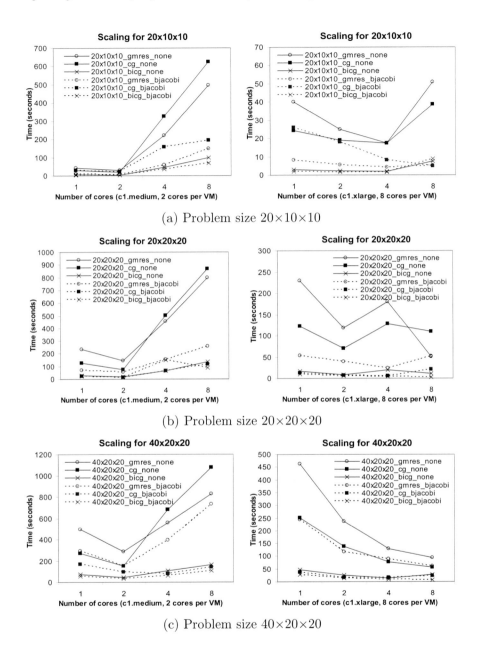

(a) Problem size $20 \times 10 \times 10$

(b) Problem size $20 \times 20 \times 20$

(c) Problem size $40 \times 20 \times 20$

FIGURE 25.10
Time to Completion for Simulations of Various Sizes with Different Solvers (GMRES, CG, BiCG) and Block–Jacobi Preconditioner. Benchmarks Ran on EC2 Nodes with MPI. Problem Size Increases Going from Left to Right. The Top Row is for a 2 Core VM and the Bottom Row is for an 8 Core VM.

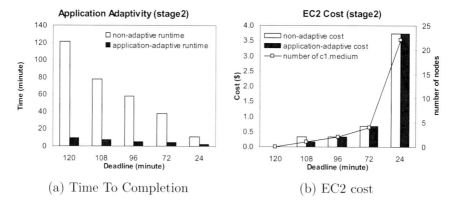

(a) Time To Completion   (b) EC2 cost

FIGURE 25.11
Experiments with Application Adaptivity. The Optimized Application Option is Used for the Application-Adaptive run. The TTC is Reduced with Application Adaptivity for Equivalent or Slightly Less Cost.

in MPI are hit hardest, affecting Gram-Schmidt orthogonalization routines adversely. Conjugate gradient solvers, on the other hand, are optimal for banded diagonally dominant systems (as is the case in this particular problem) and require less internal iterations to reach convergence tolerance.

From Figure 25.10 we see that as the problem size increases (moving from left to right), the performance profile changes. Comparison of the profiling data suggests that a smaller percentage of simulation time is spent in communication as the problem size increases. This is obviously due to the fact that there are larger domain partitions for each instance to work on. Detailed profiling of inter–instance MPI bandwidth and latency is still underway; however, early results suggest that this trend continues.

Figure 25.11 shows (a) TTC and (b) EC2 cost using application adaptivity. Applying an optimized application configuration reduces TTC considerably. Application configuration does not affect the selection of infrastructure and the number of nodes, hence, the cost depends on TTC. However, since EC2 costs are billed on an hourly basis (with a minimum of one hour), the decrease in cost does not match the decrease in TTC. For example, with a 108 minute deadline, the time difference between the non-adaptive mode and the application-adaptive mode is more than one hour. Hence, the cost of the application-adaptive mode is almost halved. However, because TTCs are all within one hour for other deadlines, EC2 costs remain constant.

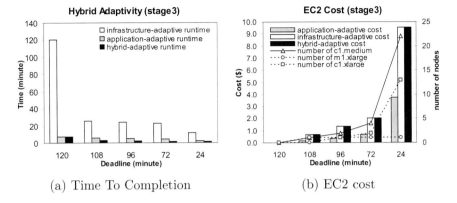

(a) Time To Completion   (b) EC2 cost

FIGURE 25.12
Experiment with Adaptivity Applied for Both Infrastructure and Application. The TTC is Reduced Further than with Application or Infrastructure Adaptivity on its Own. The Cost is Similar to that in Infrastructure Adaptivity for Durations less than one hour Since EC2 Usage is Billed Hourly with a One Hour Minimum.

25.6.6 Track 3: Adaptations at the Application and Infrastructure Levels

In this experiment set, we explore both infrastructure as well as application-level adaptivity. We try infrastructure-level adaptivity in the first stage, followed by application-level adaptivity in the second stage and hybrid adaptivity in the third stage. In principle, the final performance should not be sensitive to the ordering. We will compare hybrid-adaptivity to application-adaptivity, as well as infrastructure-adaptivity.

Figure 25.12 shows TTCs and EC2 costs; the white columns correspond to infrastructure adaptivity TTC (Stage 1), the light blue columns correspond to application adaptivity TTC (Stage 2) and the dark blue columns correspond to hybrid adaptivity TTC. As mentioned earlier, the application spends most of its time in the iterative solver routine. The application also runs twenty time-steps for each of the 128 simulations. Therefore, the potential for improvement in TTC from solver/preconditioner selection is substantial. As we expect, the TTC of infrastructure-adaptive runs are larger than those of application-adaptive runs, especially since infrastructure adaptivity occurs once per every stage and application adaptivity influences every solver iteration. As is evident from Figure 25.12 (a), both infrastructure and application adaptivity result in TTC reduction, and even more so when used simultaneously.

The cost for using infrastructure adaptivity is higher than that of using application adaptivity as seen in Figure 25.12 (b). This is due to the simple fact that application adaptivity improves the efficiency of the application without the need for an increase in resources. It is worth mentioning that ours is a special case as the application depends heavily on a sparse matrix solve with an iterative solver. Other applications that use explicit methods cannot make use of application adaptivity (no solvers).

25.7 Conclusion

Given that production computational infrastructures will soon provide a hybrid computing environment that integrates traditional HPC Grid services with on-demand Cloud services, understanding the potential usage modes of such hybrid infrastructures is important. In this chapter, we experimentally investigated, from an application's perspective, possible usage modes for integrating HPC Grids and Clouds as well as how autonomic computing can support these modes. Specifically, we used an EnKF workflow with the Comet-Cloud autonomic Cloud engine on a hybrid platform, to investigate three usage modes (i.e., autonomic objectives) — acceleration, conservation and resilience. Note that our objective in the experiments presented was to understand each of the usage scenarios and their feasibility, behaviors, and benefits, and not to optimize the performance of any one scenario (or experiment). We then focused on acceleration usage modes and explored application or/and infrastructure adaptivity. We showed that application and infrastructure adaptivity affect performance and cost.

The results of experiments demonstrate that a policy driven autonomic substrate, such as the autonomic application workflow management framework and *pull-based* scheduling supported by CometCloud, can effectively support the usage modes (objectives) investigated here. The approach and infrastructure used can manage the heterogeneity and dynamics in the behaviors and performance of the platforms (i.e., EC2 and TeraGrid). Our results also demonstrate that defining appropriate policies that govern the specifics of the integration is critical to realizing the benefits of the integration. The experimental results for investigating adaptivity show that time-to-completion decreases with both system or application-level adaptivity. We also observe that the time-to-completion decreases further when applying both the system and application-level adaptivity. Furthermore, while EC2 cost decreases when application adaptively is applied, it increases when infrastructure adaptivity is applied. This is despite a reduced time-to-completion, and is attributed to the use of more expensive instance types.

While our work so far has been focused on the EnKF inverse problem workflow which is a fairly straightforward, linear workflow, complex workflows such as parameter or model space survey workflows can be similarly scheduled. It will be interesting to extend the concepts and insight gained from this work to high-throughput Grids where the individual resources involved will have similar performance to EC2. Additionally the objectives for high-throughput

Grids would also be different and so would be the policies required to support the objective.

Acknowledgments

The research presented in this paper is supported in part by National Science Foundation via grants numbers IIP 0758566, CCF-0833039, DMS-0835436, CNS 0426354, IIS 0430826, and CNS 0723594, by Department of Energy via grant numbers DE-FG02-06ER54857 DE-FG02- 04ER46136 (UCoMS), by a grant from UT Battelle, and by an IBM Faculty Award, and was conducted as part of the NSF Center for Autonomic Computing at Rutgers University. Experiments on the Amazon Elastic Compute Cloud (EC2) were supported by a grant from Amazon Web Services and CCT CyberInfrastructure Group grants.

26

RestFS: The Filesystem as a Connector Abstraction for Flexible Resource and Service Composition

Joseph Kaylor

Department of Computer Science, Loyola University, Chicago, Illinois

Konstantin Läufer

Department of Computer Science, Loyola University, Chicago, Illinois

George K. Thiruvathukal

Department of Computer Science, Loyola University, Chicago, Illinois

CONTENTS

The broader context for this chapter comprises business scenarios requiring resource and/or service composition, such as (intra-company) enterprise application integration (EAI) and (inter-company) web service orchestration. The resources and services involved vary widely in terms of the protocols they support, which typically fall into remote procedure call (RPC) [129], resource-oriented (HTTP [270] and WEBDAV [758]) and message-oriented protocols. By recognizing the similarity between web-based resources and the kind of resources exposed in the form of filesystems in operating systems, we have found it feasible to map the former to the latter using a uniform, configurable connector layer. Once a remote resource has been exposed in the form of a local filesystem, one can access the resource programmatically using the operating system's standard filesystem application programming interface (API). Taking this idea one step further, one can then aggregate or otherwise orchestrate two or more remote resources using the same standard API. Filesystem APIs are available in all major operating systems. Some of those, most notably, all flavors of UNIX including GNU/Linux, have a rich collection of small, flexible command-line utilities, as well as various inter-process communication (IPC) mechanisms. These tools can be used in scripts and programs that compose the various underlying resources in powerful ways.

Further explorations of the role of a filesystem-based connector layer in the enterprise application architecture have lead us to the question whether one can achieve a fully compositional, arbitrarily deep hierarchical architecture by re-exposing the aggregated resources as a single, composite resource that, in turn, can be accessed in the same form as the original resources. This is indeed possible in two flavors: 1) the composite resource can be exposed internally as a filesystem for further local composition; 2) the composite resource is exposed externally as a restful resource for further external composition. We expect the ability hierarchically to compose resources to facilitate the construction of complex, robust resource- and service-oriented software systems, and we hope that concrete case studies will further substantiate our position.

Leveraging our prior work on the Naked Objects Filesystem (NOFS) [419], which exposes object-oriented domain model functionality as a Linux filesystem in user space (FUSE) [685], we have implemented RestFS [418], a (dynamically re)configurable mechanism for exposing remote restful resources and as local filesystems. Several sample adapters specific to well-known services such as Yahoo! Placefinder and Twitter are already available. Authentication poses a challenge in that it cannot always be automated; in practice, when systems such as OAuth are used, it is often only the initial granting of authentication that must be manual, and the resulting authentication token can then be included in the connector configuration. As future work, we plan to develop plugins to support resources across a broader range of protocols, such as FTP, SFTP, or SMTP.

26.1 Related Work

There are various lines of related work, which we will discuss in this section.

26.1.1 Representational State Transfer (ReST)

Partly in response to the complexity of the W3C's WS-* web service specifications [183], resource-oriented approaches such as the representational state transfer (ReST) architectural style [271] have received growing attention during the second half of this decade. In ReST, addressable, interconnected resources, each with one or more possible representations, are usually exposed through the HTTP protocol, which is itself stateless, so that all state is located within the resources themselves. These resources share a uniform interface, where resource-specific functionality is mapped to the standard HTTP request methods GET, PUT, POST, DELETE, and several others. Clients of these resources can access them directly through HTTP, use a language-specific framework with ReST client support, or rely on resource- and language-specific client-side bindings.

26.1.2 Inter-Process Communication through the Filesystem

Most methods of IPC can be represented in the filesystem namespace in many operating systems. Pipes, domain sockets and memory-mapped files can exist in the filesystem in UNIX [427]. While pipes are uni-directional, allowing one program to connect at each end point, other IPC methods such as UNIX domain sockets allow for multiple client connections and permit data to be written in both directions. With this capability, it is possible for output from several programs to be aggregated by one program instead of a 1:1 model as is allowed by pipes. Other methods of IPC, such as memory-mapped and regular files, allow several programs to collaborate through a common, named store of data.

Composition of the files in filesystems is also possible through layered or stackable filesystems. Mechanisms for this differ amongst operating systems. In 4.4BSD-Lite, Union Mounts [574] allowed for filesystems to be mounted in a linear hierarchy. Changes to files lower in the hierarchy would override files in the higher part of the hierarchy. The Plan 9 distributed operating system allowed for the filesystem namespace to be manipulated through the mount, unmount, and bind system calls [580, 581]. In our own research, we have implemented a layered filesystem, OLFS, which allowed for a flexible layering and inheritance scheme through folder manipulation [417]. Each of these approaches manipulates the filesystem namespace and consequently allows for changes in configuration and how IPC resources are located. This capability can help provide for new and interesting ways to share data between programs. Although not as widespread, some operating systems implement more advanced IPC such as network connections, specific protocols such as HTTP or FTP, and other services through the filesystem namespace. An excellent example of this is the Plan 9 operating system. Plan 9's filesystem layer, the 9P protocol, is used to represent user interface windows, processes, storage files, and network connections. In Plan 9, it is possible through filesystem calls to engage in IPC in a more uniform way on a local machine and across separate machines.

In terms of inter-machine file-based IPC, it has been possible for many years to coordinate and share data among processes by writing to files on network filesystems. As long as the network filesystem has adequate locking mechanisms and an adequate solution to the cache coherency problem, it is possible to perform IPC through file-based system calls over a network filesystem.

Other than coordination through network filesystems or specialized operating system mechanisms like 9P, much inter-machine IPC has been through abstractions on top of the network socket. Remote procedure call approaches such as RPC or RMI have provided a standard way for processes to share data and coordinate with each other. Other socket-based approaches include the HTTP protocol and abstractions on top of HTTP, such as SOAP and REST.

26.1.3   Recent Developments in File-Based IPC

Some more recent advances have been made in terms of inter-machine IPC over the filesystem. Application filesystems are being built on top of FUSE (File Systems in Userspace) to act as clients for web services such as Flickr, IMAP email services, Amazon S3, and others. Instead of using the socket as the basis for IPC with these services, it has become possible to be able to interact with them through filesystem calls.

IPC through the filesystem offers some advantages. Although in UNIX-like operating systems, it is possible to redirect output to a socket through a program like socat, netcat, or nc, there are many network options and issues like datagram versus streaming to consider. File-based IPC often presents a simpler interface to work with, and leaves many of the networking and protocol questions to the implementing filesystem. Another important advantage that it offers processes that interact with these application filesystems is transparency. These processes that interact with these application filesystems do not need to be aware of which service they are interacting with, which URL it is located at or what types of SOAP messages it requires to communicate with. With a Flickr filesystem, it is possible to use programs that simply interact with images aside from a web browser to interact with the Flickr photo service.

26.1.4 The Shift from Kernel Mode to User Mode Filesystem Development

In very early systems, development of new filesystem code was a challenge because of high coupling with storage device architecture and kernel code.

In the 1970s, with the introduction of MULTICS, UNIX, and other systems of the time, more structured systems with separated layers became more common. UNIX used a concept of i-nodes, which were a common data structure that described structures on the filesystem [704]. Different filesystem implementations within the same operating system kernel could share the i-node structure; this included on-disk and network filesystems. Early UNIX operating systems shared a common disc and filesystem cache and other structures related to making calls to the I/O layer that managed the discs and network interfaces.

Newer UNIX-like systems such as 4.2 BSD and SunOS included an updated architecture called v-nodes [437]. The goal was to split the filesystem's implementation-independent functionality in the kernel from the filesystem's implementation-dependent functionality. Mechanisms like path parsing, buffer cache, i-node tables, and other structures became more shareable. Also, operations based on v-nodes became reentrant, thereby allowing new behavior to be stacked on top of other filesystem code or to modify existing behavior. V-nodes also helped to simplify systems design and to make filesystems implementations more portable to other UNIX-like systems. Many modern UNIX-like systems have a v-nodes-like layer in their filesystems code.

With the advent of micro-kernel architectures, filesystems being built as user-mode applications became more common and popular even in operating systems with monolithic kernel architectures. Several systems with different design philosophies have been built. We describe three of these systems that are most closely related to NOFS: FUSE [685], ELFS [414], and Frigate [432].

The Extensible File System (ELFS hereafter) is an object-oriented framework built on top of the filesystem that is used to simplify and enhance the performance of the interaction between applications and the filesystem. ELFS uses class definitions to generate code that takes advantage of pre-fetching and caching techniques. ELFS also allows developers to automatically take advantage of parallel storage systems by using multiple worker threads to perform reads and writes. Also, since ELFS has the definition of the data structures, it can build efficient read and write plans. The novelty of ELFS is that the developer can use an object-oriented architecture and allow ELFS to take care of the details.

Frigate is a framework that allows developers to inject behavioral changes into the filesystem code of an operating system. Modules built in Frigate are run as user-mode servers that are called to by a module that exists in the operating system's kernel. Frigate takes advantage of the reentrant structure of vnodes in UNIX-like operating systems to allow the Frigate module developer to layer behavior on top of existing filesystem code. Frigate also allows the developer to tag certain files with additional metadata so that different Frigate modules can automatically work with different types of files. The novelty of Frigate is that developers do not need to understand operating-systems development to modify the capabilities of filesystem code, and they can test and debug their modules as user-mode applications. But they still need to be aware of the UNIX filesystem structures and functions.

File Systems in Userspace (FUSE hereafter) is a user mode filesystems framework. FUSE is supported by many UNIX-like operating systems such as Linux, FreeBSD, NetBSD, OpenSolaris, and Mac OSX. The interface supported by FUSE is very similar to the set of UNIX system calls that are available for file and folder operations. Aside from the ability to make calls into the host operating system, there is less sharing with the operating system than with v-nodes such as path parsing. FUSE has helped many filesystem implementations such as NTFS and ZFS to be portable to many operating systems. Since FUSE filesystems are built as user-land programs, they can be easier to develop in languages other than C or C++, easier to unit test, and easier to debug. Accordingly, FUSE has become a popular platform for implementing application-specific filesystems.

26.2   Composition of Web Services through the Filesystem

Filesystems can play different roles in the composition of web-based resources and services. We will now study these in more detail.

26.2.1   Commonalities between Web Resources and the Filesystem

We believe that there are clear commonalities between web services and the filesystem. Both systems have a concept of a URI. In web services, this can be an HTTP URL. In the filesystem this can be a file or folder path. In both systems there are protocol actions that can be used to send and retrieve data. In web services this can be accomplished through HTTP GET and POST. In filesystems, this can be accomplished through read() and write() system calls. In both systems it is possible to invoke executable elements. In web services this can be performed with GET and POST calls and the use of SOAP messages to web service URLs. On a local filesystem, executable services can be invoked by loading and executing programs from the local filesystem.

In our exploration we believe that there are three candidates for how to build the filesystem layer to expose resources from the web. The first way is through application filesystems built with the Naked Object Filesystem (NOFS) framework. The second way is to use the filesystem as a connector layer to abstract and re-expose web resources to the local system. The third way is to use a combination of the filesystem as a connector layer and the filesystem as an application. We have explored this second route with RestFS, which has been implemented using the NOFS framework. In each of these methodologies we demonstrate how to map concepts from web services onto the filesystem. We will also explain the advantages and disadvantages to each approach.

26.2.2   The Filesystem as a Connector Layer

In our exploration of filesystems, we questioned whether a filesystem could be used as a connector layer for web services. We also questioned whether that connector layer could be used to compose web services with local and other web services and then expose those web services externally as a new web service. RestFS is our attempt to implement such a filesystem.

RestFS is an application filesystem implemented with the NOFS framework. RestFS uses files to model interaction with web services. When a file is created in RestFS, two files are created: a configuration file and a resource file. The configuration file contains an XML document that can be updated to contain a web service URI, web method, authentication information, and a triggering filesystem method. Once configured, the resource file can be interacted with on the local machine to interact with a web service.

One example of the usage of RestFS is to create a file that can perform a Google Search. In this example, the file is configured with the Google APIs server and the web search service. Web requests are sent with the GET HTTP method and are triggered by the utime filesystem call. When a user of the filesystem issues a 'touch' command on the resource file, a GET request is issued by RestFS to the Google API server and the response from that server is written back to the resource file, which will be available for subsequent reads. In this example, the task of configuring the resource, triggering the request, and parsing the results are left to a Bash shell script.

Another example usage of RestFS is with the Yahoo! PlaceFinder service. This example is similar to the Google search example. The configuration file is setup with the URI for the web service, and the utime system call is used to trigger the web request. Also, in this example, a shell script is used to configure the RestFS file, trigger the web service call, and to parse the results. With our implementation of resource files in RestFS, remote web resources can be interacted with in a similar way as other local file based IPC. The local nature of the resource files allows for programs that read from and write to the resource files to be unaware of the web service that RestFS is communicating

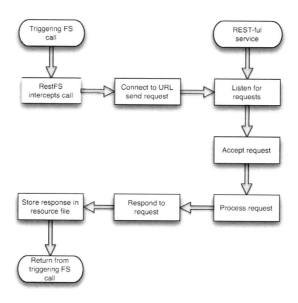

FIGURE 26.1
The Timeline of a RestFS Web Service Call.

with. For example, it is possible to use programs such as grep, sed, or perl to search, transform, and manipulate the data in the resource file. In each of these cases, these programs do not need to be aware that the data they are working with has been transparently read from or written to a remote web service.

Because RestFS acts as only a connector layer and provides no additional interpretation or filtering of requests or responses, external programs are required to read and write the structured data that is necessary for interact with configured web services. In the Google Search and Yahoo! PlaceFinder examples, the task of writing a structured request and parsing the response was left to a shell script that took advantage of UNIX command line tools like sed, grep, and others. These scripts had to be aware of the structure of both the requests and response needed by the web service. It is possible to filter, translate, and load data from the resource files with any local program that can accept data from a file or a UNIX pipe. As a consequence, it is possible to augment the value added of the web service with local programs in several possible combinations.

The connector model presented by RestFS in combination with other IPC mechanisms on the local operating system makes it possible to compose the data from several web services with each other in a flexible and reconfigurable way. One possible example of this would be to set up several resource files for RSS news feeds across the internet. A script could be implemented to

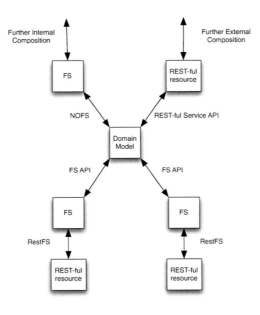

FIGURE 26.2
The Flexible Internal and External Composition Possible with RestFS.

parse each of those news sources for specific topics, aggregate them, and then write them to another resource file that could represent a submission form and service for creating articles on a blog. The same system then could have several resource files set up to watch Twitter accounts for comments on the article and post responses on Twitter to the blog site. If new news sources become important or new Twitter accounts are necessary, new resource files and alterations to scripts can be made to expand and reconfigure the system. It is possible to do all of this with a series of scripts and small programs on a UNIX operating system that use RestFS as a connector layer.

There are some instances where the connection layer concept has some difficulties in our exploration. When trying to compose some web services that are built around human interaction through rich user interfaces, it can be difficult to create a program that can interact with these services in a simple way.

One example of this is the CAPTCHA human test. To reduce "spam" in the form of email and as entries on blogs, many websites incorporate a form that requests the user perform a small test such as recognizing a sound or interpreting letters on an image to prove to the system that the user of the web service is in fact a human. Often, after these initial interactions, it is possible for simple interaction with RestFS, but because of them it is not always straightforward to automate the entire interaction with a web service. Other forms of non machine readable interactions such as the use of images, sounds, or video can present complications for composing web services with RestFS.

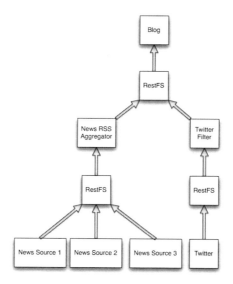

FIGURE 26.3
A Sample Composition of a Blog, News Sources, and Twitter.

Another example would be web services that make use of the user interface for complex validation or additional business rules. While not an ideal design, such web services still exist on the internet. Because local programs will interact with the application tier and not the presentation tier of a web service, any logic that exists in that presentation tier that is necessary for proper communication with the application tier must be duplicated in whatever local composition is made of the web service.

26.2.3   The Filesystem as an Application and Abstraction

While exploring the possibilities for using filesystems to interact with web services, we observed the emergence of application oriented filesystems such as WikipediaFS, IMAPFS, and FlickrFS. Each of these filesystems demonstrate different web services represented as different components on filesystems. In several email oriented filesystems, folders available in IMAP accounts are represented as folders on the local filesystem and individual email messages as files. In photo-sharing-oriented filesystems such as FlickrFS, photos are categorized into folders and exposed as standard image files. In each of these application filesystems, normal file operations work as expected. Copying and deleting files in FlickrFS completes the expected operation of downloading and uploading photos with a user's Flickr account.

After our own experiences with implementing storage oriented filesystems in FUSE, we felt that application filesystems would benefit from a different abstraction than what is presented by FUSE. To that end, we implemented the Naked Objects Filesystem (NOFS). NOFS allows a developer to implement an application filesystem by annotating Java classes in an application domain model. Through inspection of these domain objects and associated annotations, NOFS presents a filesystem composed of files, folders, and executable scripts to the user through FUSE to interact with the domain model. We will explore in detail the architecture and internal workings of NOFS in a later section.

With the NOFS framework, we were able to implement application filesystems in a more rapid fashion with less filesystem glue code needed. This helped reduce the necessary components to expose a web service such as the Flickr photo service as a filesystem (Figures 26.4, 26.5) to the interaction with the REST-ful web service and the construction of an adequate domain model to represent the structure of the service and filesystem. Our implementation of a simple Flickr filesystem took 484 lines of Java code. An existing Python implementation of the Flickr filesystem that uses FUSE directly took 2144 lines of code. About half of the Python implementation was code used to glue FUSE to the Flickr photo service. The remainder of the code was related to handling the Flickr photo service.

Another example of an application filesystem built with NOFS is the Yahoo! Finance stock ticker filesystem. We were able to implement the entire filesystem with just 155 lines of code in two Java classes (see Figures 26.6, 26.7)

Application filesystems like those that can be built with NOFS are very useful for user interaction. Actions that make sense in a photo library service have excellent mappings to filesystem actions. The fundamental unit in the service, the photo, maps well to a file. Collections and categories of photos map well to folder structures. In this particular case, for the sake of user interaction, the structure of the web service calls and their mapping into a connector layer like RestFS would not be a convenient structure for user interaction. The application filesystem allows for a better mapping of the business unit / domain model that is presented by the web service.

Application filesystems built through NOFS also are able to handle action validation and interaction in a simpler way than is possible with RestFS-like systems. If an action on the domain model for an NOFS filesystem is in some way invalid, an exception can be raised so that the filesystem call that triggered the action can return an error code. In this way, NOFS domain models can restrict copy, delete, read, write or other filesystem operations to those that are considered valid by the domain model. Resource files in RestFS expect that data written to and read from the resource files is in a valid format.

Application filesystems are not as well suited for simple re-configuration or changes in composition as RestFS is. To introduce changes in an application filesystem, either facilities for dynamically adding plugins must be introduced, or the system must be unmounted, modified and mounted as a filesystem again.

```
@DomainObject(CanWrite=false)
public class FlickrPhoto implements IProvidesUnstructuredData {
   private byte[] _data;
   public void setData(byte[] data) {
      _data = data;
   }

   public FlickrPhoto() {}

   private String _name;
   @ProvidesName
   public String getName() { return _name; }
   @ProvidesName
   public void setName(String name) { _name = name; }

   public boolean Cacheable() { return false; }
   public long DataSize() { return _data.length; }
   public void Read(ByteBuffer buffer, long offset, long length) {
      for(long i = offset; i < offset + length && i < _data.length;
            i++) {
         buffer.put(_data[(int)i]);
      }
   }
   public void Truncate(long length) { }
   public void Write(ByteBuffer buffer, long offset,
      long length) { }
}
```

FIGURE 26.4
The FlickrPhoto Domain Object from FlickrFS.

```
@FolderObject(CanAdd=false, CanRemove=false)
@DomainObject
public class FlickrUser {
   private List<FlickrPhoto> _photos =
      new LinkedList<FlickrPhoto>();
   public FlickrUser() {}

   private String _name;
   @ProvidesName
   public String getName() { return _name; }
   @ProvidesName
   public void setName(String name) { _name = name; }

   private IDomainObjectContainerManager _manager;
   @NeedsContainerManager
   public void setContainerManager(IDomainObjectContainerManager
         manager) {
      _manager = manager;
   }

   private long _lastGet = 0;
   @FolderObject(CanAdd=false, CanRemove=false)
   public List<FlickrPhoto> getPhotos() throws Exception {
      if(_lastGet == 0 || System.currentTimeMillis() - 10000 >
            _lastGet) {
         UpdatePhotos();
         _lastGet = System.currentTimeMillis();
      }
      return _photos;
   }

   private void UpdatePhotos() throws Exception {
      _photos = new LinkedList<FlickrPhoto>();
      FlickrFacade facade = new FlickrFacade();
      for(PhotoSet set : facade.getPhotoSets(_name)) {
         for(Photo photo : facade.getPhotosInASet(set, 100)) {
            FlickrPhoto newPhoto = _manager
               .GetContainer(FlickrPhoto.class)
               .NewPersistentInstance();
            newPhoto.setName(photo.getTitle() +".jpg");
            newPhoto.setData(facade.getDataForPhoto(photo));
            _photos.add(newPhoto);
            _manager.GetContainer(FlickrPhoto.class)
               .ObjectChanged(newPhoto);
         }
      }
      _manager.GetContainer(FlickrUser.class).ObjectChanged(this);
   }
}
```

FIGURE 26.5
The FlickrUser Domain Object from FlickrFS.

```
@RootFolderObject
@DomainObject
@FolderObject(CanAdd=false, CanRemove=false)
public class Portfolio {
   private IDomainObjectContainerManager _manager;
   private List<Stock> _stocks = new LinkedList<Stock>();

   @NeedsContainerManager
   public void setContainerManager(IDomainObjectContainerManager
         manager) {
     _manager = manager;
   }

   @FolderObject(CanAdd=true, CanRemove=true)
   public List<Stock> getStocks() throws Exception {
      UpdateStockData();
      return _stocks;
   }

   private void UpdateStockData() throws Exception {
      String url = BuildURL();
      List<String> dataLines = getDataFromURL(url);
      for(Stock stock : _stocks) {
         String dataLine = null;
         for(String line : dataLines) {
            if(line.startsWith("\"" + stock.getTicker())) {
               dataLine = line;
                break;
            }
         }
         if(dataLine != null) {
            stock.UpdateData(dataLine);
         }
      }
   }

   private String BuildURL() { .... }
   private List<String> getDataFromURL(String url) { .... }

   @Executable
   public void AddAStock(String ticker) throws Exception {
      Stock stock = _manager.GetContainer(Stock.class)
         .NewPersistentInstance();
      stock.setTicker(tocker);
      _stocks.add(stock);
      _manager.GetContainer(Stock.class).ObjectChanged(stock);
      _manager.GetContainer(Portfolio.class).ObjectChanged(this);
   }
}
```

FIGURE 26.6
The Portfolio Class for the Stock Ticker Filesystem.

```
@DomainObject(CanWrite=false)
public class Stock {
   private String _ticker;
   private string _data;

   public Stock(String ticker) {
      _ticker = ticker;
   }

   @ProvidesName
   public String getTicker() { return _ticker; }

   public void UpdateData(String data) { _data = data; }

   public String getPrice() {
      return _data.split(",")[1];
   }

   public String getDate() {
      return _data.split(",")[2];
   }

   public String getTime() {
      return _data.split(",")[3];
   }
}
```

FIGURE 26.7
The Stock Class for the Stock Ticker Filesystem.

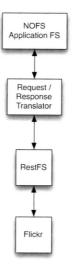

FIGURE 26.8
FlickrFS with both RestFS and NOFS.

26.2.4   Combining the Approaches: Using the RestFS Connector Layer in a NOFS Application Filesystem

It is also possible to use the filesystem as an application and the filesystem as a connector layer to form service compositions. The positive aspects of both approaches can be combined to derive the advantages of each system.

One of the important disadvantages of a filesystem as an application is that extra code must be added to the implementation to accommodate changing configurations and compositions of external resources. If this extra code is not present, then to realize changes, a filesystem must be unmounted, modified and then mounted again. With the filesystem as a connector layer, adding complex validation and advanced user interaction semantics is difficult. When both approaches are combined, these disadvantages are no longer present.

To demonstrate a possible use of both technologies, consider a photo service such as Flickr that you wish to represent as a filesystem. One possible way to construct a filesystem is to use both RestFS and an application filesystem built with NOFS. A domain model similar to the one in the FlickrFS example discussed earlier can be constructed. In this case, instead of using a library to interact with Flickr in the application filesystem, the application filesystem could use a RestFS resource file and a small script that translates requests and replies from the Flickr photo service into representations that conform to the domain model of the application filesystem.

This composition is more flexible to change than it would be implemented only as an application filesystem. For example, if an additional photo service

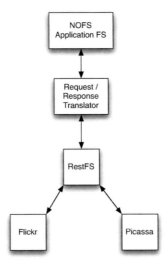

FIGURE 26.9
A Photo Filesystem Composed of Multiple Photo Services.

were added, it would involve creating a second resource file in RestFS that the
NOFS application filesystem would interact with. All that would be needed
is to implement a small script that could translate requests and replies from
the new web service into a form that could be consumed by the application
filesystem's domain model.

26.3   Building Application Filesystems with the Naked Object Filesystem (NOFS)

The capabilities, role and development process of the filesystem have evolved
throughout the years. Early on, filesystems were developed as tightly integrated operating system kernel components. Kernel mode filesystems require
a complex understanding of systems programming, systems programming languages, and the underlying operating system. There are fewer people who have
this skill set as object-oriented frameworks and languages are becoming more
and more popular. As user mode programs are more suited for loading and
launching programs dynamically, a kernel mode component often has to take
additional steps to support being unloadable or configurable at run time.
Also because operating system kernels cannot easily depend upon user mode
libraries, it is difficult to reuse software components within the operating system and by extension in filesystem implementations. Because of this, there is

much code that has already been developed using the patterns available and common to enterprise application frameworks that either cannot be used or are difficult to reuse in systems development. Two important advancements needed over kernel mode filesystems development are the ability to implement filesystems as user-mode programs and frameworks that allow enterprise development techniques and patterns to be applied to filesystems development. The answer to the user mode problem has been user-mode filesystem frameworks such as FUSE for UNIX-like operating systems and Dokan for the Windows operating systems. Our answer to provide an enterprise-patterns-friendly framework is the NOFS framework.

26.3.1   An Explanation of Naked Objects

Naked Objects [572] is the term used to describe the design philosophy of using plain object-oriented domain models to build entire applications. In the realm of desktop applications, Naked Object frameworks remove the concern of the developer in implementing user interfaces, model-view-controller patterns, and persistence layers. These components are generated for the domain model by the Naked Objects framework automatically, either through the use of reflection or through additional metadata supplied with the domain model.

A characteristic feature of Naked Object frameworks is that they present an object-oriented user interface. Applications where the user is treated more as a problem solver than as a process follower benefit from an object oriented user interface [572, p41]. For many applications, processes are very important and an object-oriented user interface is not the best fit. We believe that the interface presented to the programmer and to the user of a filesystem is also object-oriented. In a filesystem, the components are not exposed to the user to facilitate the moving, reading, writing, creation, or deletion of files and folders. These actions are accomplished with external programs and references to the actual objects as command line parameters. The user interaction with filesystems is a noun-verb style of interaction and not a verb-noun interaction, which is more common with typical desktop applications. Like the Naked Object user interfaces, filesystems "present the user with a set of tools with which to operate and allow a business system to be designed that does not dictate the users sequence of actions" [572, p41].

26.3.2   The Naked Object Filesystem (NOFS)

There are three important contributions made by the NOFS framework. The first is that NOFS demonstrates the filesystem can be used as an object-oriented user interface in a Naked Objects framework and that the Naked Objects design principle can be applied successfully to filesystems development. The second contribution is that NOFS inverts and simplifies the normal filesystem development contract. In FUSE and operating system kernels, there are a series of functions to implement and data structures to work with.

With the NOFS framework, a domain model is inspected to produce a filesystem user interface. Domain models for NOFS do not implement filesystem contracts or work with filesystem structures. Instead, they are described with metadata that is used by NOFS to allow the domain model to interact with the FUSE filesystem framework. In this way, NOFS follows the dependency inversion principle in that the higher level domain model does not depend upon the lower level filesystem model. The third contribution made by the NOFS framework is that by providing an object-oriented framework to develop filesystems, we allow developers who are unfamiliar with systems or UNIX programming to more easily and rapidly implement experimental or lightweight filesystems. With this object-oriented framework, it becomes easier to unit test a filesystem implementation because details of the operating system do not need to be stubbed or mocked out; only the domain model needs to be verified.

26.3.3 Implementing a Domain Model with NOFS

Here we will explore developing a domain model with NOFS. We will explore three domain models: an address book domain model that was developed for presentation purposes, a Flickr domain model for manipulating photos on the Flickr photo service, and a stock ticker tracking filesystem for Yahoo! Finance.

26.3.3.1 Implementing Files and Folders in NOFS

In NOFS, files are modeled as plain classes that are described with metadata. The methods on the class are not constrained to any specific interface but are used to model the structure of the data in a file. There are two ways for classes to expose their data: through translation of the return values of public methods to structured XML files or by defining the structure of these files by implementing an interface with read and write methods.

In the example in Figure 26.10, the class Contact marks itself as a file object by using the @DomainObject Java annotation. The class also tells NOFS that it manages its own file name with the @ProvidesName annotation on the getName accessor and the setName mutator methods. The persistence mechanism of NOFS is injected upon construction of the Contact class through the setContainer method, which is marked by the @NeedsContainer method. An example representation of the Contact class as a file in the NOFS filesystem is as follows in Figure 26.11.

In this example the class FlickrPhoto (Figure 26.4) marks itself as a file object by using the @DomainObject Java annotation. It tells NOFS that it is immutable by setting the CanWrite member of the DomainObject annotation to false. IFlickrPhoto's responsibility is to model a graphical image from the Flickr photo sharing website. Since it is convenient to expose to the filesystem these photos as an image file and not as an XML file, FlickrPhoto provides

```
@DomainObject
public class Contact {
   private String _name;
   private String _phoneNumber;
   private IDomainObjectContainer<Contact> _container;

   @ProvidesName
   public String getName() { return _name; }

   @ProvidesName
   public void setName(String name) { _name = name; }

   public String getPhoneNumber() { return _phoneNumber; }
   public void setPhoneNumber(String value) {
      _phoneNumber = value;
   }

   @NeedsContainer
   public void setContainer(IDomainObjectContainer<Contact>
         container) {
      _container = container;
   }
}
```

FIGURE 26.10
The Contact NOFS Domain Object.

```
<?xml version="1.0"?>
<Contact>
   <PhoneNumber>555-5555</PhoneNumber>
<Contact>
```

FIGURE 26.11
Representation on the Filesystem of the Contact Domain Object.

```
@DomainObject
@FolderObject(CanAdd=true, CanRemove=true)
public class Category extends LinkedList<Contact> {
   private String _name;

   @ProvidesName
   public void setName(String name) { _name = name; }

   @ProvidesName
   public String getName() { return _name; }
}
```

FIGURE 26.12
The Category NOFS Domain Object.

read and write methods as defined by the IProvidesUnstructuredData NOFS interface.

In the example in Figure 26.6, the class Portfolio marks itself as a folder object by using the @DomainObject and the @FolderObject Java annotations. The FolderObject annotation sets CanAdd and CanRemove to false, to tell NOFS that the user of the filesystem cannot add or remove files from the folder. The Portfolio class exposes two objects to NOFS, a folder called Stocks through the getStocks() method and an executable script through the AddAStock method. NOFS can tell that getStocks() is a folder because its return type is a collection and because of the FolderObject annotation on the method declaration. NOFS can tell that the AddAStock method is to be exposed as an executable script because of the Executable annotation on the method declaration. The script that will appear in the Portfolio object's folder will be an automatically generated Perl script that will accept one argument and pass it back to NOFS, which will in turn pass it to the correct domain object instance based upon path. In this way, NOFS domain objects can expose additional executable behavior to the filesystem interface.

Another way to implement a folder is through extending a collection type such as LinkedList. The Category class in Figure 26.12, which is a part of the address-book filesystem, takes advantage of this approach. Instead of statically defining the components of a folder as was done in the Portfolio example, the Category folder's components will be defined by what is present in the collection.

26.3.4   Architecture of NOFS

There are two important aspects to the architecture of NOFS. The first is its place and role in the filesystem architecture and the second is how domain objects are mapped to FUSE calls. First, the overall architecture of FUSE is

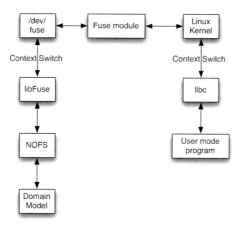

FIGURE 26.13
The Relationship between NOFS, FUSE, and the Linux Kernel.

not changed by NOFS. NOFS exists as an additional layer on top of FUSE.
A diagram of this relationship is available in Figure 26.13.

The existing context switches between user-mode programs with the kernel
and between filesystem implementations with FUSE still exist with NOFS.
No new context switches are created by the NOFS framework. The reader
is encouraged to consult literature and documentation on FUSE to explore
additional details of FUSE and its implementations (see also 26.1.4 above).

The way domain models are mapped to FUSE calls can be split into two
important parts: how paths are translated to domain objects and how domain
objects are translated to different file object types.

Domain objects are translated to files, folders, root-folders, and executable
scripts through the use of Java annotations. Depending upon the annotation,
classes or methods are scanned to see if there are matching annotations. If
a class or method is marked as a file, then that class instance or the return
value of that method is exposed as a file on the filesystem. The same is true
of folders. If a class is marked as a folder and if it is also a list, then the class
is exposed as a folder and the contained objects in the list are exposed as
children of that folder. If the class is marked as a folder and is not also a list,
then the member methods of the class are exposed as children of the folder. If a
particular method is encountered and marked as executable, NOFS generates
a Perl script that accepts as arguments a list matching the parameters of the
method. Executable methods will be explored in more detail soon.

Paths are translated with the algorithms in Figures 26.14 and 26.15. The
algorithm basically finds the root of the filesystem by searching for an object
instance of type root and then traverses the path from that instance until

```
translate_path(path) {
   current = find_root();
   for-each(segment in path) {
      if(current IsA folder) {
         if(current IsA list) {
            current = current[segment];
         } else if(current HasA member whose name matches
               segment) {
            current = current.members[segment];
         } else {
            raise exception "invalid path";
         }
      } else {
         raise exception "invalid path";
      }
   }
   return current;
}
```

FIGURE 26.14
The NOFS Path Translation Algorithm.

```
find_root() {
   List roots = new List();
   for-each(instance in all_instances) {
      if(instance IsA root-folder) {
         roots.add(instance);
      }
   }
   if(roots.count() == 0)  {
      raise exception "no roots found";
   } else if(roots.count() > 1) {
      raise exception "more than one root found";
   }
   return roots[0];
}
```

FIGURE 26.15
The NOFS Root Discovery Algorithm.

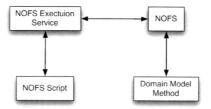

FIGURE 26.16
The Communication Path for Executable Scripts in NOFS.

```
translate_arguments(arg_list, method) {
   for(int i = 0; i < arg_list.length; i++) {
      if(method.parameters[i] IsA NOFS-domain-object) {
         args_list[i] = translate_path(arg_list[i]);
      }
   }
}
```

FIGURE 26.17
The NOFS Argument Translation Algorithm.

it encounters a mismatch or runs out of segments in the path and returns a matching object.

Additional path and type translation is involved in methods that are exposed as executable scripts in NOFS. If a method has as parameters just primitive or string types, then NOFS has no additional translation work to perform and just passes values as they are to a method from the script. If a method parameter is of one of the domain model's types, then the script will accept a path as a valid argument and NOFS will translate the path to an object reference that is then passed to the method (see Figure 26.17). In this way, it is possible to pass by value or by reference to methods on NOFS domain classes.

With path to object translation, filesystem calls like getdir(), mkdir(), mknod(), unlink() and similar calls map pretty well into path translation and object creation and deletion actions. Next, we will discuss how calls such as read(), write(), open(), and close() work.

In NOFS, there are three ways that a file object's data is managed. The first way is if the file happens to be an executable script. If a method is determined to be an executable script, NOFS will generate Perl code to wrap a call back into NOFS and make file that the Perl code is placed in read-only. The second way data is managed is through the IProvidesUnstructuredData interface. This interface was mentioned earlier in the FlickrPhoto example. If NOFS encounters a file object that implements this interface, it will pass read and

```
represent_as_xml(object) {
   for-each(member in object.class_definition) {
       if(member IsA primitive) {
          emit element with value of primitive;
       } else {
          represent_as_xml(member);
       }
   }
}
```

FIGURE 26.18
The NOFS XML Serialization Algorithm.

write calls directly to the object. The final way data is managed is if the domain object exposes public members. In this case, NOFS will examine the members and translate all primitive members into XML elements. If a non primitive type is encountered, an element will be emitted and it will also be serialized into XML. The algorithm is available in Figure 26.18.

In the case of XML files being written back to, all writes are cached by NOFS until the file handle is closed. When the file handle is closed, NOFS will perform a similar algorithm as represent_as_xml except to deserialize the XML back into the domain object. If there is a mismatch in the XML structure with respect to the domain object or if the deserialization process causes the domain object to throw an exception, the change to the domain is rolled back entirely and the contents of the XML file are reverted to their state before any write occurred. The cache management algorithm can be found in Figure 26.19.

The final set of calls mapped to FUSE by NOFS are metadata calls such as getxattr, getattr, chown, chmod, and other related calls. There are two ways that these are managed. The first way is if a method has any of the Provides-GID, ProvidesUID, ProvidesMode, ProvidesLastAccessTime, ProvidesLast-ModifiedTime, or ProvidesCreateTime annotations. For any class that has methods with these annotation, NOFS assumes that the domain object maintains this metadata. For each case where one of these annotations is not encountered, NOFS will provide a default implementation and store appropriate metadata in a small db4o database for each instance of a domain object.

It is sometimes useful for domain models to manage this additional metadata in a non-default way. One important reason is if the data is a legitimate part of the domain model. One good example would be a web service that provides online document editing. The domain object that models a document should also retrieve attributes like creation, modification, and access times from the server. For other domain models, such as the stock ticker domain model presented earlier, this information is less important to the domain model and

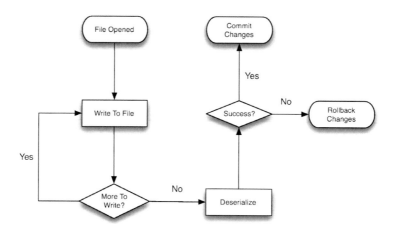

FIGURE 26.19
The NOFS Cache and Serialization Relationship.

can be adequately handled by the NOFS default implementation. These two possibilities allow the creator of the domain model to model only attributes that they are concerned with and nothing more.

The domain object persistence mechanism used in NOFS is straightforward and natural in the way it maps annotated class definitions to XML elements at run time. A thorough evaluation of this approach and its alternatives is still needed. One alternative is our earlier work on simple XML data bindings and linearized external representations of XML data [133]. Other choices include more complex, schema-based XML data binding frameworks such as JAXB [268] and XStream [767], as well as non-XML formats such as JSON [209]. In addition, we plan to allow domain classes in future versions of NOFS to choose alternate representations through their own serializers or XSLT transformations.

26.4   Architecture and Details of RestFS

Our work on RestFS was inspired by two other bodies of work: Plan 9's 9P protocol and netfs [580], and Representational State Transfer or REST [271]. While exploring REST, we realized that the GET, PUT, POST, and DELETE HTTP methods mapped well into filesystem operations and that there were a few ways that we might map REST-ful services onto the filesystem. Another important observation that we made at the time is how other forms of interprocess communication and especially sockets have been the basis for composing programs and services. We felt after our exploration of layered filesystems research with the OLFS filesystem that the filesystem held the possibility to mediate the composition of web services. With these observations in hand, and with the NOFS filesystem framework, we set about developing a filesystem to support communication with and composition of web services.

In Plan 9, network communication is not performed through the use of system calls like accept, connect, listen, send or recv. Network communications are performed through file operations in netfs under a special folder '/net' in the Plan 9 filesystem. In addition to folders separating types of network connections into UDP and TCP, there are two types of folders in netfs: connection / configuration files and stream files. Connection / configuration files contained details about IP addresses, port numbers, and socket options. Once fully configured, it is possible to read from and write to the special stream files in netfs to send and receive data from a remote computer.

26.4.1   RestFS's Approach

The use of files for networking and the separation of files into configuration and streams offer very important advantages over the family of calls used in UNIX and other operating systems for networking. The first advantage is that no additional system calls other than the ones necessary for filesystem interaction are needed to work with the network. Calls like connect, listen, send, recv, accept, and others are not necessary when the network can be managed through the filesystem. The other important advantage is in the separation of responsibility between the files. With the separation, it is possible for one process to manage configuration of the network connection while another process is responsible for reading and writing to the connection as if it were a normal file. In this way, software that is capable of working with just file I/O calls does not need to be extended to support networking code; it need only be supplemented with some prior configuration. Another important advantage of using the filesystem for network communication is that it allows for network connections to be named in a namespace that has a longer lifetime than programs that may take advantage of a network connection. For example, a program may read from and write to a network file and work correctly for some time. If that program crashes, it can be re-launched and resume working with the network file without having to re-establish any connections. This capability also allows the programs on either end point of the connection to change over time without resetting the connection.

26.4.1.1   Configuration Files in RestFS

In RestFS, when a file is created, it is created as a pair consisting of a resource and a configuration file that are bound to each other. For example, if a file called "GoogleSearch" is created, then a companion configuration file called ".GoogleSearch" will also be created in skeleton form.

```xml
<?xml version="1.0" encoding="UTF-8"?>
<RestfulSetting>
  <FsMethod>utime</FsMethod>
  <WebMethod>get</WebMethod>
  <FormName></FormName>
  <Resource>ajax/services/search/web?v=1.0&amp;q=Brett%Favre
  </Resource>
  <Host>ajax.googleapis.com</Host>
  <Port>80</Port>
  <OAuthTokenPath></OAuthTokenPath>
</RestfulSetting>
```

FIGURE 26.20
An Example RestFS Configuration File for a Google Search.

Next, this skeleton is populated manually to contact a specific web service. In the example shown in 26.20, the resource file has been configured to contact the Google search service and perform a GET HTTP request when the utime filesystem call is performed on the GoogleSearch file. When this occurs, RestFS will make a call to the web service and place the results in the resource file.
The Web Application Description Language (WADL) [330] has been proposed as a REST-ful counterpart to the Web Service Definition Language (WSDL) [183]. We are currently investigating ways to use WADL in conjunction with RestFS, in particular, to populate RestFS configuration files from WADL service descriptions.

26.4.1.2   Implementation of Configuration Files in RestFS

Since RestFS is implemented as a NOFS application filesystem, implementing files that are represented as XML is straightforward. The individual elements are implemented as accessors and mutators in a Java class called RestfulSetting in Figure 26.21. These settings objects are managed by the resource files that we will discuss shortly.

26.4.1.3   Resource Files in RestFS

As stated before, resource files in RestFS contain the state of a current request or response with a web service. Resource files can be configured to be triggered to respond to web service calls upon being opened, before deletion, when the resource file's timestamp is updated, before the resource file is read from, and after the resource file has been written to. This triggering capability is accomplished through the implementation of the NOFS IListensToEvents interface. With this interface, the RestFS resource file is notified by NOFS when actual calls to FUSE are encountered. Once a triggering call is encountered, the algorithm in Figure 26.22 is run.
When the triggering call is made on the resource file, RestFS will check

```
@DomainObject
public class RestfulSetting extends BaseFileObject {
   private String _method;
   public String getMethod() { return _method; }
   public void setMethod(String value) { _method = value; }

   private String _formName;
   public String getFormName() { return _formName; }
   public void setFormName(String value) { _formName = value; }

   private String _port = "";
   public String getPort() { return _port; }
   public void setPort(String value) { _port = value; }

   private String _host = "";
   public String getHost() { return _host; }
   public void setHost(String value) { _host = value; }

   private String _resource = "";
   public String getResource() { return _resource; }
   public void setResource(String value) { _resource = value; }

   private String _oauthTokenPath = "";
   public String getOAuthTokenPath() { return _oauthTokenPath; }
   public void setOAuthTokenPath(String value) {
      _oauthTokenPath = value;
   }
}
```

FIGURE 26.21
The RestfulSetting NOFS Domain Object.

```
RespondToEvent(event_type, settings, current_file_data) {
   if(settings.triggering_call == event_type) {
      response = IssueWebRequest(settings.URI,
         settings.WebMethod, current_file_data);
      SetCurrentFileData(response);
   }
}
```

FIGURE 26.22
RestFS Resource File Triggering Algorithm.

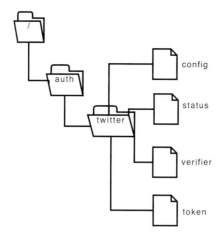

FIGURE 26.23
An Example of an OAuth Configuration in RestFS.

the current contents of the file. If the file contains a JSON object, the object will be parsed and passed as arguments to the web service call. For example, the JSON object {"description" : "student", "name": "Joe"} would translate to the URI http://host/service?description=student &name=joe.

26.4.1.4   Authentication in RestFS

As many REST-ful web services support the OAuth authentication model, we decided to add special OAuth file and folder types to assist in establishing authorization for web services. In RestFS, there is one special folder '/auth' in the root of every mounted RestFS filesystem. When a folder is created in the '/auth' folder, a config, status, verifier, and token file are created. The config file takes the OAuth API-Key, secret, and set of URLs to communicate with to establish an authorization token. These fields are typically provided by the service provider for a REST-ful web service.

Once all of the appropriate fields are written to the configuration file, RestFS will contact the web service to obtain authorization. Depending upon the implementation there are a few possibilities. If the service requires human interaction to accept a PIN or pass a CAPTCHA test, the URL for that step will be written to the 'status' file. If the service provides a PIN, it should be written to the 'verifier' file. Once this process is complete, the 'token' file will be populated with the OAuth access and request tokens for use in further communications. An example of this token file can be seen in Figure 26.25.

Once authorization is successful, the token file can be referred to in any configuration file by path reference in the OAuthTokenPath element. If the configuration file contains a valid token file, RestFS will handle any call to the

```
<?xml version="1.0" encoding="UTF-8"?>
<OAuthConfigFile>
  <Key>asdf3244dsf</Key>
  <AccessTokenURL>https://api.twitter.com/oauth/access_token
  </AccessTokenURL>
  <UserAuthURL>https://api.twitter.com/auth/authorize
  </UserAuthURL>
  <RequestTokenURL>https://api.twitter.com/oauth/request_token
  </RequestTokenURL>
  <Secret>147sdfkek</Secret>
</OAuthConfigFile>
```

FIGURE 26.24
An Example OAuth Configuration File for Twitter.

```
<OAuthTokenFile>
  <AccessToken>2534534asdf2348</AccessToken>
  <RequestToken>aql2343</RequestToken>
  <TokenSecret>adfjds124522</TokenSecret>
</OAuthTokenFile>
```

FIGURE 26.25
An Example OAuth Token File.

resource file using the appropriate OAuth token. The user of the resource file then, does not need to worry about authentication any further. This process is summarized by figure 26.26.

26.4.1.5   Putting it All Together

With these three types of files: authentication, configuration, and resource, it is possible to connect to and work with a web service through filesystem calls. If several resource files are created, it is possible to work with several web services and to send multiple requests and compose multiple responses locally using UNIX command line tools or through small programs.

26.5   Summary

With RestFS and NOFS, we have demonstrated how web services can be abstracted and composed in an arbitrarily deep hierarchy through the implementation and use of filesystems. We have shown how the filesystem can be used as a connector layer to translate filesystem calls into web service calls and

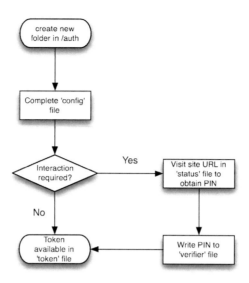

FIGURE 26.26
The RestFS Authentication Process.

how this can allow for local and external composition of web services. We have also shown how application filesystems can be used to provide a user-friendly interface for web services to provide validation and more complex structure. Finally, we have shown how the two approaches can be combined to provide effective representations of web services through the filesystem interface.

In our deeper exploration of NOFS, we discussed how the Naked Objects design principles can be used to build filesystems and how the dependency inversion approach simplifies filesystem design. We also explored several example filesystems and explained how NOFS handles translating requests from FUSE to operations on a domain model.

While exploring RestFS, we discussed the challenges of translating web service authentication to the filesystem interface, how configuration and resource files are separated, and how best to use RestFS to expose web services through external programs or scripts.

27

Aneka Cloud Application Platform and Its Integration with Windows Azure

Yi Wei

Manjrasoft Pty. Ltd., Melbourne, Victoria, Australia

Karthik Sukumar

Manjrasoft Pty. Ltd., Melbourne, Victoria, Australia

Christian Vecchiola

Cloud Computing and Distributed Systems (CLOUDS) Laboratory, Department of Computer Science and Software Engineering, The University of Melbourne, Australia

Dileban Karunamoorthy

Cloud Computing and Distributed Systems (CLOUDS) Laboratory, Department of Computer Science and Software Engineering, The University of Melbourne, Australia

Rajkumar Buyya

Cloud Computing and Distributed Systems (CLOUDS) Laboratory, Department of Computer Science and Software Engineering, The University of Melbourne, Australia

CONTENTS

Aneka is an Application Platform-as-a-Service (Aneka PaaS) for Cloud Computing. It acts as a framework for building customized applications and deploying them on either public or private Clouds. One of the key features of Aneka is its support for provisioning resources on different public Cloud providers such as Amazon EC2, Windows Azure and GoGrid. In this chapter, we will present the Aneka platform and its integration with one of the public Cloud infrastructures, Windows Azure, which enables the usage of Windows Azure Compute Service as a resource provider of Aneka PaaS. The integration of the two platforms allows users to leverage the power of Windows Azure Platform for Aneka Cloud Computing, employing a large number of compute instances to run their applications in parallel. Furthermore, customers of the

Windows Azure platform can benefit from the integration with Aneka PaaS by embracing the advanced features of Aneka in terms of multiple programming models, scheduling and management services, application execution services, accounting and pricing services, and dynamic provisioning services. Finally, in addition to the Windows Azure Platform, Aneka PaaS has integrated with other public Cloud platforms such as Amazon EC2 and GoGrid, and virtual machine management platforms such as Xen Server. The new support of provisioning resources on Windows Azure once again proves the adaptability, extensibility, and flexibility of Aneka.

27.1 Introduction

Current industries have seen Clouds [96] [156] as an economic incentive for expanding their IT infrastructure with less total cost of ownership (TCO) and higher return of investment (ROI). By supporting virtualization and dynamic provisioning of resources on demand, the Cloud Computing paradigm allows any business, from small and medium enterprise (SMEs) to large organizations, to more wisely and securely plan their IT expenditures. They will able to respond rapidly to variations in the market demand for their Cloud services. IT cost savings are realized by means of the provision of IT "subscription-oriented" infrastructure and services on a pay-as-you-go-basis. There is no more need to invest in redundant and highly fault tolerant hardware or expensive software systems, which will lose their value before they will be paid for by the generated revenue. Cloud Computing now allows paying for what the business needs at the present time and releasing it when the resources are no longer needed. The practice of renting IT infrastructures and services has become so appealing that it is not only leveraged to integrate additional resources and elastically scale existing software systems into hybrid Clouds, but also to redesign the existing IT infrastructure in order to optimize the usage of the internal IT, thus leading to the birth of private Clouds. To effectively and efficiently harness Cloud Computing, service providers and application developers need to deal with several challenges, which include: application programming models, resource management and monitoring, cost-aware provisioning, application scheduling, and energy efficient resource utilization. The Aneka Cloud Application platform, together with other virtualization and Cloud Computing technologies, aims to address these challenges and to simplify the design and deployment of Cloud Computing systems.

Aneka is a .NET-based application development Platform-as-a-Service (PaaS), which offers a runtime environment and a set of APIs that enable developers to build customized applications by using multiple programming models such as Task Programming, Thread Programming and MapReduce Program-

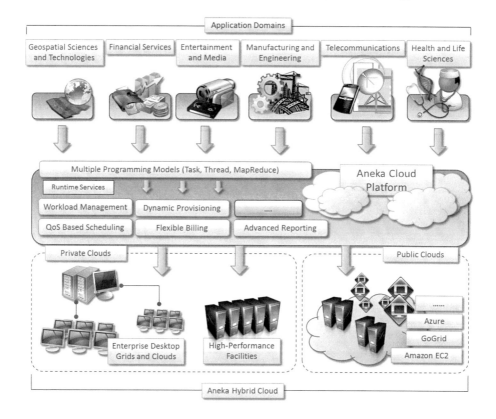

FIGURE 27.1
Aneka Cloud Application Platform.

ming, which can leverage the compute resources on either public or private Clouds [725]. Moreover, Aneka provides a number of services that allow users to control, auto-scale, reserve, monitor and bill users for the resources used by their applications. One of key characteristics of Aneka PaaS is to support provisioning of resources on public Clouds such as Windows Azure, Amazon EC2, and GoGrid, while also harnessing private Cloud resources ranging from desktops and clusters, to virtual datacenters when needed to boost the performance of applications, as shown in Figure 27.1. Aneka has successfully been used in several industry segments and application scenarios to meet their rapidly growing computing demand.

In this chapter, we will introduce Aneka Cloud Application Platform (Aneka PaaS) and describe its integration with public Cloud platforms particularly focusing on the Windows Azure Platform. We will show in detail, how an adaptable, extensible and flexible Cloud platform can help enhance the performance and efficiency of applications by harnessing resources from private, public or hybrid Clouds with minimal programming effort. The Windows Azure Platform is a Cloud Services Platform offered by Microsoft [467]. Our goal is to integrate the Aneka PaaS with Windows Azure Platform so that Aneka PaaS can leverage the computing resource offered by Windows Azure Platform. The integration supports two types of deployments. In the first case, our objective was to deploy Aneka Worker Containers as instances of Windows Azure Worker Role while the Aneka Master Container runs locally on-premises, enabling users of Aneka PaaS to use the computing resources offered by Windows Azure Platform for application execution. And in the second case, the entire Aneka Cloud is deployed on Windows Azure so that Aneka users do not have to build or provision any computing resource to run Aneka PaaS. This chapter reports the design and implementation of the deployment of Aneka containers on Windows Azure Worker Role and the integration of two platforms.

The remainder of the chapter is structured as follows: in Section 27.2, we present the architecture of Aneka PaaS, provide an overview of the Windows Azure Platform and Windows Azure Service Architecture, and list the advantages of integrating the two platforms along with the limitations and challenges we faced. Section 27.3 demonstrates our design in detail on how to integrate the Aneka PaaS with Windows Azure Platform. Next, we will discuss the implementation of the design in Section 27.4. Section 27.5 presents the experimental results of executing applications on the two integrated environments. In Section 27.6 and 27.7, we list related work and sample applications of Aneka. Finally, we present the conclusions and future directions.

27.2 Background

In this section, we present the architecture of Aneka PaaS, and then depict the overall view on Windows Azure Platform and Windows Azure Service Architecture. We also discuss the advantages brought by the integration, along with the limitation and challenges faced.

27.2.1 Overview of Aneka Cloud Application Platform

Figure 27.2 shows the basic architecture of Aneka. The system includes four key components, including Aneka Master, Aneka Worker, Aneka Management Console, and Aneka Client Libraries [725].

The Master and Worker are both software Containers which represent the basic deployment unit of Aneka based Clouds. Aneka Containers host different kinds of services depending on their role. For instance, in addition to mandatory services, the Master runs the Scheduling, Accounting, Reporting, Reservation, Provisioning, and Storage services, while the Worker nodes run Execution services. For scalability reasons, some of these services can be hosted

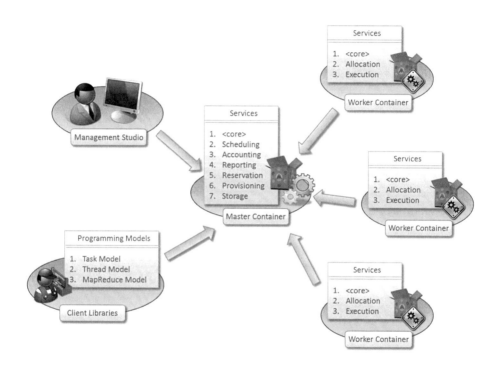

FIGURE 27.2
The Basic Architecture of Aneka PaaS.

on separate Containers with different roles. For example, it is ideal to deploy a Storage Container for hosting the Storage service, which is responsible for managing the storage and transfer of files within the Aneka Cloud. The Master Container is responsible for managing the entire Aneka Cloud, coordinating the execution of applications by dispatching the collection of work units to the compute nodes, while the Worker Container is in charge of the executing the work units, monitoring the execution, and collecting and forwarding the results.

The Management Studio and client libraries help in managing the Aneka Cloud and developing applications that utilize resources on Aneka Cloud. The Management Studio is an administrative console that is used to configure Aneka Clouds; install, start or stop Containers; set up user accounts and permissions for accessing Cloud resources; and access monitoring and billing information. The Aneka client libraries, are Application Programming Interfaces (APIs) used to develop applications which can be executed on the Aneka Cloud. Three different kinds of Cloud programming models are available for the Aneka PaaS, to cover different application scenarios: *Task Programming*, *Thread Programming*, and *MapReduce Programming* These models represent common abstractions in distributed and parallel computing and provide developers with familiar abstractions to design and implement applications.

1. **Fast and Simple: Task Programming Model.** It provides developers with the ability of expressing applications as a collection of independent tasks. Each task can perform different operations, or the same operation on different data, and can be executed in any order by the runtime environment. This is a scenario in which many scientific applications fit in and a very popular model for Grid Computing. Also, Task programming allows the parallelization of legacy applications on the Cloud.

2. **Concurrent Applications: Thread Programming Model.** It offers developers the capability of running multithreaded applications on the Aneka Cloud. The main abstraction of this model is the concept of thread, which mimics the semantics of the common local thread but is executed remotely in a distributed environment. This model offers finer control on the execution of the individual components (threads) of an application but requires more management when compared to Task Programming, which is based on a *"submit and forget"* pattern. The Aneka Thread supports almost all of the operations available for traditional local threads. More specifically, an Aneka thread has been designed to mirror the interface of the *System.Threading.Thread* .NET class, so that developers can easily move existing multi-threaded applications to the Aneka platform with minimal changes. Ideally, applications can be transparently ported to Aneka just by substituting local threads with Aneka Threads and introducing minimal changes to the code. This

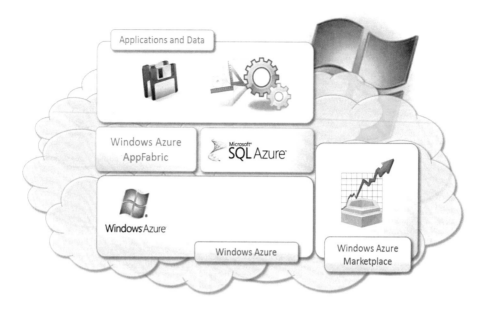

FIGURE 27.3
The Components of Windows Azure Platform.

model covers all the application scenarios of the Task Programming and solves the additional challenges of providing a distributed runtime environment for local multi-threaded applications.

3. **Data Intensive Applications: MapReduce Programing Model.** [396] It is an implementation of the MapReduce model proposed by Google [226], in .NET on the Aneka platform. MapReduce has been designed to process huge quantities of data by using simple operations that extract useful information from a dataset (the *map* function) and aggregate this information together (the *reduce* function) to produce the final results. Developers provide the logic for these two operations and the dataset, and Aneka will do the rest, making the results accessible when the application is completed.

27.2.2 Overview of Windows Azure Platform

Generally speaking, Windows Azure Platform is a Cloud platform which provides a wide range of Internet Services [177]. Currently, it involves four components (Figure 27.3). They are *Windows Azure*, *SQL Azure*, *Windows Azure AppFabric*, and *Windows Azure Market Place* respectively.

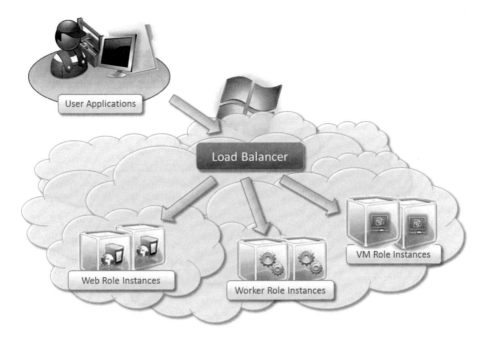

FIGURE 27.4
Windows Azure Service Architecture.

Windows Azure, which we will introduce in detail in Section 27.2.3, is a Windows based Cloud services operating system providing users with on-demand compute service for running applications, and storage services for storing data in Microsoft data centers. The second component, *SQL Azure* offers a SQL Server environment in the Cloud, whose features include supporting Transact-SQL and support for the synchronization of relational data across SQL Azure and an on-premises SQL Server. *Windows Azure AppFabric* is a Cloud-based infrastructure for connecting Cloud and on-premise applications, which are accessed through HTTP REST API. The newly born *Windows Azure Marketplace* is an online service for making transactions on Cloud-based data and Windows Azure Applications.

27.2.3 Overview of Windows Azure Service Architecture

Windows Azure is a PaaS (Platform-as-a-Service) solution for developing Cloud applications. Therefore, it does not provide any IaaS supporting, thus restricting users from direct access with administrative privileges to underlying virtual infrastructure. Users can only use the Web APIs exposed by Windows Azure to configure and use Windows Azure services [444].
A role on Windows Azure refers to a discrete scalable component built with

managed code. Windows Azure currently supports three kinds of roles [178], as shown in Figure 27.4.

1. **Web Role:** a Web role is a role that is customized for running Web application as is supported by IIS 7.

2. **Worker Role:** a worker role is a role that is useful for generalized Windows application development. It is designed to run a variety of Windows-based code.

3. **VM Role:** a virtual machine role is a role that runs a user-provided Windows Server 2008 R2 image.

A Windows Azure service must include at least one role of either type, but may consist of any number of Web roles, worker roles and VM roles. Furthermore, we can launch any number of instances of a particular role. Each instance will be run in an independent Virtual Machine(VM) and share the same binary code and configuration file of the role.

In terms of communication support, there are two types of endpoints that can be defined: input and internal. Input endpoints are those are exposed to the Internet; internal endpoints are used for communication inside the applications within the Azure environment. A Web role can define a single HTTP endpoint and a single HTTPS endpoint for external users, while a Worker Role and a VM role may assign up to five internal or external endpoints using HTTP, HTTPS or TCP. There exists a built-in load balancer on top of each external endpoint which is used to spread incoming requests across the instances of the given role. Besides, all the role instances can make outbound connections to Internet resources via HTTP, HTTPS or TCP.

Under this circumstance, we can deploy Aneka Container as instances of Windows Azure Worker Role which gets access to resources on the Windows Azure environment via the *Windows Azure Managed Library*.

27.2.4 Advantages of Integration of Two Platforms

Inevitably, the integrated Cloud environment will combine features from the two platforms together, enabling the users to leverage the best of both platforms such as access to cheap resources, easy programming, and management of Cloud Computing services.

27.2.4.1 Features from Windows Azure

For the users of Aneka Platform, the integration of the Aneka PaaS and Windows Azure resources means they do not have to build or provision the infrastructure needed for Aneka Cloud. They can launch any number of instances on the Windows Azure Cloud Platform to run their application in parallel to gain more efficiency.

FIGURE 27.5
Multiple Programming Models of the Aneka PaaS.

27.2.4.2   Features from Aneka Cloud Application Platform

For the users of Windows Azure Application, the integration of Aneka PaaS and Windows Azure Platform allows them to embrace the advanced features from Aneka PaaS:

1. **Multiple Programming Models.** As discussed in Section 27.2.1, the Aneka PaaS provides users with three different kinds of Cloud programming models, which involve *Task Programming*, *Thread Programming*, and *MapReduce Programming* to cover different application scenarios, dramatically decreasing the time needed in developing Cloud-aware applications, as shown in Figure 27.5.

2. **Scheduling and Management Services.** The Aneka PaaS Scheduling Service can dispatch the collection of jobs that compose an Aneka Application to the compute nodes in a completely transparent manner. The users do not need to take care of the scheduling and the management of the application execution.

3. **Execution Services.** The Aneka PaaS Execution Services can perform the execution of distributed application and collect the results on the Aneka Worker Container runtime environment.

4. **Accounting and Pricing Services.** Accounting and Pricing services of the Aneka PaaS enable billing the final customer for using the Cloud by keeping track of the application running and providing flexible pricing strategies that are of benefit to both the final users of the application and the service providers.

5. **Dynamic Provisioning Services.** In the current pricing model for Windows Azure, customers will be charged at an hourly rate depending on the size of the compute instance. Thus it makes sense to dynamically add instances to a deployment at runtime according to the load and requirement of the application. Similarly instances can be dynamically decreased or the entire deployment can be deleted when not being actively used to avoid charges. One of the key fea-

tures of Aneka is its support for dynamic provisioning which can be used to leverage resources dynamically for scaling up and down Aneka Clouds, controlling the lifetime of virtual nodes.

27.2.5   Limitations for the Integration

Although the integration of two platforms will generate numerous benefits for both Aneka users and Windows Azure customers, running Aneka Container as instances of Windows Azure Worker Role has some limitations.

The current version of Windows Azure does not provide administrative privileges on Windows Azure Worker Role instances. Deployments are prepared using the *Windows Azure Managed Library*, and the prepared Windows Azure Service Package is uploaded and run.

Under these circumstances, we cannot use the Aneka Management Studio to install Aneka Daemons and Aneka Containers on Windows Azure VMs directly. Furthermore, other third party software that is needed on the Worker nodes such as PovRay and Maya, cannot be run on Windows Azure Worker Role instances because of the need for administrative privileges. This limits the task execution services that Azure Aneka Worker offers to XCopy deployment applications.

27.2.6   Challenges for the Integration

Due to the access limitations and service architecture of Windows Azure, we encountered some implementation issues that required changes to some parts of the design and implementation of the Aneka framework.

27.2.6.1   Administration Privileges

The Azure applications in both Web role and Worker role instances do not have administrative privileges and does not have write access to files under the "E:\approot\" where the application code is deployed. On possible solution is to use *LocalResource* to define and use the local resource of Windows Azure VM disk.

Technically speaking, we need to dynamically change the path of files which are to be written to the local file system, to the path under the *RootPath* Property returned by the *LocalResource* object at runtime.

27.2.6.2   Routing in Windows Azure

Each Windows Azure Worker Role can define up to five input endpoints using HTTP, HTTPS or TCP, each of which is used as external endpoints to listen on a unique port.

One of the several benefits of using Windows Azure is that all the requests connected to an input endpoint of a Windows Azure Role will be connected

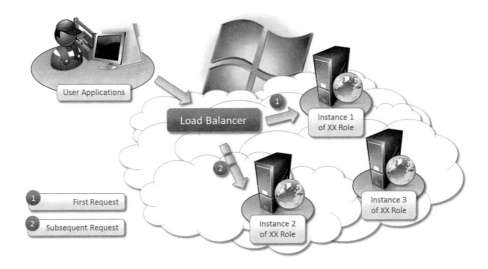

FIGURE 27.6
Routing in Windows Azure.

to a load balancer which automatically forwards the requests to all instances that are declared to be part of the role, on a round robin basis.

As depicted in Figure 27.6, instances from the same Role will share the same defined input endpoints and be behind the same Load Balancer. It is the responsibility of the Load Balancer to dispatch the incoming external traffic to instances behind it following a round robin mechanism. For instance, the first request will be sent to the first instance of the given worker role, the second will be sent to the second available instance, and so forth.

As we plan to deploy Aneka Container as instances of the Windows Azure Worker Role, there exists a situation where the Aneka Master is outside the Windows Azure Cloud and tries to send messages to a specific Aneka Worker inside the Windows Azure Cloud. Since the load balancer is responsible for forwarding these messages, there is a good possibility that the message may be sent to a Aneka Worker Container other than the specified one. Hence, in order to avoid the message being transferred to the wrong Aneka Worker Container, two possible solutions are available:

- **Forward Messages among Aneka Worker Containers.** When a Container receives a message that does not belong to it, it will forward the message to the right Container according to the InternalEndpoint address encoded in the *NodeURI* of Target Node of the Message. The advantage of this solution is the consistency of the architecture of Aneka PaaS, since no new components are introduced to the architecture. The disadvantage, however, is that the performance of Aneka Worker Containers will be hindered due to the overhead of forwarding messages.

- **Deploy a message proxy between the Aneka Worker Containers and Master for the purpose of dispatching incoming external messages.** The Message Proxy is a special kind of Aneka Worker Container which does not host any execution services. When the Windows Azure Aneka Cloud starts up, all Aneka Worker Containers in Windows Azure encode the internal endpoint address into the *NodeURI*. When the Message Proxy receives a message from the Master, it dispatches the message to the right Aneka Worker Container according to the encoded *NodeURI* specified in the Target Node of Message. The disadvantage of this solution is that it costs extra since Windows Azure charges according to the number of instances launched. However, in view of possible performance issues, the second solution is preferred. More details on the deployment of the Message Proxy Role are introduced in Section 27.3.1.3.

27.2.6.3 Dynamic Environment

As mentioned in Section 27.2.6.2, each Windows Azure Web Worker Role can define an external endpoint to listen on a unique port. As a matter of fact, endpoints are defined as ports on the load balancer. The Windows Azure Fabric will dynamically assign a random port number to each instance of the given role to listen on and requests will be forwarded from the endpoints defined on the load balancer to the instances. Consequently, before starting the Container, we need to get the dynamically assigned endpoint via *RoleEnvironment.CurrentRoleInstance.InstanceEndpoints* Property defined in the *Windows Azure Managed Library* and save it to the *Aneka Configuration File* so that the Container can bind the TCP channel to the right port.

Another changed required by the dynamic environment of Windows Azure is that we need to set the *NodeURI* of an Aneka Worker Container to the URL of the Message Proxy and encode the internal endpoint of the Container into the URL. When the Aneka Master sends a message to the *NodeURI* of an Aneka Worker Container, the Message Proxy receives the message and forwards it to the right Aneka Worker Container according to the internal endpoint address encoded in the *NodeURI*.

Furthermore, due to the dynamic nature of Windows Azure, we also need to guarantee that the Load Balancer sends the message to the instance of Message Proxy Role only if the message channel of the instance is ready and all the instances of Aneka Worker Role start to send Heartbeat Message to the Aneka Master located on-premises, after the deployment of the Message Proxy Role is finished.

27.2.6.4 Debugging

Debugging a Windows Azure Application is a bit different from debugging other Windows .Net applications.

In general, after we install *Windows Azure Tools for Visual Studio* we can debug a Windows Azure application locally when it is running in the Development Fabric during the development stage. However, after the application has been deployed on Windows Azure public Cloud, we cannot remotely debug the deployed application since we do not have direct access and administrative privilege on Windows Azure VMs. Fortunately, in June 2010 Windows Azure Tools + SDK, Windows Azure provides us with a new feature that enables us to debug issues that occur in the Cloud via *IntelliTrace*. With *IntelliTrace* debugging we can log extensive debugging information for a role instance while it is running in Windows Azure. Subsequently, we can use the *IntelliTrace* logs to step through the code from Visual Studio.

27.3 Design

In this section, we will discuss the design decisions for deploying Aneka Containers on Windows Azure as instances of Worker Role, how to integrate and leverage the dynamic provisioning service of Aneka, and how to exploit the Windows Azure Storage as a file storage system for Aneka PaaS in detail. The deployment includes two different types. The first type is to deploy Aneka Worker Containers on Windows Azure while the Aneka Master Container is run on local or on-premise resource. The second type is to deploy the entire Aneka PaaS including Aneka Master Container and Aneka Worker Containers on Windows Azure.

27.3.1 Deploying Aneka Workers on Windows Azure

27.3.1.1 Overview

Figure 27.7 provides an overall view of the deployment of Aneka Worker Containers as instances of Windows Azure Worker Role.

As shown in the figure, there are two types of Windows Azure Worker Roles used. These are the *Aneka Worker Role* and *Message Proxy Role*. In this case, we deploy one instance of Message Proxy Role and at least one instance of Aneka Worker Role. The maximum number of instances of the Aneka Worker Role that can be launched is limited by the subscription offer of Windows Azure Service that a user selects. In the first stage of the project, the *Aneka Master Container* will be deployed in the on-premises private Cloud, while *Aneka Worker Containers* will be run as instances of Windows Azure Worker

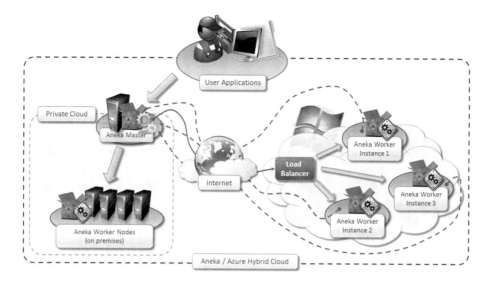

FIGURE 27.7
The Deployment of Aneka Worker Containers as Windows Azure Worker Role Instances.

Role. The instance of the Message Proxy Role is used to transfer the messages sent from the Aneka Master to the given Aneka Worker.

In this deployment scenario, when a user submits an application to the Aneka Master, the job units will be scheduled by the Aneka Master by leveraging on-premises Aneka Workers, if they exist, and Aneka Worker instances on Windows Azure simultaneously. When Aneka Workers finish the execution of Aneka work units, they will send the results back to Aneka Master, and then Aneka Master will send the result back to the user application.

27.3.1.2   Aneka Worker Deployment

Basically, we can deploy Aneka Containers of the same configuration as an Azure Worker Role, since they share the same binary code and the same configuration file. We can set up the number of instances of an Azure Worker Role to be launched in the *Windows Azure Service Configuration* file, which represents the number of Aneka Containers that will be deployed on the Windows Azure Cloud. And also we need to set up the Aneka Master URI and the shared security key in the Windows Azure Service Configuration file. When the instances of Aneka Worker Role are started up by Windows Azure Role Hosting Process, we first update the configuration of the Aneka Worker Container and start the Container program. After the container starts successfully, it will connect to the Aneka Master directly.

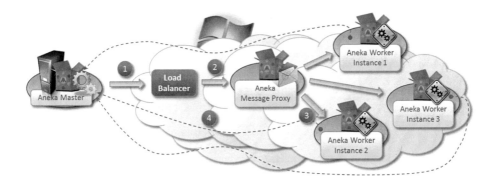

FIGURE 27.8
How the Message Proxy Works.

27.3.1.3 Message Proxy

As for the issue we discussed in Section 27.2.6.2, in order to guarantee that messages are transferred to the right target node specified by the Aneka Master, we need a mechanism to route messages to a given instance. Therefore, we introduce a Message Proxy between the Load Balancer and Aneka Worker instances. As shown in the Figure 27.8, all the messages that are sent to the Aneka Worker Containers in Windows Azure Public Cloud will be transferred to the external input endpoint of Message Proxy Role. All the messages will be transferred to the load balancer of the input endpoint(1). The load balancer will transfer the messages to the instance of Message Proxy Role(2). In this case, we only launch one instance for Message Proxy Role. The Message Proxy picks the incoming message, and parses the *NodeURI* of the target node to determine the internal address of the target node, and then forwards the messages to the given Aneka Worker(3). The Aneka Worker will handle the message and send a reply message to Aneka Master directly(4).

27.3.1.4 Dynamic Provisioning

Windows Azure provides us with programmatic access to most of the functionality available through the Windows Azure Developer Portal via the *Windows Azure Service Management REST API*. Using the Service Management API, we can manage our storage accounts and hosted services.

Hence, by using the completely extensible *Dynamic Resource Provisioning Object Model* [727] of Aneka PaaS and *Windows Azure Service Management REST API*, we can integrate Windows Azure Cloud resources into Aneka's infrastructure and provide support for dynamically provisioning and releasing Windows Azure resource on demand.

Specifically, the Aneka APIs offer the *IResourcePool* interface and the *ResourcePoolBase* class as extension points for integrating new resource pools. By implementing the interface and extending the abstract base class, we can support provisioning of Aneka Worker Containers on Windows Azure by following these steps:

1. Use the *CSPack Command-Line Tool* to programmatically packet all the binaries and the service definition file to be published to the Windows Azure fabric, into a service package file;

2. Use the *Windows Azure Storage Services REST API* to upload the service package to Windows Azure Blob;

3. Use the *Windows Azure Service Management REST API* to create, monitor and update the status of the deployment of the Windows Azure Hosted Service.

4. Use the *Windows Azure Service Management REST API* to increase or decrease the number of instances of Aneka Worker Containers to scale out or scale in on demand;

5. Use the *Windows Azure Service Management REST API* to delete the whole deployment of the Windows Azure Hosted Service when the provisioning service is shutting down.

27.3.2 Deploying Aneka Cloud on Windows Azure

27.3.2.1 Overview

In the second deployment scenario, we deploy the Aneka Master Container as an instance of the Windows Azure Worker Role. After finishing this step, we can run the whole Aneka Cloud infrastructure on the Windows Azure Cloud Platform, as can be seen from Figure 27.9.

In this scenario, users submit Aneka applications outside of the Windows Azure Cloud and receive the result of the execution from Windows Azure Cloud. The advantage of this structure is that it can dramatically decrease message transfer delay since all the messages between the Aneka Master and Aneka Workers are transferred within the same data center of Windows Azure, and the cost of data transfer charged by Windows Azure will reduce greatly as well.

Furthermore, for data persistence requirements, the Aneka Master Container, can directly use the Relational Data Service Provided by SQL Azure which would have higher data transfer rates and of higher security since they are located in the same Microsoft data center.

27.3.2.2 Aneka Cloud Deployment

Figure 27.9 shows two types of roles being deployed on the Windows Azure Cloud: one instance of the Aneka Master Role hosting the Aneka Master Container, and at least one instance of the Aneka Worker Role hosting the Aneka

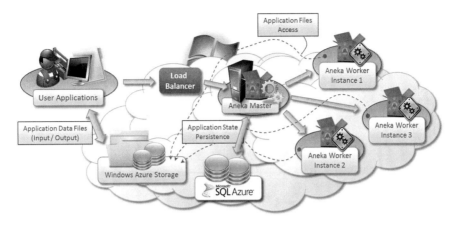

FIGURE 27.9
The Deployment of Aneka Master Container.

Worker Container. The Aneka Master Container and Aneka Worker Containers interact with each other via an internal endpoint, while the client and Aneka Master Container interact via an external endpoint of Aneka Master instance.

27.3.2.3 File Transfer System

In the current version of Aneka Cloud, the FTP protocol is used to transfer data files from the client to the Aneka Master (or a separate Storage Container) and between the Aneka Master and Aneka Worker Containers. However, due to the limitation of a maximum of five networking ports allowed on each Windows Azure Role instance, we can no longer use the FTP service to support file transfers on the Windows Azure Cloud. Instead, we can leverage Windows Azure Storage to support file transfers in Aneka.

In general, as illustrated in Figure 27.10, two types of Windows Azure Storage will be used to implement the Aneka File Transfer System: Blobs and Queues. Blobs will be used for transferring data files, and Queues for the purpose of notification. When Aneka users submit the application, if the transfer of input data files is needed, the *FileTransferManager* component will upload the input data files to the Windows Azure Blob and notify the start and end of the file transfer to Aneka's Storage Service via Windows Azure Queue. Similarly, the Aneka Worker will download the related input data file from Windows Azure Blob, and the start and end of the file transfer will be notified via Windows Azure Queue. When the execution of the work unit is completed in the Aneka Worker, if the transfer of output data files is needed, the *FileTransferManager* component of Aneka PaaS will upload the output data files to the Windows Azure Blob to enable Aneka users to download from it.

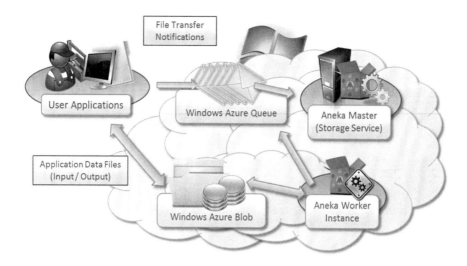

FIGURE 27.10
Using Windows Azure Storage Blob and Queue for Implementation of Aneka
File Transfer System.

27.4 Implementation

In this section, we will explore the implementation details of the design we
presented in Section 27.3. Section 27.4.1 displays the class diagrams of the
new and changed components in the Aneka PaaS. Next, Section 27.4.2 illustrates the configuration setting of the deployments, while Section 27.4.3
demonstrates the designed life cycle of deployments.

27.4.1 Class Diagrams

27.4.1.1 Windows Azure Aneka Container Deployment

Technically, in order to start an Aneka Container on Windows Azure Role
instance, we need to extend the *RoleEntryPoint* class which provides a callback to initialize, run, and stop instances of the role. We override the *Run()*
method to implement our code to start the Aneka Container which will be
called by Windows Azure Runtime after the role instance has been initialized.
Also worth noting is that due to the dynamic nature of the Windows Azure
environment, the configuration of Aneka Worker Containers must be updated
using the information obtained from the *CloudServiceConfiguration* Class.

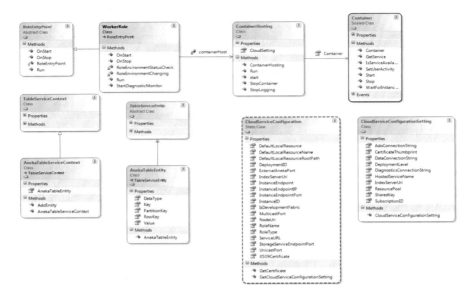

FIGURE 27.11
Class Diagram for Windows Azure Aneka Container Hosting Component.

27.4.1.2 Windows Azure Provisioning Resource Pool

The extendable and customizable *Dynamic Resource Provisioning Object Model* of Aneka PaaS enables us to provide new solutions for dynamic provisioning in Aneka. Specifically, in Azure Integration scenario, the *WindowsAzureResourcePool* class extends the *ResourcePoolBase* class and implements the *IResourcePool* interface to integrate Windows Azure as a new resource pool. The class *WindowsAzureOperation* provides all the operations that are needed to interact with the *Windows Azure Service Management REST API* via Windows Azure Service Management component.

27.4.1.3 Windows Azure Service Management

The *DeploymentOperation* component is used to interact with the *Windows Azure Service Management REST API* to manage the Windows Azure Hosted Services in terms of creating a deployment, updating the status of a deployment (such as from Suspended to Running or vice versa), upgrading the deployment, querying the deployment and deleting the deployment. The *Windows Azure Service Managment* component is used by the Resource Provisioning Service to manage the Windows Azure resource pool, and is also used during the stage of Windows Azure Role Deployment to monitor the status of deployment.

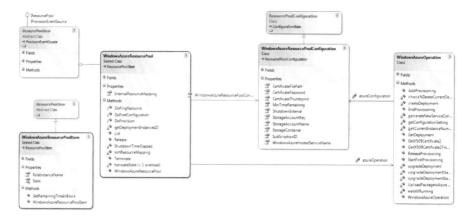

FIGURE 27.12

Class Diagram for Windows Azure Aneka Provisioning Resource Pool Component.

27.4.1.4   File Transfer System

The File Transfer System Component is used to transfer data files which are used in applications between clients and Aneka Cloud deployed on top of Windows Azure. The class *AzureFileChannelController* which implements the *IFileChannelController* interface represents the server component of the communication channel. It is responsible for providing the connection string for the client component to gain access to the Windows Azure Storage Service, providing a way to upload and retrieve a specific file. The class *AzureFileHandler* which implements the *IFileHandler* interface is in charge of retrieving a single file or a collection of files from the server component of the communication channel and uploading a single file or a collection of files to the server component of the communication channel.

27.4.2   Configuration

27.4.2.1   Provisioning Aneka Workers from Management Studio

In order to enable the Provisioning Service of Aneka to provision resources on Windows Azure, we need to configure it via the Aneka Cloud Management Studio, while configuring the services of the Aneka Master Container. This requires configuring the Scheduling Service and Resource Provisioning Service. For the Scheduling Service, we need to select a proper scheduling algorithm for the *TaskScheduler* and *ThreadScheduler*. Currently, only two algorithms are available for dynamic provisioning: the *FixedQueueProvisioningAlgorithm* and the *DeadlinePriorityProvisioningAlgorithm* as shown in Figure 27.15. The configuration required for the Resource Provisioning Service in order to

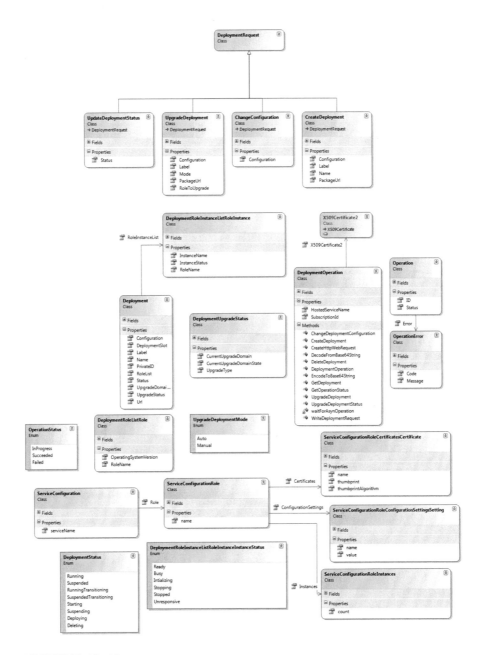

FIGURE 27.13
Class Diagram for Windows Azure Service Management Component.

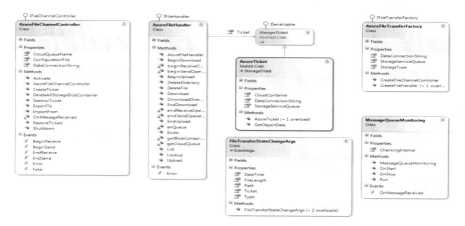

FIGURE 27.14

Class Diagram for Windows Azure Aneka Storage Service Implementation using Windows Azure Storage.

acquire resources from the Windows Azure Cloud Service Providers is depicted in Figure 27.15. For setting up a Windows Azure Resource Pool, we need the following information:

1. **Capacity:** identifies the number of instances that can be managed at a given time by the pool. This value is restricted to the maximum number of instances that the user is allowed to launch on Windows Azure, based on the subscription.

2. **Certificate File Path:** specifies the file path of an X509 certificate that is used to interact with the Windows Azure Service Management REST API. This certificate must also be uploaded to the Windows Azure Development Portal.

3. **Certificate Password:** designates the password of the X509 Certificate.

4. **Certificate Thumbprint:** assigns the thumbprint of the X509 Certificate.

5. **Hosted Service Name:** identifies the name of the Windows Azure Hosted Service; the service must have been created via the Windows Azure Development Portal.

6. **Subscription ID:** specifies the Subscription ID of Windows Azure Account.

7. **Storage Account Name:** designates the name of Windows Azure Storage account that is under the same subscription.

8. **Storage Account Key:** specifies the key of the storage account.

FIGURE 27.15
Aneka Provisioning Service Configuration.

9. **Storage Container:** defines the name of storage container which is used to store the Windows Azure Hosted Service Package File.

27.4.2.2 Deploying Aneka Cloud on Windows Azure

In order to deploy an Aneka Cloud on Windows Azure, before uploading the Windows Azure Aneka Cloud Package into Windows Azure Cloud, we need to configure the Windows Azure Service Configuration file related to the Windows Azure Aneka Cloud Package. To be more specific, as shown in the Figure 27.16, we need to specify the values below:

1. **DiagnosticsConnectionString:** the connection string for connecting to the Windows Azure Storage Service which is used to store diagnostics data.

2. **DataConnectionString:** the connection string for connecting to Windows Azure Storage Service which is used to implement the File Transfer System.

3. **SharedKey:** the security key shared between Aneka Master and Aneka Worker.

4. **SubscriptionID:** the Subscription ID of Windows Azure Account.

5. **HostedServiceName:** the name of the Windows Azure Hosted Service.

6. **CertificateThumbprint:** the thumbprint of the X509 Certificate which has been uploaded to Windows Azure Service Portal. The value of thumbprint in the Certificate Property should also be set.

7. **AdoConnectionString:** the connection string used to connect to

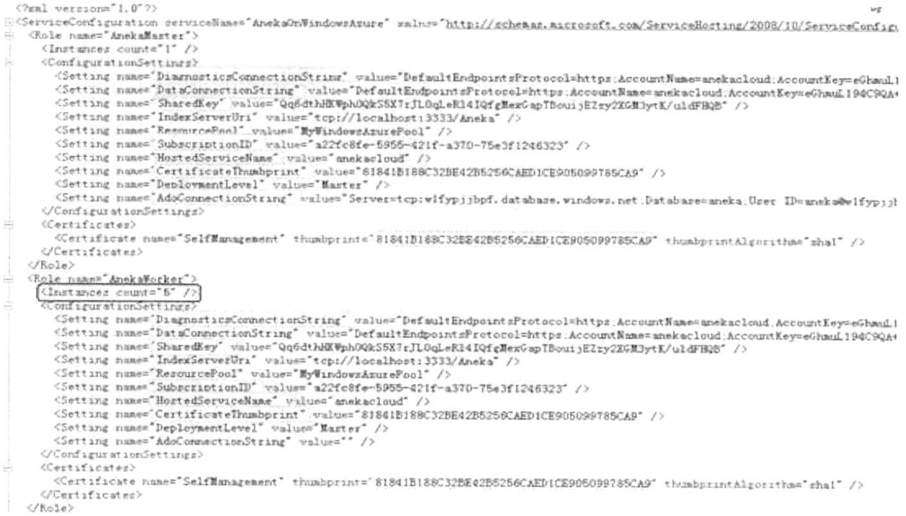

```
<?xml version="1.0"?>
<ServiceConfiguration serviceName="AnekaOnWindowsAzure" xmlns="http://schemas.microsoft.com/ServiceHosting/2008/10/ServiceConfig
  <Role name="AnekaMaster">
    <Instances count="1" />
    <ConfigurationSettings>
      <Setting name="DiagnosticsConnectionString" value="DefaultEndpointsProtocol=https;AccountName=anekacloud;AccountKey=eGhauL1
      <Setting name="DataConnectionString" value="DefaultEndpointsProtocol=https;AccountName=anekacloud;AccountKey=eGhauL194C9QA
      <Setting name="SharedKey" value="Qq6dthHKWph0QkS5X7rJL0qLeR14IQfgMexGapTBouijEZzy2XGM3ytK/uldFHQB" />
      <Setting name="IndexServerUri" value="tcp://localhost:3333/Aneka" />
      <Setting name="ResourcePool" value="MyWindowsAzurePool" />
      <Setting name="SubscriptionID" value="a22fc8fe-5955-421f-a370-75e3f1246323" />
      <Setting name="HostedServiceName" value="anekacloud" />
      <Setting name="CertificateThumbprint" value="81841B188C32BE42B5256CAED1CE905099785CA9" />
      <Setting name="DeploymentLevel" value="Master" />
      <Setting name="AdoConnectionString" value="Server=tcp:w1fypjjbpf.database.windows.net;Database=aneka;User ID=aneka@w1fypjjt
    </ConfigurationSettings>
    <Certificates>
      <Certificate name="SelfManagement" thumbprint="81841B188C32BE42B5256CAED1CE905099785CA9" thumbprintAlgorithm="sha1" />
    </Certificates>
  </Role>
  <Role name="AnekaWorker">
    <Instances count="5" />
    <ConfigurationSettings>
      <Setting name="DiagnosticsConnectionString" value="DefaultEndpointsProtocol=https;AccountName=anekacloud;AccountKey=eGhauL1
      <Setting name="DataConnectionString" value="DefaultEndpointsProtocol=https;AccountName=anekacloud;AccountKey=eGhauL194C9QA
      <Setting name="SharedKey" value="Qq6dthHKWph0QkS5X7rJL0qLeR14IQfgMexGapTBouijEZzy2XGM3ytK/uldFHQB" />
      <Setting name="IndexServerUri" value="tcp://localhost:3333/Aneka" />
      <Setting name="ResourcePool" value="MyWindowsAzurePool" />
      <Setting name="SubscriptionID" value="a22fc8fe-5955-421f-a370-75e3f1246323" />
      <Setting name="HostedServiceName" value="anekacloud" />
      <Setting name="CertificateThumbprint" value="81841B188C32BE42B5256CAED1CE905099785CA9" />
      <Setting name="DeploymentLevel" value="Master" />
      <Setting name="AdoConnectionString" value="" />
    </ConfigurationSettings>
    <Certificates>
      <Certificate name="SelfManagement" thumbprint="81841B188C32BE42B5256CAED1CE905099785CA9" thumbprintAlgorithm="sha1" />
    </Certificates>
  </Role>
```

FIGURE 27.16
Windows Azure Service Configuration File related to Windows Azure Aneka Cloud Package.

an ADO relational database if relational database is used to store persistent data.

More importantly, we need to define the *Instance Number* of Aneka Workers running on Windows Azure Cloud, which is specified in the "count" attribute of "Instance" property.

27.4.3 Life Cycle of Deployment

27.4.3.1 Aneka Worker Deployment

Figure 27.17 shows the whole life cycle of deployment of Aneka Worker Containers on Windows Azure Cloud.

Generally speaking, the whole life cycle of Aneka Worker Container deployment on Windows Azure involves five steps. They are *Configuration*, *First Time Resource Provisioning*, *Subsequent Resource Provisioning*, *Resource Release*, and *Termination of Deployment* respectively.

27.4.3.2 Aneka Cloud Deployment

Figure 27.18 shows the life cycle for deploying an entire Aneka Cloud on top of Windows Azure.

In general, the whole life cycle of Aneka Cloud deployment on Windows Azure

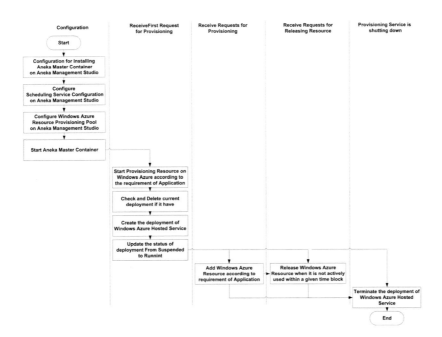

FIGURE 27.17
The Life Cycle of Aneka Worker Container Deployment on Windows Azure.

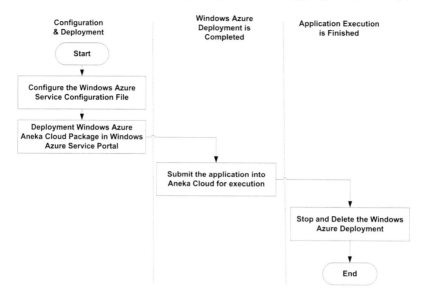

FIGURE 27.18
The Life Cycle of Aneka Cloud Deployment on Windows Azure.

involves three steps. They are *Configuration and Deployment, Application Execution*, and *Deployment Termination* respectively.

27.5   Experiments

In this section, we will present the experimental results for application execution on the Aneka Windows Azure Deployments including Aneka Worker Deployment and Aneka Cloud Deployment. The testing application we selected is Mandelbrot (Figure 27.19) which is developed on top of Aneka Thread Model to determine the suitability of Aneka Windows Azure Deployment for running parallel algorithms.

Figure 27.20 and Figure 27.21 display the experimental results for executing the Mandelbrot application using different input problem sizes, running on both Aneka Worker Deployment and Aneka Cloud Deployment on Windows Azure when the number of Aneka Workers being launched is 1, 5, and 10. The compute instance size of the Azure Instance selected to run the Aneka Worker Containers is *small computer instance* which is a virtual server with dedicated resources (CPU 1.6 GHz and Memory 1.75 GB) and specially tailored for Windows Server 2008 Enterprise operating system as the guest OS. The instance size for deploying the Aneka Master Container is *middle com-*

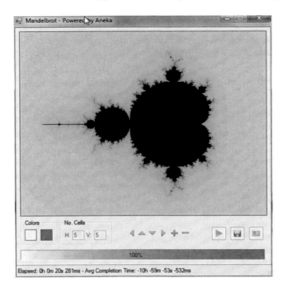

FIGURE 27.19
Mandelbrot Application Developing on Top of Aneka Thread Model.

puter instance with machine configuration CPU 2*1.6 GHz and Memory 3.5
GB.

From both Figure 27.20 and Figure 27.21, we can see that for the same input
problem size, there is a decrease in the execution time as a result of employing
more Aneka Workers to process the work units. The elapsed time used to
execute application on Aneka Worker Deployment is also much larger than
on Aneka Cloud Deployment due to the communication overhead between
the Aneka Master and Aneka Workers with Aneka Workers deployed inside
Windows Azure Cloud, while Aneka Master is deployed outside.

In the next experiment, we measure the scalability of Aneka Cloud Deployment. In this experiment, we use up to 16 small size instances. All the instances
are allocated statically. The result of the experiment is summarized in Figure
27.22. We see that the throughput of the Mandelbrot application running on
Azure Cloud Deployment increases when the number of instances ascends.

Furthermore, we can see from Figure 27.23 that the jobs are evenly distributed
across all the available Aneka Workers, whose number is 10 in this case.

27.6 Related Work

Windows Azure has been adopted in many projects to build high performance computing applications [103] [481] [468]. Augustyn and Warchal [103]

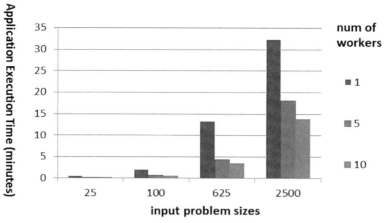

FIGURE 27.20
Experimental Result Showing the Execution Time for Running the Mandelbrot Application on Aneka Worker Deployment.

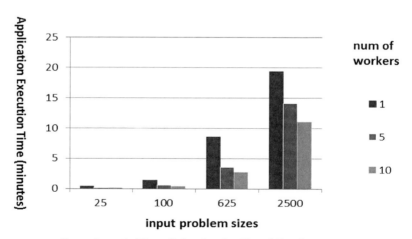

FIGURE 27.21
Experimental Result Showing the Execution Time for Running the Mandelbrot Application on Aneka Cloud Deployment.

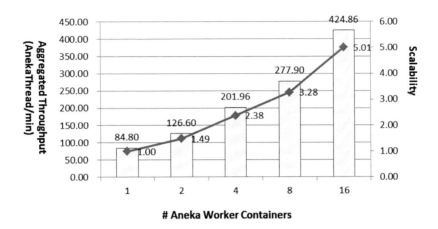

FIGURE 27.22
Scalability Diagram for Aneka Cloud Deployment.

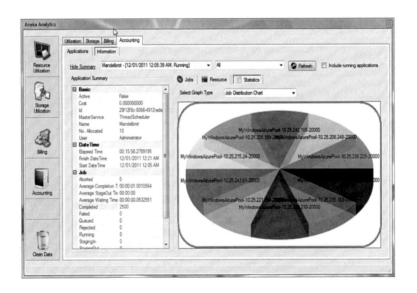

FIGURE 27.23
The Job Distribution Chart Shown on the Aneka Analytics Tool.

presented an idea and implementation on how to use Windows Azure computing service to solve the N-body problem using the Barnes-Hut Algorithm. All computation is operated in parallel on Windows Azure Worker Role instances. Lu et al. [481] delivered a case study of developing AzureBlast, a parallel BLAST engine running on Windows Azure Platform, which can be used to run the BLAST [56], a well-known and both data intensive and computational intensive bioinformatics application. Li et al. [465] demonstrated how to build the MODIS Satellite Data Reprojection and Reduction Pipeline on Windows Azure. In these cases, the whole implementation is started from scratch, which means the developers need to handle application administration, task scheduling, the communication and interaction between role instances, and the storage service access. The Aneka PaaS integration with Windows Azure Platform can speed up the entire development for a high performance application running on top of Windows Azure by using the programming models powered by Aneka.

Besides, similar to Aneka, Lokad.Cloud [479] is an execution framework which provides build-in features such task scheduling, queue processing and application administration, and allows users to define a set of services to be run in Windows Azure Worker Role instances. Nevertheless, differently from the Aneka PaaS, Lokad.Cloud is only designed to run applications on top of Windows Azure. It is worth mentioning that Aneka PaaS is designed to run applications on private Clouds as well as on public Clouds such as Windows Azure and Amazon EC2. Aneka Paas can be leveraged to integrate private Clouds with public Clouds by dynamically provisioning resources on public Clouds such as Windows Azure when local resources cannot meet the computing requirement. Moreover, Aneka supports three types of programming models, the *Task Programming Model*, *Thread Programming Model* and *MapReduce Programming Model*, to meet the requirements of different application domains.

27.7   Sample Applications of Aneka

Differently from other Cloud platforms or Cloud applications running on top of Windows Azure we introduced in Section 27.6, Aneka allows seamless integration of public Clouds and private Clouds to leverage their resources to executing applications. Specifically, a wide range of applications from scientific applications, business services, to entertainment and media, or manufacturing and engineering applications, have benefited from Aneka PaaS. A list of application types that utilized Aneka is shown in Table 27.1.

TABLE 27.1

Sample Applications of Aneka

Industry Sectors

Challenges and Issues	Aneka PaaS Usage
1. Geospatial Sciences and Technologies	
More geospatial and non-spatial data is involved due to the increase in the number of data sources and advancement of data collection methodologies.	- Enables a new approach to complex analyses of massive data and computationally intensive environments. - Builds a high-performance and distributed GIS environment over the public, private and hybrid Clouds.
2. Health and Life Sciences	
High volume and density of data require for processing.	- Enables faster execution and massive data computation. - Suitable for life science R&D, clinical simulation, and business intelligence tools.
3. Financial Services	
Applications such as portfolio and risk analysis, credit fraud detection, option pricing require the use of high-performance computing systems and complex algorithms.	- Simplifies the application development life cycle. - Reduces hardware investment. - Lowers ongoing operational expenditure. - Brings a breakthrough in industry standard tools for financial modeling such as Microsoft Office Excel by solving its computational performance barrier.
4. Telecom Industry	
The majority of Telecom providers have several disparate systems and they don't have enough capacity to handle the utilization and access information to optimize their use.	- Helps telecom providers to realize system utilization strategies in a cost effective, reliable, scalable and tightly integrated manner. - Helps mission-critical applications by automating their initiation across a shared pool of computational resources, by breaking the executions into many parallel workloads that produce results faster in accordance with agreed upon SLAs and policies.
5. Manufacturing and Engineering	
Manufacturing organizations are faced with a number of computing challenges as they seek to optimize their IT environments, including high infrastructure costs and complexity to poor visibility into capacity and utilization.	- Enables organizations to perform process simulation, modelling, and optimization at a highly increased rate so that the time-to-market of key products is faster, by effectively leveraging Cloud technologies.
6. Entertainment and Media	
Business solutions involving digital media transcoding to HD video, 3D image rendering, and gaming, require plenty of time to process and utilize vast amounts of computing capacity to encode and decode the media.	- Optimizes networked computers as a private Cloud or leverage public Cloud such as Windows Azure, Amazon EC2 and Go-Grid. - Allows scaling applications linearly. - Better utilizes the Cloud farm providing best efficiency and speed possible using Cloud scalability.

27.8 Conclusions and Future Directions

In this chapter, we have introduced the Aneka Cloud Application Platform (Aneka PaaS), presented and discussed the background, design and implementation of the integration of the Aneka PaaS and Windows Azure Platform.

The Aneka PaaS is built on a solid .NET service oriented architecture allowing seamless integration between public Clouds and mainstream applications. The core capabilities of the framework are expressed through its extensible and flexible architecture as well as its powerful application models featuring support for several distributed and parallel programming paradigms. These features enhance the development experience of software developers allowing them to rapidly prototype elastically scalable applications. Applications ranging from the media and entertainment industry, to engineering, education, health and life sciences, and several others have been proven to be appropriate to the Aneka PaaS.

Admittedly, the integration of two platforms would give numerous benefits to not only the users of Aneka PaaS but also the customers of Windows Azure Platform, enabling them to embrace the advantages of Cloud Computing in terms of more computing resources, easier programming model, and more efficiency on application execution at lower expense and lower administration overhead.

In the first stage, we deployed the Aneka Worker Container as instances of Windows Azure Worker Role, as well as support for dynamic provisioning of Aneka Workers on Windows Azure. In the second step, we deployed the Aneka Master Container on Windows Azure as an instance of Worker Role and the entire Aneka PaaS ran completely on the Windows Azure Platform. This allows users to run Aneka Cloud applications without requiring any local infrastructure. The message transfer overhead and the transfer cost will decrease dramatically. This is beneficial to both Service Providers who uses Aneka PaaS to deliver their services and the final users who consume the services.

On the whole, in addition to the integration with Windows Azure Platform, presently, Aneka PaaS has already supported the integration of Amazon EC2, GoGrid, and Xen Server. The support of provisioning resources on Windows Azure Platform once again illustrates the adaptability, flexibility, mobility and extensibility of the Aneka PaaS. In the next stage, the Aneka PaaS will continue to integrate with other public Cloud platforms and virtual machine management platforms such as VMWare, Microsoft HyperV and so forth, to help users to exploit more power of Cloud Computing.

Acknowledgments

This work is partially supported by a grant from the Australian Department of Innovation, Industry, Science and Research via its COMET (Commercialising Emerging Technologies) Program.

Bibliography

[1] eXtensible Access Control Markup Language (XACML): http://www.oasisopen.org/committees/tc_home.php?wg_abbrev=xacml.

[2] SAML V2.0 Technical Overview, OASIS, http://www.oasis-open.org/specs/index.php#saml.

[3] Forum of Federations: http://www.forumfed.org/en/index.php.

[4] Amazon Elastic Compute Cloud (Amazon EC2): http://aws.amazon.com/ec2/.

[5] OpenQRM, "the next generation, open-source Data-center management platform," http://www.openqrm.com/.

[6] Resources and Services Virtualisation without Barriers (RESERVOIR) European Project, http://www.reservoir-fp7.eu/.

[7] OASIS Web Services Resource Framework (WSRF): http://www.oasis-open.org/committees/tc_home.php?wg_abbrev=wsrf.

[8] GridShib: Bridging SAML/Shibboleth and X.509 PKI for campus and grid interoperability, http://gridshib.globus.org/.

[9] Extensible Messaging and Presence Protocol (XMPP): Core, http://xmpp.org/rfcs/rfc3920.html.

[10] Extensible Messaging and Presence Protocol (XMPP): Instant Messaging and Presence, http://xmpp.org/rfcs/rfc3921.html.

[11] StartSSL The Swiss Army Knife of Digital Certificates & PKI: http://www.startssl.com/.

[12] inContext project: Unleash Team Power, http://www.in-context.eu, 2007.

[13] inContext project WP2 Deliverable 2.1: Analysis and specification and context modeling techniques, http://www.in-context.eu/page.asp?PageRef=10. 2007.

[14] W3C. Web services activity.

[15] Xml-rpc specification.

[16] Software as a service:strategic backgrounder. *Software and Information Industry Association*, 2001.

[17] Worldwide server power and cooling expense 2006-2010. IDC Document #203598, 2006.

[18] What is cloud computing. *Jrnl of Parallel and Distributed Computing*, 2008.

[19] Amazon web services cloud watch Web Site, November 2010.

[20] Cloudkick Web Site, November 2010.

[21] Enstratus Web Site, December 2010.

[22] Eucalyptus monitoring Web Site, November 2010.

[23] Gogrid exchange Web Site, November 2010.

[24] Gogrid load balancing service Web Site, November 2010.

[25] Makara Web Site, December 2010.

[26] J. Seltzer, D. Read, and D.D. Clark. End-to-end arguments in system design. *Second International Conference on Distributed Computing Systems*, pages 509–512, 1981.

[27] Microsoft azure appfabric Web Site, November 2010.

[28] Msdn forum - microsoft azure load balancing service Web Site, November 2010.

[29] Msdn library - azure monitoring and diagnostics Web Site, November 2010.

[30] Multi-scale modeling and simulation, 2010. [Online], http://www.math.princeton.edu/multiscale/.

[31] Nagios project Web Site, November 2010.

[32] Opennebula information manager Web Site, November 2010.

[33] Opennebula scheduling policies Web Site, November 2010.

[34] OpenNebula Service Management Project, December 2010.

[35] Paraleap windows azure dynamic scaling Web Site, November 2010.

[36] Rackspace cloud tools Web Site, November 2010.

[37] RightScale Web Site, December 2010.

[38] Scalr Web Site, December 2010.

[39] Rackspace cloud sites load balancing service Web Site, January 2011.

[40] Grid Computer Operating System For Web Applications - AppLogic, 3tera.: http://www.3tera.com/AppLogic/.

[41] J.C. Mouriño, J. López, F.G. G. Castaño, D. Rodríguez, L. Domínguez, D. González, J. Pena, F. Gómez, D. G. Castaño, M. Pombar, A. Gómez, C. Fernández. Monte Carlo Verification of IMRT treatment plans on Grid. In *HealthGrid*, pages 105–114, 2007.

[42] Aalto Venture Garage seed accelerator http://www.aaltovg.com.

[43] Agnar Aamodt and Enric Plaza. Case-based reasoning: Foundational issues, methodological variations, and system approaches. *AI Communications*, 7:39–59, 1994.

[44] Daniel J. Abadi. Data Management in the Cloud: Limitations and Opportunities. *IEEE Data Eng. Bull.*, 32(1):3–12, 2009.

[45] Karl Aberer and Zoran Despotovic. Managing trust in a peer-2-peer information system. In *CIKM*, pages 310–317, 2001.

[46] A. Abouzeid, K. Bajda-Pawlikowski, D. Abadi, A. Rasin, and A. Silberschatz. Hadoopdb: An architectural hybrid of mapreduce and dbms technologies for analytical workloads. *PVLDB*, 2(1):922–933, 2009.

[47] ActiveMQ. Messaging and integration pattern provider. http://activemq.apache.org/.

[48] Atul Adya, William J. Bolosky, Miguel Castro, Gerald Cermak, Ronnie Chaiken, John R. Douceur, Jon Howell, Jacob R. Lorch, Marvin Theimer, and Roger P. Wattenhofer. Farsite: federated, available, and reliable storage for an incompletely trusted environment. In *Proceedings of the 5th Symposium on Operating Systems Design and Implementation (OSDI)*, OSDI '02, pages 1–14, New York, NY, USA, 2002. ACM.

[49] R. Agarwal and M. Gort. First Mover Advantage and the Speed of Competitive Entry: 1887-1986. *Journal of Law and Economics*, 44(1):161–178, 2002.

[50] International Atomic Energy Agency. IAEA (2009). Directory of Radiotherapy Centres (DIRAC). http://www-naweb.iaea.org/nahu/dirac/query3.asp.

[51] Gagan Aggarwal, Mayank Bawa, Prasanna Ganesan, Hector Garcia-Molina, Krishnaram Kenthapadi, Rajeev Motwani, Utkarsh Srivastava, Dilys Thomas, and Ying Xu 0002. Two can keep a secret: A distributed architecture for secure database services. In *CIDR*, pages 186–199, 2005.

[52] Gagan Aggarwal, Mayank Bawa, Prasanna Ganesan, Hector Garcia-Molina, Krishnaram Kenthapadi, Rajeev Motwani, Utkarsh Srivastava, Dilys Thomas, and Ying Xu. Two Can Keep A Secret: A Distributed Architecture for Secure Database Services. In *CIDR*, pages 186–199, 2005.

[53] Divyakant Agrawal, Amr El Abbadi, Fatih Emekçi, and Ahmed Metwally. Database Management as a Service: Challenges and Opportunities. In *ICDE*, pages 1709–1716, 2009.

[54] Ahronovitz, M. et al. Cloud Computing Use Cases White Paper, Version 4.0. `http://opencloudmanifesto.org/Cloud_Computing_Use_Cases_Whitepaper-4_0.pdf`.

[55] S. Ajuja, N. Carriero, and D. Gelernter. Linda and friends. *Computer*, 19(8):26–34, 1986.

[56] S. F. Altschul, W. Gish, W. Miller, E. W. Myers, and D. J. Lipman. Basic Local Alignment Search Tool. *Journal of Molecular Biology*, 215(3):403–410, October 1990.

[57] Amazon. Amazon elastic compute cloud (amazon ec2). *Jrnl of Parallel and Distributed Computing*, 2008.

[58] Amazon. Amazon simple storage service (amazon s3). *Jrnl of Parallel and Distributed Computing*, 2008.

[59] Amazon. Amazon elastic compute cloud: http://aws.amazon.com/ec2/, August 2009.

[60] Amazon. Amazon ec2 high performance computing: http://aws.amazon.com/ec2/hpc-applications/, September 2010.

[61] Amazon. Amazon ec2 instance types: `http://aws.amazon.com/ec2/instance-types/`, September 2010.

[62] Amazon. Amazon ec2 service level agreement: `http://aws.amazon.com/ec2-sla/`, November 2010.

[63] Amazon Web Services. http://aws.amazon.com, Dec. 2010.

[64] amazon.com. Amazon Elastic Compute Cloud. http://aws.amazon.com/ec2/.

[65] Amazon Elastic Compute Cloud (Amazon EC2): http://aws.amazon.com/ec2/.

[66] AMD Virtualization (AMD-V) Technology. http://sites.amd.com/us/business/it-solutions/virtualization/Pages/amd-v.aspx.

[67] D. W. O. Rogers B. A. Faddegon G. X. Ding C. M. Ma W. J. Mackie. TR. BEAM: a Monte Carlo code to simulate radiotherapy treatment units. In *Med Phys.*, pages 503–524, 1995.

[68] Novell BEA Systems. *Complete Transition of TUXEDO to BEA.* http://www.bea.com/framework.jsp?CNT=pr00002.htm&FP= /content/news_events/press_releases/1996, 1996.

[69] J. Pena, D. M. González-Castaño, F. Gómez, A. Gago-Arias, F. J. González-Castaño, D. Rodríguez-Silva, A. Gómez, C. Mouriño, M. Pombar, M. Sánchez. eIMRT: a web platform for the verification and optimization of radiation treatment plans. In *J Appl Clin Med Phys*, volume 10, 2009.

[70] J. Pena, D. M. González-Castaño, F. Gómez, F F. Sánchez-Doblado, G. H. Hartmann. Automatic determination of primary electron beam parameters in Monte Carlo simulation. In *Med Phys.*, pages 1076–1084, 2007.

[71] R. R. Patel and D. W. Arthur. The emergence of advanced brachytherapy techniques for common malignancies. In *Hematology/Oncology Clinics of North America*, pages 97–118, 2006.

[72] B. R. B. Walters and D. W. O. Rogers. *DOSXYZnrc Users Manual*, 2002.

[73] M. P. Coleman et. al. Responding to the challenge of cancer in Europe. *Institute of Public Health of the Republic of Slovenia*, 2008.

[74] U. Amaldi G. Kraft. Particle accelerators take up the fight against cancer. *CERN Courier*, 46(10):17–20, 2006.

[75] L. E. Taylor and M. Ding. A review of intensity-modulated radiation therapy. In *Current Oncology Reports*, pages 294–299, 2008.

[76] A. Taylor and M. E. Powell. Intensity–modulated radiotherapy–what is it? In *Cancer Imaging*, pages 68–73, 2004.

[77] M. Brada, M. Pijls-Johannesma, D. De Ruysscher. Proton therapy in clinical practice: Current clinical evidence. In *Journal of Clinical Oncology*, pages 965–970, 2007.

[78] D. Schulz-Ertner, O. Jäkel, W. Schlegel. Radiation therapy with charged particles. In *Seminars in Radiation Oncology*, pages 249–259, 2006.

[79] B. D. Kavanagh, R. D. Timmerman. Stereotactic radiosurgery and stereotactic body radiation therapy: An overview of technical considerations and clinical applications. In *Hematology/Oncology Clinics of North America*, pages 87–95, 2006.

[80] T. S. Lawrence, R. K. Ten Haken, A. Giaccia. *Cancer: Principles and Practice of Oncology*, chapter Principles of Radiation Oncology. Philadelphia: Lippincott Williams and Wilkins, 2008.

[81] B. Sotomayor, R. S. Montero, I. M. Llorente, I. Foster. Virtual Infrastructure Management in Private and Hybrid Clouds, *IEEE Internet Computing*, vol. 13, no. 5, pp. 14–22, Sep./Oct. 2009.

[82] S. E. Noda, T. Lautenschlaeger, M. R. Siedow. Technological advances in radiation oncology for central nervous system tumors. In *Seminars in Radiation Oncology*, pages 179–186, 2008.

[83] David G. Andersen, Jason Franklin, Michael Kaminsky, Amar Phanishayee, Lawrence Tan, and Vijay Vasudevan. Fawn: a fast array of wimpy nodes. In *Proceedings of the 22nd ACM Symposium on Operating Systems Principles 2009, SOSP 2009, Big Sky, Montana, USA, October 11-14, 2009*, pages 1–14, 2009.

[84] Janna Anderson, Elon University, and Lee Rainie. Technical seminar report on cloud computing. *Intel Executive Summary*, 2005.

[85] Juan M. Andrade. *The TUXEDO System: Software for Constructing and Managing Distributed Business Applications*, 1996. Addison-Wesley Professional.

[86] Andrew. Wordpress. *3rd International Conference on Grid and Pervasive Computing-gpc-workshops*, 2008.

[87] A. Andrieux, K. Czajkowski, A. Dan, K. Keahey, H. Ludwig, T. Nakata, J. Pruyne, J. Rofrano, S. Tuecke, and M. Xu. *Web Services Agreement Specification (WS-Agreement)*. Open Grid Forum, March 14, 2007.

[88] A. Andrieux, K. Czajkowski, A. Dan, K. Keahey, H. Ludwig, J. Pruyne, J. Rofrano, S. Tuecke, and M. Xu. Web services agreement specification (WS-Agreement). In *Global Grid Forum*, 2004.

[89] S. Androulidakis, T. Doukoglou, G. Patikis, and D. Kagklis. Service differentiation and traffic engineering in ip over wdm networks. *Communications Magazine, IEEE*, 46(5):52 –59, May 2008.

[90] Animoto: automatic production of video pieces from photos, video clips and music. based on AWS and rightscale http://www.animoto.com.

[91] J. Anselmi, D. Ardagna, and P. Cremonesi. A QoS-based selection approach of autonomic grid services. In *Proceedings of the 2007 workshop on Service-oriented computing performance: aspects, issues, and approaches*, pages 1–8. ACM, 2007.

[92] Google App. Multitenancy, 2010.

[93] Appistry. Cloud Taxonomy: Applications, Platform, Infrastructure : http://www.appistry.com/blogs/sam/cloud-taxonomyapplications-platform-infrastructure, 2008.

[94] Sam Charrington Appistry. Cloud taxonomy: Applications, platform, infrastructure, December 2008.

[95] M. Armbrust, A. Fox, R. Griffith, A.D. Joseph, R.H. Katz, A. Konwinski, G. Lee, D.A. Patterson, A. Rabkin, I. Stoica, et al. Above the clouds: A Berkeley view of cloud computing. *EECS Department, University of California, Berkeley, Tech. Rep. UCB/EECS-2009-28*, 2009.

[96] Michael Armbrust, Armando Fox, Rean Griffith, Anthony D. Joseph, Randy Katz, Andy Konwinski, Gunho Lee, David Patterson, Ariel Rabkin, Ion Stoica, and Matei Zaharia. A view of cloud computing. *Communications of the ACM*, 53:50–58, April 2010.

[97] Michael Armbrust, Armando Fox, Rean Griffith, Anthony D. Joseph, Randy H. Katz, Andrew Konwinski, Gunho Lee, David A. Patterson, Ariel Rabkin, Ion Stoica, and Matei Zaharia. Above the clouds: A Berkeley view of cloud computing. Technical Report UCB/EECS-2009-28, EECS Department, University of California, Berkeley, Feb 2009.

[98] Michael Armbrust, Anthony D Joseph, Randy H Katz, and David A Patterson. Above the Clouds : A Berkeley View of Cloud Computing. *University of California, Berkeley, Technical Report*, 2009.

[99] Arvind. Decomposing a program for multiple processor system. In *Proceedings of the 1980 International Conference on Parallel Processing*, pages 7–14, 1980.

[100] Everything as a Service. Gathering clouds of xaas. *Second International Symposium on Information Science and Engineering*, 2010.

[101] G. Atanassov et al. Interval valued intuitionistic fuzzy sets. *Fuzzy sets and Systems*, 31(3):343–349, 1989.

[102] L. Atzori and T. Onali. Operators challenges toward bandwidth management in diffserv-aware traffic engineering networks. *Communications Magazine, IEEE*, 46(5):154 –160, May 2008.

[103] Dariusz Rafal Augustyn and Lukasz Warchal. Cloud service solving n-body problem based on windows azure platform. In Andrzej Kwiecieï£¡ï£¡, Piotr Gaj, and Piotr Stera, editors, *Computer Networks*, volume 79 of *Communications in Computer and Information Science*, pages 84–95. Springer Berlin Heidelberg, 2010.

[104] Stefan Aulbach, Torsten Grust, Dean Jacobs, Alfons Kemper, and Jan Rittinger. Multi-tenant databases for software as a service: schema-mapping techniques. In *SIGMOD '08: Proceedings of the 2008 ACM SIGMOD international conference on Management of data*, pages 1195–1206, New York, NY, USA, 2008. ACM.

[105] T. Auld, A. W. Moore, and S. F. Gull. Bayesian neural networks for internet traffic classification. *Neural Networks, IEEE Transactions on*, 18(1):223–239, Jan 2007.

[106] AutoI project. Project Website: `http://ist-autoi.eu/autoi/index.php`.

[107] AuverGrid. Project website. http://www.auvergrid.fr.

[108] R. Aversa, B. Di Martino, N. Mazzocca, and S. Venticinque. A skeleton based programming paradigm for mobile multi-agents on distributed systems and its realization within the magda mobile agents platform. *Mob. Inf. Syst.*, 4:131–146, April 2008.

[109] Azure service platform. http://www.microsoft.com/azure/.

[110] J. S. Bagla and T Padmanabhan. Cosmological n-body simulations. *Pramana - Journal of Physics, 49 (2)*, pages 161–192, 2008.

[111] Satish Balay, Kris Buschelman, William D. Gropp, Dinesh Kaushik, Matthew G. Knepley, Lois Curfman McInnes, Barry F. Smith, and Hong Zhang. PETSc Web page, 2001. http://www.mcs.anl.gov/petsc.

[112] Magdalena Balazinska, Hari Balakrishnan, and Mike Stonebraker. Contract-based load management in federated distributed systems. In *1st Symposium on Networked Systems Design and Implementation (NSDI)*, pages 197–210, San Francisco, USA, March 2004. USENIX Association.

[113] K. Ballinger, P. Brittenham, A. Malhorta, W. A. Nagy, and S. Pharies. *Web Services Inspection Language (WS-Inspection)*, Nov. 2001.

[114] Paul Baran. On distributed communications, rm-3420. Technical report, `http://www.rand.org/about/history/baran.list.html`, 1964.

[115] Paul Barham, Boris Dragovic, Keir Fraser, Steven Hand, Tim Harris, Alex Ho, Rolf Neugebauer, Ian Pratt, and Andrew Warfield. Xen and the art of virtualization. In *SOSP '03: Proceedings of the nineteenth ACM symposium on Operating systems principles*, pages 164–177, New York, NY, USA, 2003. ACM.

[116] Paul Barham, Boris Dragovic, Keir Fraser, Steven Hand, Tim Harris, Alex Ho, Rolf Neugebauer, Ian Pratt, and Andrew Warfield. Xen and the art of virtualization. *SIGOPS Oper. Syst. Rev.*, 37:164–177, October 2003.

[117] Paul Barham, Boris Dragovic, Keir Fraser, Steven Hand, Tim Harris, Alex Ho, Rolf Neugebauer, Ian Pratt, and Andrew Warfield. Xen and the Art of Virtualization. In *Proceedings of the nineteenth ACM symposium on Operating systems principles*, pages 164 – 177, 2003.

[118] Paul Barham, Boris Dragovic, Keir Fraser, Steven Hand, Tim Harris, Alex Ho, Rolf Neugebauer, Ian Pratt, and Andrew Warfield. Xen and the art of virtualization. In *Proceedings of the nineteenth ACM symposium on Operating systems principles*, SOSP '03, pages 164–177, New York, NY, USA, 2003. ACM.

[119] Luiz André Barroso and Urs Hölzle. The case for energy-proportional computing. *IEEE Computer*, 40(12):33–37, 2007.

[120] J. Basney. *A Distributed Implementation of the C-Linda Programming Language*. PhD thesis, Oberlin College, June 1995.

[121] Christian Baun, Marcel Kunze, Jens Nimis, and Stefan Tai. *Cloud Computing: Web-basierte dynamische IT-Services*. Springer, 2010.

[122] B. Benatallah, M. Dumas, M.C. Fauvet, and F.A. Rabhi. *Towards patterns of web services composition*. Springer-Verlag, 2003.

[123] Hal Berenson, Phil Bernstein, Gray Jim, Jim Melton, Elizabeth O'Neil, and Patrick O'Neil. A critique of ANSI SQL isolation levels. In *Proceedings of the 1995 ACM SIGMOD international Conference on Management of Data*, pages 1–10. ACM, 1995.

[124] Berkeley Millenium Project. Project Website: `https://www.millennium.berkeley.edu`.

[125] Laurent Bernaille, Renata Teixeira, and Kave Salamatian. Early application identification. In *Proceedings of the 2006 ACM CoNEXT conference*, CoNEXT '06, pages 6:1–6:12, New York, NY, USA, 2006. ACM.

[126] D. Bernstein, E. Ludvigson, K. Sankar, S. Diamond, and M. Morrow. Blueprint for the Intercloud – Protocols and Formats for Cloud Computing Interoperability. In *Proceedings of the 4th International Conference on Internet and Web Applications and Services (ICIW)*, Venice/Mestre, Italy, May 24–28 2009.

[127] David Bernstein and Erik Ludvigson. Networking challenges and resultant approaches for large scale cloud construction. *Workshops at the Grid and Pervasive Computing Conference*, 2009.

[128] WoShun Luk Bin Zhou, Jian Pei. A brief survey on anonymization techniques for privacy preserving publishing of social network data. *ACM SIGKDD Explorations Newsletter*, 10(Issue 2), 2008.

[129] Andrew D. Birrell and Bruce Jay Nelson. Implementing remote procedure calls. *ACM Transactions on Computer Systems*, 2:39–59, 1984.

[130] Tomas Bittman. The evolution of the cloud computing market. *Gartner Blog Network, http://blogs.gartner.com/thomas_bittman/2008/11/03/the-evolution-of-the-cloud-computing-market/*, November 2008.

[131] Blobstore API from Google's App Engine: http://code.google.com/intl/it-IT/appengine/docs/java/blobstore/.

[132] Harold Boley. The ruleml family of web rule languages. In José Júlio Alferes, James Bailey, Wolfgang May, and Uta Schwertel, editors, *PP-SWR*, volume 4187 of *Lecture Notes in Computer Science*, pages 1–17. Springer, 2006.

[133] Matt Bone, Peter F. Nabicht, Konstantin Läufer, and George K. Thiruvathukal. Taming XML: Objects first, then markup. In *Proc. IEEE Intl. Conf. on Electro/Information Technology (EIT)*, May 2008.

[134] M. Boniface, S. C. Phillips, A. Sanchez-Macian, and M. Surridge. Dynamic service provisioning using GRIA SLAs. In *International Workshops on Service-Oriented Computing (ICSOC'07)*, 2007.

[135] O. J. Boxma, J. W. Cohen, and N. Huffel. Approximations of the mean waiting time in an $M/G/s$ queueing system. *Operations Research*, 27:1115–1127, 1979.

[136] Ivona. Brandic. Towards self-manageable cloud services. In *33rd Annual IEEE International Computer Software and Applications Conference (COMPSAC'09)*, 2009.

[137] Ivona Brandic, Dejan Music, Philipp Leitner, and Schahram Dustdar. Vieslaf framework: Enabling adaptive and versatile sla-management. In *Proceedings of the 6th International Workshop on Grid Economics and Business Models*, GECON '09, pages 60–73, 2009.

[138] J. P. Brans, P. Vincke, and B. Mareschal. How to select and how to rank projects: The PROMETHEE method. *European Journal of Operational Research*, 24(2):228–238, 1986.

[139] Matthias Brantner, Daniela Florescu, David A. Graf, Donald Kossmann, and Tim Kraska. Building a database on S3. In *SIGMOD Conference*, pages 251–264, 2008.

[140] Brein. Business objective driven reliable and intelligent grids for real business. http://www.eu-brein.com/.

[141] G. Breiter and M. Behrendt. Life cycle and characteristics of services in the world of cloud computing. *IBM Journal of Research and Development*, 53(4):3:1 –3:8, July 2009.

[142] John Brevik, Daniel Nurmi, and Rich Wolski. Predicting bounds on queuing delay for batch-scheduled parallel machines. In *PPoPP '06: Proceedings of the eleventh ACM SIGPLAN symposium on Principles and practice of parallel programming*, pages 110–118, New York, NY, USA, 2006. ACM.

[143] Eric A. Brewer. Towards robust distributed systems (abstract). In *PODC*, page 7, 2000.

[144] James Broberg, Rajkumar Buyya, and Zahir Tari. Metacdn: Harnessing 'storage clouds' for high performance content delivery. *J. Network and Computer Applications*, 32(5):1012–1022, 2009.

[145] M. Brock and A. Goscinski. Attributed publication and selection for web service-based distributed systems. In *Proceedings of the 3rd International Workshop on Service Intelligence and Computing (SIC 2009) in conjunction with the 7th IEEE International Conference on Web Services (ICWS 2009)*, pages 732–739. IEEE Los Angeles, CA, USA, 2009.

[146] M. Brock and A. Goscinski. Toward ease of discovery, selection and use of clusters within a cloud. *IEEE 3rd International Conference on Cloud Computing (CLOUD 2010)*, July 5–10 2010.

[147] Chris Bunch, Navraj Chohan, Chandra Krintz, Jovan Chohan, Jonathan Kupferman, Puneet Lakhina, Yiming Li, and Yoshihide Nomura. An Evaluation of Distributed Datastores Using the AppScale Cloud Platform. In *IEEE Cloud10: International Conference on Cloud Computing*, July 2010.

[148] Michael Burrows. The Chubby Lock Service for Loosely-Coupled Distributed Systems. In *OSDI*, pages 335–350, 2006.

[149] R. Buyya. *Economic-based distributed resource management and scheduling for grid computing.* PhD thesis, Citeseer, 2002.

[150] R. Buyya, C. S. Yeo, S. Venugopal, J. Broberg, and I. Brandic. Cloud computing and emerging IT platforms: Vision, hype, and reality for delivering computing as the 5th utility. *Future Generation Computer Systems*, 25(6):599–616, 2009.

[151] R. Buyya, Chee Shin Yeo, and S. Venugopal. Market-oriented cloud computing: Vision, hype, and reality for delivering it services as computing utilities. In *High Performance Computing and Communications, 2008. HPCC '08. 10th IEEE International Conference on*, pages 5–13, Sep 2008.

[152] R. Buyya, C.S. Yeo, S. Venugopal, J. Broberg, and I. Brandic. Cloud computing and emerging IT platforms: Vision, hype, and reality for delivering computing as the 5th utility. *Future Generation Computer Systems*, 25(6):599–616, 2009.

[153] Rajkumar Buyya. Cloud computing: The next revolution in information technology. In *Parallel Distributed and Grid Computing (PDGC), 2010 1st International Conference on*, pages 2–3, October 2010.

[154] Rajkumar Buyya, Chee S. Yeo, and Srikumar Venugopal. Market-Oriented Cloud Computing: Vision, Hype, and Reality for Delivering IT Services as Computing Utilities. In *HPCC '08: Proceedings of the 2008 10th IEEE International Conference on High Performance Computing and Communications*, pages 5–13. IEEE Computer Society, September 2008.

[155] Rajkumar Buyya, Chee Shin Yeo, Srikumar Venugopal, James Broberg, and Ivona Brandic. Cloud computing and emerging it platforms: Vision, hype, and reality for delivering computing as the 5th utility. *Future Gener. Comput. Syst.*, 25:599–616, June 2009.

[156] Rajkumar Buyya, Chee Shin Yeo, Srikumar Venugopal, James Broberg, and Ivona Brandic. Cloud computing and emerging it platforms: Vision, hype, and reality for delivering computing as the 5th utility. *Future Generation Computer Systems*, 25:599–616, June 2009.

[157] Cactus Framework. http://www.cactuscode.org.

[158] R. N. Calheiros, R. Buyya, and C. A. F. De Rose. Building an automated and self-configurable emulation testbed for grid applications. *Software: Practice and Experience*, 40(5):405–429, 2010.

[159] Andrea Calì, Georg Gottlob, and Thomas Lukasiewicz. A general datalog-based framework for tractable query answering over ontologies. In *Proceedings of the twenty-eighth ACM SIGMOD-SIGACT-SIGART symposium on Principles of database systems*, PODS '09, pages 77–86, New York, NY, USA, 2009. ACM.

[160] Roy Campbell, Indranil Gupta, Michael Heath, Steve Ko, Michael Kozuch, Marcel Kunze, Thomas Kwan, Kevin Lai, Hing Yan Lee, Martha Lyons, Dejan Milojicic, David O'Hallaron, and Yeng Chai Soh. Open Cirrus^{TM} Cloud Computing Testbed: Federated Data Centers for Open Source Systems and Services Research. In *HotCloud'09: Workshop in Hot Topics in Cloud Computing*, 2009.

[161] K. Selcuk Candan, Wen-Syan Li, Thomas Phan, and Minqi Zhou. Frontiers in information and software as services. *Data Engineering, International Conference on*, 0:1761–1768, 2009.

[162] C. Carlsson and R. Fullér. OWA operators for decision support. *Proceedings of EUFIT*, 97:8–11, 1997.

[163] Jason Carolan and Steve Gaede. Introduction to cloud computing architecture. *Sun Microsystems Inc white paper*, 2009.

[164] Nicholas Carr and Yan Yu. It is no longer important:the internet great change of the high ground - cloud computing. *The Big Switch:Rewiring the World, from Edison to Google, CITIC Publishing House, October,* 2008.

[165] N. Carriero and D. Gelernter. *How to Write Parallel Programs - A First Course.* The MIT Press, Cambridge, MA, 1990.

[166] Nicholas Carriero and David Gelernter. Linda in context. *Commun. ACM,* 32(4):444–458, 1989.

[167] Chuck Cavaness. *Quartz Job Scheduling Framework: Building Open Source Enterprise Applications.* Prentice Hall PTR, Upper Saddle River, NJ, USA, 2006.

[168] Nimbus Science Cloud: http://workspace.globus.org/clouds/nimbus.html.

[169] Antonio Celesti, Francesco Tusa, Massimo Villari, and Antonio Puliafito. How to enhance cloud architectures to enable cross-federation. *Cloud Computing, IEEE International Conference on,* 0:337–345, 2010.

[170] Memorial Sloan-Kettering Cancer Center and the University of Toronto. Pathway commons::sif interaction rules: `http://www.pathwaycommons. org/pc/sif_interaction_rules.do`, 26 December 2010.

[171] Damien Cerbelaud, Shishir Garg, and Jeremy Huylebroeck. Opening the clouds: qualitative overview of the state-of-the-art open source vm-based cloud management platforms. In *Proceedings of the 10th ACM/IFIP/USENIX International Conference on Middleware,* Middleware '09, pages 22:1–22:8, New York, NY, USA, 2009. Springer-Verlag New York, Inc.

[172] J. Cerviño, P. Rodríguez, J. Salvachúa, F. Escribano, and G. Huecas. Marte 3.0: Una videoconferencia 2.0. In *JITEL 2008,* pages 209–216, 2008.

[173] Rohit Chandra. *Parallel Programming in OpenMP.* Morgan Kauffman, 2001.

[174] M. Chandy and L. Lamport. Distributed snapshots: Determining global states of distributed systems. *ACM Transactions on Computing Systems,* 3(1):63 –75, 1985.

[175] Elizabeth Chang, Farookh Hussain, and Tharam Dillon. *Trust and Reputation for Service-Oriented Environments: Technologies For Building Business Intelligence And Consumer Confidence.* John Wiley & Sons, 2005.

[176] Fay Chang, Jeffrey Dean, Sanjay Ghemawat, Wilson C. Hsieh, Deborah A. Wallach, Michael Burrows, Tushar Chandra, Andrew Fikes, and Robert E. Gruber. Bigtable: A Distributed Storage System for Structured Data. *ACM Trans. Comput. Syst.*, 26(2), 2008.

[177] David Chappell. Introducing the windows Azure platform. David Chappell & Associates, October 2010.

[178] David Chappell. Introducing Windows Aazure. David Chappell & Associates, October 2010.

[179] S. Chen, S. Nepal and B. Yan. Sdsi: Simple distributed storage interface, csiro ict centre technical report, no. 06/322., 2006.

[180] Shiping Chen, Surya Nepal, Jonathan Chan, David Moreland, and John Zic. Virtual storage services for dynamic collaborations. In *WETICE*, pages 186–191, 2007.

[181] Yu Cheng, A. Leon-Garcia, and I. Foster. Toward an autonomic service management framework: A holistic vision of soa, aon, and autonomic computing. *Communications Magazine, IEEE*, 46(5):138 –146, May 2008.

[182] S. Cheshire and M. Krochmal. DNS-Based Service Discovery, Oct. 25, 2010. Internet-Draft.

[183] R. Chinnici, J-J Moreau, A. Ryman, and S. Weerawarana. Web services description language (WSDL) version 2.0 part 1: Core language. W3C Recommendation, June 2007. Available from http://www.w3.org/TR/wsdl20.

[184] C.Hoff. Cloud Taxonomy and Ontology : http://rationalsecurity.typepad.com/blog/2009/01/cloud-computing-taxonomy-ontology.html, 2009.

[185] Navraj Chohan, Chris Buch, Sydney Pang, Chandra Krintz, Nagy Mostafa, Sunil Soman, and Rich Wolski. AppScale Design and Implementation. Technical report, UCSB, January 2009.

[186] Frederick Chong, Gianpaolo Carraro, and Roger Wolter. Multi-tenant data architecture, 2006.

[187] Online, available at http://msdn.microsoft.com/en-us/library/aa479086.aspx

[188] T.C.K. Chou and J.A. Abraham. Load balancing in distributed systems. *Software Engineering, IEEE Transactions on*, SE-8(4):401–412, July 1982.

[189] Richard Chow, Philippe Golle, Markus Jakobsson, Elaine Shi, Jessica Staddon, Ryusuke Masuoka, and Jesus Molina. Controlling data in the cloud: outsourcing computation without outsourcing control. In *Proceedings of the 2009 ACM workshop on Cloud computing security*, CCSW '09, pages 85–90, New York, NY, USA, 2009. ACM.

[190] J. Christiansen. COPYING Y COMBINATOR – A framework for developing Seed Accelerator Programmes. *MBA Thesis*, 2009.

[191] C. Clark, K. Fraser, S. Hand, J. G. Hanseny, E. July, C. Limpach, I. Pratt, and A. Warfield. Live Migration of Virtual Machines. In *Proceedings of the 2nd Symposium on Networked Systems Design and Implementation (NSDI)*, Boston, USA, May 2–4, 2005.

[192] Christopher Clark, Keir Fraser, Steven Hand, Jacob Gorm Hansen, Eric Jul, Christian Limpach, Ian Pratt, and Andrew Warfield. Live migration of virtual machines. In *Proceedings of the 2nd conference on Symposium on Networked Systems Design & Implementation - Volume 2*, NSDI'05, pages 273–286, Berkeley, CA, USA, 2005. USENIX Association.

[193] S. Clayman, A. Galis, and L. Mamatas. Monitoring virtual networks with lattice. In *Network Operations and Management Symposium Workshops (NOMS Wksps), 2010 IEEE/IFIP*, pages 239 –246, April 2010.

[194] Cloud Security Alliance. Top threats to cloud computing v 1.0. http://www.cloudsecurityalliance.org/topthreats/csathreats.v1.0.pdf, 2010. (accessed July 20, 2010).

[195] Cloud4SOA. http://www.cloud4soa.eu/, Dec. 2010.

[196] CloudSigma. Cloud hosting interface: http://www.cloudsigma.com/en/platform-details/intuitive-web-interface, October 2010.

[197] C.A.C. Coello. A comprehensive survey of evolutionary-based multiobjective optimization techniques. *Knowledge and Information systems*, 1(3):129–156, 1999.

[198] C.A.C. Coello and M.S. Lechuga. MOPSO: A proposal for multiple objective particle swarm optimization. *Proceedings of the Evolutionary Computation on*, pages 1051–1056, 2002.

[199] M. Commuzzi, C. Kotsokalis, G. Spanoudakis, and R. Yahyapour. Establishing and Monitoring SLAs in Complex Service Based Systems. In *Proceedings of the International Conference on Web Services (ICWS)*, Los Angeles, USA, July 6–10 2009.

[200] Adaptive Computing. Adaptive operating environment: http://www.adaptivecomputing.com/solutions/solution-architecture.php, November 2010.

[201] Penguin Computing. Penguin computing on demand (pod): `http://www.penguincomputing.com/POD`, 16 November 2010.

[202] Penguin Computing. Security on penguin computing on demand (pod) `http://www.penguincomputing.com/files/datasheets/Security_on_POD.pdf`, 10 April 2010.

[203] Penguin Computing. Security `http://www.penguincomputing.com/POD/Security`, 16 November 2010.

[204] M. Comuzzi, C. Kotsokalis, G. Spanoudkis, and R. Yahyapour. Establishing and monitoring SLAs in complex service based systems. In *IEEE International Conference on Web Services 2009*, 1009.

[205] Brian F. Cooper, Raghu Ramakrishnan, Utkarsh Srivastava, Adam Silberstein, Philip Bohannon, Hans-Arno Jacobsen, Nick Puz, Daniel Weaver, and Ramana Yerneni. PNUTS: Yahoo!'s hosted data serving platform. *PVLDB*, 1(2):1277–1288, 2008.

[206] Brian F. Cooper, Adam Silberstein, Erwin Tam, Raghu Ramakrishnan, and Russell Sears. Benchmarking cloud serving systems with YCSB. In *ACM SoCC*, pages 143–154, 2010.

[207] Intel Corp. Intel xeon processor x5570. `http://ark.intel.com/Product.aspx?id=37111`.

[208] R. S. Cox, J. G. Hansen, S. D. Gribble, and H. M. Levy. A safety-oriented platform for Web applications. *2006 IEEE Symposium on Security and Privacy*, 2006.

[209] D. Crockford. The application/json Media Type for JavaScript Object Notation (JSON). RFC 4627 (Informational), July 2006.

[210] S. Crosby, R. Doyle, M. Gering, M. Gionfriddo, S. Grarup, S. Hand, M. Hapner, D. Hiltgen, M. Johanssen, L.J. Lamers, J. Leung, F. Machida, A. Maier, E. Mellor, J. Parchem, S. Pardikar, S.J. Schmidt, R. W. Schmidt, A. Warfield, M.D. Weitzel, and J. Wilson. Open Virtualization Format Specification (OVF). Technical Report DSP0243, 2009.

[211] Cloud security alliance: Security guidance for critical areas of focus in cloud computing `http://www.cloudsecurityalliance.org/guidance`.

[212] Cloud security alliance: Top threats to cloud computing http://www.cloudsecurityalliance.org/topthreats/csathreats.v1.0.pdf.

[213] Carlo Curino, Evan Jones, Yang Zhang, Eugene Wu, and Sam Madde. Relational Cloud: The Case for a Database Service. In *CIDR*, 2011.

[214] CycleComputing. Cyclecloud: `http://www.cyclecomputing.com/products/cyclecloud/overview`, November 2010.

[215] John Daintith. *Oxford Dictionary of Computing*. Oxford University Press, 4th edition, 2004.

[216] A. D'Ambrogio and P. Bocciarelli. A model-driven approach to describe and predict the performance of composite services. In *6th International Workshop on Software and Performance (WOSP'07)*, 2007.

[217] Ernesto Damiani, De Capitani di Vimercati, Stefano Paraboschi, Pierangela Samarati, and Fabio Violante. A reputation-based approach for choosing reliable resources in peer-to-peer networks. In *Proceedings of the 9th ACM conference on Computer and communications security*, CCS '02, pages 207–216, New York, NY, USA, 2002. ACM.

[218] A. Dan, D. Davis, R. Kearney, A. Keller, R. King, D. Kuebler, H. Ludwig, M. Polan, M. Spreitzer, and A. Youssef. Web services on demand: Wsla-driven automated management. *IBM Syst. J.*, 43:136–158, January 2004.

[219] C. Darwin. *The origin of species by means of natural selection of the preservation of favoured races in the struggle for life*. Signet Classics, 2003.

[220] A.V. Dastjerdi, S.G.H. Tabatabaei, and R. Buyya. An Effective Architecture for Automated Appliance Management System Applying Ontology-Based Cloud Discovery. In *Cluster, Cloud and Grid Computing (CCGrid), 2010 10th IEEE/ACM International Conference on*, pages 104–112. IEEE, 2010.

[221] D. Baran. Cloud computing basics. *ICWS*, 2007.

[222] S. K. De, R. Biswas, and A. R. Roy. An application of intuitionistic fuzzy sets in medical diagnosis. *Fuzzy Sets and Systems*, 117(2):209–213, 2001.

[223] Marcos Dias de Assunção, Alexandre di Costanzo, and Rajkumar Buyya. Evaluating the cost-benefit of using cloud computing to extend the capacity of clusters. In *HPDC*, pages 141–150, 2009. http://doi.acm.org/10.1145/1551609.1551635.

[224] J. de Bruijn, H. Lausen, R. Krummenacher, A. Polleres, L. Predoiu, M. Kifer, and D. Fensel. The web service modeling language WSML. *WSML Final Draft D*, 16, 2005.

[225] J.C. de Oliveira, C. Scoglio, I.F. Akyildiz, and G. Uhl. New preemption policies for diffserv-aware traffic engineering to minimize rerouting in mpls networks. *Networking, IEEE/ACM Transactions on*, 12(4):733 – 745, Aug 2004.

[226] Jeffrey Dean and Sanjay Ghemawat. Mapreduce: Simplified data processing on large clusters. In *Proceedings of the 6th Conference on Symposium on Opearting Systems Design and Implementation*, pages 10–10, Berkeley, CA, USA, 2004. USENIX Association.

[227] K. Deb, A. Pratap, S. Agarwal, and T. Meyarivan. A fast and elitist multiobjective genetic algorithm: NSGA-II. *IEEE transactions on evolutionary computation*, 6(2):182–197, 2002.

[228] Giuseppe DeCandia, Deniz Hastorun, Madan Jampani, Gunavardhan Kakulapati, Avinash Lakshman, Alex Pilchin, Swaminathan Sivasubramanian, Peter Vosshall, and Werner Vogels. Dynamo: Amazon's highly available key-value store. In *SOSP*, pages 205–220, 2007.

[229] Ewa Deelman, Gurmeet Singh, Miron Livny, Bruce Berriman, and John Good. The cost of doing science on the Cloud: The montage example. In *2008 ACM/IEEE Conference on Supercomputing (SC 2008)*, pages 1–12, Piscataway, NJ, USA, 2008. IEEE Press.

[230] Chrysanthos Dellarocas. Reputation mechanism design in online trading environments with pure moral hazard. *Info. Sys. Research*, 16:209–230, June 2005.

[231] J. B. Dennis. Data flow supercomputers. *Computer*, pages 48–56, 1980.

[232] Department of Defense. *Trusted Computer System Evaluation Criteria*, Dec. 26, 1985. DoD 5200.28-STD.

[233] Todd Deshane, Zachary Shepherd, Jeanna N. Matthews, Muli Ben-Yehuda, Amit Shah, and Balaji Rao. Quantitative comparison of Xen and KVM. *Xen summit*, June 2008.

[234] N. A. Detorie. Helical tomotherapy: A new tool for radiation therapy. In *Journal of the American College of Radiology*, pages 63–66, 2008.

[235] Amazon DevPay http://aws.amazon.com/devpay.

[236] Prashant Dewan and Partha Dasgupta. Pride: peer-to-peer reputation infrastructure for decentralized environments. In *WWW (Alternate Track Papers & Posters)*, pages 480–481, 2004.

[237] M. D. Dikaiakos, D. Katsaros, P. Mehra, G. Pallis, and A. Vakali. Cloud computing: Distributed internet computing for it and scientific research. *Internet Computing, IEEE*, 13(5):10–13, Sep 2009.

[238] Tham Dillon, Chen Wu, and Elizabeth Chang. Cloud computing : Issues and challenges. *24th IEEE International Conference on Advanced Information Networking and Applications*, 2010.

[239] Distributed Management Task Force. *Interoperable Clouds – A White Paper from the Open Clouds Standards Incubator*, Nov. 2009. DSP-IS0101.

[240] Distributed Management Task Force. *Open Virtualization Format White Paper*, Feb. 2009. DSP2017 Version 1.0.0.

[241] Distributed Management Task Force. *Open Virtualization Format Specification*, Jan. 12, 2010. DSP0243 Version 1.1.0.

[242] G. Dobson and A. Sanchez-Macian. Towards unified QoS/SLA ontologies. In *IEEE Services Computing Workshops (SCW'06)*, 2006.

[243] Dropbox: cloud based online backup `http://www.dropbox.com`.

[244] J. J. Dujmovic. Mixed Averaging by Levels (MAL)–A System and Computer Evaluation Method. In *Proceedings of the Informatica Conference, Bled, Yugoslavia*, 1973.

[245] E. Cristensen, F. Curbera, G. Meredith, S. Weerawarana. Web Services Description Language (WSDL) 1.1: http://www.w3.org/TR/2001/NOTE-wsdl-20010315, 2001.

[246] Amazon elastic compute cloud. http://aws.amazon.com/ec2/.

[247] C. Perkins (ed.). *IP Mobility Support for IPv4*, Aug. 2002. RFC 3344.

[248] Vincent. C. Emeakaroha, I. Brandic, M. Maurer, and S. Dustdar. Low level metrics to high level SLAs - LoM2HiS framework: Bridging the gap between monitored metrics and SLA parameters in cloud environments. In *High Performance Computing and Simulation Conference (HPCS'10)*, 2010.

[249] Vincent C. Emeakaroha, Rodrigo N. Calheiros, Marco A. S. Netto, Ivona Brandic, and César A. F. De Rose. DeSVi: An architecture for detecting SLA violations in cloud computing infrastructures. In *Proceedings of the 2nd International ICST Conference on Cloud Computing (Cloud-Comp'10)*, 2010.

[250] Patrícia Takako Endo, Glauco Estácio Gonçalves, Judith Kelner, and Djamel Sadok. A Survey on Open-Source Cloud Computing Solutions. *Brazilian Symposium on Computer Networks and Distributed Systems*, May 2010.

[251] ENISA – cloud computing: Benefits, risks and recommendations for information security `http://www.enisa.europa.eu/act/rm/files/deliverables/cloud-computing-risk-assessment`.

[252] U. Villano, E. P. Mancini, M. Rak. PerfCloud: GRID Services for Performance-oriented Development of Cloud Computing Applications. In *Proceedings of WETICE*. IEEE Computer Society, July 2009.

[253] ESPER. Event stream processing. `http://esper.codehaus.org/`.

[254] A. Goscinski et al. The Cloud Miner: Moving Data Mining to Computational Clouds. Grid and Cloud Database Management. Springer, 2010.

[255] I. Khalil et al. The potential of biologic network models in understanding the etiopathogenesis of ovarian cancer. *Gynecol Oncol. 116(2):282-5*, 2004.

[256] J. Shendure et al. Next-generation dna sequencing. *Nat Biotech 26*, pages 1135–1145, 2008.

[257] P. Shannon et al. Cytoscape: a software environment for integrated models of biomolecular interaction networks. *Genome Research, 13(11):2498-504*, 2003.

[258] Eucalyptus Documentation Page. http://www.eucalyptus.com/.

[259] C. Evangelinos and C. Hill. Cloud computing for parallel scientific hpc applications: Feasibility of running coupled atmosphere-ocean climate models on amazon's ec2. *Cloud Computing and its Applications*, 2008.

[260] M. Factor, K. Meth, D. Naor, O. Rodeh, and J. Satran. Object Storage: the Future Building Block for Storage Systems. In *Proceedings of the 2nd International Symposium on Mass Storage Systems and Technology*, Sardinia, Italy, June 20–24, 2005.

[261] Ming Fan, Yong Tan, and Andrew B. Whinston. Evaluation and design of online cooperative feedback mechanisms for reputation management. *IEEE Trans. on Knowl. and Data Eng.*, 17:244–254, February 2005.

[262] Xiaobo Fan, Wolf-Dietrich Weber, and Luiz André Barroso. Power provisioning for a warehouse-sized computer. In *34th International Symposium on Computer Architecture (ISCA 2007), June 9-13, 2007, San Diego, California, USA*, pages 13–23.

[263] D. Farinacci, V. Fuller, D. Meyer, and D. Lewis. *Locator/ID Separation Protocol (LISP)*, March 2 2009. Internet-Draft.

[264] D. Fensel and C. Bussler. The web service modeling framework WSMF. *Electronic Commerce Research and Applications*, 1(2):113–137, 2002.

[265] J. M. Fernández Salido and S. Murakami. Extending Yager's orness concept for the OWA aggregators to other mean operators. *Fuzzy Sets and Systems*, 139(3):515–542, 2003.

[266] D. F. Ferraiolo and D. R. Kuhn. Role-Based Access Controls. In *Proceedings of the 15th National Computer Security Conference*, Baltimore, USA, Oct 13-16 1992.

[267] Tiago C. Ferreto, Cesar A. F. de Rose, and Luiz de Rose. Rvision: An open and high configurable tool for cluster monitoring. *Cluster Computing and the Grid, IEEE International Symposium on*, 0:75, 2002.

[268] Joe Fialli and Sekhar Vajjhala. Java architecture for XML binding (JAXB) 2.0. Java Specification Request (JSR) 222, October 2005.

[269] R. Fielding. Representational state transfer (rest). *Architectural Styles and the Design of Network-based Software Architectures. University of California, Irvine*, 2000.

[270] R. Fielding, H. Frystyk, Tim Berners-Lee, J. Gettys, and J. C. Mogul. Hypertext transfer protocol - HTTP/1.1, 1996.

[271] Roy T. Fielding. *Architectural Styles and the Design of Network-based Software Architectures*. PhD thesis, University of California, 2000.

[272] J. Figueira, S. Greco, and M. Ehrgott. *Multiple criteria decision analysis: state of the art surveys*. Springer Verlag, 2005.

[273] Flexiscale. http://www.flexiant.com/products/flexiscale/, Dec. 2010.

[274] Daniela Florescu and Donald Kossmann. Rethinking cost and performance of database systems. *SIGMOD Record*, 38(1):43–48, 2009.

[275] Michael Flynn. Some computer organizations and their effectiveness. *IEEE Trans. Comput.*, c(21):948, 1972.

[276] MESSAGE-PASSING INTERFACE FORUM. *MPI: A message-passing interface standard.* http://www.mpi.org, 1994.

[277] FoSII. Foundations of self-governing infrastructures. http://www.infosys.tuwien.ac.at/linksites/FOSII/index.html.

[278] I. Foster. What's faster - a supercomputer or ec2?: http://ianfoster.typepad.com, 2009.

[279] I. Foster, C. Kesselman, J. M. Nick, and S. Tuecke. The physiology of the grid. *Grid computing: making the global infrastructure a reality*, pages 217–250, 2003. Wiley and Sons.

[280] I. Foster, C. Kesselman, and S. Tuecke. The anatomy of the grid: Enabling scalable virtual organizations. *International Journal of High Performance Computing Applications*, 15(3):200, 2001.

[281] I. Foster, Y. Zhao, I. Raicu, and S. Lu. Cloud computing and grid computing 360-degree compared. In *Grid Computing Environments Workshop, 2008. GCE'08*, pages 1–10. IEEE, 2009.

[282] I. Foster, Yong Zhao, I. Raicu, and S. Lu. Cloud computing and grid computing 360-degree compared. In *Grid Computing Environments Workshop, 2008. GCE '08*, pages 1–10, Nov 2008.

[283] Ian Foster, Yong Zhao, Ioan Raicu, and Shiyong Lu. Cloud computing and grid computing 360-degree compared. *3rd International Conference on Grid and Pervasive Computing-gpc-workshops*, 2008.

[284] Ian Foster, Yong Zhao, Ioan Raicu, and Shiyong Lu. Cloud computing and grid computing 360-degree compared. 2009.

[285] J. Franks, P. Hallam-Baker, J. Hostetler, S. Lawrence, P. Leach, A. Luotonen, and L. Stewart. *HTTP Authentication Basic and Digest Access Authentication*, June 1999. RFC 2617.

[286] FreeCBR. Online. http://freecbr. Sourceforge.net/.

[287] H. M. Frutos and I. Kotsiopoulos. BREIN: Business objective driven reliable and intelligent grids for real business. *International Journal of Interoperability in Business Information Systems*, 3(1):39–42, 2009.

[288] Kevin Fu, M. Frans Kaashoek, and David Mazières. Fast and secure distributed read-only file system. *ACM Trans. Comput. Syst.*, 20:1–24, February 2002.

[289] W. Fu and Q. Huang. GridEye: A service-oriented grid monitoring system with improved forecasting algorithm. In *International Conference on Grid and Cooperative Computing Workshops*, 2006.

[290] Xiaodong Fu, Ping Zou, Ying Jiang, and Zhenhong Shang. Qos consistency as basis of reputation measurement of web service. In *Proceedings of the The First International Symposium on Data, Privacy, and E-Commerce*, pages 391–396, Washington, DC, USA, 2007. IEEE Computer Society.

[291] B. Furht. Cloud computing fundamentals. In Borko Furht and Armando Escalante, editors, *Handbook of Cloud Computing*, pages 3–19. Springer US, 2010.

[292] Galen Gruman and Eric Knorr. What cloud computing really means. InfoWorld : http://www.infoworld.com/article/08/04/07/15FE-cloud-computing-reality 1.html, 2008.

[293] Ganglia Project. Project Website: `http://ganglia.sourceforge.net/`.

[294] J. M. Garcıa, D. Ruiz, and A. Ruiz-Cortes. On User Preferences and Utility Functions in Selection: A Semantic Approach? In *Service-Oriented Computing-ICSOC 2007 Workshops: ICSOC 2007 International Workshops, Vienna, Austria, September 17, 2007, Revised Selected Papers*, page 105. Springer Verlag, 2009.

[295] J. M. Garcia, D. Ruiz, A. Ruiz-Cortes, and J.A. Parejo. Qos-aware semantic service selection: An optimization problem. In *Services-Part I, 2008. IEEE Congress on*, pages 384–388. IEEE, 2008.

[296] J. M. García, I. Toma, D. Ruiz, and A. Ruiz-Cortes. A service ranker based on logic rules evaluation and constraint programming. In *Proc. of 2nd Non Functional Properties and Service Level Agreements in SOC Workshop (NFPSLASOC)*, 2008.

[297] Francisco Garcia-Sanchez, Eneko Fernandez-Breis, Rafael Valencia-Garcia, Enrique Jimenez, Juan M. Gomez, Javier Torres-Niño, and Daniel Martinez-Maqueda. Sitio: semantic business processes based on software-as-a-service and cloud computing. In *Proceedings of the 8th WSEAS International Conference on E-Activities and information security and privacy*, E-ACTIVITIES'09/ISP'09, pages 130–135, Stevens Point, Wisconsin, USA, 2009. World Scientific and Engineering Academy and Society (WSEAS).

[298] Gartner. Gartner highlights five attributes of cloud computing: `http://www.gartner.com/it/page.jsp?id=1035013`, June 2009.

[299] S. I. Gass and T. Rapcsák. Singular value decomposition in AHP. *European Journal of Operational Research*, 154(3):573–584, 2004.

[300] J. Geelan. Twenty-one experts define cloud computing: `http://cloudcomputing.sys-con.com/node/612375`, 24 January 2009.

[301] F. Gens. A key driver of new growth. idc forecast. it cloud services forecast, 2008.

[302] Sanjay Ghemawat, Howard Gobioff, and Shun-Tak Leung. The google file system. *SIGOPS Oper. Syst. Rev.*, 37:29–43, October 2003.

[303] Sanjay Ghemawat, Howard Gobioff, and Shun-Tak Leung. The Google file system. In *SOSP*, pages 29–43, 2003.

[304] Seth Gilbert and Nancy A. Lynch. Brewer's conjecture and the feasibility of consistent, available, partition-tolerant web services. *SIGACT News*, 33(2):51–59, 2002.

[305] A. S. Glassner et al. *An introduction to ray tracing*. Academic Press London, 1989.

[306] Rich Mogull and Glenn Brunette. Security guidance for critical areas of focus in cloud computing. *Cloud Security Alliance*, December 2009.

[307] Gogrid. http://www.gogrid.com.

[308] Eu-Jin Goh, Hovav Shacham, Nagendra Modadugu, and Dan Boneh. Sirius: Securing remote untrusted storage. In *NDSS*, 2003.

[309] D. E. Goldberg. *Genetic algorithms in search, optimization, and machine learning*. Addison-Wesley, 1989.

[310] R. P. Goldberg. Architecture of virtual machines. In *Proceedings of the workshop on virtual computer systems*, pages 74–112, New York, NY, USA, 1973. ACM.

[311] R. P. Goldberg. Survey of virtual machine research. *IEEE Computer*, 7(6):34–45, June 1974.

[312] Google App Engine: http://code.google.com/appengine/.

[313] P. Goyal. Enterprise usability of cloud computing environments: Issues and challenges. In *Enabling Technologies: Infrastructures for Collaborative Enterprises (WETICE), 2010 19th IEEE International Workshop on*, pages 54–59, June 2010.

[314] Bernardo Cuenca Grau, Ian Horrocks, Boris Motik, Bijan Parsia, Peter Patel-Schneider, and Ulrike Sattler. Owl 2: The next step for owl. *Web Semant.*, 6:309–322, November 2008.

[315] Jim Gray. Distributed computing economics. Microsoft Research Technical Report MSRTR- 2003-24, Microsoft Research, 2003.

[316] R. Grønmo and M. C. Jaeger. Model-driven methodology for building qos-optimised web service compositions. In *Distributed Applications and Interoperable Systems*, pages 68–82. Springer, 2005.

[317] William Gropp and Ewing Lusk. Fault tolerance in mpi programs. *Special Issue, Journal of High Performance Computing Applications*, 1(18):363–372, 2002.

[318] Yaqing Gu and Dean S. Oliver. The ensemble kalman filter for continuous updating of reservoir simulation models. *Journal of Engineering Resources Technology*, 128(1):79–87, 2006.

[319] Yaqing Gu and Dean S. Oliver. An iterative ensemble kalman filter for multiphase fluid flow data assimilation. *SPE Journal*, 12(4):438–446, 2007.

[320] Saikat Guha, Kevin Tang, and Paul Francis. Noyb: privacy in online social networks. In *Proceedings of the first workshop on Online social networks*, WOSP '08, pages 49–54, New York, NY, USA, 2008. ACM.

[321] T. Guha and S. A. Ludwig. Comparison of Service Selection Algorithms for Grid Services: Multiple Objective Particle Swarm Optimization and Constraint Satisfaction Based Service Selection. In *Tools with Artificial Intelligence, 2008. ICTAI'08. 20th IEEE International Conference on*, volume 1, pages 172–179. IEEE, 2008.

[322] D. Gunter, B. Tierney, B. Crowley, M. Holding, and J. Lee. Netlogger: A toolkit for distributed system performance analysis. In *8th International Symposium on Modeling, Analysis and Simulation of Computer and Telecommunication Systems (MASCOTS'00)*, 2000.

[323] Chang Jie Guo, Wei Sun, Ying Huang, Zhi Hu Wang, and Bo Gao. A framework for native multi-tenancy application development and management. *E-Commerce Technology, IEEE International Conference on, and Enterprise Computing, E-Commerce, and E-Services, IEEE International Conference on*, 0:551–558, 2007.

[324] GWOS. Project Website: `http://www.groundworkopensource.com/products`.

[325] H9Labs. Cloud computing explained - part 2. *3rd International Conference on Grid and Pervasive Computing-gpc-workshops*, 2008.

[326] Volker Haarslev and Ralf MÃüller. Description of the racer system and its applications. In *Proc of DL2001 Workshop on Description Logics*, pages 132–141, 2001.

[327] H. Hacigumus, B. Iyer, and S. Mehrotra. Providing database as a service. In *Data Engineering, 2002. Proceedings. 18th International Conference on*, pages 29–38, 2002.

[328] Hakan Hacigümüs, Balakrishna R. Iyer, Chen Li, and Sharad Mehrotra. Executing SQL over encrypted data in the database-service-provider model. In *SIGMOD Conference*, pages 216–227, 2002.

[329] Hakan Hacigümüs, Sharad Mehrotra, and Balakrishna R. Iyer. Providing Database as a Service. In *ICDE*, 2002.

[330] Marc J. Hadley. Web application description language (WADL). Technical report, Sun Microsystems, Inc., Mountain View, CA, USA, 2006.

[331] Yan Hai and Zhao Chong. Security policies based on security requirements of city emergency management information system multi-layer structure. In *IT in Medicine Education, 2009. ITIME '09. IEEE International Symposium on*, volume 1, pages 351–354, August 2009.

[332] James Hamilton. On designing and deploying internet-scale services. In *LISA'07: Proceedings of the 21st conference on Large Installation System Administration Conference*, pages 1–12, Berkeley, CA, USA, 2007. USENIX Association.

[333] James R. Hamilton. Cooperative expendable micro-slice servers (CEMS): Low cost, low power servers for internet-scale services. In *Proceedings of the 4th Biennial Conference on Innovative Data Systems Research (CIDR)*, 2009.

[334] M. Handley, H. Schulzrinne, E. Schooler, and J. Rosenberg. SIP: Session Initiation Protocol. RFC 2543 (Proposed Standard), March 1999.

[335] Steve Hanna. Cloud computing: Finding the silver lining. `http://www.ists.dartmouth.edu/docs/HannaCloudComputingv2.pdf`, 2009. (accessed March 12, 2009).

[336] J. Hartigan. *Clustering Algorithms*. John Wiley and Sons, New York, 1975.

[337] M. M. Hassan, B. Song, S. M. Han, E. N. Huh, C. Yoon, and W. Ryu. Multi-objective Optimization Model for Partner Selection in a Market-Oriented Dynamic Collaborative Cloud Service Platform. In *Tools with Artificial Intelligence, 2009. ICTAI'09. 21st International Conference on*, pages 637–644. IEEE, 2009.

[338] Philipp C. Heckel. Hybrid Clouds: Comparing Cloud Toolkits. Technical report, University of Mannheim, 2010.

[339] Mark Hefke. A framework for the successful introduction of KM using CBR and semantic web technologies. *Journal of Universal Computer Science*, 10(6), 2004.

[340] Alexander Heitzmann, Bernardo Palazzi, Charalampos Papamanthou, and Roberto Tamassia. Efficient integrity checking of untrusted network storage. In *Proceedings of the 4th ACM international workshop on Storage security and survivability*, StorageSS '08, pages 43–54, New York, NY, USA, 2008. ACM.

[341] N. Hemsoth. Hpc in the cloud: An intelligent public cloud for hpc: Computing shares news of cluster compute involvement: `http://www.hpcinthecloud.com/features/99334309.html`, 27 July 2010.

[342] N. Hemsoth. Intel lays groundwork to fulfill 2015 cloud vision, hpc in the cloud: `http://www.hpcinthecloud.com/features/106350403.html`, 29 October 2010.

[343] N. Hemsoth. Renting hpc - what's cloud got to do with it?, hpc in the cloud: `http://www.hpcinthecloud.com/features/97452994.html`, 29 October 2010.

[344] N. Hemsoth. Virtualization and performance: Brocade's dr. maria iordache, hpc in the cloud: `http://www.hpcinthecloud.com/features/104296594.html`, 14 December 2010.

[345] N. Hemsoth. Will public clouds ever be suitable for hpc?, hpc in the cloud: `http://www.hpcinthecloud.com/features/97269804.html`, 29 October 2010.

[346] Nicole Hemsoth. Microsoft's Dan Reed on new paradigms for scientific discovery, hpc in the cloud: `http://www.hpcinthecloud.com/news/106298163.html`, October 2010.

[347] Heroku: a ruby cloud platform `http://www.heroku.com`.

[348] Tony Hey, Stewart Tansley, and Kristin Tolle, editors. *The Fourth Paradigm: Data-Intensive Scientific Discovery*. Microsoft Research, October 2009.

[349] Z. Hill and M. Humphrey. A quantitative analysis of high performance computing with amazon's ec2 infrastructure: The death of the local cluster? In *10th IEEE/ACM International Conference on Grid Computing. Banff*, pages 26–33. SIAM Press, 2009.

[350] D. Hilley. Cloud Computing: A Taxonomy of Platform & Infrastructure-level Offerings. Technical report, Georgia Institute of Technology, 2009.

[351] H. Lohninger. *Teach/Me Data Analysis*. Springer-Verlag, Berlin-New York-Tokyo, ISBN 3-540-14743-8, 1999.

[352] Christina Hoffa, Gaurang Mehta, Timothy Freeman, Ewa Deelman, Kate Keahey, Bruce Berriman, and John Good. On the Use of Cloud Computing for Scientific Workflows. *SWBES 2008*, December 2008.

[353] P. Hokstad. Approximations for the $M/G/m$ queues. *Operations Research*, 26:510–523, 1978.

[354] X. Hu and R. Eberhart. Multiobjective optimization using dynamic neighborhood particle swarm optimization. *Proceedings of the Evolutionary Computation on*, pages 1677–1681, 2002.

[355] Mei Hui, Dawei Jiang, Guoliang Li, and Yuan Zhou. Supporting database applications as a service. *Data Engineering, International Conference on*, 0:832–843, 2009.

[356] S. F. Hummel, E. Schonberg, and L. E. Flynn. Factoring: A method for scheduling parallel loops. *CACM*, 35(8):90–101, August 1992.

[357] Trung Dong Huynh, Nicholas R. Jennings, and Nigel R. Shadbolt. An integrated trust and reputation model for open multi-agent systems. *Autonomous Agents and Multi-Agent Systems*, 13:119–154, September 2006.

[358] C. L. Hwang and K. Yoon. *Multiple attribute decision making: methods and applications: a state-of-the-art survey*, volume 13. Springer-Verlag New York, 1981.

[359] Kai Hwang. Massively distributed systems: From grids and p2p to clouds. In *Proceedings of The 3rd International Conference on Grid and Pervasive Computing - gpc-workshops*, page xxii, 2008.

[360] Kai Hwang. Massively distributed systems:from grids and p2p to clouds. *3rd International Conference on Grid and Pervasive Computing-gpc-workshops*, 2008.

[361] Hyperic Community. Project Website: `http://www.hyperic.com/community`.

[362] Hyperic HQ. Project Website: `http://www.hyperic.com/`.

[363] Hyperic HQ. Hq inventory model. Project Website: `http://www.hyperic.com/`, 2009.

[364] I. Horrocks, P. F. Patel-Schneider, H. Boley, S. Tabet, B. Grosof, M. Dean. SWRL: A Semantic Web Rule Language Combining OWL and RuleML: http://www.w3.org/Submission/SWRL/.

[365] IBM. *WebSphere: MQ V6 Fundamentals*, 2005.

[366] IBM. *MC91: High Availability for WebSphere MQ on Unix Platforms, Vol.,7*, 2008.

[367] IBM. IBM corp, cloud computing. *Journal of Object Technology*, 2009.

[368] IBM. IBM point of view: Security and cloud computing. cloud computing white paper, 2009.

[369] I. Foster, C. Kesselman, and S. Tuecke. The anatomy of the grid:enabling scalable virtual organization. *The Intl Jrnl of High Performance Computing Applications 15(3):200-222*, 2001.

[370] I. Foster, C. Kesselman, and S. Tuecke. The physiology of the grid:an open grid services architecture for distributed systems integration. *Globus Project*, 2002.

[371] I. Foster, J. Vockler, M. Wilde, and Y. Zhao. Chimera: a virtual data system for representing,querying and automating data derivation. *SSDBM*, 2002.

[372] R. Ihaka and R. Gentleman. R: A language for data analysis and graphics. *Journal of Computational and Graphical Statistics*, pages 299–314, 1996.

[373] Silicon Graphics International. Hpc cloud computing - cyclone: `http://www.sgi.com/products/hpc_cloud/cyclone/`, 11 November 2010.

[374] Silicon Graphics International. Hpc cloud computing: Cyclone applications: `http://www.sgi.com/products/hpc_cloud/cyclone/applications.html`, 11 November 2010.

[375] Silicon Graphics International. Sgi announces cyclone cloud computing for technical applications: `http://www.sgi.com/company_info/newsroom/press_releases/2010/february/cyclone.html`, 16 November 2010.

[376] A. Iosup, H. Li, M. Jan, S. Anoep, C. Dumitrescu, L. Wolters, and D.H.J. Epema. The grid workloads archive. *Future Generation Computer Systems*, 2008.

[377] Alexandru Iosup, Dick H. J. Epema, Todd Tannenbaum, Matthew Farrellee, and Miron Livny. Inter-operating Grids through delegated matchmaking. In *2007 ACM/IEEE Conference on Supercomputing (SC 2007)*, pages 1–12, New York, USA, November 2007. ACM Press.

[378] I. Raicu, Y. Zhao, C. Dumitrescu, I. Foster, and M. Vilde. Falkon: A fast and light-weight task execution framework. *IEEE/ACM SuperComputing*, 2007.

[379] IT Infrastructure Library. http://www.itil-officialsite.com, Dec. 2010.

[380] Jabber Software Foundation. Extensible messaging and presence protocol (xmpp): Core. 2004. Website available at http://xmpp.org/rfcs/rfc3920.html

[381] Keith R. Jackson, Lavanya Ramakrishnan, Karl J. Runge, and Rollin C. Thomas. Seeking supernovae in the clouds: a performance study. In *Proceedings of the 19th ACM International Symposium on High Performance Distributed Computing*, HPDC '10, pages 421–429, New York, NY, USA, 2010. ACM.

[382] Dean Jacobs and Stefan Aulbach. Ruminations on multi-tenant databases. In Alfons Kemper, Harald Schöning, Thomas Rose, Matthias Jarke, Thomas Seidl, Christoph Quix, and Christoph Brochhaus, editors, *BTW*, volume 103 of *LNI*, pages 514–521. GI, 2007.

[383] M. Jaeger and G. Rojec-Goldmann. SENECA–simulation of algorithms for the selection of web services for compositions. *Technologies for E-Services*, pages 84–97, 2006.

[384] M. C. Jaeger, G. Muhl, and S. Golze. QoS-aware composition of web services: A look at selection algorithms. In *Web Services, 2005. ICWS 2005. Proceedings. 2005 IEEE International Conference on*. IEEE, 2005.

[385] M. C. Jaeger, G. Rojec-Goldmann, and G. Muhl. Qos aggregation for web service composition using workflow patterns. In *Enterprise Distributed Object Computing Conference, 2004. EDOC 2004. Proceedings. Eighth IEEE International*, pages 149–159. IEEE, 2005.

[386] Ravi Chandra Jammalamadaka, Roberto Gamboni, Sharad Mehrotra, Kent E. Seamons, and Nalini Venkatasubramanian. gvault: a gmail based cryptographic network file system. In *Proceedings of the 21st annual IFIP WG 11.3 working conference on Data and applications security*, pages 161–176, Berlin, Heidelberg, 2007. Springer-Verlag.

[387] Ravi Chandra Jammalamadaka, Roberto Gamboni, Sharad Mehrotra, Kent E. Seamons, and Nalini Venkatasubramanian. . a middleware approach for building secure network drives over untrusted internet data storage. In *Proc. ACM conference on Extending database technology: Advances in database technology*, pages 710–714, 2008.

[388] Jeffery F. Rayport and Andrew Heyward. Envisioning the cloud: The next computing paradigm. 2009. *Technical Report*, available at http://www.hp.com/hpinfo/analystrelations/marketspace-090320-Envisioning-the-cloud.pdf

[389] Meiko Jensen and Nils Gruschka. Flooding attack issues of web services and service-oriented architectures. In *GI Jahrestagung (1)*, pages 117–122, 2008.

[390] Meiko Jensen, Jorg Schwenk, Nils Gruschka, and Luigi Lo Iacono. On technical security issues in cloud computing. *IEEE International Conference on Cloud Computing*, 2009.

[391] Jeremy Geelan. Twenty-one experts define cloud computing. Virtualization : http://virtualization.sys-con.com/node/612375, 2008.

[392] J. Heiser and M. Nicolett. Assessing the security risks of cloud computing. *Gartner Report*, 2009.

[393] Jim Gray. The dangers of replication and a solution. In *ACM SIGMOD International Conference on Management of Data Archive*, pages 173 – 182, Montreal, Quebec, Canada, 1996.

[394] Enrique Jimenez Domingo, Javier Torres Nino, Angel Lagares Lemos, Miguel Lagares Lemos, Ricardo Colomo Palacios, and Juan Miguel Gomez Berbis. Cloudio: A cloud computing-oriented multi-tenant architecture for business information systems. *Cloud Computing, IEEE International Conference on*, 0:532–533, 2010.

[395] Chris Dyer Jimmy Lin. *Data-IntensiveText Processing with MapReduce*. Morgan & Claypool, 2010.

[396] Chao Jin and Rajkumar Buyya. Mapreduce programming model for .net-based cloud computing. In Henk Sips, Dick Epema, and Hai-Xiang Lin, editors, *Euro-Par 2009 Parallel Processing*, volume 5704 of *Lecture Notes in Computer Science*, pages 417–428. Springer Berlin / Heidelberg, 2009.

[397] JMS. Java messaging service. `http://java.sun.com/products/jms/`.

[398] D. Johnson, C. Perkins, and J. Arkko. *Mobility Support in IPv6*, June 2004. RFC 3775.

[399] Don W Jones. *The Definitive Guide to Monitoring the Data Center, Virtual Environments, and the Cloud*. Nimsoft, 2010.

[400] E. Jopling and D. Neill. Vno phenomenon could shake up the world's telecom market. Gartner Research Report g00131283, 2005.

[401] Joulemeter. Joulemeter: VM, server, client, and software energy usage. `http://research.microsoft.com/en-us/projects/joulemeter/`.

[402] Gueyoung Jung, M.A. Hiltunen, K.R. Joshi, R.D. Schlichting, and C. Pu. Mistral: Dynamically managing power, performance, and adaptation cost in cloud infrastructures. In *Distributed Computing Systems (ICDCS), 2010 IEEE 30th International Conference on*, pages 62 –73, june 2010.

[403] Fu K. Group sharing and random access in cryptographic storage file system. master's thesis. MIT, 1999.

[404] Avinash W. Kadam. Information security policy development and implementation. *Inf. Sys. Sec.*, 16:246–256, September 2007.

[405] Mahesh Kallahalla, Erik Riedel, Ram Swaminathan, Qian Wang, and Kevin Fu. Plutus: Scalable secure file sharing on untrusted storage. In *Proceedings of the 2nd USENIX Conference on File and Storage Technologies*, pages 29–42, Berkeley, CA, USA, 2003. USENIX Association.

[406] R. E. Kalman. A new approach to linear filtering and prediction problems. *Transactions of the ASME–Journal of Basic Engineering*, 82(82 (Series D)):35–45, 1960.

[407] Aditya Kalyanpur, Bijan Parsia, Evren Sirin, Bernardo C. Grau, and James Hendler. Swoop: A Web Ontology Editing Browser. *Web Semantics: Science, Services and Agents on the World Wide Web*, 4(2):144–153, June 2006.

[408] Sepandar D. Kamvar, Mario T. Schlosser, and Hector Garcia-Molina. The eigentrust algorithm for reputation management in p2p networks. In *Proceedings of the 12th international conference on World Wide Web*, WWW '03, pages 640–651, New York, NY, USA, 2003. ACM.

[409] Aman Kansal, Feng Zhao, Jie Liu, Nupur Kothari, and Arka A. Bhattacharya. Virtual machine power metering and provisioning. In *Proceedings of the 1st ACM symposium on Cloud computing*, SoCC '10, pages 39–50, New York, NY, USA, 2010. ACM.

[410] Murat Kantarcioglu and Chris Clifton. Security Issues in Querying Encrypted Data. In *DBSec*, pages 325–337, 2005.

[411] Thomas Karagiannis, Konstantina Papagiannaki, and Michalis Faloutsos. Blinc: multilevel traffic classification in the dark. *SIGCOMM Comput. Commun. Rev.*, 35:229–240, August 2005.

[412] A. Karaman. Constraint-based routing in traffic engineering. In *Computer Networks, 2006 International Symposium on*, pages 1–6, 2006.

[413] David R. Karger, Eric Lehman, Frank Thomson Leighton, Rina Panigrahy, Matthew S. Levine, and Daniel Lewin. Consistent Hashing and Random Trees: Distributed Caching Protocols for Relieving Hot Spots on the World Wide Web. In *STOC*, pages 654–663, 1997.

[414] John F. Karpovich, Andrew S. Grimshaw, and James C. French. Extensible file system (ELFS): an object-oriented approach to high performance file I/O. In *OOPSLA '94: Proceedings of the ninth annual conference on Object-oriented programming systems, language, and applications*, pages 191–204, New York, NY, USA, 1994. ACM.

[415] G. Katsaros, G. Kousiouris, S. Gogouvitis, D. Kyriazis, and T. Varvarigou. A service oriented monitoring framework for soft real-time applications. In *Service-Oriented Computing and Applications*, 2010.

[416] Harry Katzan. Cloud software service: Concepts, technology, economics. *Service Science*, 1:256–269, 2009.

[417] Joe Kaylor, Konstantin Läufer, and George K. Thiruvathukal. Online layered file system (OLFS): A layered and versioned filesystem and performance analysis. In *Proc. IEEE Intl. Conf. on Electro/Information Technology (EIT)*, May 2010.

[418] Joe Kaylor, Konstantin Läufer, and George K. Thiruvathukal. RestFS: A FUSE filesystem to expose REST-ful services. http://restfs.googlecode.com/, 2010–2011.

[419] Joe Kaylor, George K. Thiruvathukal, and Konstantin Läufer. Naked object file system (NOFS): A framework to expose an object-oriented domain model as a filesystem. Technical report, Loyola University Chicago, May 2010.

[420] K. Keahey, R. Figueiredo, J. Fortes, T. Freeman, and M. Tsugawa. Science Clouds: Early Experiences in Cloud Computing for Scientific Applications, August 2008.

[421] Katarzyna Keahey and Tim Freeman. Contextualization: Providing One-Click Virtual Clusters. *eScience 2008*, December 2008.

[422] Chiaw K. Kee. Security policy roadmap - process for creating security policies. http://www.sans.org/reading_room/whitepapers/policyissues/security-policy-roadmap-process-creating-security-policies_494, 2001. (accessed August 13, 2010).

[423] Bettina Kemme, Ricardo Jiménez-Peris, and Marta Patiño-Martínez. *Database Replication.* Synthesis Lectures on Data Management. Morgan & Claypool Publishers, 2010.

[424] J. Kennedy, R.C. Eberhart, et al. Particle swarm optimization. In *Proceedings of IEEE international conference on neural networks*, volume 4, pages 1942–1948. Perth, Australia, 1995.

[425] J. O. Kephart and D. M. Chess. The vision of autonomic computing. *IEEE Computer*, 36(1):41–50, 2003.

[426] J.O. Kephart and W.E. Walsh. An artificial intelligence perspective on autonomic computing policies. In *Policies for Distributed Systems and Networks, 2004. POLICY 2004. Proceedings. Fifth IEEE International Workshop on*, pages 3–12. IEEE, 2004.

[427] Brian W. Kernighan and Rob Pike. *The UNIX Programming Environment.* Prentice Hall Professional Technical Reference, 1983.

[428] Roy W. Keyes, Christian Romano, Dorian Arnold, and Shuang Luan. Radiation therapy calculations using an on-demand virtual cluster via cloud computing. Technical Report arXiv:1009.5282, University of New Mexico, Sep 2010. Comments: 12 pages, 4 figures.

[429] Vishal Kher and Yongdae Kim. Building trust in storage outsourcing: Secure accounting of utility storage. In *Proceedings of the 26th IEEE International Symposium on Reliable Distributed Systems*, SRDS '07, pages 55–64, Washington, DC, USA, 2007. IEEE Computer Society.

[430] Hyunjoo Kim, Shivangi Chaudhari, Manish Parashar, and Christopher Marty. Online risk analytics on the cloud. In *Cluster Computing and the Grid, 2009. CCGRID '09. 9th IEEE/ACM International Symposium on*, pages 484–489, May 2009.

[431] Hyunjoo Kim, M. Parashar, D.J. Foran, and Lin Yang. Investigating the use of autonomic cloudbursts for high-throughput medical image registration. In *Grid Computing, 2009 10th IEEE/ACM International Conference on*, pages 34–41, October 2009.

[432] Ted H. Kim and Gerald J. Popek. Frigate: an object-oriented file system for ordinary users. In *COOTS'97: Proceedings of the 3rd conference on USENIX Conference on Object-Oriented Technologies (COOTS)*, pages 9–9, Berkeley, CA, USA, 1997. USENIX Association.

[433] T. Kimura. Diffusion approximation for an $M/G/m$ queue. *Operations Research*, 31:304–321, 1983.

[434] T. Kimura. Optimal buffer design of an $M/G/s$ queue with finite capacity. *Communications in Statistics Ũ Stochastic Models*, 12(6):165–180, 1996.

[435] T. Kimura. A transform-free approximation for the finite capacity $M/G/s$ queue. *Operations Research*, 44(6):984–988, 1996.

[436] Avi Kivity, Yaniv Kamay, Dor Laor, Uri Lublin, and Anthony Liguori. kvm: the linux virtual machine monitor. In *Proceedings of the Linux Symposium*, pages 225–230, Ottawa, Ontario, Canada, 2007.

[437] S. R. Kleiman. Vnodes: An architecture for multiple file system types in Sun UNIX. In *Proc. Summer USENIX Technical Conf.*, pages 238–247, 1986.

[438] L. Kleinrock. *Queueing Systems, volume 1, Theory*. Wiley-Interscience, 1975.

[439] B. Koller and L. Schubert. Towards autonomous sla management using a proxy-like approach. *Multiagent Grid Systems*, 3(3):313–325, 2007.

[440] J. Koomey. Estimating total power consumption by servers in the U.S. and the world. `http://hightech.lbl.gov/documents/DATA_CENTERS/svrpwrusecompletefinal.pdf`, 2007.

[441] Donald Kossmann, Tim Kraska, and Simon Loesing. An evaluation of alternative architectures for transaction processing in the cloud. In *SIGMOD Conference*, pages 579–590, 2010.

[442] D. Krafzig, K. Banke, and D. Slama. *Enterprise SOA: Service-Oriented Architecture Best Practices (The Coad Series)*. Prentice Hall PTR Upper Saddle River, NJ, USA, 2004.

[443] Tim Kraska, Martin Hentschel, Gustavo Alonso, and Donald Kossmann. Consistency Rationing in the Cloud: Pay only when it matters. *PVLDB*, 2(1):253–264, 2009.

[444] Sriram Krishnan. *Programming Windows Azure: Programming the Microsoft Cloud*. O'REILLY, Sebastopol, CA, USA, May 2009.

[445] K. Kritikos and D. Plexousakis. Semantic qos metric matching. 2006.

[446] John Kubiatowicz, David Bindel, Yan Chen, Steven Czerwinski, Patrick Eaton, Dennis Geels, Ramakrishan Gummadi, Sean Rhea, Hakim Weatherspoon, Westley Weimer, Chris Wells, and Ben Zhao. Oceanstore: an architecture for global-scale persistent storage. *SIGPLAN Not.*, 35:190–201, November 2000.

[447] P. Lairds. Cloud Computing Taxonomy. In *Procs. Interop09*, pages 201–206. IEEE Computer Society, May 2009.

[448] Avinash Lakshman and Prashant Malik. Cassandra: structured storage system on a p2p network. In *PODC*, page 5, 2009.

[449] Avinash Lakshman and Prashant Malik. Cassandra: a decentralized structured storage system. *Operating Systems Review*, 44(2):35–40, 2010.

[450] S. Lamparter, A. Ankolekar, R. Studer, and S. Grimm. Preference-based selection of highly configurable web services. In *Proceedings of the 16th international conference on World Wide Web*, pages 1013–1022. ACM, 2007.

[451] S. Lamparter, A. Ankolekar, R. Studer, D. Oberle, and C. Weinhardt. A policy framework for trading configurable goods and services in open electronic markets. In *Proceedings of the 8th international conference on Electronic commerce: The new e-commerce: innovations for conquering current barriers, obstacles and limitations to conducting successful business on the internet*, pages 162–173. ACM, 2006.

[452] Leslie Lamport. Time, clocks, and the ordering of events in a distributed system. *CACM*, 21(7):558–565, 1978.

[453] Lattice Monitoring Framework. Project Website: `http://clayfour.ee.ucl.ac.uk/lattice`.

[454] F. Le Faucheur. RFC 4124: Protocol Extensions for Support of Diffserv-aware MPLS Traffic Engineering. Technical report, IETF, June 2005.

[455] N. Leavitt. Is cloud computing really ready for prime time? *Growth*, 27:5, 2009.

[456] Neal Leavitt. Is Cloud Computing Really Ready for Prime Time? *Computer*, 42:15–20, 2009.

[457] Cynthia Bailey Lee and Allan Snavely. On the user-scheduler dialogue: Studies of user-provided runtime estimates and utility functions. *Int. J. High Perform. Comput. Appl.*, 20(4):495–506, 2006.

[458] Kevin Lee, Norman W. Paton, Rizos Sakellariou, and A. A. Fernandes Alvaro. Utility driven adaptive worklow execution. In *CCGRID '09: Proceedings of the 2009 9th IEEE/ACM International Symposium on Cluster Computing and the Grid*, pages 220–227, Washington, DC, USA, 2009. IEEE Computer Society.

[459] Tsang-yean Lee, Huey-ming Lee, Homer Wu, and Jin-shieh Su. Data Transmission Encryption and Decryption Algorithm in Network Security. *Proceedings of the 6th WSEAS International Conference on Simulation, Modelling and Optimization*, pages 417–422, 2006.

[460] Y. C. Lee and A. Y. Zomaya. Energy conscious scheduling for distributed computing systems under different operating conditions. *IEEE Trans. Parallel Distrib. Syst.*, in press.

[461] Young Choon Lee, Chen Wang, Albert Y. Zomaya, and Bing Bing Zhou. Profit-driven service request scheduling in clouds. In *Proceedings of the 2010 10th IEEE/ACM International Conference on Cluster, Cloud and Grid Computing*, CCGRID '10, pages 15–24, Washington, DC, USA, 2010. IEEE Computer Society.

[462] Zhou Lei, Bofeng Zhang, Wu Zhang, Qing Li, Xuejun Zhang, and Junjie Peng. Comparison of Several Cloud Computing Platforms. *Second International Symposium on Information Science and Engineering*, pages 23–27, 2009.

[463] Alexander Lenk, Markus Klems, Jens Nimis, Stefan Tai, and Thomas Sandholm. What's inside the cloud? an architectural map of the cloud landscape. In *Proceedings of the 2009 ICSE Workshop on Software Engineering Challenges of Cloud Computing*, CLOUD '09, pages 23–31, Washington, DC, USA, 2009. IEEE Computer Society.

[464] Alexander Lenk, Markus Klems, Jens Nimis, Stefan Tai, and Thomas Sandholm. What's inside the cloud? an architectural map of the cloud landscape. In *Proceedings of the 2009 ICSE Workshop on Software Engineering Challenges of Cloud Computing*, CLOUD '09, pages 23–31, Washington, DC, USA, 2009. IEEE Computer Society.

[465] Ang Li, Xiaowei Yang, Srikanth Kandula, and Ming Zhang. Cloudcmp: comparing public cloud providers. In *Proceedings of the 10th annual conference on Internet measurement*, IMC '10, pages 1–14, New York, NY, USA, 2010. ACM.

[466] Feifei Li, Marios Hadjieleftheriou, George Kollios, and Leonid Reyzin. Dynamic authenticated index structures for outsourced databases. In *In SIGMOD*, pages 121–132. SIGMOD, 2006.

[467] Henry Li. *Introducing Windows Azure*. Apress, Berkely, CA, USA, 2009.

[468] Jie Li, Marty Humphrey, Deb Agarwal, Keith Jackson, Catharine van Ingen, and Youngryel Ryu. escience in the cloud: A modis satellite data reprojection and reduction pipeline in the windows azure platform. In *Parallel Distributed Processing (IPDPS), 2010 IEEE International Symposium on*, pages 1 –10, 2010.

[469] W. Li, L. Ping, and X. Pan. Use Trust Management Module to Achieve Effective Security Mechanisms in Cloud Environment. In *Proceedings of the International Conference On Electronics and Information Engineering (ICEIE)*, Kyoto, Japan, Aug. 1–3, 2010.

[470] Xin Li, Christopher White, Zhou Lei, and Gabrielle Allen. Reservoir model updating by ensemble kalman filter-practical approaches using grid computing technology. In *Petroleum Geostatistics 2007*, Cascais,Portugal, August 2007.

[471] Zhen Li and Manish Parashar. A computational infrastructure for grid-based asynchronous parallel applications. In *HPDC '07: Proceedings of the 16th international symposium on High performance distributed computing*, pages 229–230, New York, NY, USA, 2007. ACM.

[472] Libvirt. Project Website: http://www.libvirt.org.

[473] M. Lieberman and D. Montgomery. First-Mover Advantages. *Strategic Management Journal*, 9(Special Issue: Strategy Content Research):41–58, 1988.

[474] John Little. A proof for the queueing formula: $l = \lambda w$. *Operations Research: A Journal of the Institute for Operations Research and the Management Sciences*, pages 383–387, 1961.

[475] Xiaoqing (Frank) Liu, Lijun Dong, and Hungwen Lin. High order object-oriented modeling technique for structured object-oriented analysis. *ACIS Int. J Comp. Inf. Sci.*, 2(2):74–96, 2001.

[476] Yan Liu, Mingguang Zhuang, Qingling Wang, and Guannan Zhang. A new approach to web services characterization. In *Asia-Pacific Services Computing Conference, 2008. APSCC '08. IEEE*, pages 404 –409, Dec 2008.

[477] Yan Liu, Mingguang Zhuang, Biao Yu, Guannan Zhang, and Xiaojing Meng. Services characterization with statistical study on existing web services. In *Web Services, 2008. ICWS '08. IEEE International Conference on*, pages 803–804, Sep 2008.

[478] Z. Liu, T. Liu, L. Cai, and G. Yang. Quality Evaluation and Selection Framework of Service Composition Based on Distributed Agents. In *2009 Fifth International Conference on Next Generation Web Services Practices*, pages 68–75. IEEE, 2009.

[479] Lokad.Cloud Project. http://code.google.com/p/lokad-cloud/.

[480] D. A. Low, W. B. Harms, S. Mutic, and J. A. Purdy. A technique for the quantitative evaluation of dose distributions. *Medical Physics*, 25(5):656–661, 1998.

[481] Wei Lu, Jared Jackson, and Roger Barga. Azureblast: a case study of developing science applications on the cloud. In *Proceedings of the 19th ACM International Symposium on High Performance Distributed Computing*, HPDC '10, pages 413–420, New York, NY, USA, 2010. ACM.

[482] H. Ludwig, A. Keller, A. Dan, R. P. King, and R. Franck. Web service level agreement (WSLA) language specification. *IBM Corporation*, 2003.

[483] H. Ludwig, A. Keller, A. Dan, R. P. King, and R. Franck. Web service level agreement (WSLA) language specification. *IBM Corporation*, 2003.

[484] S. A. Ludwig and SMS Reyhani. Selection algorithm for grid services based on a quality of service metric. In *High Performance Computing Systems and Applications, 2007. HPCS 2007. 21st International Symposium on*, page 13. IEEE, 2007.

[485] S.A. Ludwig and SMS Reyhani. Selection algorithm for grid services based on a quality of service metric. In *High Performance Computing Systems and Applications, 2007. HPCS 2007. 21st International Symposium on*, page 13. IEEE, 2007.

[486] Ian Lumb, Eunmi Choi, and Bhaskar Prasad Rimal. Virtualization for dummies. *Second International Symposium on Information Science and Engineering*, 2010.

[487] Lutz Schubert. The Future of Cloud Computing - Opportunities for European Cloud Computing Beyond 2010. `http://cordis.europa.eu/fp7/ict/ssai/docs/cloud-report-final.pdf`.

[488] B. N. W. Ma and J. W. Mark. Approximation of the mean queue length of an $M/G/c$ queueing system. *Operations Research*, 43:158–165, 1998.

[489] Dan Ma. The Business Model of "Software-As-A-Service." *In Proceedings of the International Conference on Service Computing*, July 2007.

[490] G. S. Machado, D. Hausheer, and B. Stiller. Considerations on the Interoperability of and between Cloud Computing Standards. In *Proceedings of the 27th Open Grid Forum (OGF27)*, Banff, Canada, Oct. 12–15, 2009.

[491] Huajian Mao, Nong Xiao, Weisong Shi, and Yutong Lu. Wukong: Toward a cloud-oriented file service for mobile devices. In *Proceedings of the 2010 IEEE International Conference on Services Computing*, SCC '10, pages 498–505, Washington, DC, USA, 2010. IEEE Computer Society.

[492] Maplesoft, Inc. *Maple 13*. CRC Press, Waterloo, ON, Canada, 2009.

[493] Sojan Markose, Xiaoqing (Frank) Liu, and Bruce McMillin. A systematic framework for structured object-oriented security requirements analysis in embedded systems. In *EUC '08: Proceedings of the 2008 IEEE/IFIP International Conference on Embedded and Ubiquitous Computing*, pages 75–81, Washington, DC, USA, 2008. IEEE Computer Society.

[494] M. Armbrust, A. Fox, R. Griffith, A. Joseph, R. Katz, A. Konwinski, G. Lee, D. Patterson, A. Rabkin, and I. Stoica. Above the clouds: a Berkeley view of cloud computing. *EECS Department, University of California, Berkley, Tech Rep*, 2009.

[495] Paul Marshall, Kate Keahey, and Tim Freeman. Elastic Site: Using Clouds to Elastically Extend Site Resources. *IEEE/ACM International Symposium on Cluster, Cloud and Grid Computing (CCGrid 2010)*, May 2010.

[496] Paul Marshall, Kate Keahey, and Tim Freeman. Elastic site: Using clouds to elastically extend site resources. *Cluster Computing and the Grid, IEEE International Symposium on*, 0:43–52, 2010.

[497] Paul Marshall, Kate Keahey, and Tim Freeman. Elastic site: Using clouds to elastically extend site resources. *Cluster Computing and the Grid, IEEE International Symposium on*, 0:43–52, 2010.

[498] D. Martin, M. Burstein, J. Hobbs, O. Lassila, D. McDermott, S. McIlraith, S. Narayanan, M. Paolucci, B. Parsia, T. Payne, et al. OWL-S: Semantic markup for web services. *W3C Member Submission*, 22:2007–04, 2004.

[499] David Martin, Massimo Paolucci, Sheila Mcilraith, Mark Burstein, Drew Mcdermott, Deborah Mcguinness, Bijan Parsia, Terry Payne, Marta Sabou, Monika Solanki, Naveen Srinivasan, and Katia Sycara. Bringing Semantics to Web Services: The OWL-S Approach. In J. Cardoso and A. Sheth, editors, *SWSWPC 2004*, volume 3387 of *LNCS*, pages 26–42. Springer, 2004.

[500] M. L. Massie, B. N. Chun, and D. E. Culler. The Ganglia distributed monitoring system: Design, implementation and experience. *Parallel Computing*, 30(7):817–840, 2004.

[501] M. Maurer, I. Brandic, V. C. Emeakaroha, and S. Dustdar. Towards knowledge management in self-adaptable clouds. In *4th International Workshop of Software Engineering for Adaptive Service-Oriented Systems (SEASS'10)*, 2010.

[502] E. M. Maximilien and M. P. Singh. A framework and ontology for dynamic web services selection. *Internet Computing, IEEE*, 8(5):84–93, 2004.

[503] D. L. McGuinness, F. Van Harmelen, et al. OWL web ontology language overview. *W3C recommendation*, 10:2004–03, 2004.

[504] McGuinness, D. L., van Harmelen, F. OWL Web Ontology Language Overview. W3C Recommendation: http://www.w3.org/TR/2004/REC-owl-features-20040210/, 2004.

[505] Michael Mcintosh, Paula Austel, and Yorktown Heights. XML Signature Element Wrapping Attacks and countermeasures. *IBM research division*, 23691, 2005.

[506] Marvin McNett, Diwaker Gupta, Amin Vahdat, and Geoffrey M. Voelker. Usher: An Extensible Framework for Managing Clusters of Virtual Machines. In *Proceedings of the 21st Large Installation System Administration Conference (LISA)*, November 2007.

[507] David Meisner, Brian T. Gold, and Thomas F. Wenisch. Powernap: eliminating server idle power. In *Proceeding of the 14th international conference on Architectural support for programming languages and operating systems*, ASPLOS '09, pages 205–216, New York, NY, USA, 2009. ACM.

[508] P. Mell and T. Grance. The nist definition of cloud computing: `http://csrc.nist.gov/groups/SNS/cloud-computing/`, 18 December 2009.

[509] P. Mell and T. Grance. The NIST Definition of Cloud Computing. Technical report, National Institute of Standards & Technology, Oct. 7, 2009.

[510] Members of EGEE-II. An EGEE comparative study: Grids and clouds - evolution or revolution. Technical report, Enabling Grids for E-sciencE Project : https://edms.cern.ch/document/925013/, 2008.

[511] D. A. Menascé. QoS issues in Web services. *Internet Computing, IEEE*, 6(6):72–75, 2002.

[512] D. A. Menascé, E. Casalicchio, and V. Dubey. On optimal service selection in Service Oriented Architectures. *Performance Evaluation*, 2009.

[513] T. Metsch, A. Edmonds, and V. Bayon. Using Cloud Standards for Interoperability of Cloud Frameworks. Technical report, RESERVOIR and SLA@SOI, Apr. 2010.

[514] A. Michael, F. Armando, G. Rean, D. J. Anthony, K. Randy, K. Andy, L. Gunho, P. David, R. Ariel, S. Ion, et al. Above the clouds: A berkeley view of cloud computing. *EECS Department, University of California, Berkeley, Tech. Rep. UCB/EECS-2009-28*, 2009.

[515] Microsoft. Azure: `http://www.microsoft.com/azure/default.mspx`, 5 May 2009.

[516] Microsoft. Microsoft pushes its technical computing initiative forward with windows hpc server 2008 r2: `http://www.microsoft.com/Presspass/press/2010/sep10/09-17WindowsHPCServerPR.mspx`, 22 October 2010.

[517] Microsoft. *SQL Server Online Book.* `http://msdn.microsoft.com/en-us/library/ms187956.aspx`, 2010.

[518] Dejan Milojicic. Cloud computing: Interview with russ daniels and franco travostino. *IEEE Internet Computing*, 12(5):7–9, 2008.

[519] M. Isard, M. Budiu, Y. Yu, A. Birrell, and D. Fetterly. Dryad: distributed data-parallel programs from sequential building blocks. *Euro. Conf on computer systems (EuroSys)*, 2007.

[520] M. Miyazawa. Approximation of the queue-length distribution of an $M/GI/s$ queue by the basic equations. *Journal of Applied Probability*, 23:443–458, 1986.

[521] MONIT. Project Website: `http://monit.com/monit`.

[522] Andrew W. Moore and Denis Zuev. Internet traffic classification using bayesian analysis techniques. *SIGMETRICS Perform. Eval. Rev.*, 33:50–60, June 2005.

[523] J. N. Morse. Reducing the size of the nondominated set: Pruning by clustering. *Computers & Operations Research*, 7(1-2):55–66, 1980.

[524] Angela Moscaritolo. Most organizations falling short on cloud security policies. `http://www.scmagazineus.com/most-organizations-falling-short-on-cloud-security-policies/article/167415/`, 2010. (accessed September 17, 2010).

[525] Francesco Moscato, Beniamino Di Martino, Salvatore Venticinque, and Angelo Martone. Overfa: a collaborative framework for the semantic annotation of documents and websites. *Int. J. Web Grid Serv.*, 5:30–45, March 2009.

[526] Hamid R. Motahari-Nezhad, Bryan Stephenson, and Sharad Singhal. Outsourcing business to cloud computing services: Opportunities and challenges. *Development*, 10(4):1–17, 2009.

[527] M. Turner, D. Budgen, and P. Brereton. Turning software into a service. *IEEE Computer, Vol 36*, 2008.

[528] Ahuva W. Mu'alem and Dror G. Feitelson. Utilization, predictability, workloads, and user runtime estimates in scheduling the IBM SP2 with backfilling. *IEEE Transactions on Parallel and Distributed Systems*, 12(6):529–543, 2001.

[529] MySQL. mysqlslap - load emulation client. `http://dev.mysql.com/doc/refman/5.1/en/mysqlslap.html`.

[530] Nagios. Project Website: `http://www.nagios.org/`.

[531] Nagios Enterprises, LLC. Nagios XI - Features. `http://assets.nagios.com/datasheets/nagiosxi/Nagios%20XI%20-%20Features.pdf`.

[532] Nagios Plugins Development Team. Nagios plug-in development guidelines. `http://nagiosplug.sourceforge.net/developer-guidelines.html`.

[533] Ripal Nathuji and Karsten Schwan. Virtualpower: coordinated power management in virtualized enterprise systems. In *Proceedings of twenty-first ACM SIGOPS symposium on Operating systems principles*, SOSP '07, pages 265–278, New York, NY, USA, 2007. ACM.

[534] National Institute of Standards and Technology. Cloud computing. `http://csrc.nist.gov/groups/SNS/cloud-computing/index.html`, 2009. (accessed October 03, 2009).

[535] Alon Naveh, Efraim Rotem, Avi Mendelson, Simcha Gochman, Rajshree Chabukswar, Karthik Krishnan, and Arun Kumar. Power and thermal management in the intel core duo processor. *Intel Technology Journal*, 10(2):109–122, 2006.

[536] Surya Nepal, Zaki Malik, and Athman Bouguettaya. Reputation propagation in composite services. In *Proceedings of the 2009 IEEE International Conference on Web Services*, ICWS '09, pages 295–302, Washington, DC, USA, 2009. IEEE Computer Society.

[537] SuitCloud NetSuite. Suitecloud platform, 2010.

[538] Patrick Nicolas. Introduction to multi-tenant web applications, 2006.

[539] Nimbus Home Page. http://www.nimbusproject.org/.

[540] National Institute of Standards and Technology (NIST), Cloud Standards: http://csrc.nist.gov/groups/SNS/cloud-computing/.

[541] NIST definition of cloud computing `http://www.nist.gov/itl/cloud`.

[542] Natalya F. Noy, Michael Sintek, Stefan Decker, Monica Crubézy, Ray W. Fergerson, and Mark A. Musen. Creating semantic web contents with protégé-2000. *IEEE Intelligent Systems*, 16:60–71, March 2001.

[543] S. A. Nozaki and S. M. Ross. Approximations in finite-capacity multi-server queues with poisson arrivals. *Journal of Applied Probability*, 15:826–834, 1978.

[544] D. Nurmi, R. Wolski, C. Grzegorczyk, G. Obertelli, S. Soman, L. Youseff, and D. Zagorodnov. The Eucalyptus open-source cloud-computing system. In *9th International Symposium on Cluster Computing and the Grid (CCGRID'09)*, 2009.

[545] Daniel Nurmi, Rich Wolski, Chris Grzegorczyk, Graziano Obertelli, Sunil Soman, Lamia Youseff, and Dmitrii Zagorodnov. Eucalyptus : A Technical Report on an Elastic Utility Computing Architecture Linking Your Programs to Useful Systems. Technical report, Computer Science Department, University of California, Santa Barbara, 2008.

[546] Daniel Nurmi, Rich Wolski, Chris Grzegorczyk, Graziano Obertelli, Sunil Soman, Lamia Youseff, and Dmitrii Zagorodnov. Eucalyptus: A Technical Report on an Elastic Utility Computing Architecture Linking Your Programs to Useful Systems. In *UCSB Computer Science Technical Report Number 2008-10*, August 2008.

[547] Daniel Nurmi, Rich Wolski, Chris Grzegorczyk, Graziano Obertelli, Sunil Soman, Lamia Youseff, and Dmitrii Zagorodnov. The Eucalyptus Open-source Cloud-computing System. In *CCA'08: Proceedings of Cloud Computing and Its Applications workshop*, October 2008.

[548] Daniel Nurmi, Rich Wolski, Chris Grzegorczyk, Graziano Obertelli, Sunil Soman, Lamia Youseff, and Dmitrii Zagorodnov. The Eucalyptus Open-source Cloud-computing System. *9th IEEE/ACM International Symposium on Cluster Computing and the Grid*, pages 124–131, 2009.

[549] O. Lassila and R. Swick. Resource Description Framework (RDF) Model and Syntax Specification.: http://www.w3.org/TR/REC-rdf-syntax/, 1998.

[550] University of Manchester. Dataflow research project, 1997.

[551] Open Grid Forum: Open Cloud Computing Interface (OCCI): http://forge.ogf.org/sf/projects/occi-wg.

[552] OpenNebula Home Page. http://www.opennebula.org/.

[553] Open Cloud Manifesto, Spring 2009 : http://www.opencloudmanifesto.org.

[554] Oracle. *Oracle 11g Online Library.* `http://download.oracle.com/docs/cd/B28359_01/server.111/b28313/usingpe.htm#i1007101`, 2010.

[555] Oracle Corporation. http://www.virtualbox.org/manual/ UserManual.html.

[556] Organization for the Advancement of Structured Information Standards (OASIS). *Web Services Dynamic Discovery (WS–Discovery)*, July 1, 2009. Version 1.1.

[557] S. Ostermann, R. Prodan, and T. Fahringer. Extending grids with cloud resource management for scientific computing. In *Grid Computing, 2009 10th IEEE/ACM International Conference on*, pages 42 –49, oct. 2009.

[558] Dustin Owens. Securing elasticity in the cloud. *Queue*, 8(5):10–16, 2010.

[559] M. Tamer Özsu and Patrick Valduriez. *Principles of Distributed Database Systems, Second Edition*. Prentice-Hall, 1999.

[560] E. Page. Tables of waiting times for $M/M/n$, $M/D/n$ and $D/M/n$ and their use to give approximate waiting times in more general queues. *J. Operational Research Society*, 33:453–473, 1982.

[561] Mayur R. Palankar, Adriana Iamnitchi, Matei Ripeanu, and Simson Garfinkel. Amazon s3 for science grids: a viable solution? In *Proceedings of the 2008 international workshop on Data-aware distributed computing*, DADC '08, pages 55–64, New York, NY, USA, 2008. ACM.

[562] Mayur R. Palankar, Adriana Iamnitchi, Matei Ripeanu, and Simson Garfinkel. Amazon S3 for science Grids: a viable solution? In *International Workshop on Data-aware Distributed Computing (DADC'08) in conjunction with HPDC 2008*, pages 55–64, New York, NY, USA, 2008. ACM.

[563] G. Pallis. Cloud computing: The new frontier of internet computing. *Internet Computing, IEEE*, 14(5):70 –73, No.v 2010.

[564] I.V. Papaioannou, D.T. Tsesmetzis, I.G. Roussaki, and M.E. Anagnostou. A QoS ontology language for Web-services. In *Advanced Information Networking and Applications, 2006. AINA 2006. 20th International Conference on*, volume 1, page 6. IEEE, 2006.

[565] P. Papakos, L. Capra, and D.S. Rosenblum. VOLARE: context-aware adaptive cloud service discovery for mobile systems. In *Proceedings of the 9th International Workshop on Adaptive and Reflective Middleware*, pages 32–38. ACM, 2010.

[566] Michael P. Papazoglou, Paolo Traverso, Schahram Dustdar, and Frank Leymann. Service-oriented computing: State of the art and research challenges. *Computer*, 40:38–45, November 2007.

[567] A. V. Parameswaran and A. Chaddha. Cloud Interoperability and Standardization. *SETLabs Briefings*, 7(7):19–27, 2009.

[568] K. E. Parsopoulos and M. N. Vrahatis. Particle swarm optimization method in multiobjective problems. In *Proceedings of the 2002 ACM symposium on applied computing*, pages 603–607. ACM, 2002.

[569] Gopi Kannan Parthiban. Cloud computing use cases white paper version 1.0. *3rd International Conference on Grid and Pervasive Computing-gpc-workshops*, 2008.

[570] Mukaddim Pathan, James Broberg, and Rajkumar Buyya. Maximizing utility for content delivery clouds. In *Proceedings of the 10th International Conference on Web Information Systems Engineering*, WISE '09, pages 13–28, Berlin, Heidelberg, 2009. Springer-Verlag.

[571] Paul McFedries. The cloud is the computer. IEEE Spectrum Online, : http://www.spectrum.ieee.org/aug08/6490, 2008.

[572] R. Pawson. *Naked Objects*. PhD thesis, Trinity College, Dublin, Ireland, 2004.

[573] Vern Paxson. Bro: a system for detecting network intruders in real-time. *Comput. Netw.*, 31:2435–2463, December 1999.

[574] Jan-Simon Pendry and Marshall Kirk McKusick. Union mounts in 4.4BSD-lite. In *TCON'95: Proc. of the USENIX 1995 Technical Conf.*, pages 3–3, Berkeley, CA, USA, 1995. USENIX Association.

[575] Somesh Jha Peng Ning, Paul Syverson. Robust Defenses for Cross-Site Request Forgery . *Proceedings of the 15th ACM conference on Computer and communications security*, october 2008.

[576] Donald H. Perkins. *Introduction to High Energy Particle Physics*. Cambridge University Press, ISBN 978-0521621960, 2000.

[577] R. Perlman. The ephemerizer: Making data disappear, sun microsystems whitepaper, 2005.

[578] Gunnar Peterson. Don't trust. and verify: A security architecture stack for the cloud. *IEEE Security and Privacy*, 8:83–86, September 2010.

[579] P. Goyal, R. Mikkilineni, and M. Ganti. The fcaps in the business services fabric model. *in proceedings of WETICE 2009:18th IEEE International Workshops on Enabling Technologies:Infrastructure for Collaborative Enterprises*, 2009.

[580] Rob Pike, Dave Presotto, Sean Dorward, Bob Flandrena, Ken Thompson, Howard Trickey, and Phil Winterbottom. Plan 9 from Bell Labs. *Computing Systems*, 8(3):221–254, Summer 1995.

[581] Rob Pike, Dave Presotto, Ken Thompson, Howard Trickey, and Phil Winterbottom. The use of name spaces in Plan 9. *SIGOPS Oper. Syst. Rev.*, 27(2):72–76, 1993.

[582] D. Pisinger. *Algorithms for knapsack problems*. Datalogisk Institut, Københavns Universitet, 1995.

[583] C. Polychronopoulos and D. Kuck. Guided self-scheduling: A practical scheduling scheme for parallel computers. *IEEE Transactions on Computers*, C-36(12):1425–1439, December 1987.

[584] F. Pop, A. Costan, C. Dobre, C. Stratan, and V. Cristea. Monitoring of complex applications execution in distributed dependable systems. In *Parallel and Distributed Computing, 2009. ISPDC '09. Eighth International Symposium on*, pages 241 –244, July 2009.

[585] Raluca Ada Popa, Jay Lorch, David Molnar, Helen J. Wang, and Li Zhuang. Enabling security in cloud storage slas with cloudproof. technical report, microsoft research, 2010.

[586] Gerald J. Popek and Robert P. Goldberg. Formal requirements for virtualizable third generation architectures. *Commun. ACM*, 17:412–421, July 1974.

[587] Gerald J. Popek and Robert P. Goldberg. Formal requirements for virtualizable third generation architectures. *Communications of the ACM*, 17(7):412–421, July 1974.

[588] R. Prodan and S. Ostermann. A Survey and Taxonomy of Infrastructure as a Service & Web Hosting Cloud Providers. In *Proceedings of the Intl. Conference on Grid Computing*, Banff, Canada, Oct. 13–15, 2009.

[589] FAWN Project. FAWN: A fast array of wimpy nodes. `http://www.cs.cmu.edu/~fawnproj/`.

[590] Nimbus project. http://www.nimbusproject.org/.

[591] J. Quemada, T. de Miguel, S. Pavon, G. Huecas, T. Robles, J. Salvachua, D.a.a. Ortiz, V. Sirvent, F. Escribano, and J. Sedano. Isabel: An Application for real time Collaboration with a flexible Floor Control. In *2005 International Conference on Collaborative Computing: Networking, Applications and Worksharing*, pages 1–9. IEEE, 2005.

[592] Juan Quemada, Gabriel Huecas, Tomás de Miguel, Joaquín Salvachúa, Blanca Fernandez, Bernd Simon, Katherine Maillet, and Efiie Lai-Cong. Educanext: a framework for sharing live educational resources with isabel. In *WWW (Alternate Track Papers & Posters)*, pages 11–18, 2004.

[593] R. Akkiraju, J. Farrell, J.Miller, M. Nagarajan, M. Schmidt, A. Sheth, K. Verma. Web Service Semantics WSDL-S. A joint UGA-IBM Technical Note, version 1.0: http://lsdis.cs.uga.edu/projects/METEOR-S/WSDL, 2005.

[594] Rackspace. http://www.rackspacecloud.com/index.php, Dec. 2010.

[595] Sandro Rafaeli and David Hutchison. A survey of key management for secure group communication. *ACM Comput. Surv.*, 35:309–329, September 2003.

[596] Barath Raghavan, Kashi Vishwanath, Sriram Ramabhadran, Kenneth Yocum, and Alex C. Snoeren. Cloud control with distributed rate limiting. *ACM SIGCOMM Computer Communication Review*, 37(4):337, oct 2007.

[597] Arcot Rajasekar, Michael Wan, and Reagan Moore. Mysrb & srb: Components of a data grid. In *Proceedings of the 11th IEEE International Symposium on High Performance Distributed Computing*, HPDC '02, pages 301–, Washington, DC, USA, 2002. IEEE Computer Society.

[598] S. Ran. A model for web services discovery with QoS. *ACM SIGecom Exchanges*, 4(1):1–10, 2003.

[599] Omer Rana, Martijn Warnier, Thomas B. Quillinan, and Frances Brazier. Monitoring and reputation mechanisms for service level agreements. In *Proceedings of the 5th international workshop on Grid Economics and Business Models*, GECON '08, pages 125–139, Berlin, Heidelberg, 2008. Springer-Verlag.

[600] EPA Technical Report. Report to congress on server and data center energy efficiency. http://www.energystar.gov/ia/partners/prod_development/downloads/EPA_Datacenter_Report_Congress_Final1.pdf, 2007.

[601] RESERVOIR project. Project Website: http://www.reservoir-fp7.eu.

[602] Cluster Resources. Moab access portal: http://www.clusterresources.com/products/moab-cluster-suite/access-portal.php, November 2010.

[603] Cluster Resources. Moab adaptive hpc suite: http://www.clusterresources.com/products/adaptive-hpc-suite.php, November 2010.

[604] Cluster Resources. Moab cluster manager: http://www.clusterresources.com/products/moab-cluster-suite/cluster-manager.php, November 2010.

[605] Cluster Resources. Moab workload manager: http://www.clusterresources.com/products/moab-cluster-suite/workload-manager.php, November 2010.

[606] B. P. Rimal, E. Choi, and I. Lumb. A Taxonomy and Survey of Cloud Computing Systems. In *Proceedings of the 5th International Conference on Networked Computing*, Seoul, Korea, Aug. 25–27 2009.

[607] Bhaskar Prasad Rimal, Eunmi Choi, and Ian Lumb. A Taxonomy and Survey of Cloud Computing Systems. *Fifth International Joint Conference on INC, IMS and IDC*, pages 44–51, 2009.

[608] Bhaskar Prasad Rimal, Eunmi Choi, and Ian Lumb. A taxonomy and survey of cloud computing systems. *Fifth International Joint Conference on INC, IMS and IDC*, 2009.

[609] Bhaskar Prasad Rimal, Eunmi Choi, and Ian Lumb. A taxonomy and survey of cloud computing systems. *Networked Computing and Advanced Information Management, International Conference on*, 0:44–51, 2009.

[610] Thomas Ristenpart, Eran Tromer, Hovav Shacham, and Stefan Savage. Hey, you, get off of my cloud: exploring information leakage in third-party compute clouds. In *CCS'09: Proceedings of the 16th ACM conference on Computer and communications security*, pages 199–212. ACM, 2009.

[611] B. Rochwerger, D. Breitgand, E. Levy, A. Galis, K. Nagin, I. Llorente, R. Montero, Y. Wolfsthal, E. Elmroth, J. Caceres, M. Ben-Yehuda, W. Emmerich, and F. Galan. The RESERVOIR Model and Architecture for Open Federated Cloud Computing. *IBM Journal of Research and Development*, 53(4):4:1–4:11, July 2009.

[612] Benny Rochwerger, David Breitgand, Eliezer Levy, Alex Galis, Kenneth Nagin, Ignacio M. Llorente, Ruben Montero, Yaron Wolfsthal, Erik Elmroth, Juan Caceres, Muli Ben-Yehuda, Wolfgang Emmerich, and Fermin Galan. The RESERVOIR Model and Architecture for Open Federated Cloud Computing. *IBM Systems Journal*, 2008.

[613] Luis. Rodero-Merino, Luis M. Vaquero, Víctor Gil, Fermín Galán, Javier Fontán, Rubén Montero, and Ignacio M. Llorente. From infrastructure delivery to service management in clouds. *Future Generation Computer Systems*, 26:1226–1240, October 2010.

[614] Pedro Rodríguez, Daniel Gallego, Javier Cerviño, Fernando Escribano, Juan Quemada, Salvachú, and Joaquin A. VaaS : Videoconference as a Service. In *CollaborateCom*, Washington, 2009.

[615] D. Roman, U. Keller, H. Lausen, J. de Bruijn, R. Lara, M. Stollberg, A. Polleres, C. Feier, C. Bussler, and D. Fensel. Web service modeling ontology. *Applied Ontology*, 1(1):77–106, 2005.

[616] F. Rosenberg, C. Platzer, and S. Dustdar. Bootstrapping performance and dependability attributes of web services. In *IEEE International Conference on Web Services (ICWS'06)*, 2006.

[617] J. Rosenberg, H. Schulzrinne, G. Camarillo, A. Johnston, J. Peterson, and R. Spark. SIP: Session Initiation Protocol. RFC 3261 (Proposed Standard), June 2002.

[618] Antony Rowstron and Peter Druschel. Storage management and caching in past, a large-scale, persistent peer-to-peer storage utility. *SIGOPS Oper. Syst. Rev.*, 35:188–201, October 2001.

[619] Roy Bragg. Cloud computing: When computers really rule: http://www.technewsworld.com/story/63954.html, 2008.

[620] RSoft Design. *Artifex v.4.4.2.* RSoft Design Group, Inc., San Jose, CA, 2003.

[621] A. Ruiz-Cortés, O. Martin-Diaz, A. Duran, and M. Toro. Improving the automatic procurement of web services using constraint programming. *International Journal of Cooperative Information Systems*, 14(4):439–467, 2005.

[622] J. Rumbaugh, I. Jacobson, and G. Booch. *The Unified Modeling Language Reference Manual.* Erewhon: Addison-Wesley, 1999.

[623] P. Ruth, X. Jiang, D. Xu, and S. Goasguen. Virtual distributed environments in a shared infrastructure. *Computer*, 38(5):63 – 69, May 2005.

[624] Matt Ryan, Sojan Markose, Xiaoqing (Frank) Liu, and Ying Cheng. Structured object-oriented co-analysis/co-design of hardware/software for the facts power system. In *COMPSAC '05: Proceedings of the 29th Annual International Computer Software and Applications Conference*, pages 396–402, Washington, DC, USA, 2005. IEEE Computer Society.

[625] S. Battle, A. Bernstein, H. Boley, B. Grosof, M. Gruninger, R. Hull, M. Kifer, D. Martin, S. Mcllraith, D. McGuinness, J. Su, S. Tabet. Semantic Web Services Language (SWSL): http://www.w3.org/Submission/SWSF-SWSL/, 2005.

[626] T. L. Saaty. *Fundamentals of decision making and priority theory with the analytic hierarchy process*, volume 6. RWS Publications USA, 1994.

[627] Jordi Sabater and Carles Sierra. Reputation and social network analysis in multi-agent systems. In *Proceedings of the first international joint conference on Autonomous agents and multiagent systems: part 1*, AAMAS '02, pages 475–482, New York, NY, USA, 2002. ACM.

[628] F. D. Sacerdoti, M. J. Katz, M. L. Massie, and D. E. Culler. Wide area cluster monitoring with ganglia. In *Cluster Computing, 2003. Proceedings. 2003 IEEE International Conference on*, pages 289–298, December 2003.

[629] A. Sahai, V. Machiraju, M. Sayal, A. Van Moorsel, and F. Casati. Automated SLA monitoring for web services. *Management Technologies for E-Commerce and E-Business Applications*, pages 28–41, 2002.

[630] Usman Sait. The future of cloud computing. *Intel Executive Summary*, 2005.

[631] Usman Sait. Welcome to www.cloudtutorial.com. *Intel Executive Summary*, 2005.

[632] Salesforce. Appexchange: `http://sites.force.com/appexchange/home`, 17 November 2010.

[633] SAP. Business by design. sap, 2010.

[634] SAX. Simple API for XML. `http://sax.sourceforge.net/`.

[635] Cristina Schmidt and Manish Parashar. Squid: Enabling search in dht-based systems. *J. Parallel Distrib. Comput.*, 68(7):962–975, 2008.

[636] D. Schmidt and R. Guida. Hpc in the cloud - achieving ultra high performance in the cloud: `http://www.hpcinthecloud.com/features/101386074.html`, 24 August 2010.

[637] W3C School. MathML3 Manual. 2011.

[638] N. Seitz. Itu-t qos standards for ip-based networks. *Communications Magazine, IEEE*, 41(6):82 – 89, June 2003.

[639] Peter Sempolinski and Douglas Thain. A Comparison and Critique of Eucalyptus, OpenNebula and Nimbus. *2nd IEEE International Conference on Cloud Computing Technology and Science*, December 2010.

[640] Sequoia capital `http://www.sequoiacap.com/`.

[641] Amazon Web Service. Amazon web services (aws), 2010.

[642] A. ShaikhAli, O.F. Rana, R. Al-Ali, and D.W. Walker. Uddie: An extended registry for web services. In *Applications and the Internet Workshops, 2003. Proceedings. 2003 Symposium on*, pages 85–89. IEEE, 2003.

[643] Jin Shao, Hao Wei, Qianxiang Wang, and Hong Mei. A runtime model based monitoring approach for cloud. In *Cloud Computing (CLOUD), 2010 IEEE 3rd International Conference on*, pages 313 –320, July 2010.

[644] R. Shearer, B. Motik, and I. Horrocks. Hermit: A highly-efficient owl reasoner. In *Proceedings of the 5th International Workshop on OWL: Experiences and Directions (OWLED 2008)*, 2008.

[645] Justin Shi and Suntian Song. Apparatus and Method of Optimizing Database Clustering with Zero Transaction Loss. (Pending), 2007.

[646] Justin Y. Shi. *Synergy v3.0 Manual.* `http://spartan.cis.temple.edu/synergy/`, 1995.

[647] Justin Y. Shi. Fault tolerant self-optimizing multiprocessor system and method thereof. 2007.

[648] Justin Y. Shi. Decoupling as a foundation for large scale parallel processing. In *Proceedings of 2009 High Performance Computing and Communications*, Seoul, Korea, 2009.

[649] Y. Shi. Heterogeneous computing for graphics applications. In *National Conference on Graphics Applications*, April 1991.

[650] Y. Shi. A distributed programming model and its applications to computation intensive applications for heterogeneous environments". In *International Space Year Conference on Earth and Space Information Systems*, pages 10–13, Pasadena, CA., February 1992.

[651] Y. Shi. Program scalability analysis. In *International Conference on Distributed and Parallel Processing*, Geogetown University, Washington D.C., October 1997.

[652] J. Shiers. The worldwide lhc computing grid (worldwide lcg). *Computer Physics Communications, 177 (1-2)*, pages 219–223, 2007.

[653] M. Shultz, G. Bronevetsky, R. Fernandes, D. M. K. Pingali, and P. Stodghill. Implementation and evaluation of a scalable application-level checkpoint-recovery scheme for mpi programs. In *Proceedings of Supercomputing 2004 Conference*, Pittsburgh, PA., November 2004.

[654] Guttorm Sindre. Mal-activity diagrams for capturing attacks on business processes. In *REFSQ'07: Proceedings of the 13th international working conference on Requirements engineering*, pages 355–366, Berlin, Heidelberg, 2007. Springer-Verlag.

[655] Guttorm Sindre and Andreas L. Opdahl. Eliciting security requirements with misuse cases. *Requir. Eng.*, 10(1):34–44, 2005.

[656] Aameek Singh and Ling Liu. Sharoes: A data sharing platform for outsourced enterprise storage environments. In *ICDE*, pages 993–1002, 2008.

[657] R. O. Sinnott, J. Jiang, J. Watt, and O. Ajayi. Shibboleth-based Access to and Usage of Grid Resources. In *Proceedings of the 7th International Conference on Grid Computing*, Barcelona, Spain, Sep. 28–29, 2006.

[658] Radu Sion. Query execution assurance for outsourced databases. In *Proceedings of the 31st international conference on Very large data bases*, VLDB '05, pages 601–612. VLDB Endowment, 2005.

[659] E. Sirin, B. Parsia, B. Grau, A. Kalyanpur, and Y. Katz. Pellet: A practical OWL-DL reasoner. *Web Semantics: Science, Services and Agents on the World Wide Web*, 5(2):51–53, June 2007.

[660] K. Sivashanmugam, K. Verma, and A. Sheth. Discovery of Web Services in a Federated Registry Environment. In *Proceedings of the International Conference on Web Services (ICWS)*, San Diego, USA, June 6–9, 2004.

[661] A. Skonnard. A developer's guide to the microsoft .net service bus, white paper, pluralsight: http://go.microsoft.com/fwlink/?LinkID=150834, May 2009.

[662] J. M. Smith. $M/G/c/K$ blocking probability models and system performance. *Perform. Eval.*, 52:237–267, May 2003.

[663] Suntian Song. Method and apparatus for database fault tolerance with instant transaction replication using off-the-shelf database servers and low bandwidth networks. (#6,421,688), 2002.

[664] B. Sotomayor, R. S. Montero, I. M. Llorente, and I. Foster. Virtual Infrastructure Management in Private and Hybrid Clouds. *Internet Computing, IEEE*, 13(5):14–22, Sept.-Oct. 2009.

[665] B. Sotomayor, R. S. Montero, I. M. Llorente, and I. Foster. Virtual infrastructure management in private and hybrid clouds. *IEEE Internet Computing*, 13(5):14–22, 2009.

[666] Borja Sotomayor, Ian Foster, Rubén S. Montero, and Ignacio M. Llorente. Virtual Infrastructure Management in Private and Hybrid Clouds. *IEEE Internet Computing*, 2009.

[667] Borja Sotomayor, Rubén S. Montero, Ignacio M. Llorente, and Ian Foster. Virtual Infrastructure Management in Private and Hybrid Clouds. *IEEE Internet Computing*, 13(5):14–22, September 2009.

[668] Matthew J. Sottile and Ronald G. Minnich. Supermon: A high-speed cluster monitoring system. *Cluster Computing, IEEE International Conference on*, 0:39, 2002.

[669] L. Sotto, B. Treacy, and M. McLellan. Privacy & Data Security Risks in Cloud Computing. In *Electronic Commerce & Law Report*, Feb. 2010.

[670] Scott Spetka, Haris Hadzimujic, Stephen Peek, and Christopher Flynn. High Productivity Languages for Parallel Programming Compared to MPI. *HPCMP Users Group Conference*, 0:413–417, 2008.

[671] Amazon EC2 Spot Instances. http://aws.amazon.com/ec2/spot-instances/.

[672] V. Springel. The cosmological simulation code gadget-2. *MNRAS, astro-ph/0505010*, 2005.

[673] SPRUCE: Special PRiority and Urgent Computing Environment. http://spruce.teragrid.org/.

[674] Shekhar Srikantaiah, Aman Kansal, and Feng Zhao. Energy aware consolidation for cloud computing. In *Proceedings of the 2008 Conference on Power Aware Computing and Systems*, HotPower'08, pages 10–10, Berkeley, CA, USA, 2008. USENIX Association.

[675] N. Srinivas and K. Deb. Muiltiobjective optimization using nondominated sorting in genetic algorithms. *Evolutionary computation*, 2(3):221–248, 1994.

[676] Ion Stoica, Robert Morris, David Liben-Nowell, David R. Karger, M. Frans Kaashoek, Frank Dabek, and Hari Balakrishnan. Chord: A scalable peer-to-peer lookup protocol for internet applications. In *ACM SIGCOMM*, pages 149–160, 2001.

[677] Michael Stonebraker. The case for shared nothing architecture. *Database Engineering*, 9(1), 1986.

[678] Feijian Sun. *Automatic Program Parallelization Using Stateless Parallel Processing Architecture*. PhD thesis, Temple University, 2004.

[679] Wei Sun, Xin Zhang, Chang J. Guo, Pei Sun, and Hui Su. Software as a Service: Configuration and Customization Perspectives. In *2008 IEEE Congress on Services Part II (services-2 2008)*, pages 18–25. IEEE, September 2008.

[680] Yuanhui Sun, Zongshui Xiao, Dongmei Bao, and Jie Zhao. An architecture model of management and monitoring on cloud services resources. In *Advanced Computer Theory and Engineering (ICACTE), 2010 3rd International Conference on*, volume 3, pages V3–207 –V3–211, August 2010.

[681] Sun Microsystems. *Introduction to Cloud Computing Architecture*, 1st edition, June 2009. White Paper.

[682] Sonesh Surana, Brighten Godfrey, Karthik Lakshminarayanan, Richard Karp, and Ion Stoica. Load balancing in dynamic structured peer-to-peer systems. *Performance Evaluation*, 63(3):217–240, 2006.

[683] Swedish seed accelerator `http://www.swedacc.com`.

[684] Symantec. *Veritas Cluster Server (VCS)*. `http://eval.symantec.com/mktginfo/products/Datasheets/High_Availability/cluster_server_datasheet.pdf`, 2006.

[685] M. Szeredi. Filesystem in userspace. http://fuse.sourceforge.net, February 2005.

[686] E. Szmidt and J. Kacprzyk. Intuitionistic fuzzy sets in group decision making. *Notes on IFS*, 2(1):11–14, 1996.

[687] Tom Szuba. *Safeguarding Your Technology.* Washington: National Center for Education Statistics, 1998.

[688] B. Szymanski, Y. Shi, and N. Prywes. Synchronized distributed termination. *IEEE Transactions on Software Engineering,* SE11(10):1136–1140, 1985.

[689] S.G.H. Tabatabaei, A.V. Dastjerdi, W.M.N.W. Kadir, S. Ibrahim, and E. Sarafian. Security conscious AI-planning-based composition of semantic web services. *International Journal of Web Information Systems,* 6(3):203–229, 2010.

[690] H. Takagi. *Queueing Analysis,* volume 1: Vacation and Priority Systems. North-Holland, Amsterdam, The Netherlands, 1991.

[691] Y. Takahashi. An approximation formula for the mean waiting time of an $M/G/c$ queue. *J. Operational Research Society,* 20:150–163, 1977.

[692] Tingxi Tan and Cameron Kiddle. An Assessment of Eucalyptus Version 1.4. Technical report, Grid Research Centre, University of Calgary, Canada, 2009.

[693] Andrew S. Tanenbaum and Maarten van Steen, editors. *Distributed Systems: Principles and Paradigms.* Prentice Hall, 2002.

[694] Cloud Computing Interoperability Forum, Cloud taxonomy : http://groups.google.com/group/cloudforum/web/ccif-cloud-taxonomy.

[695] Cloud Computing Use Case Discussion Group: Cloud Computing Use Cases, White Paper v0.1, Section 2: Definitions and Taxonomies: http://groups.google.com/group/cloud-computing-use-cases, 2009.

[696] W. T. Luke Teacy, Jigar Patel, Nicholas R. Jennings, and Michael Luck. Coping with inaccurate reputation sources: experimental analysis of a probabilistic trust model. In *Proceedings of the fourth international joint conference on Autonomous agents and multiagent systems,* AAMAS '05, pages 997–1004, New York, NY, USA, 2005. ACM.

[697] Akamai Technologies. Fast internet content delivery with freeflow. technical report, 2000.

[698] Intel Virtualization Technology. http://www.intel.com/technology/ virtualization/technology.htm.

[699] TeraGrid. http://www.teragrid.org/.

[700] Douglas Thain, Todd Tannenbaum, and Miron Livny. Condor and the Grid. In *Grid computing: making the global infrastructure a reality,* pages 299–350. John Wiley and Sons, 2003.

[701] Douglas Thain, Todd Tannenbaum, and Miron Livny. Distributed computing in practice: the Condor experience. *Concurrency - Practice and Experience*, 17(2-4):323–356, 2005.

[702] Douglas Thain, Todd Tannenbaum, and Miron Livny. Distributed computing in practice: The condor experience. *Concurrency and Computation: Practice and Experience*, 17:2–4, 2005.

[703] Wolfgang Theilmann, Ramin Yahyapour, and Joe Butler. Multi-level sla management for service-oriented infrastructures. In *Proceedings of the 1st European Conference on Towards a Service-Based Internet*, ServiceWave '08, pages 324–335, 2008.

[704] K Thompson. *UNIX implementation*, pages 26–41. Prentice-Hall, Inc., Upper Saddle River, NJ, USA, 1986.

[705] H. C. Tijms. Heuristics for finite-buffer queues. *Probability in the Engineering and Informational Sciences*, 6:277–285, 1992.

[706] H. C. Tijms, M. H. V. Hoorn, and A. Federgru. Approximations for the steady-state probabilities in the $M/G/c$ queue. *Advances in Applied Probability*, 13:186–206, 1981.

[707] I. Toma, D. Foxvog, and M. C. Jaeger. Modeling QoS characteristics in WSMO. In *Proceedings of the 1st workshop on Middleware for Service Oriented Computing (MW4SOC 2006)*, pages 42–47. ACM, 2006.

[708] I. Toma, D. Foxvog, and M. C. Jaeger. Modeling QoS characteristics in WSMO. In *Proceedings of the 1st workshop on Middleware for Service Oriented Computing (MW4SOC 2006)*, pages 42–47. ACM, 2006.

[709] I. Toma, D. Roman, D. Fensel, B. Sapkota, and J. Gomez. A multi-criteria service ranking approach based on non-functional properties rules evaluation. *Service-Oriented Computing–ICSOC 2007*, pages 435–441, 2010.

[710] TOP500 Supercomputers . Supercomputer's application area share. http://www.top500.org/stats/list/33/apparea (2009).

[711] J. Torres, D. Carrera, K. Hogan, R. Gavalda, V. Beltran, and N. Poggi. Reducing wasted resources to help achieve green data centers. In *Parallel and Distributed Processing, 2008. IPDPS 2008. IEEE International Symposium on*, pages 1 –8, April 2008.

[712] V. Tosic, K. Patel, and B. Pagurek. WSOLÜWeb Service Offerings Language. *Web Services, E-Business, and the Semantic Web*, pages 57–67, 2002.

[713] V. X. Tran, H. Tsuji, and R. Masuda. A new QoS ontology and its QoS-based ranking algorithm for Web services. *Simulation Modelling Practice and Theory*, 17(8):1378–1398, 2009.

[714] D. Tsarkov and I. Horrocks. Fact++ description logic reasoner: System description. In *Proc. of the Int. Joint Conf. on Automated Reasoning (IJCAR 2006)*, volume 4130 of *Lecture Notes in Artificial Intelligence*, pages 292–297. Springer, 2006.

[715] D. Tsesmetzis, I. Roussaki, and E. Sykas. Modeling and simulation of QoS-aware Web service selection for provider profit maximization. *Simulation*, 83(1):93, 2007.

[716] Alan Turing. On computable numbers, with an application to the entscheidungsproblem. *Proceedings of the London Mathematical Society*, Series 2(42):230–265, 1936.

[717] Francesco Tusa, Maurizio Paone, Massimo Villari, and Antonio Puli-afito. CLEVER: a CLoud Enabled Virtual EnviRonment. In *Computers and Communications (ISCC), 2010 IEEE Symposium on*, pages 477 –482, June 2010.

[718] Use Case Discussion Group. Cloud Computing Use Cases v2.0, 2009.

[719] D.A. Van Veldhuizen. Multiobjective evolutionary algorithms: classifications, analyses, and new innovations. In *Evolutionary Computation*. Citeseer, 1999.

[720] L. M. Vaquero, L. Rodero-Merino, J. Caceres, and M. Lindner. A break in the clouds: towards a cloud definition. *SIGCOMM Comput. Commun. Rev.*, 39:50–55, December 2008.

[721] Luis M. Vaquero, Luis Rodero-Merino, Juan Caceres, and Maik Lindner. A Break in the Clouds : Towards a Cloud Definition. *Computer Communication Review*, 39(1):50–55, 2009.

[722] Luis M. Vaquero, Luis Rodero-Merino, Juan Caceres, and Maik Lindner. A break in the clouds: towards a cloud definition. *SIGCOMM Comput. Commun. Rev.*, 39(1):50–55, 2009.

[723] C. Vazquez, E. Huedo, R. S. Montero, and I. M. Llorente. Dynamic provision of computing resources from grid infrastructures and cloud providers. In *Grid and Pervasive Computing Conference, 2009. GPC '09. Workshops at the*, pages 113–120, May 2009.

[724] Christian Vecchiola, Xingchen Chu, and Rajkumar Buyya. *High Speed and Large Scale Scientific Computing*, chapter Aneka: A Software Platform for .NET-based Cloud Computing. IOS Press, Amsterdam, Netherlands, 2009.

[725] Christian Vecchiola, Xingchen Chu, and Rajkumar Buyya. *Aneka: A Software Platform for .NET-based Cloud Computing.* IOS Press, Amsterdam, Netherlands, 2010.

[726] Christian Vecchiola, Xingchen Chu, Michael Mattess, and Rajkumar Buyya. *Cloud Computing: Principles and Paradigms*, chapter Aneka – Integration of Private and Public Clouds. John Wiley & Sons, August 2010.

[727] Christian Vecchiola, Xingchen Chu, Michael Mattess, and Rajkumar Buyya. *Cloud Computing: Principles and Paradigms*, chapter Aneka Integration of Private and Public Clouds. Wiley Press, New York, USA, February 2011.

[728] Kim H. Veltman. Syntactic and Semantic Interoperability: New Approaches to Knowledge and the Semantic Web. *New Review of Information Networking*, 7(1):159–183, 2001.

[729] VMWare. VMWare vCloud.

[730] VMware. Vmware vsphere: `http://www.vmware.com/products/vsphere/mid-size-and-enterprise-business/features.html`, December 2010.

[731] Werner Vogels. Eventually consistent. *Commun. ACM*, 52(1):40–44, 2009.

[732] Mladen A Vouk. Cloud computing-issues,research and implementations. *Second International Symposium on Information Science and Engineering*, 2010.

[733] J. Broberg, W. Voorsluys, and R. Buyya. *Introduction to Cloud Computing, Cloud Computing: Principles and Paradigms.* John Wiley and Sons (in print), 2011.

[734] Luay A. Wahsheh and Jim Alves-Foss. Security policy development: Towards a life-cycle and logic-based verification model. *American Journal of Applied Sciences*, 5(9):1117–1126, 2008.

[735] Phil Wainewright. Many degrees of multi-tenancy, 2008. *Technical Report*, available at http://www.zdnet.com/blog/saas/many-degrees-of-multitenancy/533

[736] E. Walker. Benchmarking amazon ec2 for high-performance scientific computing: `http://www.usenix.org/publications/login/2008-10/openpdfs/walker.pdf`, 2008.

[737] N. Wancheng, H. Lingjuan, L. Lianchen, and W. Cheng. Commodity-Market Based Services Selection in Dynamic Web Service Composition. In *Asia-Pacific Service Computing Conference, The 2nd IEEE*, pages 218–223. IEEE, 2007.

[738] Cong Wang, Student Member, Sherman S Chow, Qian Wang, Kui Ren, and Wenjing Lou. Privacy-Preserving Public Auditing for Secure Cloud Storage. *INFOCOM, 2010 Proceedings IEEE*, pages 1–12, 2009.

[739] Cong Wang, Qian Wang, Kui Ren, and Wenjing Lou. Ensuring data storage security in Cloud Computing. *2009 17th International Workshop on Quality of Service*, pages 1–9, July 2009.

[740] G. Wang, A. Chen, C. Wang, C. Fung, and S. Uczekaj. Integrated quality of service (QoS) management in service-oriented enterprise architectures. In *Enterprise Distributed Object Computing Conference, 2004. EDOC 2004. Proceedings. Eighth IEEE International*, pages 21–32. IEEE, 2005.

[741] Haiqin Wang, Guijun Wang, Changzhou Wang, A. Chen, and R. Santiago. Service level management in global enterprise services: from qos monitoring and diagnostics to adaptation, a case study. In *EDOC Conference Workshop, 2007. EDOC '07. Eleventh International IEEE*, pages 44 –51, Oct 2007.

[742] L. Wang, G. V. Laszewski, A. Younge, X. He, M. Kunze, J. Tao, and C. Fu. Cloud computing: a perspective study. *New Generation Computing*, 28:137–146, 2010.

[743] Lizhe Wang, Gregor Von Laszewski, Marcel Kunze, and Jie Tao. Cloud Computing: a Perspective Study. *Proceedings of the Grid Computing Environments (GCE) workshop*, November 2008.

[744] Lizhe Wang, Gregor von Laszewski, Andrew Younge, Xi He, Marcel Kunze, Jie Tao, and Cheng Fu. Cloud computing: a perspective study. *New Generation Computing*, 28:137–146, 2010. 10.1007/s00354-008-0081-5.

[745] P. Wang. QoS-aware web services selection with intuitionistic fuzzy set under consumer's vague perception. *Expert Systems with Applications*, 36(3):4460–4466, 2009.

[746] X. Wang, T. Vitvar, M. Kerrigan, and I. Toma. A QoS-aware selection model for semantic web services. *Service-Oriented Computing–ICSOC 2006*, pages 390–401, 2006.

[747] X. Wang, K. Yue, J.Z. Huang, and A. Zhou. Service selection in dynamic demand-driven Web services. In *Web Services, 2004. Proceedings. IEEE International Conference on*, pages 376–383. IEEE, 2004.

[748] Yan Wang and Kwei-Jay Lin. Reputation-oriented trustworthy computing in e-commerce environments. *IEEE Internet Computing*, 12:55–59, July 2008.

[749] Yung-Terng Wang and Robert J. T. Morris. Load sharing in distributed systems. *IEEE Transactions on Computers*, C-34(3):204–217, March 1985.

[750] M.S. Ware, J.B. Bowles, and C.M. Eastman. Using the common criteria to elicit security requirements with use cases. In *Proceedings of the IEEE SoutheastCon*, pages 273 –278. 10.1109/second.2006.1629363, 2006.

[751] web.mit.edu. StarCluster. http://web.mit.edu/stardev/cluster/index.html.

[752] Yi Wei and M.B. Blake. Service-oriented computing and cloud computing: Challenges and opportunities. *Internet Computing, IEEE*, 14(6):72 –75, Nov. 2010.

[753] S. Weibel, J. Kunze, C. Lagoze, and M. Wolf. Dublin core metadata for resource discovery. *Internet Engineering Task Force RFC*, 2413, 1998.

[754] M.H. Weik. Computer science and communications dictionary. Springer, 2000.

[755] J. Weise. *Public Key Infrastructure Overview*. Sun Microsystems, August 2001.

[756] Aaron Weiss. Computing in the clouds. *netWorker*, 11:16–25, December 2007.

[757] Craig D. Weissman and Steve Bobrowski. The design of the force.com multitenant internet application development platform. In *Proceedings of the 35th SIGMOD international conference on Management of data*, SIGMOD '09, pages 889–896, New York, NY, USA, 2009. ACM.

[758] J. Whitehead and Y. A. Goland. WebDAV: A network protocol for remote collaborative authoring on the web. In *ECSCW 1999*, 1999.

[759] Wikipedia. Wikipedia, free encyclopedia. *3rd International Conference on Grid and Pervasive Computing-gpc-workshops*, 2008.

[760] Wikipedia. Wikipedia, free encyclopedia. *3rd International Conference on Grid and Pervasive Computing-gpc-workshops*, 2008.

[761] Wikipedia. Wikipedia, free encyclopedia. 2009.

[762] Wikipedia. Cloud computing. *Second International Symposium on Information Science and Engineering*, 2010.

[763] D. Winer et al. Xml-rpc specification, 1999.

[764] W.Kim. Cloud computing:today and tomorrow. *Journal of Object Technology*, 2009.

[765] Rich Wolski, James S. Plank, John Brevik, and Todd Bryan. Analyzing market-based resource allocation strategies for the computational Grid. *The International Journal of High Performance Computing Applications*, 15(3):258–281, Fall 2001.

[766] Rich Wolski, Neil T Spring, and Jim Hayes. The network weather service: a distributed resource performance forecasting service for metacomputing. *Future Generation Computer Systems*, 15(5-6):757 – 768, 1999.

[767] Eugene Y. C. Wong, Alvin T. S. Chan, and Hong Va Leong. Xstream: A middleware for streaming XML contents over wireless environments. *IEEE Trans. Softw. Eng.*, 30:918–935, December 2004.

[768] T. Wood, P. J. Shenoy, A. Venkataramani, and M. S. Yousif. Sandpiper: Black-box and gray-box resource management for virtual machines. *Computer Networks*, 53(17):2923–2938, 2009.

[769] xCAT Extreme Cloud Administration Toolkit. http://xcat.sourceforge.net/.

[770] xen.org. Xen. http://www.xen.org/.

[771] Fanfan Xiong. *Resource Efficient Parallel VLDB with Customizable Degree of Redundancy*. PhD thesis, Temple University, 2009.

[772] K. Xiong and H. Perros. Service performance and analysis in cloud computing. In *Proceedings of the 2009 Congress on Services - I*, pages 693–700, Los Alamitos, CA, USA, 2009.

[773] Li Xiong and Ling Liu. Peertrust: Supporting reputation-based trust for peer-to-peer electronic communities. *IEEE Trans. Knowl. Data Eng.*, 16(7):843–857, 2004.

[774] Li Xiong, Ling Liu, and Mustaque Ahamad. Countering feedback sparsity and manipulation in reputation systems. In *Proceedings of the 2007 International Conference on Collaborative Computing: Networking, Applications and Worksharing*, pages 203–212, Washington, DC, USA, 2007. IEEE Computer Society.

[775] B. Yang, F. Tan, Y. Dai, and S. Guo. Performance evaluation of cloud service considering fault recovery. In *Cloud Computing*, volume 5931 of *Lecture Notes in Computer Science*, pages 571–576. Springer Berlin Heidelberg, 2009.

[776] Yijian Yang. *Fault Tolerance Protocol for Multiple Dependent Master Protection in a Stateless Parallel Processing Framework*. PhD thesis, Temple University, August 2007.

[777] Yin Yang, Stavros Papadopoulos, Dimitris Papadias, and George Kollios. Authenticated indexing for outsourced spatial databases. *The VLDB Journal*, 18:631–648, June 2009.

[778] D. D. Yao. Refining the diffusion approximation for the $M/G/m$ queue. *Operations Research*, 33:1266–1277, 1985.

[779] Y Combinator seed accelerator `http://ycombinator.com`.

[780] Chee Shin Yeo and Rajkumar Buyya. Pricing for utility-driven resource management and allocation in clusters. *Int. J. High Perform. Comput. Appl.*, 21(4):405–418, 2007.

[781] Yieldex: advanced forecasting and delivery simulation algorithms to manage digital business for publishing companies. `http://www.yieldex.com`.

[782] N. Yigitbasi, A. Iosup, D. Epema, and S. Ostermann. C-meter: A framework for performance analysis of computing clouds. In *CCGRID '09: Proceedings of the 2009 9th IEEE/ACM International Symposium on Cluster Computing and the Grid*, pages 472–477, Washington, DC, USA, 2009.

[783] L. Youseff, M. Butrico, and D. Da Silva. Toward a Unified Ontology of Cloud Computing. In *Grid Computing Environments Workshop, 2008. GCE '08*, pages 1–10, Nov 2008.

[784] H. Q. Yu and S. Reiff-Marganiec. A method for automated web service selection. In *Services-Part I, 2008. IEEE Congress on*, pages 513–520. IEEE, 2008.

[785] H.Q. Yu and S. Reiff-Marganiec. Non-functional property based service selection: A survey and classification of approaches. In *Proc. of 2nd Non Functional Properties and Service Level Agreements in SOC Workshop (NFPSLASOCŠ08)*. Citeseer, 2008.

[786] Shucheng Yu, Cong Wang, Kui Ren, and Wenjing Lou. Achieving Secure, Scalable, and Fine-grained Data Access Control in Cloud Computing. *2010 Proceedings IEEE INFOCOM*, pages 1–9, March 2010.

[787] Y.Zhao, I.Raicu, and I.Foster. Scientific workflow systems for 21st century,new bottle or new wine. *IEEE Workshop on Scientific Workflows*, 2008.

[788] Zabbix. Project Website: `http://www.zabbix.com`.

[789] Giorgos Zacharia and Pattie Maes. Trust management through reputation mechanisms. *Applied Artificial Intelligence*, 14(9):881–907, 2000.

[790] Michal Zalewski. *Browser Security Handbook*. Google, 2008.

[791] Zencoder: video encoding on the cloud `http://www.zencoder.com`.

[792] L. Zeng, B. Benatallah, A.H.H. Ngu, M. Dumas, J. Kalagnanam, and H. Chang. QoS-aware middleware for web services composition. *Software Engineering, IEEE Transactions on*, 30(5):311–327, 2004.

[793] W. Zeng, Y. Zhao, and J. Zeng. Cloud service and service selection algorithm research. In *Proceedings of the first ACM/SIGEVO Summit on Genetic and Evolutionary Computation*, pages 1045–1048. ACM, 2009.

[794] Zenoss. Zenoss Core - Open Source IT Management. `http://community.zenoss.org/docs/DOC-2614`.

[795] Ya-Qin Zhang. The future of computing in the cloud - client. *The Economic Observer*, 2008.

[796] Yingqian Zhang, Bin Sun, and Jia Liu. A markup language for parallel programming model on multi-core system. *SCALCOM-EMBEDDEDCOM 2009 International Conference*, pages 640–643, Sept. 2009.

[797] Ben Y. Zhao, John D. Kubiatowicz, and Anthony D. Joseph. Tapestry: a fault-tolerant wide-area application infrastructure. *SIGCOMM Comput. Commun. Rev.*, 32:81–81, January 2002.

[798] C. Zhou, L.T. Chia, and B.S. Lee. DAML-QoS ontology for web services. In *Web Services, 2004. Proceedings. IEEE International Conference on*, pages 472–479. IEEE, 2004.

[799] Runfang Zhou and Kai Hwang. Powertrust: A robust and scalable reputation system for trusted peer-to-peer computing. *IEEE Trans. Parallel Distrib. Syst.*, 18:460–473, April 2007.

[800] Dakai Zhu, Rami G. Melhem, and Bruce R. Childers. Scheduling with dynamic voltage/speed adjustment using slack reclamation in multiprocessor real-time systems. *IEEE Trans. Parallel Distrib. Syst.*, 14(7):686–700, 2003.

[801] H.J. Zimmermann. Fuzzy programming and linear programming with several objective functions* 1. *Fuzzy sets and systems*, 1(1):45–55, 1978.

[802] E. Zitzler and L. Thiele. Multiobjective evolutionary algorithms: A comparative case study and the strength pareto approach. *evolutionary computation, IEEE transactions on*, 3(4):257–271, 2002.

[803] Zope. Project Website: `http://www.zope.org`.

Index